Abiotic Stress and Physiological Process in Plants

NIPA® GENX ELECTRONIC RESOURCES & SOLUTIONS P. LTD.
New Delhi-110 034

Abiotic Stress and Physiological Process in Plants

A. Bhattacharya

Former Principal Scientist (Plant Physiology)
Indian Institute of Pulses Research
Kanpur, India

NIPA® GENX ELECTRONIC RESOURCES & SOLUTIONS P. LTD.
New Delhi-110 034

NIPA® GENX ELECTRONIC
RESOURCES & SOLUTIONS P. LTD.

101,103, Vikas Surya Plaza, CU Block
L.S.C. Market, Pitam Pura, New Delhi-110 034
Ph : +91 11 27341616, 27341717, 27341718
E-mail: newindiapublishingagency@gmail.com
www: www.nipabooks.com

For customer assistance, please contact
Phone: + 91-11-27 34 17 17
Fax: + 91-11- 27 34 16 16
E-Mail: feedbacks@nipabooks.com

ISBN 978-81-96075-47-7

Composed & Designed by NIPA

Preface

Abiotic stresses are serious threats to agriculture and the environment which have been exacerbated in the current century by global warming and industrialization. According to FAO statistics, more than 800 million hectares of land throughout the world are currently salt-affected, including both saline and sodic soils equating to more than 6% of the world's total land area. Continuing salinization of arable land is expected to have overwhelming global impact, resulting in a 30% loss of agricultural land over the next 25 years and up to 50% loss by 2050. Overall, it has been estimated that the world is losing at least 3 ha of arable land every minute due to soil salinity. Some of the most serious effects of abiotic stresses occur in the arid and semiarid regions where lower rainfall, higher evaporation, native rocks, saline irrigation water, and poor water management all contribute in agricultural areas.

In the current millennium, an international initiative focusing on increasing our productivity per unit of water has been launched to achieve **"More crop per drop"**. Efforts to improve the efficiency of agricultural water use while simultaneously reducing adverse environmental impacts will need to draw on results of extensive and diverse research in several areas.

As we move forward, emerging information and novel approaches must continuously be applied in a timely and effective manner by both the research and applied agricultural communities. One promising approach to improving the ability of plants to cope with abiotic stress is to combine utilization of the vast biodiversity of crop plants and their wild relatives with the rapidly emerging genetic and molecular techniques. Global programs, such as the Global Partnership Initiative for Plant Breeding Capacity Building (GIPB), aim to select and distribute seed crops and cultivars with tolerance to abiotic stresses in order to facilitate sustainable use of plant genetic resources for food and agriculture.

Abiotic stress leads to a series of morphological, physiological, biochemical and molecular changes in plants that adversely affect growth and productivity. A frequent result is protein dysfunction. Understanding the mechanisms of protein folding stability and how this knowledge can be utilized is one of the most challenging strategies for aiding organisms undergoing stress conditions. Stresses also affect the biosynthesis, concentration, transport, and storage of primary and secondary metabolites. As a more comprehensive view of these processes evolves, applications to reducing plant stress are emerging.

Under natural conditions plants can suffer from various stress combinations at different development stages and during different time periods. Many of the gene products differentially expressed under stress, such as dehydrins, enzymes for the synthesis of osmolytes, and enzymes for the removal of reactive oxygen species (ROS), protect plant cells from damage. The production of these functional proteins is widely regulated by specific transcription factors. Use of transcription factors is now under development as an additional biotechnological approach to improving plant response to abiotic stresses.

Abiotic stress factors frequently constrain the growth and productivity of major crop species such as cereals. The single greatest abiotic stress factor that limits crop growth worldwide is water availability. While genetic increases in yield potential are best expressed in optimum environments, they are also associated with enhanced yields under drought and nitrogen deficiency. These gains are especially relevant given that further large increases in the area under irrigation are not expected, and land deterioration associated with intensive agriculture threatens those areas already irrigated.

This book is intended to cover all known abiotic stresses and their effects on the various aspects physiological processes in plants. Since to cover all the physiological processes has become a voluminous work, the present book has been divided in different parts. The present volume has eight chapters. The first chapter deals with introduction and the introductory remarks on abiotic stress and overall effects of different factors on crop. Second chapter deals with aspects of seed germination, seedling growth and root development under stressful condition for different factors of abiotic stress. Third chapter is a contributory chapter and deals with different aspects of biological nitrogen fixation under abiotic stress. Fourth chapter deals with growth, development and phenology of crop plants under environmental stress condition. Various aspects of photosynthesis under abiotic stress have been discussed in the fifth chapter. Mineral nutrition under abiotic stress condition and roles of different nutrient to overcome abiotic stress has been discussed in Sixth chapter. Seventh chapter discussions have been provided regarding transgenic approach to enhance drought tolerance and it is a contributory chapter. Eighth chapter is about the effects of different factors of abiotic stress on respiration processes of plant have been discussed.

These review chapters addresses how knowledge of the physiological mechanisms of crops can improve crop yield. The study of crop physiology can assist cereal breeding by: i) improving our understanding of the factors that determine crop yield and adaptation through the syncretic concept of ideotype and, consequently, improve crop simulation models; ii) defining particular "secondary" traits to select (analytical breeding) when choosing: iii) indicating the kind of genetically modified organisms (GMOs) with potential for development and how to test them; and iv) phenotyping associated with marker-assisted selection.

Suggestions and advices are most welcome for the improvements in structural and for additions of current technologies as well as methodologies. A feeling of successfulness would prevail in my efforts in publishing this book if it is known that how this text has become meaningful to our readers. Suggestions from researchers, teachers and students, who use the book, would be highly appreciated. Hope this book shall serve the need of students, faculty members and researchers. The suggestions and advice are most welcome for the improvements in structural and for additions of current techniques. I sincerely hope the review chapters of this part of the book shall meet the requirement of post graduate students as well as faculty members of plant physiology and will be glad to accept constructive criticism and suggestions from the faculty.

A. Bhattacharya

Former Principal Scientist (Plant Physiology)
Indian Institute of Pulses Research
Kanpur, India

Contents

1

Abiotic Stress: An Introduction

A. Bhattacharya

Plants are bound to places. They, therefore, have to be considerably more adaptable to stressful environments and must acquire greater tolerance to multiple stresses than animals and humans. This is shown very clearly by the limitations in the distribution of particular types of vegetation, for example, the tree line on mountains. Extremely high light intensities, mechanical stress by wind, frequent periods of frost, and a winter period of many months have left their marks on these spruces and pines at about 3000 m altitude. The natural environment for plants is composed of a complex set of abiotic stresses and plant responses to these stresses are equally complex. Systems biology approaches facilitate a multi-targeted approach by allowing one to identify regulatory hubs in complex networks. Systems biology takes the molecular parts (transcripts, proteins and metabolites) of an organism and attempts to fit them into functional networks or models designed to describe and predict the dynamic activities of that organism in different environments.

Abiotic stress has been defined as the negative impact of non-living factors on the living organisms in a specific environment. The non-living variable must influence the environment beyond its normal range of variation to adversely affect the population performance or individual physiology of the organism in a significant way. Whereas a biotic stress would include such living disturbances as fungi or harmful insects, abiotic stress factors, or stressors, are naturally occurring, often intangible, factors such as intense sunlight or wind that may cause harm to the plants and animals in the area affected. Abiotic stress is essentially unavoidable. Abiotic stress affects animals, but plants are especially dependent on environmental factors, so it is particularly constraining. Abiotic stress is the most harmful factor concerning the growth and productivity of crops worldwide. Research has also shown that abiotic stressors are at their most harmful when they occur together, in combinations of abiotic stress factors.

The environment affects an organism in many ways, at any time. To understand the reactions of a particular organism in a certain situation, so-called environmental factors, are usually considered separately, if at all possible. Environmental factors can be of abiotic and biotic nature. Biotic environmental factors, resulting from interactions with other organisms, are, for example, infection or mechanical damage by herbivory or trampling, as well as effects of symbiosis or parasitism. Abiotic environmental factors include temperature, humidity, light intensity, the supply of water and minerals, and CO_2; these are the parameters and resources that determine the growth of a plant. Many other influences, which are only rarely beneficial to the plant (wind as distributor of pollen and seeds), or not at all beneficial or are even damaging (ionising rays or pollutants), are also classified as abiotic factors. The effect of each abiotic factor depends on its quantity. With optimal quantity or intensity, as may be provided in a greenhouse, the plant grows "optimally" and thus achieves its "physiological normal type", maximizing its physiologically achievable performance. Plants almost never find the optimal quantities or intensities of all essential abiotic factors. Thus the "physiological normal type" is rather the exception and deviation from the rule. It is very important to realize that growth is only one of many reactions of a plant to its environment. Flowering and fruiting determine the plant's success in reproduction and propagation and might equally be used as a measure of the plant's reaction to the environment. The value of the factors might, in this case, change but the principal behaviour would be similar. Deviations from the physiological normal type are regarded as reactions to suboptimal or damaging quantities or intensities of environmental factors, *i.e.* situations for which we use the term stress. Thus stress and reactions caused by it (stress reactions) can be used as a measure of the strength of the stress on a scale of intensity, ranging from deficiency to excessive supply. Environmental factors deviating from the optimal intensity or quantity for the plant are called stress factors. The optimal quantity can, in fact, be zero, *e.g.* with xenobiotics.

If the dosage is inappropriate, stress is caused, as is obvious with the effects of the following factors: light (weak light, strong light), temperature (cold, heat), water (drought, flooding), nutrients (lack of ions, over-fertilisation, salt stress), carbon dioxide and oxygen (photosynthesis, respiration/photorespiration, oxidative stress, anaerobiosis); optimal intensities and concentrations of these may also differ not only for individual organisms, but also for particular organs of the same organism. Environmental noxae are stress factors which trigger stress reactions when applied in any concentration or intensity: UV-B, ozone, ionising radiation, xenobiotics, heavy metals and aluminium. In this context, electrical and strong magnetic fields can also be considered as stress factors. Endogenous stress may also occur, for example, by separating an organ from its water supply, as is the case during ripening of seeds and the desiccation of embryo and endosperm.

1.1. Specific and Unspecific Reactions to Stress

An organism that is stressed, for example, by elevated temperature, not only increases its metabolic rate, but other reactions occur which are usually not observed in the unstressed organism, or take place only to a very small degree. An example of this is the formation of "heat shock proteins". The modification of the basic metabolism could be interpreted as an unspecific reaction, whilst the production of heat shock proteins would be considered a specific stress reaction of the organism. Both types of reactions are partly "specific and partly unspecific." Contrary to plants there is, in humans, also a strong psychic-humoral stress component. The concept of both components of the stress reaction is complicated by the fact that even the specific reactions often lack specificity: The above mentioned heat shock proteins also assist the folding of proteins during synthesis and after denaturing, not only by high temperature stress, but also under other stresses. They are produced in high amounts, for example, under stress by xenobiotics (*e.g.* heavy metals). This does not exclude that there are in addition more specific responses by which an organism differentiates between stress by heat and by heavy metals. There is yet another facet to the question of specificity of stress reactions which is described by the term cross-protection. Earlier drought stress or salt stress (osmotic stress) is known to harden plants against temperature stress, and particularly cold stress. Is this an unspecific stress response? The apparent lack of specificity of the adaptation is explained, on the one hand by considering the physiological effects of salt and drought stress on cells and, on the other, the effects of frost. All three factors lead to a partial dehydration of cells. This causes problems with the stability of biomembranes in particular, as the lipid bilayers are stabilised by so-called hydrophobic interactions, which are disturbed if the availability of water, or the ion concentration at the surface of membranes, is drastically changed. If too much water is removed from the aqueous environment of the biomembranes (by evaporation or freezing), the concentration of solutes increases, *e.g.* in the cytosol or the chloroplast stroma. Increase in the ion concentrations in turn changes with the charges at the surface of membranes, and as a consequence the membrane potentials. This usually leads to destabilization of membrane structure. High charge densities, however, not only result from water deficiency, but also from excessive salt concentration. A general reaction to stress is the synthesis of hydrophilic low molecular protectants, so-called compatible solutes (sugars, sugar alcohols and cyclitols, amino acids and betaines), which replace water at the membrane surfaces and dislodge the ionic compounds upon loss of cellular water. Production of compatible solutes requires, synthesis of respective enzymes, triggered by stress. Synthesis of these enzymes is often preceded by signals transmitted by certain phytohormones

– particularly abscisic acid (ABA) or the stress hormone jasmonate, but also ethylene, may transiently change their concentration. One example of such cross-protection is induction of frost hardening in wild potatoes by salt stress. Potato plants treated with NaCl are able to tolerate lower temperatures than untreated controls. A transient increase in ABA concentration mediates this hardening reaction.

Abiotic stress, as a natural part of every ecosystem, will affect organisms in a variety of ways. Although these effects may be either beneficial or detrimental, the location of the area is crucial in determining the extent of the impact that abiotic stress will have. The higher the latitude of the area affected, the greater the impact of abiotic stress will be on that area. So, a taiga or boreal forest is at the mercy of whatever abiotic stress factors may come along, while tropical zones are much less susceptible to such stressors.

2

Seed Germination, Seedling Growth and Root Development Under Abiotic Stresses

A. Bhattacharya

The most crucial step of plant life cycle, which ensures the survival of the next generation, is seed germination. Plants have evolved mechanisms that enable seed germination to be arrested under stress conditions and then resumed when conditions are favorable. Investigations are needed on seed enhancement treatments and environmental influences after radicle protrusion to understand implications to further root system development and, ultimately, stand establishment and seedling growth. Few investigations have focused on variability of growth and development of root systems or root components. Seed development and germination are crucial processes for plant survival.

Roots usually suffer greater exposure to multiple abiotic stresses than shoots. Therefore, the root system can be as affected, or even more affected, than the aerial parts of a plant by such stresses. Changes in the shoot:root ratio are often observed when plants are subjected to various stresses and a redistribution of metabolites from shoots to roots is frequently observed under drought, salt, or sub-optimal temperature stress, as well as during some nutrient deficiencies, or elevated levels of atmospheric CO_2*. Conversely, reductions in solar radiation or excess nutrient usually cause an increased shoot:root ratio. Plants subjected to increased atmospheric* CO_2 *concentrations, or to low-moisture regimes, may develop a more extensive root system; however, the other stressescommonly inhibit root growth, and cause significant modifications to the architecture of the root system, often giving rise to more branched root systems with shorter roots. A wide variety of hormones and biochemical processes are involved in the regulation of root growth under abiotic stress. Essential regulatory functions have been attributed to abscisic acid, ethylene, reactive oxygen species, and reactive nitrogen species.*

Germination is the process by which a plant grows from a seed. The most common example of germination is the sprouting of a seedling from a seed of an angiosperm or gymnosperm. In addition, however, the growth of a sporeling from a spore, such as the growth of hyphae from fungal spores, is also germination. Thus, germination can be thought of in a general sense as anything expanding into greater being from a small existence or germ, a method that is commonly used by many seed germination projects.

Germination is the growth of a plant contained within a seed and it results in the formation of the seedling. The seed of a vascular plant is a small package produced in a fruit or cone after the union of male and female sex cells. All fully developed seeds contain an embryo and, in most plant species some store of food reserves, wrapped in a seed coat. Some plants produce varying numbers of seeds that lack embryos; these are called empty seeds and never germinate. Dormant seeds are ripe seeds that do not germinate because they are subject to external environmental conditions that prevent the initiation of metabolic processes and cell growth. Under proper conditions, the seed begins to germinate and the embryonic tissues resume growth, developing towards a seedling.

Seed germination depends on both internal and external conditions. The most important external factors include temperature, water, oxygen and sometimes light or darkness (Raven *et al.,* 2005). Various plants require different variables for successful seed germination. Often this depends on the individual seed variety and is closely linked to the ecological conditions of a plant's natural habitate. For some seeds, their future germination response is affected by environmental conditions during seed formation; most often these responses are types of seed dormancy.

Water is required for germination. Mature seeds are often extremely dry and need to take in significant amounts of water, relative to the dry weight of the seed, before cellular metabolism and growth can resume. Most seeds need enough water to moisten the seeds but not enough to soak them. The uptake of water by seeds is called imbibitions. which leads to the swelling and the breaking of the seed coat. When seeds are formed, most plants store a food reserve within the seed, such as starch, proteins or oils. This food reserve provides nourishment to the growing embryo. When the seed imbibes water, hydrolytic enzymes are activated which break down these stored food resources into metabolically useful chemicals (Raven *et al.*, 2005). After the seedling emerges from the seed coats and starts growing roots and leaves, the seedling's food reserves are typically exhausted; at this point photosynthesis provides the energy needed for continued growth and the seedling now requires a continuous supply of water, nutrients, and light.

Oxygen is used in aerobic respiration, the main source of the seedling's energy until it grows leaves (Raven *et al.*, 2005). Oxygen is an atmospheric gas that is found in soil pore spaces; if a seed is buried too deeply within the soil or the soil is waterlogged, the seed can be oxygen starved. Some seeds have impermeable seed coats that prevent oxygen from entering the seed, causing a type of physical dormancy which is broken when the seed coat is worn away enough to allow gas exchange and water uptake from the environment.

Temperature affects cellular metabolic and growth rates. Seeds from different species and even seeds from the same plant germinate over a wide range of temperatures. Seeds often have temperature range within which they will germinate, and they will not do so above or below this range. Many seeds germinate at temperatures slightly above 16-24°C, while others germinate just above freezing and others germinate only in response to alternations in temperature between warm and cool. Some seeds germinate when the soil is cool (-2 - 4°C), and some when the soil is warm 24-32°C. Some seeds require exposure to cold temperatures (vernalization) to break dormancy. Some seeds in a dormant state will not germinate even if conditions are favorable. Seeds that are dependent on temperature to end dormancy have a type of physiological dormancy. For example, seeds requiring the cold of winter are inhibited from germinating until they take in water in the fall and experience cooler temperatures. Four degrees Celsius is cool enough to end dormancy for most cool dormant seeds, but some groups, especially within the family Ranunculaceae and others, need conditions cooler than -5°C. Some seeds will only germinate after hot temperatures during a forest fire which cracks their seed coats; this is a type of physical dormancy.

Most common annual vegetables have optimal germination temperatures between $24\text{-}32^0$ C, though many species (*e.g.* radishes or spinach) can germinate at significantly lower temperatures, as low as 4°C, thus allowing them to be grown from seeds in cooler climates. Suboptimal temperatures lead to lower success rates and longer germination periods.

Light or darkness can be an environmental factor trigger for germination and is a type of physiological dormancy. Most seeds are not affected by light or darkness, but many seeds, including species found in forest settings, will not germinate until an opening in the canopy allows sufficient light for growth of the seedling (Raven *et al.*, 2005).

2.1. Seed Germination

During their life cycle, plants experience a variety of abiotic stresses such as high and low temperatures, drought, salinity, heavy metals, UV and highlight, which can have a profound effect on viability, growth, morphology, and reproduction (Deborah *et al.*, 2011). These stresses lead to a significant reduction of agricultural products around the world. Soil contamination by heavy *metals* is one of the main reasons of ecological and environmental degradation in large area (Raskin and Ensley, 2000). Various operations such as using of pesticides in agriculture, mining operations, transportation of untreated industrial waste, and metallurgical industry lead to the release of various heavy metals like Cu, Fe, Cd, Cr and Hg into the environment (Chandra *et al.*, 2009). Accumulations of chemical environmental pollutants cause interruptions in ecological processes and energy chain such as chemical weathering and reduction in plant growth and energy production. It also causes physiological damages such as genotoxicity, impaired photosynthesis, reduced uptake and transport of mineral nutrients and growth retardation in plants (Nagajyoti *et al.*, 2010). Nickel is a heavy metal that is essential in small amounts for plant growth and development because it is involved in the structure of some important enzymes such as urease. Urea accumulates in leaves when nickel is deficient, and urease activity within the cell completely disappears resulting in the leaf tip necrosis (Taiz and Zeiger, 2007). Toxic and even fatality properties of nickel for humans, animals and plants in high concentrations have been found. In the culture medium, high concentrations cause a change in absorption of necessary elements, chlorosis, reduced absorption of CO_2, irregularities in the exchange of gases, changes in water absorption and free radical production and production of any reactive oxygen that cause oxidative stress (Ali *et al.*, 2009). Nickel is a metallic element that is easily absorbed by plant roots. It is moving in plant and its excessive accumulation in plants will reduce the value of plant products. Due to the toxic effects, high concentrations of nickel can inhibit plant growth (Zdeneck, 1999). Cadmium is a very strong pollutant due to extreme toxicity at low concentrations and its high solubility in water. The availability of cadmium in soil depends on organic matter in soil, root exudates, presence of mycorrhiza, pH, and cation exchange capacity of the soil, temperature and concentration of other elements in the soil. Cadmium alters nutrient absorption by plants by competing with potassium, magnesium, calcium, manganese, copper, zinc and nickel (Moreno *et al.*, 1999). Cadmium has a negative effect on plant metabolism, such as decreased nutrient absorption, inhibition of photosynthesis through effect on chlorophyll metabolism and chloroplast structure, photosynthetic carbon metabolism enzyme, photosystem II activity and changes in metabolism of nitrogen. Cadmium causes the closing of stomata and reduces the amount of plant water in the long term. Cadmium leads to lower growth and lower biomass in plants. Damage to DNA and changes in RNA synthesis is the effect of cadmium toxicity (Liang, 2005; Karantev, 2006; Popova, 2009).

Abiotic stress, such as salinity, drought, extreme temperature and heavy metals toxicity is an important cause of limiting plant growth and reducing average yields for most crop plants. However salinity is major problem in arid and semi-arid regions (Munns, 2002). High salinity includes both an ionic (chemical) and an osmotic (physical) component. Salinity becomes a concern when an excessive amount or concentration of soluble salts occurs in the soil or water. In general NaCl is the predominant salt causing salinization. Plants have mechanisms to regulate its accumulation (Munns and Tester, 2008). Salinity induces specific changes on cell, tissue and organ levels. These changes are physiological, hormonal balance, morphological and anatomical (Foolad, 1996). The establishment stage of the crop consists of three parts : germination, emergence and early seedling growth; that are particularly sensitive to substrate salinity (Jamil *et al.*, 2005). Germination is reactivation of growth triggered by environmental stimuli as simple as availability of water and oxygen, or as complex as temperature, light, endogenous inhibitor and promoter interactions. Successful seedling establishment depends on the frequency and the amount of precipitation as well as on the ability of the seed species to seed germination and grow while soil moisture and osmotic potentials decrease (Welbaum *et al.*, 1991). The decrease in germination rate particularly under drought and salt stress conditions may be due to the fact that they prevent germination development and osmotically enforced dormancy under water stress conditions. This may be an adaptive strategy of seeds to they prevent germination under stressful environment thus ensuring proper establishment of the seedling (Goll *et al.*, 2003). Several investigation of seed germination under salinity stress have indicated that seeds of most species attain their maximum germination in distilled water and are very sensitive to elevated salinity at the germination and seeding phases of development (Gulzar *et al.*, 2003).

In many crop species seed germination and early seedling growth are the most sensitive developmental stages to salinity stress. Salinity stress greatly delays onset and reduces the rate of germination events. It has been reported that salinity limits plant growth and productivity (Ayaz *et al.*, 2000). There are big changes in morphology and anatomy of plants growing in saline soils. Salinity generally affects growth rate and it results in plants with smaller leaves, shorter stature and roots' structure by reducing their length and mass (Munns and Termaat, 1986). Tomatoes (*Solanum lycopersicum*) are one of the most important vegetable crops and is moderately sensitive to salinity (Cuartero *et al.*, 2006). Salinity reduces tomato seed germination and lengthens the time required for germination. Salinity also reduced the fresh and dry shoot and root weight of tomato (Levitt, 1980). Increased salinity over 4000 ppm led to reduction in dry weight, stem, and roots, leaf area (Omar *et al.*, 1982). The reduction of dry weights due to increased salinity may be a result of a combination of osmotic and specific ion effects of Cl and Na (Al-Rwahy, 1989). The leaf and stem dry weights of tomato were also reduced significantly in plants irrigated with saline nutrient solution in contrast with control plants (Satti and Al-Yahyai, 1995). Byari and Almaghrabi (1991) found that tomato cultivars varied greatly in their response to different salinity levels.

The detrimental effects of high salinity and drought on plants can be observed at the whole-plant as the death of plants and/or decreases in productivity (Allakhverdive *et al.*, 2000b). During the onset and development of salt and drought stress within a seed, all major processes such as protein synthesis and energy and lipid metabolism are affected. Carbohydrates and other substrates which are needed for cell growth are mainly supplied through the process of photosynthesis, and photosynthesis rates are usually lower in plants exposed to salinity and especially to NaCl. Effect of salinity and drought stress on plant or seed responses have been discussed (Hasegawa *et al.*, 2000; Zhu, 2002). Drought along with salinity stresses is the most important and effective environmental factors in agricultural practices in arid and semi-arid regions of the world. Extreme environmental conditions like high temperature and low water exerts severe abiotic stress in the microenvironment of crops. Drought is one of the major environmental factors causing low productivity and growth (Peleg *et al.*, 2011). The detrimental drought effects could be listed as the loss in yield and quality; and in certain scenarios the death of plant. Reportedly, the application of silicon is known to enhance crop tolerance against various environmental stresses by tailoring the plant water uptake and transport. Silicon, actually, induces dehydration tolerance at tissue or cellular levels by improving the water status (Gao *et al.*, 2004, 2006). This in return, facilitates the plant photosynthetic access and availability to the plant.

Crop yield losses due to water scarcity are much more adverse than other environmental stresses. Plants cannot survive without water. Water deficiency adversely has effects on number of physiological and metabolic mechanism in plants. Drought inhibits the growth at its first instance, as lack of water hampers the photosynthetic rates. Drought is the main factor hindering to reach a crop maximum theoretical yield (Mitra, 2001). It is one of the most limiting factors to the crop productivity. It has a differential effect on crop's micro and macro-environment across a common field, different plant growth stages and the plant material to be harvested. Promising future researches need to be focused on a wide range of genetic, metabolic, environmental and physiological factors for improvement of drought resistance in crops (Araus *et al.*, 2008). The drought resistance recently has drawn the future research focus due to exhausting water availability and distribution (Habash *et al.*, 2009). This goal could only be promised by complete knowledge of genetic and molecular basis of drought resistance (Nachit and Elouafi, 2004). Blum (2005) depicted that plant adopted various mechanisms to face drought *e.g.* osmotic adjustment, cell membrane stability, epicuticular wax, partitioning and stem reserve mobilization, manipulation and stability of flowering processes, seedling drought traits. Plant transpiration is the second drought victim to hamper the plant efficiency by decreasing transpiration that results in lowering wheat yield productivity to a greater extent. Carbon-dioxide assimilation and chlorophyll content were also influenced by drought (Monneveux *et al.*, 2006). Chlorophyll contents were found to increase in leaves in short term water stress during the vegetative phase, whereas the reverse has been noticed (decrease in chlorophyll) during long term stress. Chlorophyll content, stomatal conductance, transpiration rate were known to strongly influence photosynthetic rate. Drought resulted to decrease in transpiration rate, stomatal conductance and photosynthetic rate of the seedlings of four spring wheat varieties while water use efficiency of leaves was increased (Chang-juan, 2012). Therefore, drought not only influences photosynthetic and transpiration rate, but also affects chlorophyll. Therefore, drought not only influences photosynthetic and transpiration rate, but also affects chlorophyll content and CO_2 assimilation rate to a considerable degree.

2.1.1. Drought Tolerance in Plants

Drought is a condition when an area faces a lack of precipitation and high evapotranspiration resulting to non-availability of minimum amount of water. Drought tolerance means a condition when plants are adapted to drought or arid condition by modifying physio-morphic features. These features include leaf water potential, stomatal frequency, stomatal size, osmotic adjustments (*Ahmadet al.*, 2006). Water stress is a severe environmental and production constraint for a variety of food crops. Water is turning as a rare commodity and its severity has been predicted in the years to come. Drought alters all elements of the plant growth and the capability of plant to rehabilitate whereas different physiomorphic traits can amend the stress. These traits are genetically complex and are not simple to handle. Drought-induced loss in crop yield probably exceeds losses from all other causes, since both the severity and duration of the stress are critical. Drought stress affects the growth, phenology, water and nutrient relations, photosynthesis, assimilate partitioning, and respiration in plants. Proline and quaternary ammonium compounds *e.g.* glycinebetaine, choline, prolinebetaine are key osmolytes contributing towards osmotic adjustment (Kavikishore *et al.*, 2005). Furthermore, breeding for drought tolerance in crops is economical and considered to be an effective way of coping with drought stress (Blum, 1999).

2.1.2. Osmotic Adjustment

When a plant experiences water stress, certain osmotically active compounds such as amino acids, ions and sugars inside the cell accumulate, which results in lowering osmotic potential of cell. It is reported to be an important mechanism in drought conditions (Yousifi *et al.*, 2010). After the water accumulates in intercellular spaces, it will move inside the cell called as osmotic adjustment. osmotic adjustment is the possible phenomena that will maintain the turgor and plant can survive in lack of water and facilitates root growth (Gonzalez *et al.*, 2010). Osmotic adjustment was found as a positively related to grain yield in drought conditions in sorghum (Tangpremsri *et al.*, 1995), wheat (Zivack *et al.*, 2009), barley (Gonzalez *et al.*, 2009), chickpea (Moinuddin and Khanna-Chopra, 2004), and pigeon pea (Subbarao *et al.*, 2000). Verities with high osmotic adjustment are reported to have better grain yield than plants with low osmotic adjustment (Moinuddin and Khanna-Chopra, 2004). Although this adoptive mechanism seems beneficial, certain studies suggested to have negative (Kirkham, 1988) or no relation (Tangpremsri *et al.*, 1995) with the grain yield.

2.1.3. Cell membrane Stability

Cell membrane stability is ability of a cell to live in dehydration. This ability is particularly useful in osmotic adjustment. Osmotic adjustment coupled with cell membrane stability help plant to reduced water conditions during drought. Different crops have different tolerance to this stress depending upon the cell membrane stability. The tolerance depends upon cell membrane stability because it prevents electrolyte leakage in drought condition and maintains the cell activity in abiotic stresses like salinity (Ghoulam *et al.*, 2002), drought (Blum 2011, Yang *et al.*, 2009), air pollution (Neves *et al.*, 2009) and heat, cold (Arvin and Donnelly, 2008). Therefore cell membrane stability is considered to be one of the stress adaptive mechanisms under drought stress and is helpful in increasing yields as it maintains the cell activity in stress.

2.1.4. Epicuticular Wax

Epicuticular wax is the organic compounds of cuticle which covers the outer surface of plant tissues (Li *et al*, 2007). Epicuticular wax is very helpful for reducing leaf surface water losses. Species with thick cuticle layer retain more moisture and maintain leaf turgor for longer period of time in stress conditions due to reduced evapotranspiration. The wax composition of species may differ for different parts of the same plant and may vary with season, locale, and the age of the plant (Eglinton and Hamilton, 1967).

2.1.5. Partitioning and Stem Reserve Mobilization

In drought conditions the process of photosynthesis is abridged and grain yield then have to rely upon stem reserves. The stem reserves buffer in stress condition and help in maintaining the grain yield by helping in photosynthesis and nutrient uptake during grain filling stage (Tahir and Nakata, 2005). The contribution of these assimilates to seed may be effected in severe stress conditions when photosynthesis is reduced and water, mineral uptake are limited (Arduini *et al.*, 2006). In severe stress, the grain filling varies in different species, varieties and environment; and a large mobilization can occur in low soil fertility (Masoni *et al.*, 2007). The partitioning and stem reserve mobilization starts from plant senescence, which transfers stored reserves from stems to seed (Ercoli *et al.*, 2008). The previous study depicted that amount of reserve mobilization depends upon environment, genotype and cultural practices (Aynehband *et al.*, 2011). A continuous rise in temperature during grain filling can cause a severe reduction in dry matter production (Alvaro *et al.*, 2008). This mechanism encourages a decrease in gibberallic acid and increase in abscisic acid concentration (Chinnusamy *et al.*, 2004).

2.1.6. Salt Stress and Seed Germination

Shokohifard *et al.* (1989) reported that salt stress negatively affected seed germination; either osmotically through reduced water absorption or ionically through accumulation of Na and Cl causing an imbalance in nutrient uptake and a toxic effect. Furthermore, Younis *et al.* (1991) reported that low moisture content under salt stress caused a cessation of metabolism or inhibition of certain steps in metabolic sequences of germination. Conversely, salt stress increased the intake of toxic ions that may have altered certain enzymatic or hormonal activity in seeds during germination (Smith and Comb, 1991). Moreover, salinity perturbs a plant's hormone balance (Khan and Rizvi, 1994) and reduces utilization of seed reserves (Ahmad and Bano, 1992). Decreasing germination due to increased salinity can be connected to the nature of salinity that reduces imbibitions of water due to lowered osmotic potential of the medium and changes in metabolic activity (Yupsanis *et al.*, 1994). Salt induced inhibition of seed germination could be attributed to osmotic stress or to specific ion toxicity (Huang and Reddman, 1995). It appears that a decrease in germination is related to salinity-induced disturbance of the metabolic process leading to an increase of phenolic compounds (Ayaz *et al.*, 2000).

It had been demonstrated that elevated salinity slowed down water uptake by seeds, thereby inhibiting germination and root elongation. Significant reduction in seed germination was observed in canola seeds grown under salinity conditions (Puppala *et al.*, 1999). Both NaCl and KCl have been reported to reduce the final germination percentage and germination rate of two arid-land varieties of wheat (*Triticum aestivum* L.); furthermore, the salt had increased the production of abnormal seedlings (Al-Ansari, 2003). The seed germination of *Panicum turgidum* has been significantly reduced and slowed at high concentrations of both NaCl and KCl and was completely inhibited at 300 and 400 mM (El-Keblawy, 2004). In another report, four lentil (*Lens culinaris* M.) genotypes were treated with salt stress (0, 50, 100, 150 or 200 mM NaCl), and it was reported that the increasing NaCl concentration reduced the germination percentage, the growth parameters and the relative water content (Sidari *et al.*, 2007). Pratap and Sharma (2010) reported that germination percentage and all of the seedling growth parameters showed inhibition with an increase in osmotic potentials generated by polyethylene glycol (PEG) 6000 in black gram (*Phaseolus mungo*). The concentration limits for germination in *Dianthus chinensis* L. have been reported to be 150 mM for NaCl and -1.66 MPa for PEG 6000; again, it was reported that NaCl had exerted less effect than PEG on the final germination percentage (He *et al.*, 2009). Similarly, both NaCl and PEG-6000 have been shown to inhibit the germination and seedling growth in two *P. mungo* varieties, where the effect of NaCl was also shown to be less than PEG-6000 (Garg, 2010). Seed germination and the growth of young seedlings of sugar beet (*Beta vulgaris* L.) have been reported to be inhibited by NaCl treatment (Wang *et al.*, 2011). Bahrami *et al.* (2012) studied that sesame seed germination was strongly affected to 98, 95.7, 87.4, 78.5, 50.65 and 15.02% with the salinity levels of 0, 5.3, 8.5, 12.05, 14.65 and 18.45 dSm^{-1}, respectively.

2.2. Seedling Drought Traits

In drought conditions seedlings are planted deep to reach the receding water profile so that drought can be avoided and ultimately to help plant protected from high surface temperatures causing germination inhibition. Deep placed seeds will be able to capture more water from soil and exclude salts from root cells (Vadez *et al.*, 2005).

Drought stress is one of the most restrictive of seed germination. Seed germination is negatively affected by unfavorable conditions of soil moisture due to lack of rainfall and irrigation (Mwale *et al.*, 2003). At planting, inadequate moisture leads to irregular germination and seedling emergence, which influence establishment and performance of the species as a crop (Mwale *et al.*, 2003; Okcu *et al.*, 2005). In arid and semi-arid regions, water is usually one of the most important factors limiting crop production. Seed germination and emergence are critical stages for plant establishment and crop growth in arid and semi-arid regions and help determine standing crop density and yields (Hadas, 1976). High levels of salt in the soil can often create limitations for agricultural production and development. The main factors that lead to this problem are the semi-arid climate and addition of salt by water that is used for irrigation. Soil salinity may lead to deleterious effects of on plant growth and development, physiological and biochemical in plants (Munns, 2002). These effects could be due to low osmotic potential of soil solution, specific ion effects and nutritional imbalances, or a combination of these factors (Zalba and Peinemann, 1998). High levels of salt in the germination media can lead to a high osmotic pressure that prevents water absorption necessary for germination. High concentrations of salts have toxic effects on the embryo and can reduce or delay germination (Ramden, 1986). Salinity reduces germination, root length, callus size, coleoptile length and seedling growth (Lallu and Dixit, 2005; Ghannadha *et al.*, 2005; Bera *et al.*, 2006; Agnihotri *et al.*, 2006). Different stages of the plant life cycle, such as seed germination, seedling growth, vegetative growth, flowering and fruiting are adversely affected by non-biological stresses. In general, seeds and seedlings may have less tolerance than mature plants (Bhattacharya *et al.*, 2012).

Seeds are the main way through which plants propagate, and a seed contains all of the genetic material of the plant. As seed germination is the beginning of the life cycle of plants, seedling emergence is critical for the establishment of plant populations (Khan and Gulzar, 2003). Although salt stress affects all growth stages of a plant, seed germination and seedling growth stages are known to be more sensitive for most plant species (Cuartero *et al.*, 2006) and germination has been reported to decline with increasing salinity levels (Houle *et al.*, 2001). Germination failures on saline soils are often the results of high salt concentrations in the seed planting zone because of upward movement of soil solution and subsequent evaporation at the soil surface. Soil salinity affects germination by either an osmotic stress or ion toxic effect (Bewley and Black, 1982). Salinity may create an external osmotic potential limiting water absorption by the seeds, or sodium and chloride ions may accumulate in the germinating seed, resulting in a toxic effect. Under drought and salt stress, the mobilization of stored reserves is prevented or reduced in seeds (Bouaziz and Hicks, 1990), and the structural organization and synthesis of proteins are restricted in germinating embryos (Ramagopal, 1990). Seed weight and consequently viability percentage and germination rate decreases as salinity increases. Salinity causes not only differences between the mean yield and the potential yield, but also causes yield reduction from year to year.

Germination characteristics are usually essential process in seedling establishment and plant development to obtain seedling numbers those results in higher seed crop (Almansouri *et al.*, 2001, Murungu *et al.*, 2003). Seed germination and establishment are the most sensitive stages to abiotic stress in most crop species (Patade *et al* 2011). Abiotic stress may delay seed germination and reduce the rate (Patade *et al.*, 2011. Rouhi *et al.*, 2011. Ansari *et al.*, 2012). Patade *et al.* (2009) and Ashraf and Foolad (2005) reported that seed priming is one of the methods that can be taken to counteract the adverse effects of abiotic stress. The purpose of seed priming is to a partially hydrated with water, or various chemical solutions the seeds to a point where germination processes are begun but not completed (Asfraf and Foolad., 2005), followed by drying of seeds to the original moisture level. Seed priming techniques have been used to increase germination, improve germination uniformity, improve seedling establishment and stimulate vegetative growth in more field crops (Iqbal and Ashraf, 2007; Kaya *et al*, 2006b; Kaur *et al.*, 2002; Patade *et al.*, 2011; Saglam, *et al.*, 2011; Sadeghi *et al.*, 2011), under stressed conditions. Also, the priming strategies enhanced activities of free radical scavenging enzymes such as catalase (CAT) and superoxide dismutase (SOD) (Ansari *et al.*, 2012).

Germination of seed is strongly influenced by variation in temperature, water stress and light requirement and these factors often show significant interaction in their effects on germination (Hampson and Simpson, 1990). Water stress causes both reductions in the rate of protein synthesis as well as changes in the type of proteins produced. It is believed that these stress induced proteins that allow plants to make biochemical and structural adjustments that enable plants to cope with the stress. All the plants have an inbuilt ability to adjust to environmental variables. Aboitic stress negatively influences survival grain yield, biomass accumulation and production of most crops (Grover *et al.*, 2001; Bhattacharjee, 2008; Ozdener and Kutbay, 2008). In sandy soils, where moisture retention is poor, the development of stress is rapid and soil moisture deficit adversely influences the metabolism, growth and yield of crops. Vyas *et al.* (1996) reported that increasing water stress progressively decreased the activities of nitrate reductase, glutamine synthetase and glutamate synthase, and soluble protein content in moth bean leaves. Garg *et al.* (2001) also found that increasing water stress progressively decreased plant water potential, leaf area, net photosynthetic rate, starch and soluble protein contents and nitrate reductase activity in moth bean genotyps. In several arid zone crops genotypic differences in response to water stress exist (Garg *et al.*, 1998; Kuhad and Sheoran, 1986). The reactions of the plants to water stress differ significantly at various organizational levels depending upon intensity and duration of stress as well as plant species and its stage of development (Chaves *et al.*, 2003). Global climate change, in the form of rising temperature and altered soil moisture, is projected to decrease the yield of food crops over the next 50 years (Thomson *et al.*, 2005). Global changes include mainly temperature rise and water deficiency. The lack of moisture in soil directly affects seed germination and seedling growth in plant.

Plant resort to many adaptive strategies in response to different abiotic stresses such as high salt, dehydration, cold, heat and excessive osmotic pressure which ultimately affect the plant growth and productivity (Yancey *et al.*,1982). Against these stresses, plants adapt themselves by different mechanisms including change in morphological and developmental pattern as well as physiological and biochemical processes (Bohnert *et al.*, 1995). Adaptation to all these stresses is associated with metabolic adjustments that leads to the accumulation of several organic solutes like sugars, polyols, betaines and proline (Yancey *et al.*,1982). Among these accumulating solutes, sugars represent the major reserve in the seeds (Bewley and Black, 1994) which maximally synthesized during germination and mobilized to various tissues like stem and internods (Smith, 1967) in the form of sucrose, glucose and fructose, that are readily transportable to sites where they are required for growth (Mayer and Poljkoff-Mayber, 1975) and maintained the osmotic regulation of cells (Garham *et al.*, 1981). There are earlier reports on carbohydrate accumulation during various abiotic stresses in the temperate grasses and cereals from the gramineae family where long term carbohydrate storage occur during reproductive development (Meier and Reid, 1982). Accumulation of sugars in different parts of plants is enhanced in response to the variety of environmental stresses (Prado *et al.*, 2000. In case of salt (Lin and Kao, 1995) and water stress (Prado *et al.*, 2000), adaptation to these stresses has been attributed to the stress induced increase in carbohydrate level. There are few studies on carbohydrate status in germinated seeds and its early developmental stages under stress conditions. The metabolism of these compounds can be affected by a number of environmental factors such as irradiance, temperature, salinity and type of ion present (Bohnert *et al.*, 1995). Thus the variation that occurs in carbohydrate level, during early developmental stages of seedling under different abiotic stresses is not well understood and information on physiological events involved in this process is scarce. In this report, we present details on growth and status of soluble carbohydrates in the developing seedlings of sorghum under different abiotic conditions. Up to best of our knowledge it is the first report on sorghum which try to compare the results obtained from different stresses in view to describe the common physiological tolerance of this important for arid and semiarid regions crop (Aronson, 1985).

The Three Phases of Water Uptake During Seed Germination

Process of seed development in higher plants is divided into embryo and endosperm development and seed maturation (Gutierrez *et al.*, 2007; Linkies *et al.*, 2010). Seed maturation is completed when storage compounds have accumulated, water content has decreased, desiccation tolerance has been developed, phytohormones level have been established and, in the case of dormant seeds, primary dormancy have been established. The *Arabidopsis* seed is composed of the embryo surrounded by a single layer of endosperm cells and a testa (Debeaujon and Koornneef, 2000). After the seed shape and size are determined, germinating seeds start to uptake water in order to set the metabolic events essential for germination in motion (Nonogaki, 2008a). This water uptake by seeds is tri phasic (Fig. 1). During Phase I, dry seeds absorb water rapidly. Initially, imbibition results in the hydration of the cell walls and reserve polymers within the cells. The kinetics of uptake depends on the increasing number of cells within the seed, which become hydrated (Nonogaki, 2008a). Then, the seed moves to Phase II, which is called the plateau because of the stable water content. During this phase, testa rupture begins. This is followed by Phase III of water uptake when the endosperm rupture and protrusion of radicle is observed. The emerging radicle is a sign of completed germination (Bentsink and Koorneef, 2008; Weitbrecht *et al.*, 2011). During initial imbibition, massive leakage of cellular solutes is observed. Taking into account that the concentration of inhibitors, for example, phenolic compounds, decreases within seeds due to the leakage, this process accelerates seed germination (Matilla *et al.*, 2005).

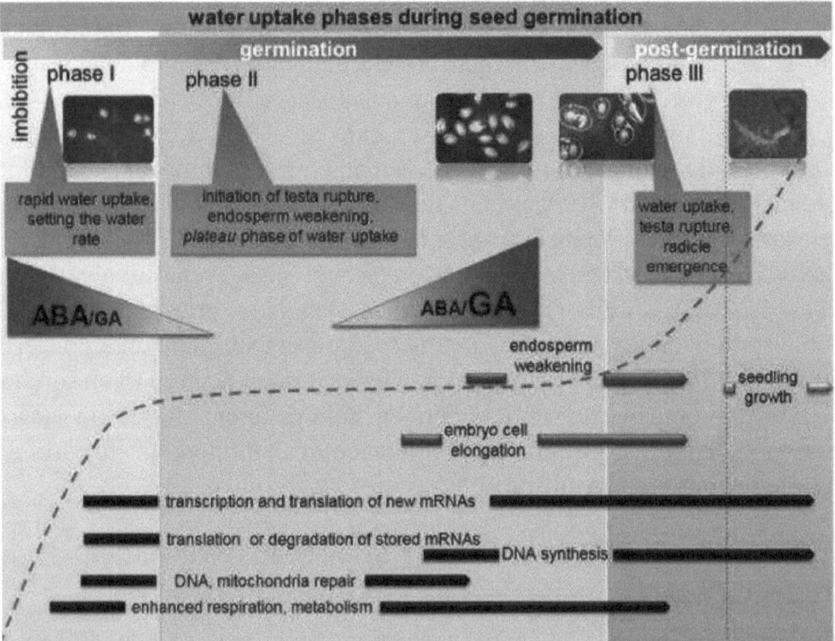

Fig. 1: Critical processes during the germination of typical endospermic eudicot seeds with a separate testa and endosperm rupture (two-step germination). Time courses of *Arabidopsis thaliana* seed water uptake (Ogawa *et al.*, 2003; Preston *et al.*, 2009).

However, leakage is also the sign of damage to the cellular membranes caused by rehydration. Cellular membrane structures, which are maintained in the gel phase during maturation, undergo a transition to a liquid crystalline phase. The processes of drying and rapid rehydration also influence DNA integrity (Matilla *et al.*, 2005). In order to overcome the consequences of the damage, seeds switch on a number of repair mechanisms during seed germination, including the repair of membranes, as well as proteins and DNA (Oge *et al.*, 2008).

Although a precise description of genes and hormones involved in the seed coat development is available for *Arabidopsis*, yet little is known about the changes in the seed coat biochemical and physical properties that lead to testa rupture (Leubner-Metzger, 2003). Endosperm rupture by the expanding radicle is closely associated with endosperm weakening. The process of endosperm weakening is linked with cell wall remodeling enzymes (CWRE), which include endo-β-mannanase (Bewley, 1997), b-1,3-glucanases (Leubner-Metzger, 2003), xyloglucan endotransglycosylase (Chen *et al.*, 2002), and others. Prior to the completion of seed germination, the embryo cells elongate. It is not clear which cells are responsible for the initial emergence of the radicle. The emergence could be facilitated by a reduction in the resistance of the tissue surrounding the radicle through specific modifi- cations of the cell walls by CWRE (Schopfer, 2006). This implies that embryo growth during germination depends on changes in the extensibility of the cell wall and is accompanied by progressive vacuolation during late water uptake in Phase II (Bethke *et al.*, 2007). Sliwinska and coworkers (2009) recently reported that embryo elongation occurs in the basal part of the hypocotyl and the hypocotyls-radicle transition zone.

2.3. Seed Germination Under Abiotic Stress

In natural environmental conditions, seeds have to overcome abiotic stresses in order to ensure the survival of the next generation. The ability to tolerate and resist salt, osmotic stress, dehydration, and cold stress during germination is essential for plant survival. When taking into account changes in water content in a seed and triphasic water uptake during imbibition prior to radicle emergence, it is obvious that seed germination cannot take place in the absence of water. When sudden abiotic stress, osmotic stress, salt, or dehydration occurs between a seed's first exposure to water and cotyledon greening, the germination process can be arrested. This process is regulated by a network of transcription factors that have both overlapping and discrete functions (Finkelstein, 2004; Finch-Savage and Leubner-Metzger, 2006; Hilhorst, 2007). In order to maintain or break the period of arrested germination and to complete germination, different metabolic pathways including phytohormones biosynthesis and signal transduction pathways, chromatin modifications, and micro-RNA posttranscriptional regulation, are involved.

Salt stress can influence many physiological processes from seed germination to plant development. The complexity and severity of a plant's response to salt stress can be explained by the fact that salinity works on two levels: ionic toxicity and osmotic stress. Different sets of genes regulate response to salt during different stages of development (Quesada *et al.*, 2002). At the cellular level, salt tolerance mechanisms function to reduce sodium accumulation in the cytoplasm. This is achieved by the active transport of sodium out of the cell through plasma membrane Na^+/H^+ antiporters, that is, salt over sensitive1 (Salt Oversensitive 1) (Shi *et al.*, 2005) or addressing it into the vacuole by tonoplast antiporter NHX1 (Na^+/H^+ Exchanger) (Blumwald *et al.*, 2000; Chung *et al.*, 2008).

In order to cope with dehydration and osmotic stresses, plants have evolved complicated machinery to synthetize and accumulate metabolites, "osmoprotectants," which help to withstand osmotic pressure and maintain turgor and the driving gradient for water uptake. Osmoprotectants include low molecular weight compounds such as amino acids, polyols, sugars, and methylamines (Hasegawa *et al.*, 2000). These compatible metabolites stabilize the enzyme structure, cellular membranes, and other cellular components during any exposure to stress (Chaves *et al.*, 2009; Zhu, 2001, 2002, 2003). In response to stress, *de novo* synthesis of dehydrins, osmotins, and LEA (Late Embryogenesis Abundant) proteins occurs. Osmotins and dehydrins stabilize the integrity of the cellular membranes and protein structure, whereas LEA are able to sequestrate ions and water in order to protect cellular components against stress (Hundertmark *et al.*, 2011; Kosova *et al.*, 2011).

ABA and its role in response to abiotic stress during seed germination

Before imbibition, over 10,000 different mRNAs are stored in the dry seeds of *Arabidopsis*(Okamoto *et al*, 2010). Among them, transcripts of genes that have ABA responsive elements within their promoter sequences are overrepresented. This implies that ABA has a significant role during seed development (Nakabayashi *et al.*, 2005). Many components of ABA signaling take part in the regulation of seed germination under abiotic stress. Kinases are important for stress response; some like SnRK2 (Sucrose Non-fermenting Related Kinases 2) are known to be signal mediators. The identification of ABA receptors—PYR/PYL/RCAR (Pyrabactin Resistance1 / PYR1-LIKE / Regulatory Component of ABA Response 1)—has provided a breakthrough in understanding the relations between some ABA signaling components

(Ma *et al.*, 2009; Nishimura *et al.*, 2010; Park *et al.*, 2009; Santiago *et al.*, 2009). These ABA receptors regulate the phosphatases PP2C (Protein Phosphatases 2 C) interacting with SnRK2 (Fujii *et al.*, 2009; Melcher *et al.*, 2009; Nishimura *et al.*, 2010; Santiago *et al.*, 2009; Yin *et al.*, 2009; reviewed in Umezawa *et al.*, 2010). SnRK2 are important for activating transcription factors that are crucial for seed germination, such as ABI5 (ABA insensitive 5) (Fujii *et al.*, 2007; Nakashima *et al.*, 2009; Wasilewska *et al.*, 2008). Three well-characterized transcription factors ABI3 (ABA insensitive 3), ABI4 (ABA insensitive 4) and ABI5 (ABA INSENSITIVE 5) were isolated in screens for ABA insensitivity during germination. They belong to different transcription factor families : B3-, APETALA2-, and bZIPdomain and regulate overlapping subsets of seed-specific and/or ABA-related genes (Finkelstein and Lynch 2000; Lopez-Molina and Chua, 2000). ABI3 and ABI4 are expressed from globular stage embryogenesis, and their products can regulate ABI5 expression, which is activated at the heart stage. ABI5 also activates its own expression, similarly to ABI4 (Bossi *et al.*, 2009; Garcia *et al.*, 2008). All three transcription factors are most highly expressed in mature seeds, but differ slightly in localization within the seeds (Brocard *et al.*, 2002; Lopez-Molina *et al.*, 2002; Penfield *et al.*, 2006). Although abi3, abi4, and abi5 mutants were initially isolated as ABA-insensitive, they have also been shown to have defects in response to glucose, NaCl, and osmotic stress during germination and seedling growth (Arenas-Huertero *et al.*, 2000; Brocard *et al.*, 2002; Laby *et al.*, 2000; Quesada *et al.*, 2002).

During seed imbibition under nonstress conditions, the level of ABI3 and ABI5 mRNAs and proteins rapidly decrease in order to enable the completion of the germination process. A decrease in the level of proteins occurs through the interaction of RING E3 ligases: AIP2 (ABI3-interacting protein 2) with ABI3 and KEG (keep on going) with ABI5 (Stone *et al.*, 2006; Zhang *et al.*, 2005). This action of the ubiquitin/26S proteasome system can be arrested or delayed in response to osmotic stress or a high exogenous ABA signal. In the presence of abiotic stresses and/or exogenous ABA, the degradation process of ABI3 and ABI5 is slowed. It has been shown that ABA promotes KEG self-degradation via the ubiquitin dependent 26S proteasome pathway (Liu and Stone, 2010). ABA and/or abiotic stresses induce the expression of ABI3 and ABI5 in order to activate their respective signal cascade transduction (Stone *et al.*, 2006; Zhang *et al.*, 2005). ABI5 action is also regulated by sumoylation, which both represses its activity and increases its stability (Miura *et al.*, 2009). Post translational modification of the ABI3, farnesylation, occurs through its interaction with ERA1 (enhanced response to ABA 1) (Brady *et al.*, 2003). Recent findings of Reeves and coworkers (2011) have shown that ABI4 is also regulated in the proteasome-mediated degradation pathway in order to prevent proteins from accumulating to high levels. The mechanisms and their components remain unknown.

Another ABA-related gene that regulates the response to salt, osmotic, and cold stress during seed germination is RAP2.6 (Related to AP*ETALA* 2), initially isolated as activated by JA, ethylene, SA, and two virulent strains (Wang *et al*, 2008). The expression of RAP2.6 is also upregulated by ABA, salt, osmotic, and cold stresses (Zhu *et al.*, 2010). Overexpression lines of RAP2.6 displayed a hypersensitivity to ABA and abiotic stresses during germination and early seedling development. Stomatal movements, drought, and salt tolerance were studied in adult wild-type and transgenic plants. No significant differences were observed, which suggests that RAP2.6 plays a role in the ABA response to abiotic stress during seed germination and early seedling growth. Further studies showed the presence of abiotic- (ABRE, DRE, MYBR) and biotic- (W-box, RAV-box) related cis-regulatory motifs within the promoter region of RAP2.6. These data suggested that RAP2.6 may serve as a common point in crosstalk, not only for hormones, but also for biotic and abiotic stress responses (Zhu *et al.*, 2010).

RACK1 (Receptor for Activated C Kinase 1) is another ABA response gene associated with salt stress response during germination. Originally, RACK1 was identified as a receptor of activated protein kinase C (PKC) in mammalian cells (Ron *et al.*, 1994). It is now known that this multifunctional protein plays a role in the regulation of many of metabolic pathways. The sequence and structure of RACK1 is highly conserved in plants (Chen *et al.*, 2006; Guo *et al.*, 2007). The response of *Arabidopsis* rack1a mutants to salt stress during germination revealed its hypersensitivity, which is consistent with its hypersensitive response to ABA. The ABA-responsive marker genes, RAB18 (Responsive to ABA 18) and RD29B (Responsive to Desiccation 29B), showed upregulation in rack1a mutant. These data suggest that RACK1 encodes a critical regulator of ABA signaling in terms of salt stress response during seed germination (Guo *et al.*, 2009).

Gibberellins—another group of phytohormones involved in stress response

Seed germination is antagonistically controlled by ABA and GAs. GAs promote seed germination by enhancing the proteasome-mediated degradation of RGL2 (RGA-LIKE2), a negative regulator of GAs signaling, primarily controlling seed germination. RGL2 has a conserved DELLA motif, which is essential for its proteasome-mediated degradation (Lee *et al.*, 2002). After binding GAs to the receptor GID1 (gibberellin insensitive dwarf1), SLY1 (sleepy 1)—an F-box protein interacts with RGL2 facilitating its ubiquitination and subsequent destruction (Feng *et al.*, 2008). There are four other DELLA factors : RGA (repressor of ga1–3), GAI (gibberellin insensitive), RGL1, RGL3. All of these are expressed during seed germination (Tyler *et al.*, 2004). The function of RGL2 in salt stress response was highlighted by Kim and coworkers (2008), who observed an elevated RGL2 transcript level in seeds germinated under salt stress and Achard *et al.* (2006), who showed that a quadruple DELLA mutant (gai, rga, rgl1, rgl2) is more sensitive to salt stress than the wild-type and DELLA single mutants.

Crosstalk between ABA and GAs under stress during seed germination

Ko and coworkers (2006) reported that RGA over accumulation up-regulates XERICO and subsequently stimulates ABA synthesis. XERICO encodes a RING-H2 zinc-finger factor, which promotes ABA accumulation in an unknown mechanism (Zentella *et al.*, 2007). An increased level of endogenous ABA activates ABI5 expression, which consequently arrests seed germination. Yuan and coworkers (2011) showed an interaction between components of ABA (ABI3 and ABI5) and GAs (RGL2) signaling pathways in response to salt during germination and early seedling development. This strongly supported the cooperation between these two phytohormones that was already known. The findings of Yuan and others (2011) have enriched knowledge about the involvement of GAs in abiotic stress response. They studied two ABA signaling mutants, well known for their salt insensitive phenotype, abi3 and abi5, and the GAs signaling mutant rgl2. The rgl2, abi3 and abi5 are resistant to high sodium chloride (salt) and mannitol (osmotic) concentrations during germination. These data suggest that both ABA and GAs are involved in salt and osmotic stress inhibition and the delay of seed germination. The level of expression of RGL2, ABI3, and ABI5 was investigated in each mutant background with similar results. After treatment of seeds with NaCl, the expression of RGL2, ABI3, and ABI5 was significantly lower in any mutant background than in the wild-type. In the wild-type genotype expression of all the genes being studied was highly induced by salt stress. These data clearly show that RGL2, ABI3, and ABI5 are required for induction of each other during salt stress and strongly also demonstrate crosstalk between ABA and GAs when seeds are exposed to abiotic stress (Yuan *et al.*, 2011) (Fig. 2).

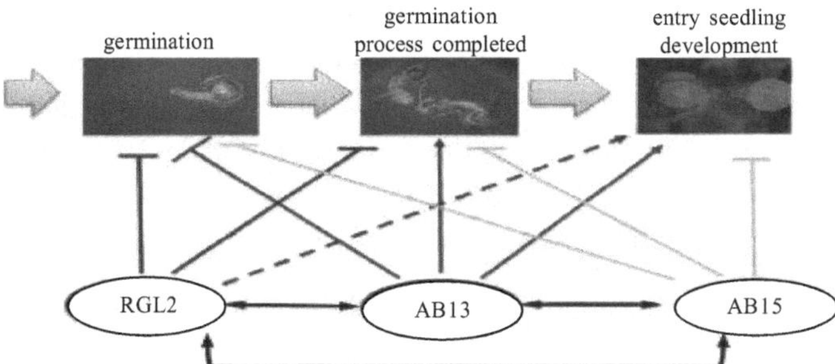

Fig. 2: The possible interaction between ABA signaling components: ABI3, ABI5, and GA signaling component: RGL2 during seed germination under salt stress (Yuan *et al.*, 2011). Lines with blunt end indicate negative regulation, arrows indicate positive regulation, dashed lines indicate hypothetical interaction.

Brassinosteroids are engaged in salt stress response in concert with ABA

Brassinosteroids are steroid hormones that play an important role in the growth and development of various organisms (Steber and McCourt, 2001). Exogenous application of brassinosteroids (BRs) was reported to improve salt stress tolerance in rice and *Arabidopsis*(Kagale *et al.*, 2007), but little is known about the impact of endogenous BRs on stress response. The components of BR signaling have been identified and characterized using BRdeficient or BR-insensitive mutants. The membrane-localized receptors BRI1 and BAK1 bind BR and transfer a signal into the cytoplasm to inhibit BIN2, a glycogen synthase kinase-3/ SHAGGY-like kinase. BIN2 negatively regulates the downstream signaling components by phosphorylating two transcription factors, BES1 and BZR1, and inhibiting their binding to BR target promoters (Li and Jin, 2006). *Arabidopsis* DET2 encodes an enzyme involved in BR biosynthesis. The defects in root and leaf development of the mutant det2-1 can be rescued by the application of brassinolide (Li *et al.*, 1996).

The involvement of endogenous BR in stress response can be studied using BR-deficient or BR-signaling mutants. Koh and others (2007) demonstrated that the rice knock-out mutant OsGSK1, which is hypersensitive to BRs, is more tolerant to salt stress. Zeng and Hua (2010) investigated the response to salt stress of *Arabidopsis* BR-deficient mutant det2-1 and BRsignaling mutant bin2-1. The seed germination and seedling growth of both, det2-1 and bin2-1, were hypersensitive to salt stress. The hypersensitivity of det2-1 was associated with an inhibited induction of stress-related genes, such as COR78 (cold regulated 78) when compared to the wild type, whereas RD22 (response to desiccation 22) the salt induced level was maintained in both det2-1 and the wild type. The proline content was lower in det-2 than in the wild-type. The same results were observed in the case of a bin2-1 mutant. Exogenous epibrassinolid (EBR) could partially rescue the growth of det2-1 under salt stress and improve NaCl-induced proline accumulation (Zeng and Hua, 2010). Recently, the inhibitory role of ABA in the modification of the phosphorylation status of BES1 and its impact on BRresponsive gene expression has been shown (Zhang *et al.*, 2009a). These data, together with det2-1 hypersensitivity to ABA (Zeng and Hua, 2010), point to a link between BR and ABA in seed germination under abiotic stress.

SA—alone and in concert with GAs and ABA— is involved in stress response during seed germination

SA is a phenolic compound. It has been demonstrated that SA is a critical signaling molecule involved in the regulation of plant responses to pathogen infections and abiotic stresses FIG. 2. The possible interaction between ABA signaling components : ABI3, ABI5, and GA signaling component : RGL2 during seed germ ination under salt stress (Yuan *et al.*, 2011). Lines with blunt end indicate negative regulation, arrows indicate positive regulation, dashed lines indicate hypothetical interaction. Seed germination under abiotic stress - through extensive signaling crosstalk with other hormones (Vlot *et al.*, 2009; Wolters and Jurgens, 2009). Recently, it was suggested that SA is engaged in the seed germination process as a key regulator (Alonso-Ramirez *et al.*, 2009). Nishimura and coworkers (2005) reported that SA inhibits seed germination in a dosage-dependent manner. A high SA concentration such as > 1 mM may delay or even inhibit germination, possibly by causing a toxic effect by inducing reactive oxygen species (ROS) biosynthesis (Rajjou *et al.*, 2006). Low doses of exogenous SA (1–10 lM), close to physiological concentrations, significantly improve seed germination under stress (Alonzo-Ramirez *et al.*, 2009). Proteomics studies have shown that SA treatment improves seed germination by promoting the synthesis of proteins that are crucial for germination, and by the mobilization or degradation of seed proteins accumulated during seed maturation. Proteomics data have also revealed that SA can induce an accumulation of proteins characteristic for ABA signaling, such as LEA. Taking into account that SA can induce a transient accumulation of ABA in wheat seedlings (Shakirova *et al.*, 2003), it can be assumed that *Arabidopsis* undergoes the same process during seed germination. This could explain the accumulation of ABA related proteins and the delay of seed germination by SA in non stress conditions (Rajjou *et al.*, 2006). It was proven that exogenous application of GA counteracted the inhibitory effects of salt, oxidative or heat stress during germination and seedling development in *Arabidopsis* (Achard *et al.*, 2006). This is closely related to an increase in SA levels as a result of GAs action (Navarro *et al.*, 2008; Robert-Seilaniantz *et al.*, 2007). Alonzo-Ramirez *et al.* (2009) reported that exogenously applied SA partially rescued seed germination in ga1–3 (GA-deficient 1– 3) mutants in the presence of NaCl. After GA3 was applied, nearly 100% of Col-0 seeds were able to complete germination and start seedling growth compared with no significant differences in the germination rate of a sid2 (SA-induction-deficient 2) mutant that was hypersensitive to salt stress. To summarize, these data together with the SA induction of LEA proteins accumulation showed a link between SA, ABA, and GAs during stress response in germinating seeds.

ABA is thought to be the main stress hormone when seed germination is considered. However, seed germination under stress conditions is regulated by various phytohormones as was revealed through the use of mutants in genes involved in phytohormone metabolism and signaling pathways. All of the possible interactions between the phytohormones involved in the seed germination process under stress are shown in Figure 3.

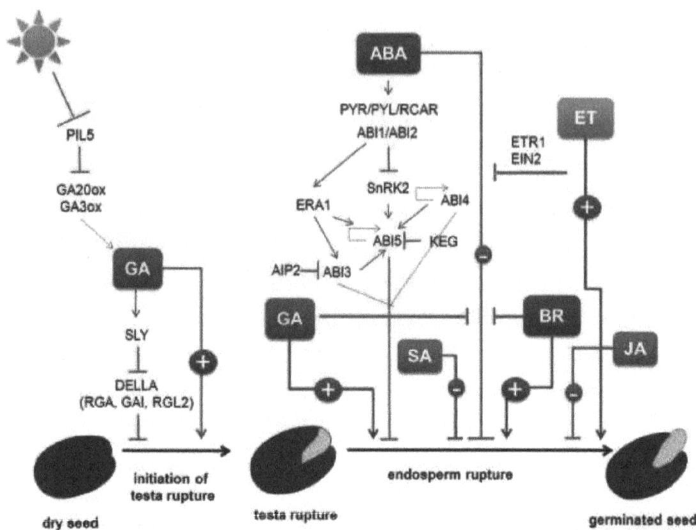

Fig. 3: Hormonal regulation of the seed germination process under both nonstress and stress conditions. Visible events during two-step germination: testa and endosperm rupture. GA, gibberellin; ABA, abscisic acid; BR, brassinosteroid; SA, salicylic acid; JA, jasmonic acid; ET, ethylene; PIL5, PHY-INTERACTING FACTOR 2; GA20ox, GIBBERELLIC ACID-20-OXIDASE; GA3ox, GIBBERELLIC ACID-3-OXIDASE, SLY1–SLEEPY1; RGA, REPRESSOR OF GA1–3; GAI. GIBBERELLIN INSENSITIVE; RGL2, RGA-LIKE 2; PYR/PYL/RCAR, PYRABACTIN RESISTANCE1 / PYR1-LIKE/REGULATORY COMPONENT OF ABA RESPONSE; ABI1/2/3/4/5, ABA INSENSITIVE 1/2/3/4/5; SnRK2, SUCROSE NONFERMENTING RELATED KINASE 2; KEG, KEEP ON GOING; AIP2, ABA INSENSITIVE 3 INTERACTING PROTEIN 2; ERA1, ENHANCED RESPONSE TO ABA 1; ETR1, ETHYLENE RECEPTOR 1; EIN2, ETHYLENE INSENTIVE 2. A plus or a minus indicates a positive or negative effect on the seed germination process. The interactions that are shown are described in detail in the text.

Epigenetics is Involved in the Regulation of Seed Germination

Abiotic stresses can regulate gene expression through the induction of changes in the DNA methylation status. One example is the action of PICKLE (PKL)-mediated chromatin remodeling events during seed germination in response to osmotic stress and ABA induction. During seed imbibition under nonstress conditions, the levels of ABI3 and ABI5 mRNAs and proteins rapidly decrease in order to enable the completion of germination process. This process was described in detail above. Using pkl mutants Perruc and others (2007) showed that pkl overexpressed both ABI3 and ABI5 in response to ABA. PICKLE encodes a putative SWI/SNF-class chromatin-remodeling factor of the CHD3 type. Chromatin immunoprecipitation (ChIP) analysis revealed that mutant pkl seeds had two- to 2.5-fold lower H3-K9 and H3-K27 methylation levels at the ABI3 and ABI5 promoter when compared to the wild type after ABA treatment. In contrast, 7-day-old wildtype and pkl seedlings displayed no differences in the methylation of H3-K9 and H3-K27 within the promoters of both ABI genes. It has been suggested that PKL is essential for repressing ABI3 and ABI5 during seed germination in the presence of exogenous ABA and osmotic stress (Perruc *et al.*, 2007).

Micro RNAs (miRNAs) are small RNAs that perform posttranscriptional gene silencing through the guided cleavage of complementary target mRNAs (Llave *et al.*, 2002) or translational repression (Brodersen *et al.*, 2008). miRNA regulation is also involved in early plant development events, such as seed germination and young seedling growth. Findings of Reyes and Chua (2007) showed the crucial role of miR159 in the negative regulation of ABA signaling during seed germination. The response to exogenously applied ABA mimics the response to abiotic stress. When ABA is applied, miR159 accumulates in seeds and down regulates MYB33 and MYB101, positive regulators of ABA signaling, which enables the completion of seed germination. Reyes and Chua (2007) also reported the involvement of ABI3 and ABI5 in the regulation of MIR159 transcription. Another miRNA involved in seed germination is miR160, which regulates ARF10 (Auxin-Responsive Factor 10). miR160 mutants exhibited a changed ABA sensitivity (Liu *et al.*, 2007). ARFs are important factors for auxin signaling; therefore, this result suggests crosstalk between auxins and ABA during seed germination (Guilfoyle and Hagen, 2007). Microarray analyses indicate upregulation of ABA-related genes in seeds of an miR160-resistant mutant mARF10. The down regulation of ARF10, an important factor for auxin signaling, may be a regulatory step to decrease ABA sensitivity in mature seeds and to switch on the germination program (Nonogaki, 2008b). Jung and Kang (2007) reported the essential role of miR417 in seed germination under the stress. The expression of miR417 was up regulated by salt, dehydration stress, and after ABA treatment. Seed germination of over expressing miR417 plants was retarded compared to the wild type in the presence of NaCl and ABA. These results imply a negative regulatory role of miR417 during seed germination.

Crosstalk between epigenetic and micro RNA regulation of gene expression in terms of abiotic stress during seed germination miRNAs are able to indirectly regulate the DNA demethylation process as an adaptive mechanism to stress conditions during seed germination in plants. An excellent example is miR402, which affects seed germination under stress conditions. It targets the DML3 (DEMETER-LIKE 3) gene that encodes a protein involved in DNA demethylation. miR402 is upregulated by various abiotic stresses such as cold, dehydration, and salt. miR402 mutants germinated earlier than the wild type under stress conditions. In seeds of the over expressing miR402 mutant DML3 showed a decreased level of expression under stress. An analysis of the dml3 and miR402 6 DASZKOWSKA-GOLE.

2.4. Salt Stress and Seed Germination

Seed germination is an essential process in plant development to obtain optimal seedling numbers that results in higher seed yield. Germination and seedling growth declined with many abiotic factors such as salt and drought stress that are perhaps two of the most important grounded abiotic stress that limit number of seedling and seedling growth (Almansouri *et al.*, 2001; Kaya *et al.*, 2006a, Ansari *et al.*, 2012; Ansari and Sharif-Zadeh, 2012; Ansari *et al.*, 2013). Salinity stress delays the onset, reduces the rate and increases the dispersion of germination events, resulting in reduced plant growth and, finally, crop yield (Ashraf and Foolad, 2005). Therefore salinity has caused to significant decrease in germination characteristics in all studied species. The general response of many plants to increased salinity is to accumulate high concentrations of Na^+ and Cl ions in their vacuoles or to distribute these ions to different parts of the plant to activate metabolic functions. In this way, the plants protect their cytoplasmic water potential (Munns, 2002). Certain biochemical strategies were used to enhance salt tolerance in plants, including the control of ion transfer from roots to leaves, the distribution of ions into cellular compartments, the synthesis of osmotic regulators, changes in photosynthesis and cell membranes, and the induction of antioxidative enzymes and certain plant hormones (Nakamura *et al.*, 2002). Enzymatic ROS-scavenging mechanisms in plants include production of superoxide dismutase (SOD), ascorbate peroxidase (APX), guaiacol peroxidase (GPX), catalase (CAT), and glutathione reductase (GR) (Skopelitis *et al.*, 2006). Stimulating of oxidative stress is the most important pathway applied by salinity for aquatic plants and may be applied for all of plants. Several studies have shown that salt tolerant species increase the activities of their antioxidant enzymes in response to salt treatment. Salinity has caused significant increase in SOD activity in all studied species. One way that plants adjust to high salt concentrations is by increasing their tissue osmotic potential. This results in the accumulation of inorganic and organic solutes. The cellular response to short- and long-term salt stress by plants involves the synthesis and accumulation of osmoprotective compounds (Banu *et al.*, 2009; Yazici *et al.*, 2007), which are small nontoxic molecules. Osmotic active compounds increase the osmotic potential of the cell (Apse and Blumwald, 2002). Proline is the major compound that protects cells by stabilizing proteins and cellular membranes (Kumar *et al.*, 2003; Martinez *et al.*, 2003; Wang and Han, 2009; Çelik and Atak, 2011).

2.5. Seedling Growth

The seedling environment is seldom optimum for uninterrupted growth. Seedlings commonly experience temporary periods of abiotic stresses, resulting in necessary root and shoot metabolic and structural adjustments to withstand stress conditions. Abiotic stresses not only have an impact on the subsequent growth and productivity of cultivated plants, but also constitute a significant barrier for introducing crop plants into areas not used in agriculture. Adverse conditions in the root environment include mechanical impedance, deficient or excess water or salts, deficit or imbalance of nutrients, and insufficient O_2 and CO_2. The canopy environment may also limit photosynthesis and assimilate translocation. In the past decade, considerable progress has been made in identifying and understanding seedling growth physiological, morphological, and physical adaptation mechanisms to drought, O_2, salt, nutrient, and light stresses. This new knowledge has enhanced our understanding of vegetable seedling growth responses during establishment. However, considerable scientific investigation remains to be accomplished. Abiotic stresses generally do not occur separately—rather, interaction of stresses are the norm in natural environments (*e.g.*, drought and high temperature).

Seedling root systems have important physical and physiological roles from the onset of seed germination (radicle protrusion) and emergence through subsequent seedling growth and development. Important root functions include shoot anchorage, water uptake, nutrient uptake, and hormone production. The size, morphology, or architecture of a root system may control the relative size and growth rate of the shoot. Optimum root systems throughout the plant life cycle can ensure optimum shoot growth, shoot development, and subsequent yields. However, abiotic stresses originating in the root zone can be expressed in the shoots, thereby affecting the dry matter partitioning and, ultimately, the productivity of the plant. Understanding the relationships of root and shoot components to stand establishment, particularly as influenced by abiotic stresses, is essential for developing optimum cultural and management techniques for field planting systems.

To establish seedling root system relationships with shoot growth or development, a clear understanding of root morphology is required. Klepper and Rickman (1991) described root systems in terms of axes and their branches: 1. vertical or main axes (or taproots), 2. horizontal or vertical axes (first-order laterals), and 3. second- or higher-order laterals. Generally, seedling root morphology differs between monocotyledonous and dicotyledonous species. For example, the seedling root systems of mungbean (*Vigna radiata* L.), corn (*Zea mays* L.), and onion (*Allium cepa* L.) are distinctly different (Fig. 4). Mungbean, a dicotyledonous species, has a defined taproot (originating from the radicle), lateral roots (initiating from the taproot, or pericycle cells), basal roots (originating from the basal part of the hypocotyl or shoot–root transition zone when pericycle cells do not originate in this area), and adventitious roots (originating from the hypocotyl). This root origin classification was first described by Zobel (1975) for tomato (*Lycopersicon esculentum* Mill.) and by Stoffella *et al.* (1979b) for beans (*Phaseolus vulgaris* L.). Onion, a monocot, has no defined taproot, but only an accumulation of adventitious roots and lateral roots (which originate from the preexisting adventitious roots). However, distinctive root morphological differences occur within monocots (Fig. 4) and dicots. Root systems have been illustrated for field grown vegetable (Weaver and Bruner, 1927) and agronomic crops (Weaver, 1926).

Root dry-weight accumulation, number, diameter, length, surface area, and distribution in a soil profile can differ within crop species. Seedling taproot, basal root, and adventitious root variables differ between genotypes of beans (Kahn *et al.*, 1985; Lynch and van Beem, 1993), peppers (*Capsicum annuum* L.) (Leskovar *et al.*, 1989; Stoffella *et al.*, 1988a), and mature tomatoes (Stoffella, 1983). In cowpea [*Vigna unguiculata* (L.) Walp.], 'California Blackeye #5' seedlings had longer taproots, more lateral roots, taller shoots, and heavier roots and shoots than 'Knuckle Purple Hull' or 'Texas Cream 40' seedlings when measured 11 days after radicle protrusion. Mean taproot growth rate (length) and lateral root production from 3 to 11 days after radicle protrusion also differed among these three cowpea cultivars.

Root architectural traits, such as secondary and tertiary root branching, angles, and radii, also should be considered in evaluating seedling root systems. Fitter (1991) reported an architectural approach to studying plant root systems and suggested that topological models and architectural root system components,

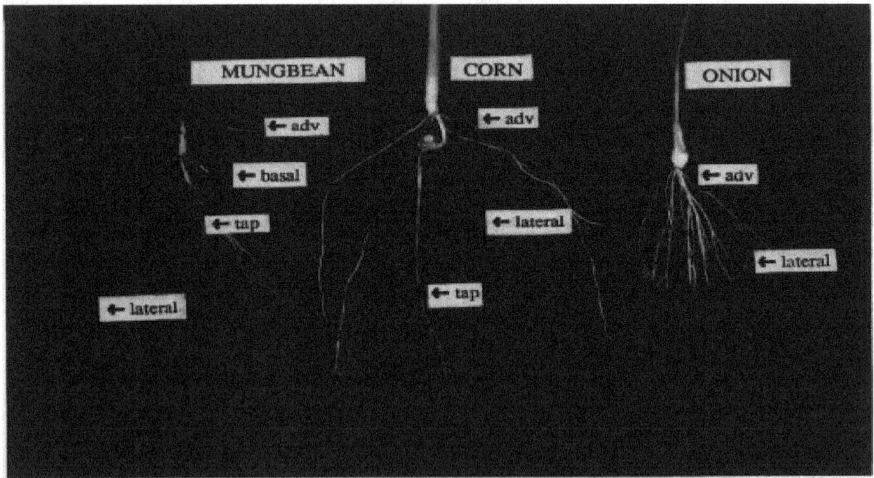

Fig. 4: Root morphological components of mungbean, corn, and onion seedlings depicting adventitious (adv), basal, lateral, and taproots (tap)

such as link lengths, branching angles, and diameters, should be used for comparative ecological studies of plant root systems. Detecting differences in root growth patterns and architecture between genotypes may offer unique genetic selection criteria for tolerance to root diseases and pests, lodging, drought, flooding, stressful root-zone temperatures, or edaphic adaptations.

Reviews of genotypic variation in root morphological characteristics and variation in root system phenotypes have been reported by O'Toole and Bland (1987) and of rhizogenetic variation in vegetable crops by Zobel (1975, 1986). Root morphological traits can be directly related to plant lodging (Ennos, 1990; Stoffella and Kahn, 1986; Stoffella *et al.*, 1979a), flooding tolerance (Kahn *et al.*, 1985), and nodule formation (Kahn and Stoffella, 1991). Zobel (1989) has reviewed studies on the effects of edaphic and environmental factors, such as moisture, temperature, mineral concentration, and gaseous atmosphere, on root morphology.

MacIsaac *et al.* (1989) reported that 2-naphthaleneacetic acid (NAA) increased the number of lateral root primordia, but kinetin prevented their spontaneous formation and inhibited the NAA-enhanced lateral root primorida of lettuce (*Lactuca sativa* L.). In corn, most primordia from the primary root completed their development and emerged as laterals within a fixed period, irrespective of their origin along the main axis (MacLeod, 1990). Cytokinins, particularly zeatin, inhibited the initiation and formation of lateral roots on decapitated primary roots of pea seedlings. Removal of adventitious root tips promoted the formation of primordia along most of the remaining axis in onion seedlings (Abadia-Fenoll *et al.*, 1986); however, normal onion roots appeared to be unbranched (Fig. 4) over distances of up to 400 mm from the apex (Melchior and Steudle, 1993).

Root developmental sequence in bell pepper has been reported and emphasized the importance of specific root morphological traits and their importance in improving or overcoming physiological stresses. Initiation, organization, and emergence of lateral roots are complex processes involving growth regulators and specific gene expression (Keller and Lamb, 1989. In peas, the first lateral roots are initiated in the radicle before germination (Rost and Toriyama, 1991). After seeding, initiation of root primordia on the primary root and hypocotyls of tomato seedlings varied between genotypes and the number of primordia on the primary root and hypocotyl was increased either by Hoagland's solution or by presoaking seeds with abscisic acid (ABA) or 6-benzylamino purine (BA) (Aung, 1982).

2.6. Stress Effects on Seedling Growth

The abiotic root–soil environment is seldom optimum for continuous growth and can have a profound influence on root anatomical characteristics, root developmental patterns, and the physiology of roots. Root systems in relation to stress tolerance have been reviewed by Miller (1986), who discussed the effects of water, O_2, mechanical impedance, and temperature stress on mature plants. Latimer (1991) reviewed vegetable seedling growth responses to mechanical stress. Adverse environments that include the interaction of factors such as nonoptimum temperature, deficiency or imbalance of nutrients, drought, and waterlogging have received less attention in seedlings.

2.6.1. Temperature

Early root elongation, branching, and function are highly temperature-dependent mechanisms (Russell, 1977). Results from root studies conducted at constant temperatures are difficult to extrapolate to estimates of root responses in the field due to fluctuating air and soil depth temperatures. No lateral roots developed for 12 legumes exposed to constant 10 or 30^0 C, and lateral and tap root elongation were highly variable between 15 and 25^0 C (Brar *et al.*, 1990). Tap root elongation rate was slower for pre germinated than for raw or primed pepper seeds incubated at 15^0 C, but similar at 25 or 35^0 C (Stoffella *et al.*, 1988b).

Root systems of tropical or subtropical origin species are sensitive to chilling temperatures. For example, root growth decreased when squash (*Cucurbita pepo* L. var. *melopepo*) and cucumber (*Cucumis sativus* L.) seedlings were exposed to 2 and 6^0 C for 48 h, and when seedlings were returned to 26^0 C, irreversible damage (inability to resume growth) occurred after 96 h (Reyes and Jennings, 1994). Short term cyclic cold stresses (2^0 C) before transplanting generally stimulated seedling height, leaf area, and shoot and root growth, but decreased chlorophyll content in tomato seedlings (Dufault and Melton, 1990). Temperatures between 10 to 13^0 C suppressed tomato root growth (Hardenburg *et al.*, 1986). However, Leskovar and Cantliffe (1991) reported that stem length, and root fresh and dry weights were not suppressed when containerized tomato transplants were stored at 5^0 C for 8 days. When containerized tomato seedlings were transplanted and mulched in beds with clear polyethylene, seedling root extension rates between 1 to 2 weeks after transplanting were significantly correlated with soil temperature during the first week (Wien *et al.*, 1993).

Temperature directly influences the nutrient status of seedlings by affecting nutrient absorption and transport from the roots to the shoots. Seedling root elongation of beans, cucumber, eggplant (*Solanum melongena* L.), pea, pepper, radish (*Raphanus sativus* L.), and spinach (*Spinacia oleracea* L.) was optimum when constant in the 16.7 to 21.1^0 C range, and a synergistic effect with phophorus nutrition occurred on beans, cucumber, eggplant, and pepper, but not on radish, spinach, and pea (Wilcox and Pfeiffer, 1990). Enhanced early root growth and nutrition uptake, particularly phosphorus, characterized tomato transplants grown with clear polyethylene mulch where soil temperatures were 5 to 7^0 C higher than soil not mulched (Wien *et al.*, 1993). Papadopoulos and Tiessen (1987) reported that leaf tissue accumulation of nutrients at high root temperatures (27 vs. 21^0 C) was greater for phosphorus than for nitrogen, with minimum or no effect on the accumulation of potassium, calcium and magnesium.

2.6.2. Nutrition, Salinity, and pH

Seedlings have a high demand for mineral nutrients, which, in part, is thought to be a result of their fast growth rates relative to mature plants (Widders and Lorenz, 1982). A strategy used to facilitate root growth, minimize transplant shock, and increase seedling survival is to enhance the seedling nutrient content before transplanting (Dufault, 1986). This technique, known as pre-transplant nutritional conditioning (PNC), recently has been reviewed by Dufault (1994). While pre-transplant treatment with phosphorus did not enhance shoot growth of tomato seedlings, root initiation and growth were stimulated by nitrogen and phosphorus pre-transplant treatments (Garton and Widders, 1990). In five bell pepper genotypes, total and basal root length increased linearly with increasing nitrogen rates from 56 to 112 mg/liter when measured 40 days after seeding (Leskovar *et al.*, 1989). Localized feeding of NO_3 stimulated lateral root initiation in corn (Granato and Raper, 1989). Ratio between shoot and root in tomato seedlings was reduced from 4.0 to 2.3 when nitrogen rate was decreased from 400 to 100 mg/liter regardless of phosphorus rate (15 to 60 mg/liter) (Weston and Zandstra, 1989). Small shoot to root ratio differences occurred on cauliflower (*Brassica oleracea* L. Botrytis Group) seedlings exposed to nitrogen rates between 0 and 155 mg/liter (Booij, 1992). Increasing nitrogen rates between 150 to 350 mg/liter increased shoot dry weight and shoot to root ratio, but decreased root dry weight in celery (*Apium graveolens* L.) seedlings (Tremblay and Gosselin, 1989). It had been reported that root morphology was found to be associated with root affinity for NO_3 or NH_4, root ability to discriminate between NO_3 or NH_4 and root detoxification of NH_4.

Root growth is generally less sensitive to salinity than shoot growth (Lauchli and Epstein, 1990), as reflected by lower shoot to root ratios when salinity is increased. However, salinity effects on root systems can be significant. For example, Snapp and Shennan (1992) reported that root-length density of two tomato cultivars (CX8303, root-rot tolerant to *Phytophtora parasitica* Dast., and UC82B, sensitive to root rot) was reduced between 21% and 39% at a moderate salinity level of 75 mM NaCl plus $CaCl_2$ (4 Na : 1 Ca molar ratio) as compared to 3 mM. Calcium plays an important role in adapting seedling plant tissue to saline conditions (Kent and Lauchli, 1985). Calcium also has a role in the root cap in regulating root growth and is translocated to the shoots mainly from young roots (Russell, 1977). Harrison-Murray and Clarkson (1973) indicated that absorption and translocation were greater within 3 cm of the root apex in taproots of marrow (*Cucurbita pepo* L.) than in more distal positions. Takahashi *et al.* (1992) reported that exogenous Ca^{2+} applied at 10 or 20 mM directly to the root cap stimulated root elongation in pea and corn seedlings. They suggested that differences in responsiveness of roots to Ca^{2+} may result from water conditions at the root tip that differ from those along the older portions. Inhibitory effects of 100 mM NaCl on maize root elongation were partially reversed by Ca^{2+} levels at 10 mM in the root medium, as evidenced by increasing root diameter and root hydraulic conductivity in the apical 4 cm of the primary roots (Azaizeh *et al.*, 1992; Evlagon *et al.*, 1992).

Root and shoot weights were higher and taproot growth rate was faster for bell pepper seedlings grown at pH 5.9 as compared with pH 4.0 or 7.3 (Stoffella *et al.*, 1991). Plant roots increased (Marschner *et al.*, 1986) or decreased rhizosphere pH (Schaller, 1987). Rhizosphere pH changes are influenced by soil buffering capacity (Schaller, 1987), mineral absorption capacity, absorption efficiency of the plant species (Marschner and Romheld, 1983), and initial soil pH (Youssef and Chino, 1989).

2.6.3. Water

Effect on root growth. Root activity throughout the soil profile has been investigated for water uptake (Klepper, 1990). Water potential of the root medium influences root growth and the water potential gradient between the medium and root. Significant root elongation restrictions occur when soil water potential is below –0.8 MPa (Klepper, 1991). Although taproots have the capacity to explore deep layers of the soil profile, under drought conditions, monocot and dicot seedlings may also highly depend for survival and subsequent growth on crown (or nodal), adventitious, basal, or lateral roots. In maize, the water potential at which growth stopped was –0.5, –1.0, and –1.4 MPa for stems, leaves, and nodal roots, respectively (Westgate and Boyer, 1985). Water flux is a function of water potential differences and hydraulic conductances (Melchior and Steudle, 1993). In dry surface soils, roots may develop strong axial and radial resistance to the pathway of water flow from root to shoot tissues (Cornish *et al.*, 1984; Steudle *et al.*, 1987). The main resistance to water flux lies in the radial pathway (from the soil solution to the tracheary elements) and not in the axial path (within the tracheary elements) (Frensch and Steudle, 1989).

McCully and Canny (1988) reported that the diameters of the largest nonoccluded tracheary element(s) are the most important characteristic influencing hydraulic conductivity. The number of first-order branch roots and the diameter of their largest tracheary element differed at nodes along the root axes of corn (Varney *et al.*, 1991).

2.6.4. Under Flooding Condition

When flooding occurs, the O_2 concentration in the soil declines gradually and roots undergo a period of hypoxia before O_2 is fully depleted from the soil and anoxia is reached. (Pradet and Bomsel, 1978). Hypoxia is a condition in which the concentration of O_2 limits ATP production in the mitochondria. Anoxia is defined as a condition in which the concentration of O_2 is so low that virtually no ATP is produced by the mitochondria. It has been observed that when roots are pre treated with low O_2 before subjecting them to anoxia, root tips survive longer than those made anoxic without a hypoxic pretreatment (Saglio *et al.*, 1988) and has been termed "hypoxic acclimation." There is a pronounced difference in the responses of acclimated and un-acclimated seminal maize (*Zea mays* L.) roots to anoxia. At the onset of anoxia, energy production declines, there is acidosis of the cytoplasm, and cell death occurs within 15 to 20 h (Johnson *et al.*, 1989; Saglio *et al.*, 1988). Thus, after 24 h of anoxia, there is little remaining cellular organization of un-acclimated root tip cells. The only recognizable organelle is the nucleus. In contrast, acclimated root tips survive 96 h of anoxia (Johnson *et al.*, 1989). Acclimated root tips subjected to anoxia appear to maintain normal cellular organization through 48 h and possess higher ATP concentrations and higher rates of anaerobic metabolism than Un-acclimated root tips (Hole *et al.*, 1992; Johnson *et al.*, 1989; Saglio *et al.*, 1988). Clearly, hypoxia induces physiological responses that give rise to improved resistance to anoxia, and the identification of those responses has been the focus of research for several years (Cobb *et al.,* 1995).

2.7. Drought Stress and Abscisic acid (ABA)

Pepper seedlings exposed to mild drought conditions had lower root dry weight, basal root number and diameter, and shoot growth as compared with well watered seedlings (Leskovar and Cantliffe, 1992). The influence of drought stress on root and shoot growth is mediated by endogenous changes in ABA concentration acting as a signal for control of growth processes (Davis *et al.*, 1986). ABA synthesis induced by water stress has been correlated with a reduction of shoot to root ratio either by shoot growth reduction (Creelman *et al.*, 1990) or by more root growth relative to shoot growth (Karmoker and Van Steveninck, 1979). ABA was studied as a substitute for drought stress to control seedling growth and improve transplant field establishment in pepper and eggplant seedlings. When ABA was applied at 10^{-4} M to pepper seedlings 3 weeks after seeding, there was a transient inhibition of leaf weight increase, but root growth was unaffected (Leskovar and Cantliffe, 1992). ABA at 10^{-3} M reduced relative growth rates of eggplant seedlings (Latimer and Mitchell, 1988). Increased seedling survival due to exogenous application of ABA was related to increased leaf resistance and leaf water potential in pepper seedlings (Berkowitz and Rabin, 1988).

***Flooding*:** Soil flooding limits the diffusion of O_2 to the root zone (Drew, 1992), affecting young plants with a high percentage of primary roots as compared to more mature roots in older plants (Letey *et al.*, 1962). Adventitious roots play a major role in the survival of many mesophytes by partially replacing the functions of the original roots (Jackson and Drew, 1984). Aloni and Rosenshtein (1982), Jackson (1979) reported that flooding-induced root injury is limited if adventitious roots develop, indicating that these roots are important for recovery of shoot growth following flooding. Kahn *et al.* (1985) reported that adventitious roots became the predominant root component (highest weight as compared to taproot or basal roots) after bean seedlings were exposed to 7 days of flooding, followed by 7 days of recovery. Similarly, in continuously flooded tomato cultivars, adventitious roots accounted for more than 50% of the total root mass (Poysa *et al.*, 1987). However, adventitious root development did not overcome flooding injury of sunflower (*Helianthus annum* L.) (Wample and Reid, 1978).

2.8. Salt Stress on Seedling Length

Root and shoot lengths are the most important indicators of salt stress because roots are in direct contact with the soil and absorb water from soil and then shoots enable its supply to the rest of the plant. For this reason, root and shoot lengths provide important indications of a plant's response to salt stress (Jamil and Rha, 2004). Root length as well as shoot length is affected by salinity. Bahrami *et al.* (2012) found that 0, 5.3, 8.5, 12.05, 14.65 and 18.45 dSm^{-1} salinity levels decreases the radicle length of 2.6, 2.34, 1.6, 1, 0.6, 0.3 cm, respectively. According to them plumule length was more sensitive than radicle length. Mean plumule length of the cultivars studied varied between 3.1 and 4.3 cm and between 6.2 and 0.42 cm for salinity level. Xiong and Zhu (2002) explained that salt stress inhibited the efficiency of translocation and assimilation of stored materials and might have caused a reduction in shoot growth. The reduction in root and shoot development may be due to toxic effects of the NaCl used as well as unbalanced nutrient uptake by the seedlings. It may be due to the ability of a root system to control entry of ions to the shoot, which is of crucial importance to plant survival in the presence of NaCl (Hajibagheri *et al.*, 1989).

2.8.1. Salt Stress on Seedling Dry Weight

Findings of EL-Melegi *et al.* (2004) on tomato under salinity stress showed that, dry weight of plants had a positive correlation with root length and these plants were more tolerate to salinity stress. Similar results were reported by François and Bahizire (2007) in canola under salinity stress. Bahrami *et al.* (2012) reported that under increasing stress levels, radicle dry weight significantly decreased from 6, 5.4, 3.2, 2.7, 2.1 and 1.5 mg and plumule dry weight also decreased from 11.2, 8.3, 6.1, 5, 2.6 and 1 mg at the salt levels of 0, 5.3, 8.5, 12.05, 14.6 and 18.45 dSm^{-1}, respectively.

2.8.2. Mechanism of Salt Stress on Germination

Salinity reduces the ability of plants to take up water, leading to growth reduction as well as metabolic change similar to those caused by the water stress (Munns, 2002). High salt concentration in root affects the growth and yield of many important crops. The salinity may reduce the crop yield by upsetting water and nutritional balance of plant (Khan *et al.*, 2007). Water availability and nutrient uptake by plant roots is limited because of high osmotic potential and toxicity of Na and Cl ions (AlKaraki, 1997).

2.8.3. Root Establishment Methods

Effect on root development

Seedling establishment in any field environment depends on the adequate development of a taproot, associated laterals, and basal roots for dicot species, or adventitious roots and associated lateral roots for most monocot species. Root growth patterns of plants established by direct seeding differ from those of plants established by transplanting. Direct-seeded pepper plants, established with primed or raw seeds, had variable root growth over time as compared with a constant rate of root growth from transplants (Leskovar *et al.*, 1990). Early root development was unaffected by seed priming in pepper (Stoffella *et al.*, 1992), and total root growth (dry-weight basis) was similar for primed or untreated pepper plants when measured at maturity (Leskovar *et al.*, 1990). Basal, lateral, and taproot dry weights accounted for 81%, 15%, and 4%, respectively, of the total root dry weights for containerized transplants and 25%, 57%, and 18%, respectively, of the total root dry weights for direct-seeded bell pepper plants (Leskovar and Cantliffe, 1993). The latter also exhibited greater root relative growth rates and total root growth than transplants. As a consequence, transplanted pepper plants may be more prone to lodging than direct-seeded plants. Cooksey *et al.* (1994) reported that the force to uproot pepper plants established by transplanting was 15% lower than for plants established by direct seeding.

Transplant production systems, including flotation and overhead irrigation, also influence root and shoot growth characteristics in bell pepper and tomato seedlings in the greenhouse (Leskovar and Cantliffe, 1993; Leskovar *et al.*, 1994). In jalapeño peppers, overhead irrigation enhanced basal root elongation and root weight as compared with seedlings grown with flotation irrigation; however, the flotation irrigation system improved water-use efficiency, reduced shoot : root ratio, and promoted seedling hardiness (Leskovar and Heineman, 1994). Providing uniform moisture levels around the hypocotyl during the last 2 weeks before field establishment is also thought to enhance basal root primordia development in transplants. The importance of basal root development in the field has been emphasized in peppers (Leskovar and Cantliffe, 1993), black beans (Stoffella *et al.*, 1979a), and tomatoes (Zobel, 1975). The ability of a transplant (bareroot or containerized) to overcome transplant shock and become established in a field environment following transplanting depends on how seedlings withstand root disturbance (McKee, 1981), and how water and nutrient uptake capacity of the roots and the capacity of the preexisting roots to rapidly regenerate new lateral, basal, or adventitious roots are affected. Root pruning before transplanting tomatoes delayed reestablishment (McKee, 1981), and the net rate of root and shoot biomass increase was limited (Leskovar *et al.*, 1994). Enhanced early tomato root growth and nutrient uptake are key factors in increasing shoot growth (Wien *et al.*, 1993).

Early root and shoot seedling response to various cultural systems and environments also depends on the physiological state of the seedlings. Under favorable field conditions, young tomato transplants had a higher relative root growth rate and greater capacity to resume shoot growth than older transplants (Leskovar and Cantliffe, 1991; Weston and Zandstra, 1989). Under windy or sandblasting conditions, older tomato seedlings survived better than young seedlings (Liptay, 1987). Older onion transplants, with more root and shoot growth, had higher survival rates compared to younger seedlings when exposed to low winter temperatures after transplanting (Leskovar and Stoffella, 1995). Survival can also depend on the cultivar and differences in root characteristics between cultivars. For example, root counts were higher, but dry weights lower, for onion cultivars Texas 1015 and Early White Supreme than for cultivars Henry Special and Red Granex.

Information on the effects of establishment methods (direct-seeding or transplanting) or seed enhancement treatments of vegetable crops on subsequent seedling root growth, development, or architecture is limited. The influence of cultural practices on mature root systems has been reported for various vegetables; *e.g.*, plant population density on root morphological characteristics in beans and peppers (Stoffella *et al.*, 1981; Stoffella and Bryan, 1988); establishment methods on tomato and pepper root growth (Leskovar and Cantliffe, 1993; Leskovar *et al.*, 1994); irrigation on lettuce root elongation, weight, and diameter (Rowse, 1974); irrigation on root length and density in tomato (Sanders *et al.*, 1989); and irrigation on root distribution of sweet corn (Phene *et al.*, 1991).

Seed reserve utilization under salinity stress

Seed germination negatively affected by stress conditions in more crops (Patade *et al.*, 2011; Ansari and Sharif-Zadeh, 2012). Seed germination is the most sensitive stage to abiotic stress (Patade *et al.*, 2011; Ansari *et al.*, 2012). Seed reserve utilization, seedling growth and weight of mobilized seed reserve decreased with increasing drought and salt intensity (Soltani *et al.*, 2006; Ansari *et al.*, 2012). Ansari *et al.* (2012) reported that seed priming can be taken to counteract the adverse effects of abiotic stress. Seed priming techniques have been used to increase germination characteristics and improve germination uniformity in more field crops under stressed conditions (Iqbal and Ashraf, 2007; Kaya *et al.*, 2006b; Patade *et al.*, 2011; Saðlam *et al.*, 2010; Ansari *et al.*, 2012). Seed priming increases seed reserve utilization, seedling dry weight and seed reserve depletion percentage in mountain rye (Ansari *et al.*, 2012) and wheat (Soltani *et al.*, 2006). In monocotyledon plants like wheat (Soltani *et al.*, 2006) and mountain rye (Ansari *et al.*, 2012), gibberellic acid after synthesis in the scutellum migrates in to the aleurone layer. The hetrotraphic seedling growth (mg per seedling) could be quantitatively described as the product of the following two components: the weight of mobilized seed reserve (WMSR; mg per seed), and the conversion efficiency of mobilized seed reserve to seedling tissue (mg mg-1) (Soltani *et al.*, 2006; Ansari *et al.*, 2012).

2.9. Root Development

Raven and Edwards (2001) defined root as "roots are axial multicelular structures of sporophytes or vascular plants which usually occurs underground, have strictly apical elongation growth, and generally have gravitropic responses which range from positive gravitropism to diagra vitropism, combined with negative phototropism". The apical meristem of one (lower vascular plants) to many (all seed plants) dividing cells produces a root cap acropetlly and initials of stele, cortex and epidermis besipetally. The branching of roots involves the endogenous origin of new root apical meristems in the pericycle (Raven and Edwards, 2001). The most conserved functions of roots present in extant plants are anchorage to substrate, and uptake of water and mineral nutrients. The evolution of multicellular organs such as roots was necessary to successful colonization of land by early plants (Pires and Dolan, 2012).

Plants have developed various mechanisms to tolerate water or salt stress, including an altered shoot:root ratio. Thus, it is known that drought may cause a greater inhibition of shoot growth compared with root growth and, in some cases, the absolute root biomass in drying soil may increase when compared to well-watered soils (Santos *et al.*, 2007). Also, illumination stress induces changes in the shoot:root ratio (Miralles *et al.*, 2011). However, it is known that the degree of the responses of roots to abiotic stresses may vary considerably within a family, a genus, and even a species (Francini and Sebastiani, 2010; Kachout *et al.*, 2011; Rahnama *et al.*, 2011).

2.9.1. Root System Development and Abiotic Stress

Typically, root system has to grow in media where the biotic and abiotic components are distributed heterogeneously. Soils are complex, a broad range of chemical a physical process occurs due to intrinsic soil characteristics and the action of biotic factors. Thus, this complexity presents several challenges to survive. As soon as the root makes contact with the soil must sense and integrate biotic and abiotic cues in order to adjust their genetic program of post-embryonic root development (PERD). This capacity to change their PERD allows them change their architecture to find the supplies of water and nutrients that could be limited and localized (Fraire-velázquez *et al.*, 2012). Environmental cues such as water, salinity and nutrient can modulate the root system architecture.

In drying soil, shoot growth can be limited as a result of hydraulic insufficiency and chemical signalling from the roots to the shoots *via* the xylem (Hsiao and Xu, 2000). Over the last few years, progress has been made in understanding the biochemical and hormonal bases for the response of roots to stress conditions (Karni *et al.*, 2010; Postaire *et al.*, 2010; Sangakkara *et al.*, 2010; Liao *et al.*, 2011). A central role has been attributed to abscisic acid (ABA), ethylene, reactive oxygen species, and reactive nitrogen species. On the other hand, root morphology determines the ability of the plant to acquire water and nutrients under water-stress conditions (Neumann, 2003; Vamerali *et al.*, 2003). A prolific root system can confer an advantage for accelerated plant growth during the early stages of crop growth and can serve to extract water from shallow soil layers which would otherwise be easily lost through evaporation. Moreover, root length is an indicator of the ability of plants to absorb water from the deeper layers of soil, and is influenced by better root penetrability. In addition to measuring root length and diameter, other root characteristics such as the structure of the xylem vessels, number of root hairs, suberisation of the exodermis, or metacutisation (*i.e.*, the suberisation of one or more root-cap cell layers which results in a resting root that is protected against adverse environmental conditions and is capable of regrowth when conditions ameliorate) are also of great interest (Franco *et al.*, 2001; 2002b; 2008; Ostonen *et al.*, 2007).

Other abiotic stresses can also affect the development and performance of roots. For example, extreme temperatures have a notable impact on the growth and development of plant root systems. The relationships between extreme temperatures and specific plant functions ranging from nutrient uptake, photosynthesis, and carbon partitioning (McMichael and Burke, 2002) have been well-documented. In addition, inappropriate fertilisation may result in an imbalance in shoot-to-root partitioning, thereby affecting the shoot:root ratio as well as root growth (Maurel *et al.*, 2010). Moreover, heavy metal toxicity is a major form of abiotic stress over a wide swathe of acidic soils. This limits root growth and increases the risk of plants succumbing to drought and nutritional deficiencies (López-Marín *et al.*, 2009). Likewise, mechanical stress is an inherent feature of natural root growth because roots need to overcome the mechanical resistance of soils in order to grow. This resistance, commonly referred to as soil strength, is influenced by numerous soil parameters which affect its compactness (Wang and Smith, 2004). Root development in plants growing in soils of increased compactness, or in a restricted rooting medium, can therefore be altered considerably (Clark *et al.*, 2003).

2.10. Root Development Under Abiotic Stress

2.10.1. Drought Stress

Root development is strongly influenced by growing conditions such as moisture deficit (Sangakkara *et al.*, 2010).This is true even for drought-tolerant species such as *Manihot esculentum* (Sriroth *et al.*, 2001). The dynamics of root growth under drought conditions might be a key factor to understanding the contribution of roots to drought tolerance. The effect of drought stress is usually greater on shoot growth than it is on root growth. This lower sensitivity of roots appears to occur as a consequence of the rapid osmotic adjustment of roots in response to a decrease in soil water-content, which allows the maintenance of water uptake, and is also due to the enhanced loosening ability of root cell walls (Sharp *et al.*, 2004). Thus, a decrease in the shoot:root ratio under drought stress is a frequently observed phenomenon, and is caused either by an increase in root growth, or by a relatively larger decrease in shoot growth than in root growth (Franco *et al.*, 2006). This was the case with *Capsicum annuum* (Kulkarni and Phalke, 2009; Shao *et al.*, 2010), *Lonicera implexa* (Navarro *et al.*, 2008), *Lotus creticus* (Franco *et al.*, 2001; Bañón *et al.*, 2004), *Lupinus havardii* (Niu *et al.*, 2007), *Myrtus communis* (Bañón *et al.*, 2002), *Nerium oleander* (Bañón *et al.*, 2006; Niu *et al.*, 2008), *Opuntia ficus-indica* and *Opuntia robusta* (Snyman, 2004), *Rhamnus alaternus* (Bañón *et al.*, 2003), two rose (*Rosa multiflora* and *Rosa odorata*) rootstocks (Niu and Rodriguez, 2009), *Rosmarinus officinalis* (Sánchez-Blanco *et al.*, 2004), *Sambucus mexicana* (Feser *et al.*, 2005), and *Silene vulgaris* (Franco *et al.*, 2008).

Not only the shoot:root ratio, but also other root characteristics *e.g.*, root length, fresh weight (FW), dry weight (DW), diameter and surface area, deep rooting and cortex thickness and root behaviour (*e.g.*, root turnover, metacutisation, hardening, and hydraulic conductivity) may be strongly affected by drought. Plants subjected to low-moisture regimes can develop a more extensive root system to capture the available water. This was the case with *Phaseolus vulgaris*, as reported by de Sousa and Lima (2010), who found an increase in root length under drought stress. Similarly, *Viburnum odoratissimum* plants growing under drought conditions exhibited greater root growth when they were irrigated every 2 d than plants receiving irrigation every 4 days (Shober *et al.*, 2009). Nevertheless, Silber *et al.* (2006) showed that, in low-irrigated *Leucadendron* sp. plants, the horizontal extent of the root volume followed the wetted zone, being almost congruent with the spatial distribution of the available water-content.

Deep rooting is also a critical factor in influencing the ability of a plant to absorb water from the deeper layers of the soil. Huang (2008) found that deep rooting was essential to maintain cellular hydration by avoiding water-deficit in turfgrass. Similarly, deep root development in *C. annuum*, *Lycopersicon esculentum*, and *Solanum khasianum* were reported to be associated with drought tolerance (Kulkarni *et al.*, 2008). Likewise, the greater number and lengths of the first roots of *Zea mays* seedlings under moisture deficits, as reported by Sangakkara *et al.* (2010), were beneficial for the plants to overcome stress. Furthermore, studies on *C. annuum* (Kulkarni and Phalke, 2009; Shao *et al.*, 2010) showed that the length of the primary root, at the expense of root thickening, was greater in drought-stressed plants than in control plants. This response may permit plants to survive drought by accessing water from deeper in the soil profile. Thinner roots, compared with non-stressed controls, were also reported for drought-stressed *S. vulgaris* (Franco *et al.*, 2008) and *L.*

esculentum (Kulkarni and Deshpande, 2007). A greater percentage of fine roots, capable of penetrating smaller soil pores, presumably optimises the exploratory capabilities of the root system as a whole, and may have an important role in the survival of plants faced with adverse edaphic factors (Koike *et al.*, 2003). In contrast, deficit-irrigation increased the percentage of thick roots, and reduced the percentage of medium and fine roots in *M. communis* and *N. oleander* plants (Bañón *et al.*, 2002; 2006). Branching of the roots and the total root length of *S. vulgaris* plants increased under moderate drought stress (Arreola *et al.*, 2008; Franco *et al.*, 2008). Conversely, Waisel and Eshel (2002) reported, in a different range of grasses, that branching of the roots usually decreased with decreasing water supply, but the ability of the seedlings of those species to grow with little water had been thought to be related to the greater degree of branching of the roots. Likewise, the root surface area of *S. vulgaris* plants was increased under moderate drought stress (Franco *et al.*, 2008). This can minimise localised water depletion around the roots, thus minimising the resistance to water transport to the root system (Franco *et al.*, 2006). Such plants had a higher water-use efficiency (WUE) than well-watered plants. Studies by Fernández *et al.* (2006) also reported that WUE improved with limited water availability in different species.

The combined effect of deficit-irrigation and low air humidity during the nursery phase reduced the mortality rate of *M. communis* (Bañón *et al.*, 2002) and *N. oleander* (Bañón *et al.*, 2006) seedlings after transplantation under drought and heat conditions. Such behavior was related to morphological changes observed in the roots (*i.e.*, roots were shorter, thicker, denser, and less ramified) and in the shoot:root ratio (*i.e.*, reduced by approx. 60% in both species) of the pre-conditioned plants. This behavior was also related to the re-ordering of the assimilate gradient as the flow of solutes towards the roots intensified. On the other hand, drought-induced changes in the root dynamics of trees may affect carbon sequestration, since fine root production and mortality (root turnover) represent an important path for organic carbon input to the soil, and fine root turnover increases with a decrease in soil moisture (Gaul *et al.*, 2008). However, root activity can decrease as a result of drought stress. This was the case with *Pyrus pyrifolia*, as reported by Lee *et al.* (2006), who found that the number and weight of fibrous nonactive roots increased under drought stress. Even though stressed plants had more roots, their activity was reduced.

The hardening of roots, as revealed by an increased percentage of brown roots, is frequent in drought stressed plants (Franco *et al.*, 2006). This change in root colour from white to brown (and, in some species, with roots showing beads) was associated with suberisation of the exodermis, and may reflect a metacutisation process (Brundrett, 2002). This was correlated with the capacity of *Limonium cossonianum* (Franco *et al.*, 2002b), *L. creticus* (Franco *et al.*, 2001), and *S. vulgaris* plants (Franco *et al.*, 2008) to grow under drought conditions. One indicator of the capacity of a plant to absorb and transport water is the density of xylematic vessels in its roots (Sperry *et al.*, 2002). Thus, a high density of vessels would improve resistance to water-deficit situations. This was the case with *L. creticus* in a study by Bañón *et al.* (2004), who reported that plants subjected to deficit irrigation showed a higher density of xylematic vessels in their roots than well-watered plants. Drought also caused a decrease in the diameter of root metaxylem vessels, increasing water flow resistance in *C. annuum* (Kulkarni and Phalke, 2009); although, a combination of larger and smaller xylem vessels provided better stress tolerance and biomass production under water deficit conditions. In addition, Romero *et al.* (2010) reported that severe drought stress substantially reduced root hydraulic conductivity in *Vitis vinifera*.

Plants affected by drought may exhibit signs of tissue dehydration, evidenced by increasing their root DW:FW ratios. This was reported in *R. officinalis* by Sánchez-Blanco *et al.* (2004) and in *N. oleander* by Bañón *et al.* (2006). Consequently, the greater cortex thickness : root radius (C : R) ratio in *S. vulgaris* plants grown under drought stress was significant to improve their resistance to dehydration (Franco *et al.*, 2008). Tissues internal to the endodermis are relatively well-protected against dehydration, at the expense of the cortex (Tyree, 2003). Therefore, changes in the C : R ratio of *S. vulgaris* seedlings could influence the capacity of the plants to obtain water and survive under adverse conditions. This hypothesis was supported by Eissenstat and Yanai (2002), who observed that the cortex, and a well-developed epidermis, prevented root-desiccation. Moreover, Bauerle *et al.* (2008), working with *V. vinifera* on the rootstock, *Vitis riparia, Vitis rupestris* and grown under drought, demonstrated that internal hydraulic redistribution prevented an appreciable reduction in root water potential, contributing to root membrane integrity and, thus, to extending the survival of roots in dry soils. Furthermore, for species tolerant to drought such as trees of the *Rosaceae* (*e.g.*, *Eriobotrya japonica*, *Malus* spp., *Prunus* spp. and *Pyrus* spp.), it has been reported that when grown under water-deficit conditions, cell wall ingrowth (*phi* thickening) in cortical tissues adjacent to the endodermis in the roots was more pronounced compared with plants grown under normal conditions. However, *phi* thickening in the cortex of the roots could not be observed in *Citrus* spp. or in *Diospyros* spp., both of which are sensitive to drought conditions. Thus, the development of *phi* thickening of the cortex in loquat roots grown under drought stress may be regarded as a defence mechanism against stress (Cui *et al.*, 2004; Nii *et al.*, 2004; Pan *et al.*, 2006).

2.10.2. Regulation of Root System Architecture by Water Availability and Salinity Stress

Water and salinity can indirectly modulate the root system architechture because they can produce unfavorable changes in the nutritional composition of the soil, the distribution of said nutrients, the density and compaction of soil, and the type of soil particles (Malamy, 2005). Those interactions complicate the dissection of specific transduction pathways involved in root growth and development (Jovanovic *et al.,* 2007*).* The root system is the first to perceive the stress signals for drought and salinity, therefore its development is deeply affected by their availability in soil. In many agriculturally important species, the whole plant growth is inhibited during water starvation, however, root system is more resistant than shoots and continues growing under low water potentials that are completely inhibitors for shoot growth (Spollen and Sharp, 1991). Notably, while growth of primary root is not appreciably affected by water deficit, the number of lateral roots and its growth are significantly reduced (Deak and Malamy, 2005). It has been suggested that the reduction of the lateral root formation may be caused by the suppression of the activation of the lateral root meristems, not because of the reduction of the initiation in the lateral root *per se*, as primordial generation is unaffected (Malamy, 2005; Derek and Malamy, 2005). Mutants with alterations in the development of lateral roots respond differently to drought stress (Derek and Malamy, 2005; Xiong *et al.*, 2006). Suppression of the growth of lateral root by drought has been widely accepted as an adaptive response to ensure the plant survival under unfavorable growing conditions (Xiong *et al.*, 2006). Another factor that plays an important role in growing and development of plants to tolerate the drought stress is the hydrotropism (Takahashi, 1997). A study showed that a gradient of moisture generated by water stress causes an immediate degradation of amyloplasts in the columella cells of plant roots, producing a minor response to gravity and an increase of hydrotropism (Takahashi *et al.*, 2003).

However, it is unknown how the gravity signals interact with other environmental signals to modulate the direction of root growth. Less known are the adaptations in root morphology and its relevance to salinity tolerance. Many halophytes have developed morphological adaptations, like the formation of specialized organs to expel salt out of their leaves, which allows them to keep the water and take out the salt in an active manner. Glycophytes have not developed permanent changes on its morphology to deal with salt, but they can adjust the root growth and its architecture in response to salinity, like in the case of *Arabidopsis* (Galvan-ampudia and Testerink, 2011). Also it has been observed that *Arabidopsis* root system exhibit a reduced gravitropism under salt stress, growing against the gravity vector (Sun *et al.*, 2008). *Arabidopsis* root system exposed to a simultaneous salinity and gravity stimuli responded to salinity with a change in growing direction in a way that apparently represents an adaptive arrangement between gravitropic and saline simulation. Control of the relation between gravitropism and hydro tropism allows plants to direct the root growing for a better water uptake, giving an advantage during development of the radical system under stress conditions. It is known that the salt stress inhibits the growth of the primary roots in *Arabidopsis* seedlings, although it has been reported that salt stress also modulates root gravitropism of primary root in young seedlings. In vertical position, five day seedlings germinate normally in MS medium (Murashige and Skoog) containing different concentrations of sodium chloride (NaCl), however the direction of root growth changes according to the increase of NaCl concentrations, and the root curves in stressed plants with 150 mM NaCl in the medium (Sun *et al.*, 2008). These results suggest that the salt stress and the induction of signal translations by stress modulate the direction of the root, despite of the gravity. Some reports suggest that the gravitropic signal and the answers in root apex are controlled, at least partially by salt over sensitive signaling pathway. Therefore, this pathway might interact with the gravity sensor system in the cells of the columella to direct root growth in a coordinated way (Sun *et al.*, 2008).

Abscisic acid (ABA) and auxins participate in a complex signal system that plays a very important role in the development of the root system architecture under drought conditions. These hormonal effects (levels) even though are considered as intrinsic (Malamy, 2008) can change in response to environmental cues. Cytokinins, gibberellins and abscisic acid are produced in roots to be transported to other tissues, where they play their roles in development and growth. Although auxins are the major determinants of root growth (Blilou *et al.*, 2005). cytokinin and especially abscisic acid (Talanova and Titov, 1994) have been proposed as potential chemical signals in response to water stress to modulate root system architecture. The decrease in water potential of roots caused by salinity is the factor that triggers the production of ABA in different species (Kefu *et al.*, 1991). A condition of mild osmotic stress also inhibits the lateral root formation in a dependent way of ABA (Deak and Malamy, 2005; Malamy, 2005; Xiong *et al.*, 2006; Qi *et al.*, 2007). In *Arabidopsis*, the reduced water availability dramatically inhibits the formation of lateral root, but not by the suppressing of initiation of lateral root at the lateral primordia. This inhibition does not occur in lateral root mutant 2 (LRD2) nor in two ABA deficient (Deak and Malamy, 2005; Malamy, 2005). Abscisic acid and a recently identified gene LRD2 are linked to repression of lateral root formation in response to osmotic stress. It is very interesting to note that these regulators are also related to the establishment of root system architecture without apparent effect of osmotic stress. The mutant LRD2 presents an altered response to exogenous application of ABA, while ABA-deficient mutants and LRD2 show an altered response to inhibitors of polar auxin transport (Ljung *et al.*, 2002; Dubrovsky *et al.*, 2008) suggesting a joint interaction of the hormonal signaling pathway in the regulation of lateral root formation. Some authors propose a model where the promotion or suppression of hormonal signaling pathway and regulators as LRD2 determine the type of lateral root primordium and coordinate the root architecture system in response to environmental stimuli (Galvan-ampudia and Testerink, 2011). In contrast, under drought stress conditions or osmotic stress, activation of the lateral root meristem is suppressed by ABA-mediated signals, producing few small lateral roots (Deak and Malamy, 2005; De Smet *et al.*, 2006). While auxins seem to be the main initialization hormone, pattern and emergence of lateral roots; ABA is the main hormone that controls the environmental effect (like drought and salt stress) over the root system architecture (Ariel *et al.*, 2010).

2.10.3. Regulation of Root System Architecture by Nutrients

In soil, nutrients such as phosphorus (P), nitrogen (N), potassium (K) and iron (Fe), are distributed in a heterogenous patching pattern. As soon as the primary root emerges from the seed, it has to grow. As growth goes on, *de novo* lateral root are formed to generate the particular root system morphology and architecture. These nutrients alter root patterning through particular signal transduction pathways. Thus, during their life plants change their PEDP in order to increase exponentially the root-soil interaction area and find the nutrient-rich regions (Kutz *et al.* 2002; Lopez-bucio *et al.*, 2002; Ashley *et al.*, 2006). The changes in *Arabidopsis* root system are specific for each nutrient. P, N and K starvation dramatically alter primary root length.

2.10.4. Phosphate Starvation

Root system in monocotyledonous and dicotyledonous plants, present a set of developmental modifications that tend to increase the exploratory capacity of the plant (Calderón-vázquez *et al.*, 2011). When *Arabidopsis* grown under limiting P conditions their root system architecture changes dramatically such as reduction in primary root length, increased formation of lateral roots and greater formation of root hairs (Lopez-bucio *et al.*, 2002). On optimal phosphorus conditions the newly formed root cells are added by the mitotic activity of primary meristem. These cells then get away from the meristem and increase their length, and the elongation process ends when the cells start to differentiate. When plants are phosphorus starved, cell division in the primary root meristems gradually reduces and the cells start to prematurely differentiate until total inhibition of cell elongation and loss of meristematic activity occur (meristem exhaustion). At the end, root tips change their physiological characteristics and the exhausted meristem becomes a structure which takes part in phosphorus uptake. In this process, root tips locally detect phosphorus deficiency, this response being mediated by at least LPR multicopper oxidase genes (Fraire-velázquez *et al.*, 2012; Sanchez-calderon *et al.*, 2005; Svistoonoff, *et al.*, 2007). Iron (Fe) has been reported to play a role as well in the control of these PED reprogramming (Ward *et al.*, 2008). This change of root architecture is due to the fact that, in both meristematic and elongation areas, the content of ROS is reduced as long as the determined PED goes on (Chacón-lópez *et al.*, 2011).

In the past decade the changes in root system architecture evocated by phosphorus availability has been widely studied, several genes that regulates the root architectural changes has been identified, transcription factor such as WRKY75, ZAT6 (ZINC FINGER 6), Pi-responsive R2R3 MYB (MYB62) and BHLH32 (BASIC HELIX_LOOP_HELIX 32) (Devaiah *et al.,* 2007a, 2007b, 2009; Yi *et al.*, 2005) are key regulators in this response. Mutants affected in the root system architecture changes induced phosphorus availability have been isolated: PDR2 (phosphate deficiency response 2), LPI (low phosphorus-insensitive) siz1 SAP (scaffold attachment factor, acinus, protein inhibitor of activated signal transducer and activator of transcription) and Miz1 (Msx2-interacting zinc finger), SIZ (Sánchez-calderón *et al.*, 2006; Miura *et al.*, 2005). It has been reported that ethylene is involved in modulating Pi-starvation-responsive root growth, it may restrict elongation of primary root, but promote elongation of lateral roots (Nagrajan and Smith, 2012) HPS4/SABRE (important regulator of cell expansion in *Arabidopsis*) antagonistically interacts with ethylene signalling to regulate plant responses to Pi starvation. Furthermore, it is shown that Pi-starved HPS4 mutants accumulate more auxin in their root tips than the wild type, which may explain the increased inhibition of their primary root growth when grown under phosphorus deficiency (Yu *et al.*, 2012). Gibberellins and ROS also trigger responses involving DELLAs proteins which control the rate and timing of cell proliferation and they will be dealt with in further sections.

2.10.5. Nitrogen

N is fundamental for biological molecules, such as nucleotides, aminoacids, and proteins. Plants need to acquire nitrogen (N) efficiently from the soil for growth and development. In soil, nitrate (NO_3^-) is one of the major nitrogen sources for higher plant and their concentrations vary in both time and space. Plants are able to sensing these variations of NO_3^-, which is one of the most important environmental signals affecting plants physiology and development (Kant *et al.*, 2011). The effects of nitrogen supply on plant development have been particularly studied in *Arabidopsis*. NO^-_3 free medium drastically reduces shoot biomass production and appears to have little effect on primary root length. However, NO^-_3 has a dual role on lateral roots. On one hand, the uniform exposure of root system to high nitrate (>10 mM) inhibits lateral root growth at a specific developmental step corresponding to the activation of the meristem in lateral root primordium after their emergence (Zhang and Forde, 2000). As a high NO^-_3 supply to only one part of the root system is able to repress lateral root growth on the whole root system, it has been proposed that nitrate accumulation in the aerial tissues is responsible for this lateral root primordium arrest, suggesting that long-distance signals to the root are involved. On the other hand, when the entire root system is exposed to low nitrate concentration (10 μM) and only one part of the root system is exposed to a high nitrate, there is local proliferation of lateral root. NO^-_3 locally promotes lateral root growth and increased lateral root growth rate due to a higher cell production in the lateral root meristem (Hodge *et al.*, 2009). The local stimulation of lateral root growth by nitrate-rich patches is a striking example of the nutrient-induced plasticity of PERDP. This stimulation could be dependent on NRT1.1 (Nitrate Transporter 1). This is partially due to the fact that NRT1.1 represses lateral root primordium emergence and growth of young lateral roots in the absence of nitrate. NRT1.1 transports nitrate and facilitates auxin transport in a concentration dependent manner. NRT1.1 represses lateral root growth at low nitrate availability by promoting basip*et al* auxin transport out of the lateral root primordium, towards the parental root (Remans *et al.*, 2006). MADSbox transcription factor NITRATE REGULATED (ANR1) and Auxin signaling F-box protein 3 (AFB3) are key regulators of root system architecture in response to nitrate availability. The Chlorate-resistant1 mutant (chl1) is ANR1 affected, and is less responsive to the localized NO^-_3rich patches similarly to transgenic plants in which ANR1 expressions down regulated. In the tips of lateral root and lateral root primordium, ANR1 is expressed and is localize with NRT1.1 (Bouguyon *et al.*, 2012). The afb3-1 mutant shows altered root development response to nitrate. AFB3 is an auxin receptor gene induced by nitrate in the primary root tip and pericycle; its mRNA is the target of miR393 that is induced by the products of NO^-_3 assimilation.

2.10.6. Potasium and Iron

Contrasting with physiological and molecular responses to low K and Fe, changes in root system architecture have been scarcely described. Potassium deficiencies arrest lateral root and primary root development in *Arabidopsis* (Ashley *et al.*, 2006). K^+ transporters play a crucial role in SRA changes in response to K^+ availability. Disruption of the root-specific K^+ -channel AKT1 in the akt1-1 *Arabidopsis* mutant causes reduced ability of plants to grow in low potassium media (100 μM) (Hirsch *et al.*, 1998). In *Arabidopsis*, changes in the gravitropic behavior of root system were also observed in low potassium media. The genes of the KUP/HAK/KT family are homologous to bacterial KUP (TrkD) potassium porters. The trh1 (tiny root-hair 1) mutant, which is disrupted in AtKUP4/TRH1 gene shows agravitropic behavior in its roots independently of K^+ concentration in the media when grown on vertical agar plates, and also, ProTRH1:GUS expression is limited to the root cap where gravity is sensed. Interestingly, agravitropic responses in trh1 are complemented by exogenous auxin. This mutation is associated with the loss of auxin pattern in the root apex. Thus, TRH1 is an important part of auxin transport system in Arabidopsis roots (Hirsch *et al.*, 1998; Vicente-agullo *et al.*, 2004; Grabov, 2007).

Typically, the root architectural changes in response to low availability of Fe include ectopic formation of root hair due to modulation in their position and abundance (Schmidt and Schikora, 2001). Giehl *et al.* (2012) analyzed the changes in lateral root architecture in response to localized Fe supply in wild type and Fe acquisition and translocation- defective mutant plants. They found that lateral root elongation is highly responsive to local Fe and that the symplastic Fe pool in lateral root favors local auxin accumulation. They identified the auxin transporter AUX1 as a major Fe-sensitive component in the auxin signaling pathway that mainly directs the rootward auxin stream into lateral roots that have access to Fe.

2.10.7. Meristematic Activity Regulation by Abiotic Stress

To cope with environmental changes, plants have to adapt their growth timing and pattern by altering rates of cell proliferation and differentiation. The expression of several cell cycle genes is increased or decreased upon external cues (Peres *et al.*, 2007) but it is poorly understood the full molecular basis supporting these transcriptional controls, and if the cell cycle control modifications happen to fall into the post-translational category, the current knowledge is also very limited. However, there have been identified several key players in stress-induced cell cycle modifications that have cast the first light over the understanding the talk between environmental signals and the mitotic or endoreplication cycle. Gibberellins (GAs), plant hormones, promote cell expansion by disrupting growth inhibitory proteins named DELLAs (Richards *et al.*, 2001) and also promote cell proliferation in *Arabidopsis* (Archard *et al.*, 2009). In the root meristem of GAdeficient mutants, cell division rate is decreased and the phenotype is rescued by GA treatment. DELLA proteins are also involved in this regulation, as non-degradable forms of DELLA inhibit cell proliferation.

Low levels of GAs in GA-deficient mutants enhance the expression of certain CDK inhibitor genes – KRP2, SIM, SMR1 and SMR2- with a DELLA-related mechanism, and cell proliferation defects shown by these mutants can be recovered by over-expressing CYCD3;1. These findings tend to indicate that GA signaling drives cell proliferation by modulating the activity of CYC-CDK complexes, at least partially mediated by the DELLA dependent expression of CDK inhibitors, and thus making DELLA a potential intermediate in the signal transduction channel connecting environmental signals and cell cycle progression. This is proposed to be a consequence of reduced cell expansion and associated division of the endodermis layer in the root apical meristem (Achard and Genschik, 2009; Ubeda-tomas *et al.*, 2009), suggesting a role for the endodermis in controlling the growth rate in the root apical meristem. Another potential link is RICE SALT SENSITIVE 1 (RSS1), controlling the cell cycle progression under various abiotic stress conditions (Ogawa, *et al.*, 2011). The rss1 mutants do not present evident growth defects under normal conditions, but they display hypersensitivity to high salinity, ionic stress and hyperosmotic stress. Under these conditions, in rss1, shoot and root meristems are severely affected, showing a reduced population of proliferating cells, leaving RSS1 as a required factor for proliferative cell status in the meristem. RSS1 is expressed during the S phase of the mitotic cycle and its protein is degraded via APC/C during the M/G1 transition.

RSS1 interacts with a Type 1 Protein Phosphatase (PP1), known in humans to inactivate Retinoblastoma (Rb) proteins through dephosphorylation, which is inhibitory to the G1/S transition (Hirschi *et al.*, 2010). Sugars can act as signaling molecules in assorted biological processes, and even that sucrose-dependent cyclin expression is known since a decade ago (Riou-khamlichi *et al.*, 2000), lateral root formation through sucrose induction is a good example of sugar-dependent reactivation of cell proliferation (Nieuwland *et al.*, 2009). This recent study shows that the expression of CYCD4;1 levels in root pericycle cells is dependent on the sucrose availability, and that reduced CYCD4;1 levels in cyca4;1 mutants or wild-type (wt) roots grown in the absence of sucrose cause lateral root density to drop. It is not clear how sucrose up-regulates CYCD4;1 in specifically in that kind of cells, but these findings suggest that the transcriptional effect has to do with sucrose-dependent regulation of lateral root density. Notably, auxin does not have an effect over the expression of CYCD4;1 in pericycle cells, and restores the reduced lateral root density phenotype of cyca4;1 mutants, suggesting that CYCD4;1 has no role in the auxin-mediated lateral root initiation pathway. CYCD3;1, is also responsive to sucrose availability, but the effects of this over CYCD3;1 activities are not clear (Planchais *et al.*, 2004). Endoreplication progress is also affected by several environmental signals. E2F3/ DEL1, an atypical E2F present in *Arabidopsis*, and that functions as a transcriptional repressor, is one of the key regulators that negatively controls the entry into the endoreplicative cycle (Lammens *et al.*, 2008). It has been suggested that the balance between the transcriptional activator E2Fb and repressor E2Fc controls light-dependent endoreplication through the antagonistic modification of the DEL1 expression (Berckmans *et al.*, 2011). E2Fb and E2Fc compete for the same DNA-binding site of the DEL1 promoter and enhances the DEL1 expression, respectively. Under light conditions, E2Fb is the preferred binding partner, enhancing DEL1 expression and consequently repressing the endoreplicative cycle (Radziejwoski *et al.*, 2011). In the dark E2Fb is degraded, allowing E2Fc to bind to the DEL1 promoter, repressing DEL1 expression. Ultra violet-B (UVB) radiation damages DNA molecules by forming cyclobutane pyrimidine dimers (CPDs) which prevent DNA transcription and translation. Plants remove CPDs by photolyases, and these enzymes are encoded by a PHOTOLYASE 1 (PHR1) (Thiagarajan *et al.*, 2011). It has been shown that in addition to CCS52A2, a known target of DEL1, DEL1 represses the transcription of the PHR1 gene and thereby coordinates DNA repair and endocycle triggering (Radziejwoski *et al.*, 2011). After UVB treatment, DEL1 expression is strongly downregulated, permitting the upregulation of PHR1 and thus leaving the cell able to repair its DNA.

Environmental and nutrient availability condition changes affect on root apical meristem organization (Lopez-bucio *et al.*, 2003). ROS and Reactive Nitrogen Species (RNS) have been reported to be rapidly induced by several kinds of environmental stresses in a variety of plant species to regulate the plant response to biotic and abiotic stresses. In particular, oxidative stress caused by drought and salinity, has been proposed that ROS production is an obligatory element of the response to induce an adequate acclimatization process (Miller *et al.* 2010). Therefore, the degree of accumulation of ROS is what determines whether it is a part of the signaling mechanism (low production) or a harmful event (high production) to plants, making the control of production and degradation of ROS the crucial element for plant resistance to stress (Miller *et al.* 2010; Cheeseman, 2007; Garg and Manchanda, 2009; Mortimer *et al.*, 2008). ROS is never completely eliminated, as it plays an important role in signaling and growth regulation (Carol and Dolan, 2006); ROS quenching inhibits the root growth (Demidchik *et al.*, 2203), and overexpression in *Arabidopsis* of a peroxidase localized mainly in the elongation zone stimulates root elongation (Passardi *et al.*, 2006). This calls for redox control of the cell cycle, which is possibly linked to A-type cyclins, shown to be differentially expressed under oxidative stress in tobacco, resulting in cell cycle arrest (Reichheld *et al.*, 1999). It is also known that low temperatures (Lee *et al.*, 2004; Aroca *et al.*, 2005), *metals* (Sharma and Dubey, 2007) and nutrient deficiency (Tyburski *et al.*, 2009) induce the presence of ROS and RNS in specific tissues. These forms of stress affect root morphology by reducing primary root growth and promoting branching, but the mechanisms of the redox generation sensing are not well understood.

The typical response of the *Arabidopsis* radical system to low phosphorous (P) availability is an example to illustrate how complex these processes are. A study showed that ROSs are involved in the developmental adaptation of the root system to low P availability (Tyburski *et al.*, 2010). Rapidly growing roots of plants within a normal P medium synthesize ROS in the elongation zone and QC on the root, whereas seedlings within low P mediums showed a slow growth of the primary root, and the ROS normally found in the QC relocate to cortical and epidermal tissues. In a previous study (Sanchez-calderon *et al.*, 2005), it has been indicated that *Arabidopsis* plants under low P conditions show a decreased number of cells in the root apical meristem, and it decreases until it is depleted. In these roots, all root apical meristem cells differentiate and the QC is almost indistinguishable. A possible cause of this response to P starvation could be the cell cycle arrest modulated by ROS and CYCAs, but it is more complicated, as the response is also modulated by auxin (Lopez-bucio *et al.*, 2003, 2005); and gibberellin-DELLA pathways (Jiang *et al.*, 2007). Interestingly, DELLAs promote survival by reducing the levels of ROS (Achard *et al.*, 2008), suggesting a link between the gibberellin-DELLA cell cycle control pathway and ROS pathway in the developmental adaptation to the root system to low P availability It requires further study to precise the way these signals crosstalk and determine the developmental adaptation of the root system to low P availability by means of cell cycle progression control, as well as additional efforts to reveal the manners by which other regulatory pathways responding to abiotic stress interact with and influence the cell cycle control mechanisms.

2.10.8. Waterlogging Stress

Water logging, which results in a decline in the availability of oxygen to the submerged organs, may negatively affect plant development. Thus, root dysfunction, as a result of inadequate oxygenation, can modify plant growth and depress crop yields through interference with water relations, mineral nutrition, and hormone balance (Horchani *et al.*, 2009; Vidoz *et al.*, 2010). Furthermore, the total absence of oxygen from soils may be a prelude to microbially-mediated anaerobic biochemical transformations of compounds in the soil, with the creation of pools of often highly phytotoxic products (Armstrong and Drew, 2002). Overwatering often causes disease problems and slows plant growth, with root-rot being a common phenomenon when plants are over-watered. Oxygen deficiency can occur in soils because of over-irrigation, excessive rainfall, or flooding, leading to root anoxia. Deficient aeration and low oxygen diffusion may then affect rhizogenesis and the morphological, anatomical, and physiological characteristics of the root system (Ochoa *et al.*, 2003). Paradoxically, the resulting decrease in hydraulic conductivity observed in certain plant species can induce water deprivation in the shoot (Maurel *et al.*, 2010).

Waterlogging stress often negatively affects root growth. Dickin and Wright (2008) reported a decrease in total root length, but not in the final depth of the root system in waterlogged *Triticum aestivum* plants. Also, the effect of continued root-zone flooding on *Prunus persica* seedlings which, among *Prunus* species, are quite sensitive to flooding, was examined by Rutto *et al.* (2002). The visual qualities of plants deteriorated with flooding, with complete defoliation occurring over a period of 3 weeks.The viable portion of the roots in these plants was only 28%, while it was 85% in non-flooded plants. Root growth in waterlogged, fruiting *L. esculentum* plants was also severely decreased (Horchani *et al.*, 2009). They have showed that, in contrast to fruit, root protein and carbohydrate contents decreased concomitantly with a significant decrease in photosynthesis. They concluded that, at the fruiting stage, root growth was probably controlled by the relative sink strength of the fruit. Franco *et al.* (1995), working with young *Amygdalus communis* trees under different drip irrigation rates, also found a clear decline in root growth, coinciding with the period during which rapid fruit development and vegetative growth overlapped.

Plant roots can develop morphological and anatomical features that facilitate oxygen use under water logging stress. These adaptive responses can include the formation of aerenchyma tissues in the root cortex (Suralta *et al.*, 2010), the development of adventitious roots near the soil surface (Vidoz *et al.*, 2010), and increases in root diameter (Munoz-Arboleda *et al.*, 2006). Although *L. esculentum* is known for its sensitivity to water logging stress, Vidoz *et al.* (2010) found that its ability to produce adventitious roots increased plant survival when the level of oxygen decreased in the rhizosphere. This response resulted in a new root system that was capable of replacing the original root system when the latter had been damaged by water logging. On the other hand, despite the fact that *Beta vulgaris* plants submitted to a period of flooding had a significant reduction in mean root diameter compared with control plants (Costa *et al.*, 2008), it was more common to find increased root diameters in waterlogged plants. This was the case with *Solanum tuberosum*, as reported by Munoz-Arboleda *et al.* (2006), who concluded that root thickening in deep apical roots suggested the formation of aerenchyma promoted by a combination of inappropriate over-irrigation and soil compaction. It is known that the development of aerenchyma effectively facilitates the internal diffusion of oxygen to root tips under waterlogged conditions (Dickin and Wright, 2008; Suralta *et al.*, 2010).

2.10.9. Salt Stress

Salinity imposes two types of stress on roots: osmotic stress resulting from the lowered water potential in the root medium; and ionic stress induced by changes in the concentrations of specific ions in the medium and inside the root tissue (Bernstein and Kafkafi, 2002). In general, root growth is inhibited by exposure to high salinity as a result of both its osmotic and toxic effects (Bañón *et al.*, 2005). In glycophytes, since root growth was usually less affected by salinity than shoot growth, a decreased shoot:root ratio was often observed when plants were subjected to salt stress (Bernstein and Kafkafi, 2002; Munns, 2002; Khayyat *et al.*, 2009). However, the opposite was found in *Pinus* spp. (Croser *et al.*, 2001) and in *Portulaca oleracea* (Franco *et al.*, 2011).

Root growth was reduced by salinity in three ornamental species, *Calceolaria hybrida*, *Calendula officinalis*, and *Petunia hybrida*, the first being the most affected (Fornes *et al.*, 2007). Similarly, root parameters such as FW, DW, and root length, decreased with irrigation solutions of increasing electrical conductivity (EC) in experiments with three varieties of *Solanum melongena* (Akinci *et al.*, 2004) and *P. oleracea* (Franco *et al.*, 2011). Furthermore, a decrease in root DW with increased salinity has been reported in *Cynara scolymus* by Mauromicale and Licandro (2002). Croser *et al.* (2001) and Franco *et al.* (2011) also found an increase in root diameter (hypertrophy) in response to salinity. However, it has been documented that the degree of response to salt stress may vary considerably within a family, within a genus, and even within a species. For example, high NaCl had a negative effect on the DW of roots in *Fragaria ananassa* cv. Selva (Khayyat *et al.*, 2009), but Turhan and Eris (2005) reported that NaCl increased root DW in the strawberry cv. Camarosa. Likewise, salt stress can cause profound modifications to root architecture. NaCl treatment of *L. esculentum* resulted in a more branched root system, compared to the untreated controls, with the roots being shorter and each main root having more lateral roots. These modifications of root growth gave rise to a larger root system (Karni *et al.*, 2010). In addition, Rose *et al.* (2010) reported more shallow root systems for plants grown under saline conditions than for plants grown with adequate rainfall.

On the other hand, the micronutrient concentrations of roots can be affected by salinity. The concentration of Cu in roots was increased by salinity in *Cucurbita pepo* (Villora *et al.*, 2000) and in *F. ananassa* plants (Khayyat *et al.*, 2009). Also, the application of NaCl significantly increased Zn, Mn, and Fe levels in roots compared to control *F. ananassa* plants (Khayyat *et al.*, 2009).

2.10.10. Temperature Stress

The exposure of root systems to extreme (high and low) temperatures during the growing season negatively affects root growth. The roots of container-grown plants may be subjected to extreme temperatures not normally found under field conditions, which alter root growth (Mathers *et al.*, 2007; Miralles *et al.*, 2009). For plants grown in black plastic containers, extremely high root zone temperatures have often been reported (Irmak *et al.*, 2004; 2005). However, exposure to low temperatures is one of the most common abiotic stresses of plants (Wu and Zou, 2010). Many of the topics regarding the effects of low and/or high temperatures on the responses of roots have been discussed by McMichael and Burke (2002). If the temperature of the soil drops significantly from optimum values, then the structure and function of root systems may be altered. In addition, lower temperatures reduced the shoot length and the shoot:root ratio, and increased the percentage of brown roots in *L. creticus* (Franco *et al.*, 2001). Nevertheless, compared with moderate temperatures, low temperature strongly reduced root length, root surface area, and root volume, but increased the shoot:root DW ratios in *Citrus tangerine* seedlings (Wu and Zou, 2010). This was also the case with *Lactuca sativa* (He *et al.*, 2009) and *M. communis* (Miralles *et al.*, 2009). Furthermore, the exposure of roots to cold reduced hydraulic conductivity and can result in a marked water stress in the whole plant (Maurel *et al.*, 2010), as Lee *et al.* (2004) also reported for *Cucumis sativus* plants. When the combined effects of deficit-irrigation and low temperature are studied (Franco *et al.*, 2001; 2002a; Bañón *et al.*, 2004), regimes involving the least water and the lowest temperatures produced plants which had lower shoot:root ratios, higher percentages of brown roots, and greater numbers of xylem vessels in their stems and roots.

On the other hand, supra-optimal soil temperatures can affect root growth and longevity, as reported by Ostonen *et al.* (2007). This was the case with *Agrostis stolonifera* (Schlossberg *et al.*, 2002), *L. esculentum* (Nakano, 2007) and *Z. mays* (Walter *et al.*, 2002). Likewise, Kadir *et al.* (2006), studying thermo-tolerance in *F. ananassa*, concluded that the optimal temperatures for vegetative growth were 30°C day/25°C night. Exposure to higher temperatures (40°C/35°C) caused a reduction in shoot and root biomass, with root growth being more responsive to temperatures above 20°C/15°C than shoot growth. Also, exposure of the roots of *L. sativa* plants to temperatures up to 40°C severely limited root and shoot growth, with decreasing root and shoot biomass, total root lengths, numbers of root tips, and root surface areas, but with increasing average root diameter, as determined by Qin *et al.* (2007). In addition, heat-stress affected the growth and anatomy of *Salvia splendes* plants, increasing root growth (Natarajan and Kuehny, 2008).

2.10.11. Illumination Stress

Reductions in solar radiation can produce significant reductions in plant growth. The morphological parameters of roots may be affected under reduced solar radiation, as showed by Ostonen *et al.* (2007), who reported that the specific root length (*i.e.*, the length:mass ratio) of fine roots responded negatively to reduced light. The DWs of *R. alaternus* plants grown under shaded environments were reduced, with roots being more affected than shoots (Miralles *et al.*, 2011).

This fact increased the shoot:root DW ratio from 3.7 (control) to 4.7 (in 65% shade) and to 7.4 (in 80% shade). Villar-Salvador *et al.* (2004) also found a significant increase in the shoot:root ratio in *Quercus ilex* seedlings under 45% shade. The reduction in evaporation due to shade affected the need for water absorption, which may have diminished root growth. Similarly, reductions in solar radiation caused by short days of different duration were investigated by MacDonald and Owens (2010) in containerized *Pseudotsuga menziesii* plants. These authors reported that plant height and shoot DW were unaffected by the duration of short days, whereas root DW was significantly reduced in plants given 6 weeks of short days compared with the controls. Also, when short days were applied to this species under moderate drought stress (MacDonald and Owens, 2006), plant height and shoot DW were unaffected by the regime, although root DW was reduced.

Yang *et al.* (2008) studied the combined effect of low light and drought on the partitioning of biomass in *Picea asperata* and concluded that, under control natural light conditions, drought increased biomass partitioning to the roots, but no prominent drought-induced differences in biomass partitioning to roots were observed under lowlight conditions. In fact, drought and shade tolerance depend on different morphological adaptations, as reported by Markesteijn and Poorter (2009). On the other hand, high-light conditions may affect root growth. An increase in turf grass root length, as a result of using reflected light in a shaded environment, was observed by Mumford *et al.* (2004). Similarly, Matsuki *et al.* (2003) found decreased shoot:root ratios in *Quercus crispula* under high-light conditions. Also, Tani *et al.* (2003) reported a decreased leaf area:root length ratio in *Pteridophyllum racemosum* in response to abrupt increases in irradiance. These authors concluded that this decreased ratio was caused by a combination of enhanced carbon allocation to the roots, together with increased root length.

2.10.12. Nutrient Deficiency or Excess

Adequate crop growth is promoted through the application of an appropriate fertilization regime. This results in a balanced growth, in terms of the shoot:root ratio, which is vital for plants growing in harsh environments where competition for water and nutrients is frequently intense. Both nutrient deficiency and nutrient excess stresses usually affect root development, but the extent of the response of roots to these stresses may vary considerably within a family and even within a genus, as reported by Francini and Sebastiani (2010), who found that excess Cu increased root DWs in *P. persica, Prunus amygdalus*, but decreased root DWs in *Prunus cerasifera* rootstocks.

Nutrient deprivation enhances root growth in order to maintain overall nutrient uptake and shoot growth (Bumgarner *et al.*, 2008; Maurel *et al.*, 2010). Phosphorus deficiency led to increased root growth, enhancing the ability of the plant to access regions of adequate soil P (Grant *et al.*, 2001). Under conditions of Fe deficiency, root DWs of *Spathiphyllum* sp. grown hydroponically were also enhanced and the roots appeared white, with increased root hair formation on the young lateral roots. This growth response facilitated the uptake of Fe through an increased root surface area (Yeh *et al.*, 2000). Furthermore, Lecompte *et al.* (2008), studying roots and nitrate distribution related to the critical N status of fertilized *L. esculentum* plants, concluded that the pattern of soil colonization by root systems depended strongly on N fertilization treatment. These authors showed that, after an initial phase of deep and wide soil colonization, the root system receded with high soil N availability, while root extension was continuous throughout the soil profile when the soil N level was low.

Likewise, the shoot:root DW ratio of *Spathiphyllum* sp. was decreased under N-, P-, and Fe-deficiency. However, Mg-deficiency did not affect root DWs (but lateral roots were longer with fewer branches than in control plants). Under Mg- and K-deficiency, shoot:root DW ratios were not significantly different from those in plants grown with complete nutrients. Similarly, B deficiency did not affect root DWs, but longer main roots and many short, brown lateral roots were observed (Yeh *et al.*, 2000). Finally, Munoz-Arboleda *et al.* (2006) found that root length and root DW were not affected by N rate in seepage-irrigated *S. tuberosum* plants.

Conversely, nutrient deprivation can also reduce root growth. This was the case with *Spathiphyllum* sp. plants, as observed by Yeh *et al.* (2000) who found that root DW was significantly reduced under N-deficiency and under Ca-deficiency, while the roots appeared short, brown, and thickened. Furthermore, the shoot:root DW ratios of plants were increased by deficiencies in nutrients with a low mobility in the phloem such as Ca and B. Bernstein *et al.* (2005) also reported larger shoots and increased shoot:root DW ratios when *Ranunculus asiaticus* was cultivated under a low N treatment compared to control plants. Equally, *Daucus carota* plants exposed to Cu-deficiency showed morphological changes and poor root development, with root FWs being reduced to 59% of control values (Khurana and Chatterjee, 2007).

On the other hand, root growth is often adversely affected by nutrient excess, which thus provokes changes in the shoot:root ratio. This was the case with *Quercus rubra* seedlings, reported by Bumgarner *et al.* (2008), who concluded that N-fertilization greatly increased aboveground biomass production, but decreased root DW. In contrast, levels of N exceeding normal concentrations resulted in decreased growth in *C. annuum* and a lower shoot:root ratio (Guzmán and Sánchez, 2003). Also, high levels of N and Ca in the growing medium resulted in a decreased shoot:root ratio in *C. annuum* plants (Guzmán and Sánchez, 2003). Zn and Cu are essential nutrients for higher plants, but they become toxic as their availability increases. Disante *et al.* (2011) showed that root length and specific root length in *Quercus suber* seedlings decreased as the availability of Zn increased. Also, *D. carota* plants exposed to excess Cu showed reduced root length (Khurana and Chatterjee, 2007). Root elongation was similarly reduced under high Cu concentrations in *Crocus sativus* (Keyhani and Keyhani, 2008), while Groppa *et al.* (2008) also reported significantly reduced root growth under Cu excess.

2.10.13. Heavy Metal Stress

Heavy metals play an important role in plant ecology and affect the performance of roots, even in small quantities as trace elements (Hagemeyer and Breckle, 2002). Exchangeable Al is the major toxic element in most acid soils, with Al toxicity being one of the most deleterious factors for plant growth in acid soils. Al cations limit root growth, thereby limiting plant growth and production (Matsumoto, 2002). Root elongation in plants susceptible to Al can be inhibited rapidly upon exposure to the element (Zheng and Yang, 2005). Al strongly inhibited root growth in hydroponically grown *Vicia faba* (Zhang *et al.*, 2009). Root growth is also limited by other elements. For example, root growth was reduced in *Pisum sativum* plants under Ni stress (Gabbrielli *et al.*, 1999). Similarly, initial root growth in *Helianthus annuus* was impaired in Cd-contaminated soils (Groppa *et al.*, 2008). Decreased root growth was observed in *Allium sativum* plants grown under Pb stress (Jiang and Liu, 2010), in *Atriplex* spp. grown in soil polluted with Ni, Pb, and Zn (Kachout *et al.*, 2009), in *Iris pseudacorus* grown under Cd and Pb treatments (Zhou *et al.*, 2010), and in *L. esculentum* under Cd stress (Zoghlami *et al.*, 2011). The bio-accumulation of heavy metals is frequently greater in roots than in leaves and stems (Naaz and Pandey, 2010; Souza *et al.*, 2011; Zoghlami *et al.*, 2011), therefore the decrease in root growth caused by the toxicity of these metals is often more severe than the decrease in shoot growth (Kachout *et al.*, 2009; Zhou *et al.*, 2010).

Adaptive changes in root architecture may be important for plants to survive in soils with high heavy met*al* contents (Hagemeyer and Breckle, 2002). López-Marín *et al.* (2009) reported that the inheritance of resistance to Al in *P. vulgaris* was related to a differential response in root architecture traits following Al exposure. Increases in both total root length and the number of root tips enabled greater acquisition of P through root exploration of a greater volume of soil under infertile, high Al and acidic soil conditions. In addition, Cd induced anatomical changes, such as the lignification of cell walls in tissues in the roots of *Genipa americana*, and caused cell death in roots and leaves (Souza *et al.*, 2011).

2.10.14. Elevated Atmospheric Levels of CO_2

The responses of plants to increasing atmospheric CO_2 concentrations have been studied intensively. However, the effects of elevated CO_2 on root dynamics have received much less attention, despite their importance for global carbon budgets, as well as for nutrient cycling in ecosystems. Therefore, data on the effects of elevated atmospheric CO_2 levels on root dynamics are insufficient to make generalizations or to draw conclusions. Ostonen *et al.* (2007) reported decreased specific root lengths under elevated levels of CO_2. In contrast, different studies indicated a general increase in root growth under elevated CO_2 compared to ambient CO_2 levels (Arnone III *et al.*, 2000; Milchunas *et al.*, 2005a). Similarly, Sindhoj *et al.* (2000) found that an elevated CO_2 treatment increased root growth in a nutrient-poor semi-natural grassland. Also, in a short grass steppe, elevated CO_2 affected root architecture through increased branching compared to ambient CO_2. Root-length growth, the number of roots, and the diameter of roots in the upper soil profile were also greater under such conditions (Milchunas *et al.*, 2005a). Furthermore, the lignin concentrations of new annual roots were lower in an elevated CO_2 treatment (Milchunas *et al.*, 2005b).

CO_2-enriched atmospheres during nursery production were used to obtain seedlings with lower shoot:root ratios, mainly due to an elevated root biomass (Biel *et al.*, 2004; Cortes *et al.*, 2004). Some authors have hypothesized that the increased conservation of water associated with elevated CO_2 levels may result in increased shoot:root ratios, and that this may be greater in wet years (Morgan *et al.*, 2004). An alternative hypothesis suggests that the shoot:root ratio may decrease with elevated CO_2, since the latter may decrease plant N concentrations, and the increased N demand would increase the need for more extensive root exploration of the soil (Milchunas *et al.*, 2005b).

2.10.15. Mechanical Stress

Root growth can be altered both by increasing the compactness of the soil (*e.g.*, by a lack of tillage, hardpans just beneath the tilled layer, or the presence of gravelly horizons or claypans), as well as by restricted rooting media (*e.g.*, small container size, or in root restricted urban settings). Roots experience mechanical impedance and decreased growth rates due to the force required to penetrate through strong soil, where the drier or denser a soil, the higher its strength. These changes in the mechanical forces on roots trigger a signaling cascade throughout the plant with distant physiological effects on shoots (Masle, 2002). As the soil strength increases, the rate of root elongation decreases due to the increasing resistance of the soil particles to displacement. As reported by Samson *et al.* (2002), mechanical impedance is one important factor which may restrict the access of roots to deeper soil layers, thereby reducing the capacity of the root system to extract water and nutrients. Furthermore, Clark *et al.* (2008) showed that gradual rather than abrupt increases in soil strength lead to enhanced root penetration of strong soil layers.

On the other hand, growing plants in a restricted rooting medium frequently causes a down-regulation in photosynthetic capacity. This was the case with *C. sativus* (Kharkina *et al.*, 1999), *Chrysanthemun* sp. (Goto *et al.*, 2002), and *L. esculentum* (Shi *et al.*, 2008). The latter study showed that root-restriction conditions resulted in reduced root DW as early as 10 d after root confinement, and that the limitation on photosynthesis by root restriction was mimicked by water stress. In contrast, using different container sizes, no differences were observed in root growth of *Shorea leprosula* (Aminah *et al.*, 2004). In addition, Shi *et al.* (2008) found that suppression of root growth by root-restriction was alleviated, to some extent, by aeration.

Under windy conditions, a root system may differ in its windward and leeward portions. Wind ward roots are continuously stressed by tensile forces, whereas leeward roots are continuously stressed by compression forces. These different forces may generate asymmetrical root systems (Nicoll *et al.*, 2008). Similarly, the biomechanics of the root system was studied in *Spartium junceum* plants growing on slopes, where the lateral roots elongated in two directions (up-slope or down-slope) thereby generating an asymmetrical root system (Chiatante *et al.*, 2003).

2.11. Biochemical Regulation of Root

Development under abiotic stress

There have been numerous studies published in the last few years describing and discussing the chemical and hormonal bases for the responses of roots to abiotic stresses. Although the morphological adaptations of plants are largely different for diverse abiotic stresses, similar hormones and biochemical processes are involved in signaling from roots to the shoot during different stress responses (Table 2).

Table 2: Main biochemical processes involved in the regulation of root development in plants subjected to abiotic stress conditions

Biochemical process	Abiotic stress	Reference
Higher production of abcisic acid	Drought	Davies *et al.* (2005); Hund *et al.* (2009); MacDonald *et al.* (2009); Sangakkara *et al.* (2010)
	Nutrient deficiency	Vysotskaya *et al.* (2008)
Ethylene synthesis	Drought	Aloni *et al.* (2006).
	Waterlogging	Vidoz *et al.* (2010).
	Salt	Ruzicka *et al.* (2007); Karni *et al.* (2010)
Changes in auxin and cytokinin concentration	Drought	Aloni *et al.* (2006).
	Waterlogging	Armstrong and Drew (2002);Vidoz *et al.* (2010)
	Salt	Ruzicka *et al.* (2007); Albacete *et al.* (2008); Karni *et al.* (2010); Pérez-Alfocea *et al.* (2010)
Changes in amino acid metabolism	Waterlogging	do Amarante *et al.* (2006)
	Salt	Kakkar *et al.* (2000); Karni *et al.* (2010)
	Heavy metal	Groppa *et al.* (2008); Zoghlami *et al.* (2011)
Changes in ROS levels	Drought	Sharp *et al.* (2004); MacDonald *et al.* (2009)
	Salt	Karni *et al.* (2010).
	Heavy metals	Kachout *et al.* (2009) ; Zhang *et al.* (2009); Naaz and Pandey (2010)
Changes of RNS (Reactive nitrogen species) levels	Heavy metals	Xiong *et al.* (2010).

The phenomenon of root growth stimulation by drying may be due to the production of root signals that regulate shoot development to promote root growth to capture water. Dehydration of fine roots may promote the production of these signals which may alter plant growth (Loveys and Davies, 2004; Santos *et al.*, 2007). Several authors (Hund *et al.*, 2009; Sangakkara *et al.*, 2010) have shown that this root signal is due to the development of higher ABA concentrations in the xylem. Ethylene also appears to promote the growth of lateral roots and root hairs through interactions with other hormones such as cytokinins and auxins (Karni *et al.*, 2010). In addition, Sharp *et al.* (2004) and Karni *et al.* (2010) suggested that reactive oxygen species (ROS) might be induced by drought stress, thereby causing the evolution of ethylene and an inhibition of root growth. Root anti-oxidants may relieve these stress effects by reducing the levels of ROS, thereby allowing a higher production of ABA and enhancing root growth (MacDonald *et al.*, 2009; Karni *et al.*, 2010).

Armstrong and Drew (2002) reviewed the conditions which can lead to oxygen deficiency in roots, and the morphological and biochemical adaptations which enable plants to cope with the consequences of excess soil wetness. Ethylene entrapment by water may represent the first warning signal to the plant to indicate waterlogging, as shown in a study on flooded *L. esculentum* plants by Vidoz *et al.* (2010).These authors observed that ethylene stimulated auxin transport and, thereby, auxin accumulation in the stem which triggered additional ethylene synthesis. This further stimulated a flux of auxin towards the root system, inducing the growth of pre-formed root initials, resulting in new roots capable of replacing the original damaged ones. On the other hand, do Amarante *et al.* (2006) determined that, in both leguminous (*P. vulgaris*, *P. sativum*, and *Lupinus albus*) and non-leguminous species (*L. esculentum*, *Z. mays*, and *H. annuus*), the water logging of plants led to a large reduction in xylem glutamine levels, attributed to impaired N assimilation, which preceded the change in the aspartate:asparagine ratio. These authors concluded that such changes in amino acid metabolism may promote the response of the root system to stress.

Although there is ample information on the involvement of hormones during root response to water stress, little is known about the role of hormones in the response of roots to salinity. For *L. esculentum* plants grown under salt-stress, Albacete *et al.* (2008) and Pérez-Alfocea *et al.* (2010) reported changes in the auxin:cytokinin ratios in leaves and roots, which may explain the shift in the allocation of biomass to the roots. Moreover, Karni *et al.* (2010) suggest that the effect of NaCl on root morphology and growth may not be associated with ethylene production in the root, but may be associated with altering the distribution of indol eacetic acid (IAA) within the root. In addition, Ruzicka *et al.* (2007)

showed that ethylene exerted its effect on root morphology by altering the distribution of auxin within the root tip, inhibiting root elongation and promoting the initiation of lateral roots. Gao *et al.* (2009) hypothesised that higher levels of arginine metabolism in the roots, especially in white roots, rather than in the leaves of *Malus domestica* might be involved in long distance signalling from roots to shoots during the stress response. Arginine was also reported to be implicated in plant responses to NaCl stress (Kakkar *et al.*, 2000). Oxidative stress can also be induced by heavy metals stress. In fact, Zhang *et al.* (2009) found that increases in anti-oxidant enzyme activity and the malondialdehyde content of roots of *V. faba* were involved in the response of plants to Al toxicity. Therefore, anti-oxidant resistance mechanisms may provide a strategy to enhance metal tolerance. The effects of heavy metals can generate anti-oxidative defence systems (*e.g.*, superoxide dismutase, ascorbate peroxidase, glutathione reductase, and catalase). This anti-oxidative activity seems to be of fundamental importance for the adaptive responses of plants to heavy metals stress (Kachout *et al.*, 2009). Naaz and Pandey (2010) reported that catalase activity in stressed *L. sativa* plants increased with increasing Ni, Cr, or Zn concentrations in the water used for irrigation. Moreover, Cd induced qualitative and quantitative changes in the amino acids contents of roots, with asparagine content being a good marker for Cd stress (Zoghlami *et al.*, 2011). In addition, Groppa *et al.* (2008) suggested that polyamines are probably involved in the signalling pathways triggered under Cd stress. Likewise, Cu stress reduced the concentrations of sugars, starch, protein, and carotene, and increased the levels of non-proteinaceous nitrogen and phenols in roots (Khurana and Chatterjee, 2007). Finally, Xiong *et al.* (2010) reported an effect of nitric oxide (NO) on alleviating heavy metal toxicity in plants. This bio-active and signal-transmitting molecule plays a crucial role in the promotion of root development (Pagnussat *et al.*, 2002; Liao *et al.*, 2011). On the other hand, Vysotskaya *et al.* (2008) reported that ABA may also have a role in shoot-to-root partitioning in nutrient-limited plants.

Decreases in photosynthesis and, consequently, reductions in the levels of metabolites and energy occur under a range of abiotic stresses. The redistribution of metabolites in favour of the roots at the expense of the shoots is frequently observed under drought, salinity, sub-optimal temperatures, some nutrient deficiencies, elevated atmospheric CO_2, or other abiotic stresses. This redistribution is probably due to the need for the plant to maintain its root surface area under stress in order to absorb water and nutrients. Nevertheless, other stresses, such as reductions in solar radiation and nutrient excess, often produce increased shoot:root ratios. Under drought stress, metabolites are preferentially partitioned to primary root elongation in order to increase the uptake of water, which is usually more readily available deep in the soil. Thus, the adaptive response of the root system to water deficit, namely promoting growth of the primary root and inhibiting the horizontal proliferation of lateral roots in the top soil, is commonly observed. Plants subjected to low moisture regimes can develop a more extensive root system; however, such drought stress can also inhibit root activity.

Oxygen deficiency in the rhizosphere, caused by waterlogging, mechanical root-restriction, or other stresses, can affect rhizogenesis and may decrease root growth. Salinity also commonly inhibits root growth, and causes profound modifications in root morphology and to the architecture of the root system, often giving rise to more branched root systems with shorter main roots. Moreover, extreme temperatures, shaded environments, heavy metals, or high soil compactness can greatly reduce root growth and affect root system architecture. Conversely, despite there being insufficient information to make generalisations, root growth is often increased and root system architecture is affected by increased branching with increasing concentrations of atmospheric CO_2.

ABA appears to play a central role in the long distance root-to-shoot signalling process which regulates root development under stress. Nevertheless, other chemical signals and metabolic processes may also be involved in that regulation. Ethylene plays an essential role through its interactions with other hormones. In addition, changes in amino acid metabolism and anti-oxidant resistance mechanisms may be involved in long-distance root-to-shoot signalling during plant responses to different stresses. On the other hand, colonisation by AMF also induces changes to the root system of the host plant. Such changes may result in more branched roots, a higher root surface area, increased root hydraulic conductivity, or enhanced water and nutrient uptake due to extra-radical hyphae, resulting in improve resistance to several abiotic stresses.

Sensing and responding to environmental cues by roots enable plants to overcome the challenges posed by their sessile lifestyle. Root system is important to plants due to a wide variety of processes, including nutrient and water uptake from soil, which is a complex medium with high spatial and temporal environmental variability. Thus, it is not surprising that root system architecture is highly influenced by environmental cues. The importance of root system architecture in plant productivity stems from the fact that many soil resources are unevenly distributed or are subject to localized depletion, so that the spatial deployment of the root system will largely determine the ability of a plant to exploit those resources. The PERDP which regulates the changes in root system architecture, can be considered as an evolutionary response to medium with high spatial and temporal variability in resource supplies. The genetic controls regarding root deployment (PERDP) are still largely unknown. A great effort has been made to understand the molecular components that regulate the formation, proliferation and maintenance of meristems, either being embryo or pericycle-originated. Nevertheless, the facts behind their regulation by environmental factors still leave many questions to be solved. Plants are important to humans, as they provide food, fuel, fibres, medicines and materials. As the global population is projected by the UN to rise to over 9 billion by 2050, the improvement of crops is becoming an increasingly pressuring issue. The new challenge arisen is to solve the current and future obstacles to the maintenance of food supply security through higher crop yields. Water and nutrient availability limit the productivity in most agricultural ecosystems. In all environments characterized by low water and nutrient availability, root system architecture is a fundamental aspect, the acquisition of soil resources by root system is therefore a subject of considerable interest in agriculture. Root system architecture and PERDP are important agronomic traits; the right architecture in a given environment allows plants to survive periods of water of nutrient deficit, and compete effectively for resources. Most of drought-resistant rice varieties have a deeper and more highly branched root system than sensitive varieties. Understanding the root system architecture and the PERDP holds potential for the exploitation and opening of new options for genetic manipulation of the characteristics of the root, in order to both increase food plant yield and optimize agricultural land use. Improved access to deep soil water, inherently reducing the need for irrigation, is one potential benefit that could be achieved by exploitation of root system architecture. Increase in root branching and root hair in crops may enable plants to make more efficient use of existing soil nutrients and increase stress tolerance, improving yields while decreasing the need for heavy fertilizer application. Understanding which structures and environmental cues that regulate proliferation and elongation of the root system cells will allow us to develop strategies to generate crops that possess greater soil exploration capacities in order of a more efficient usage of nutrients and water present in the soil.

References

Abadia-Fenoll, F., P.J. Casero, P.G Lloret, and M.R. Vidal. 1986. Development of lateral primordial in decapitated adventitious roots of *Allium cepa*. *Ann. Bot.*, 58: 103–107.

Achard, P., Cheng, H.D.E., Grauwe, L., Decat, J., Schoutteten, H., Moritz, T., Straeten, D.D., Peng, J. and Harberd, N.P. 2006. Integration of plant responses to environmentally activated phytohormonal signals. *Science*, 311: 91–94.

Achard, P., Renou, J.P., Berthome, R., Harberd, N.P. and Genschik, P. 2008. Plant DELLAs restrain growth and promote survival of adversity by reducing the levels of reactive oxygen species. *Current biology*, 18(9): 656-660.

Achard, P., Gusti, A., Cheminant, S., Alioua, M., Dhondt, S., Coppens, F., Beemster, G.T. and Genschik, P. 2009. Gibberellin signaling controls cell proliferation rate in *Arabidopsis*. *Current biology*, 9(14): 1188-1193.

Achard, P. and Genschik, P. 2009. Releasing the brakes of plant growth: how GAs shutdown DELLA proteins. *J. Expt. Bot.*, 60(4): 1085-1092.

Agnihotri, R.K., Palni, L.M.S. and Pandey, D.K. 2006. Screening of land races of rice under cultivation in Kumaun Himalayan for salinity stress during germination and early seedling growth. *Ind. J. Plant Physiol.*, 11(30): 262-272.

Ahmad, J. and Bano, M. 1992. The effect of sodium chloride on physiology of cotyledons and mobilization of reserved food in *Cicer arietinum*. *Pak. J. Bot.*, 24: 40–48.

Ahmad, M., Akram, Z., Munir, M. and Rauf, M. 2006. Physiomorphic response of wheat genotypes under rainfed conditions. *Pak J Bot.*, 38(5): 1697-1702.

Akinci, I.E.,Akinci, S.,Yilmaz, K. and Dikici, H. 2004. Response of eggplant varieties (*Solanum melongena*) to salinity in germination and seedling stages. *New Zealand Journal of Crop and Horticultural Science*, 32: 193–200.

Al-Ansari, F.M. 2003. Salinity tolerance during germination in two aridland varieties of wheat (*Triticum aestivum* L.). *Seed Sci. Technol.*, 31: 597-603.

Albacete, A., Ghanem, M.E., Jar, C.M., Acosta, M., Bravo, J.S., Lutts, V.M.S., Dodd, I.C. and Pérez-Alfocea, F. 2008. Hormonal changes in relation to biomass partitioning and shoot growth impairment in salinized tomato (*Solanum lycopersicum* L.) plants. *Journal of Experimental Botany*, 59: 4119–4131.

Al-Karaki, G.N. 1997. Barley responce to salt stress at varied levels of phosphorous. *J. Plant Nutr.*, 20:1635-1643.

Al-Rwahy, S.A. 1989. Nitrogen uptake, growth rate and yield of tomatoes under saline condition. *Arizona Univer., Tucson, PhD. Diss.*, pp. 118.

Ali, M.A., Ashraf, M. and Athar, H.R. 2009. Influence of nickel stress on growth and some important Physiological / Biochemical attributes in some diverse canola (*Brassica napus* L.) cultivars. *J Hazar Mater.*,172(2-3): 964-969.

Allakhverdiev, S. I., Sakamoto, A., Nishiyama, Y., Inaba, M. and Murata, N., 2000b. Ionic and osmotic effects of NaCl-induced inactivation of photosystems I and II in *Synechococcus* sp. *Plant Physiol.*, 123: 1047–1056.

Almansouri, M., Kinet, J.M., and S. Lutts. 2001. Effect of salt and osmotic stresses on germination in durum wheat (Triticum durum Desf.). *Plant and Soil.* 231: 243-254.

Aloni, B. and G. Rosenshtein. 1982. Effect of flooding on tomato cultivars: The relationship between proline accumulation and other morphological and physiological changes. *Physiol. Plant.*, 56: 513–517.

Aloni, R., Aloni, E., Langhans, M. and Ullrich, C.I. 2006. Role of cytokinin and auxin in shaping root architecture: regulating vascular differentiation, lateral root initiation, root apical dominance and root gravitropism. *Annals of Botany*, 97: 883–893.

Alonso-Ramý´rez, A., Rodrý´guez, D., Reyes, D., Jime´nez, J.A., Nicola´s, G., Lo´pez-Climent, M., Gómez-Cadenas, A. and Nicolas, C. 2009. Evidence for a role of gibberellins in salicylic acid-modulated early plant responses to abiotic stress in *Arabidopsis* seeds. *Plant Physiol.*, 150: 1335–1344.

Alvaro, F., Isidro, J., Villegas, D., Garcia del Moral, L.F., and Royo, C. 2008. Breeding effect on grain filling, biomass partitioning, and remobilization in Mediterranean durum wheat. *Agron J.*, 100:361-370.

Aminah, H., Ab Rasip, A.G.,Abdullah, M.Z., Khalim, A.S.A., Elias, K. and Yahya, Y. 2004. Effects of potting media and size of root trainers on the growth of *Shorea leprosula* seedlings. *Journal of Tropical Forest Science*, 16: 145–150.

Ansari O. and Sharif-Zadeh F. 2012. Osmo and hydro priming improvement germination characteristics and enzyme activity of Mountain Rye (*Secale montanum*) seeds under drought stress. *Journal of Stress Physiology & Biochemistry*. 8(4): 253-261.

Ansari, O., Choghazardi, H.R., Sharif Zadeh, F., and Nazarli, H. 2012. Seed reserve utilization and seedling growth of treated seeds of Mountain Rye (*Seecale montanum)* as affected by drought stress. *Cercetări Agronomice în Moldova*,. 2(150): 43-48.

Ansari O., Azadi M.S., Sharif-Zadeh F. and Younesi E. 2013. Effect of hormone priming on germination characteristics and enzyme activity of mountain rye (*Secale montanum*) seeds under drought stress conditions. *Journal of Stress Physiology & Biochemistry*, 9(3): 61-71.

Arnone III, J.A., Zaller, J.G., Spehn, E.M., Niklaus, P.A. and Körner, C. 2000. Dynamics of root systems in native grassland: Effects of elevated atmospheric CO_2. *New Phytologist*, 147: 73–85.

Armstrong, W. and Drew, M.C. 2002. Root growth and metabolism under oxygen deficiency. In: "*Plant Roots: The Hidden Half.* 3rd Edition", Eds. Y. Waisel, A. Eshel, and U. Kafkafi, Marcel Dekker, New York, NY, USA. pp. 729–761.

Arreola, J., Martínez-Sánchez, J.J., Conesa, E. and Franco, J.A. 2008. Effect of pre-conditioning water regimes during nursery production on seedling root system characteristics of *Silene vulgaris*. *Acta Horticulturae*, 782: 287–292.

Ashraf, M., and Foolad, M.R. 2005. Pre-sowing seed treatment a shotgun approach to improve germination, plant growth, and crop yield under saline and non-saline conditions. *Adv Agron*. 88:223-271.

Apse M.P. and Blumwald, E. 2002. Engineering salt tolerance in plants. *Curr. Opin. Biotechnol.*, 13: 146-150.

Araus, J.L., Slafer, G., Royo, C. and Serret, M.D. 2008. Breeding for yield potential and stress adaptation in cereals. *Crit Rev Plant Sci.,* 27: 377–412.

Ariel, F., Diet, A., Verdenaud, M., Gruber, V., Frugier, F., Chan, R. and Crespi, M. 2010. Environmental regulation of lateral root emergence in *Medicago truncatula* requires the HD-Zip I transcription actor HB1. *Plant Cell*, 22(7): 2171-283.

Arduini, I., Masoni, A., Ercoli, L. and Mariotti, M. 2006. Grain yield, and dry matter and nitrogen accumulation and remobilization in durum wheat as affected by variety and seeding rate. *Eur J Agron*, 25: 309-318.

Arenas-Huertero, F., Arroyo, A., Zhou, L., Sheen, J., and Leon, P. 2000. Analysis of *Arabidopsis* glucose insensitive mutants, gin5 and gin6, reveals a central role of the plant hormone ABA in the regulation of plant vegetative development by sugar. *Genes Dev.,* 14: 2085–2096.

Aroca, R., Amodeo, G., Fernandez-illescas, S., Herman, E.M., Chaumont, F. and Chrispeels, M.J. 2005. The role of aquaporins and membrane damage in chilling and hydrogen peroxide induced changes in the hydraulic conductance of maize roots. *Plant Physiol*, 137(1): 341-353.

Aronson, J. A., 1985. Economic halophytes. In : A global review, plant for arid land. Eds. S. Wicken, J. R.Gooding, D. V. G. Efields. George Allen and Unwin, London, 177–188.

Arvin, M.J. and Donnelly, D.J. 2008. Screening potato cultivars and wild species to abiotic stresses using an electrolyte leakage bioassay. *J Agric Sci Technol.*, 10: 33-42.

Aung, L.H. 1982. Root initiation in tomato seedlings. *J. Amer. Soc. Hort. Sci.*, 107: 1015–1018.

Ayaz, F.A., Kadioglu, A. and Turgut, R. 2000. Water stress effects on the content of low molecular weight carbohydrates and phenolic acids in *Ctenanthe setosa* (Rose.) Eichler, *Can. J. Plant Sci.*, 80: 373-378.

Aynehband, A,. Valipoor, M. and Fateh E. 2011. Stem reserve accumulation and mobilization in wheat (*Triticum aestivum* L.) as affected by sowing date and N-P-K levels under Mediterranean conditions. *Turk J Agric For.*, 35: 319-331.

Azaizeh, H., B. Gunse, and E. Steudle. 1992. Effects of NaCl and $CaCl_2$ on water transport across root cells of maize (*Zea mays* L.) seedlings. *Plant Physiol.*, 99: 886–894.

Bahrami, H., Razmjoo, J. and Ostadi Jafari, A. 2012. Effect of drought stress on germination and seedling growth of sesame cultivars (*Sesamum indicum* L.). *Int J. Agric. Sci.*, 2(5): 423-428.

Bañón, S., Ochoa, J., Franco, J.A.,Alarcón, J.J., Fernández, T. and Sánchez-Blanco, M.J. 2002. The influence of acclimation treatments on the morphology, water relations and survival of *Myrtus communis* L. plants. In "*Sustainable Use and Management of Soils in Arid and Semiarid Regions*", Eds. A. Faz, R. Ortiz, and A.R. Mermut, Quaderna Editorial, Murcia, Spain. pp. 275–277.

Bañón, S., Ochoa, J., Franco, J.A., Sánchez-Blanco, M.J. and Alarcón, J.J. 2003. Influence of water deficit and low air humidity in the nursery on survival of *Rhamnus alaternus* seedlings following planting. *J. Hort. Sci. & Biotech.*, 78: 518–522.

Bañón, S., Fernández, J.A., Franco, J.A., Torrecillas, A., Alarcón, J.J. and Sánchez-Blanco, M.J. 2004. Effects of water stress and night temperature pre-conditioning on water relations and morphological and anatomical changes of *Lotus creticus* plants. *Scientia Horticulturae*, 101: 333–342.

Bañón, S., Fernández, J.A.,Ochoa, J. and Sánchez-Blanco, M.J. 2005. Paclobutrazol as an aid to reduce some effects of salt stress in oleander seedlings. *European Journal of Horticultural Science*, 70: 43–49.

Bañón, S., Ochoa, J., Franco, J.A., Alarcón, J.J. and Sánchezblanco, M.J. 2006. Hardening of oleander seedlings by deficit irrigation and low air humidity. *Environmental and Experimental Botany*, 56: 36–43.

Banu, M.N.A., Hoque M.A., WatanabeSugimoto M., Matsuoka K., Nakamura Y., Shimoishi Y. and Murata, Y. 2009. Proline and glycine betaine induce antioxidant defense gene expression and suppress cell death in cultured tobacco cells under salt stress. *J. Plant. Physiol.*, 166: 146-156.

Bauerle, T.L.,Richards, J.H., Smart, D.R. and Eissenstat, D.M. 2008. Importance of internal hydraulic redistribution for prolonging the lifespan of roots in dry soil. *Plant, Cell and Environment*, 31: 177–186.

Bentsink, L., and Koorneef, M. 2008. Seed dormancy and germination. In: *Arabidopsis* Book 6: 0119. The American Society of Plant Biologists, Rockville, MD.

Bera, A.K., Pati, M.K. and Bera, A. 2006. Bassionolide ameliorates adverse effect on salt stress on germination and seedling growth of rice. *Ind. J. Plant Physiol.*, 11(2): 182-189.

Berckmans, B., Lammens, T., Van Den Daele, H., Magyar, Z., Bogre, L., De Veylder, L.. 2011. Light-dependent regulation of DEL1 is determined by the antagonistic action of E2Fb and E2Fc. *Plant Physiol.*, 157(3): 1440-1451.

Bhattacharjee, S. 2008. Triadimefon pretreatment protects newly assembled membrane system and causes up-regulation of stress proteins in salinity stressed *Amaranthus lividus* L. during early germination. *J. Environ. Biol,*. 29: 805-810.

Bhattacharya, S., Puri, S., Jamwal, A. and Sharma, S. 2012. Studies on seed germination and seedling growth in kalmegh (*Andrographis paniculata* wall. ex nees) under abiotic stress conditions. *Int J Sci Environ Technol.*, 1(3): 197–204.

Berkowitz, G.A. and J. Rabin. 1988. Antitranspirant associated abscisic acid effects on the water relations and yield of transplanted bell peppers. *Plant Physiol.*, 86: 329–331.

Bernstein, N., Ioffe, M., Bruner, M., Nishri, Y., Luria, G., Dori, I., Matan, E., Philodoph-Hadas, S., Umiel, N. and Hagiladi, A. 2005. Effects of supplied nitrogen form and quantity on growth and postharvest quality of *Ranunculus asiaticus* flowers. *Hort.Science*, 40: 1879–1886.

Bernstein, N. and Kafkafi, U. 2002. Root growth under salinity stress. In "*Plant Roots: The Hidden Half.* 3rd Edition", Eds. Y. Waisel, A. Eshel, and U. Kafkafi, Marcel Dekker, New York, NY, USA. pp.787–805.

Bethke, P.C., Libourel, I.G.L., Aoyama, N., Chung, Y.-Y., Still, D.W., and Jones, R.L. 2007. The *Arabidopsis* aleurone layer responds to nitric oxide, gibberellin, and abscisic acid and is sufficient and necessary for seed dormancy. *Plant Physiol.*, 143: 1173–1188.

Bewley, J.D. and Black, M. 1982. Physiology and biochemistry of seeds in relation to germination. Vol. 2. Viability, Dormancy and Environmental Control. Springer-Verlag. Biochemistry and Molecular Biology of Plants. American Society of Plant Physiol.

Bewlay, J. D. and Black, M. 1994. Seeds. Physiology of developmental and germination. 2nd edition, Plenum Press, New York.

Bewley, J.D. 1997. Seed germination and dormancy. *Plant Cell* 9: 1055–1066.

Biel, C., Savé, R., Habrouk, A., Espelta, J.M. and Retana, J. 2004. Effects of restricted watering and CO_2 enrichment in the morphology and performance after transplanting of nursery-grown *Pinus nigra* seedlings. *Hort. Science*, 39: 535–540.

Blilou, I., Xu, J., Wildwater, M., Willemsen, V., Paponov, I., Friml, J., Heidstra, R., Aida, M., Palme, K. and Scheres, B. 2005. The PIN auxin efflux facilitator network controls growth and patterning in *Arabidopsis* roots. *Nature*, 433(7021): 39-44.

Blum, A., Zhang, J. and Nguyen, H.T. 1999. Consistent differences among wheat cultivars in osmotic adjustment and their relationship to plant production. *Field Crops Res.*, 64: 287–291.

Blum, A. 2005. Drought resistance, water-use efficiency, and yield potential: are they compatible, dissonant, or mutually exclusive? *Aust J Agri Res.*, 56: 1159–1168.

Blum, A. 2011. Plant Breeding for Water-Limited Environments, Drought Resistance and Its Improvement. Springer New York. ISBN - 978-1-4419-7491-4.pp 152.

Blumwald, E., Aharon, G.S., and Apse, M.P. 2000. Sodium transport in plant cells. *Biochim. Biophys. Acta.* 1465:, 140–151.

Bohnert, H. J., Nelson, D.E. and Jonsen, R.G. 1995. Adaptations to environmental stresses. *Plant Cell*, 7:1099–1111.

Bouguyon, E., Gojon, A., and Nacry, P. 2012. Nitrate sensing and signaling in plants. *Seminars in Cell and Developmental Biology*, 23(6): 648-54.

Brady, S.M., Sarkar, S.F., Bonetta, D., and McCourt, P. 2003. The ABSCISIC ACID INSENSITIVE 3 (ABI3) gene is modulated by farnesylation and is involved in auxin signaling and lateral root development in *Arabidopsis*. *Plant J.*, 34: 67–75.

Bray. E.A. 2002. Classification of genes differentially expressed during water-deficit stress in *Arabidopsis thaliana*: An analysis using microarray and differential expression data. *Ann. Bot.*, 89: 803-811.

Brocard, I.M., Lynch, T.J., and Finkelstein, R.R. 2002. Regulation and role of the *Arabidopsis* abscisic acid-insensitive 5 gene in abscisic acid, sugar, and stress response. *Plant Physiol.*, 129: 1533–1543.

Brodersen, P., Sakvarelidze-Achard, L., Bruun-Rasmussen, M., Dunoyer, P., Yamamoto, Y.Y., Sieburth, L. and Voinnet, O. 2008. Widespread translational inhibition by plant miRNAs and siRNAs. *Science*, 320: 1185–1190.

Brundrett, M.C. 2002. Coevolution of roots and mycorrhizas of land plants. *New Phytologist*, 154: 275–304.

Bumgarner, M.L., Salifu, F. and Jacobs, D. F. 2008. Sub-irrigation of *Quercus rubra* seedlings: Nursery stock quality, media chemistry, and early field performance. *Hort. Science*, 43: 2179–2185.

Booij, R. 1992. Effects of nitrogen fertilization during raising of cauliflower transplants in cellular trays on plant growth. *Netherlands J. Agr. Sci.*, 40: 43–50.

Bossi, F., Cordoba, E., Dupre, P., Mendoza, M.S., Roman, C.S., and Leon, P. 2009. The *Arabidopsis* ABA-INSENSITIVE (ABI) 4 factor acts as a central transcription activator of the expression of its own gene, and for the induction of ABI5 and SBE2.2 genes during sugar signaling. *Plant J.*, 59: 359–374.

Bouaziz, A., Souty, N. and Hicks, D. 1990. Emergence force exerted by wheat seedlings. *Soil Till. Res.*, 17: 211–219.

Brar, G.S., F.G. Gomez, B.L. McMichael, A.G. Matches, and H.M. Taylor. 1990. Root development of 12 forage legumes as affected by temperature. *Agron. J.*82: 1024–1026.

Byari, S.H. and Al-Maghrabi, A.A. 1991. Effect of salt concentration on morphological and physiological traits of tomato cultivar. *Al-Azhar J. Agric. Res.*, 14: 91-11.

Calderón-vázquez, C., Sawers, R.J.H., Herrera-Estrella, L. 2011. Phosphate deprivation in maize: genetics and genomics. *Plant Physiol.,* 156(3): 1067-1077.

Carol, R.J. and Dolan, L. 2006. The role of reactive oxygen species in cell growth: lessons from root hairs. *J. Expt. Bot.,* 57(8): 1829-1834.

Çelik Ö. And Atak Ç. 2011. The effect of salt stress on antioxidative enzymes and proline content of two Turkish tobacco varieties. *Turk. J. Biol.*, 36: 339-356.

Chandra, R., Bhargava, R.N., Yadav, S. and Mohan, D. 2009. Accumulation and distribution of toxic *metals* in wheat (*Triticumaestivum* L.) and Indian mustard (*Brassica compestries* L.) irrigated with distillery and tannery effluent. *J Hazar Mater.,*162: 1514-1521.

Chacón-lópez, A., Ibarra-laclette, E., Sánchez-calderón, L., Gutiérrez-alanis, D., and Herrera-Estrella, L. 2011. Global expression pattern comparison between *low phosphorus insensitive 4* and WT *Arabidopsis* reveals an important role of reactive oxygen species and jasmonic acid in the root tip response to phosphate starvation. *Plant Signaling and Behavior,* 6(3): 382-92.

Chang-juan, S., Ju-xiang, T., Wen-ping, Y., Xin-liang, Z., Xiujuan, R. and Yi-zhuo, L. 2012. Comparison of photosynthetic characteristics of four wheat (*Triticum aestivum* L.) genotypes during jointing stage under drought stress. *Afric J Agric Res.*, 7(8): 1289-1295.

Chaves, M.M., Maroco, J.P. and Pereira, L. 2003. Understanding plant responses to drought from genes to the whole plant. *Funct. Plant Biol.*, 30: 239-264.

Chaves, M.M., Flexas, J., and Pinheiro, C. 2009. Photosynthesis under drought and salt stress: regulation mechanisms from whole plant to cell. *Ann Bot.*, 103: 551–560.

Cheeseman, J.M. 2007. Hydrogen Peroxide and Plant Stress: A Challenging Relationship, In "Global Science Book: Plant Stress", 1(1): 4-15.

Chen, F., Nonogaki, H., and Bradford, K.J. 2002. A gibberellinregulated xyloglucan Endotrans glycosylase gene is expressed in the endosperm cap during tomato seed germination. *J. Expt. Bot*, 53: 215–223.

Chen, J.G., Ullah, H., Temple, B., Liang, J., Guo, J., Alonso, J.M., Ecker, J.R. and Jones, A.M. 2006. RACK1 mediates multiple hormone responsiveness and developmental processes in *Arabidopsis*. *J Exp Bot.*, 57: 2697–2708.

Chiatante, D., Baraldi, A., Di Iorio, A., Sarnataro, M. and Scippa, S. 2003. Root response to mechanical stress in plant growing on slopes: an experimental system for morphological, biochemical and molecular analysis. In "*Roots: the Dynamic Interface between Plants and the Earth", Ed. J.* Abe, Kluwer Academic Publishers, Dordrecht, The Netherlands. p p . 427–437.

Chinnusamy, V., Schumaker, K. AND Zhu, J.K. 2004. Molecular genetic perspectives on cross-talk and specificity in abiotic stress signaling in plants. *J Exp Bot.*, 55: 225–236.

Chung, J.S., Zhu, J.K., Bressan, R.A., Hasegawa, P.M., and Shi, H. 2008. Reactive oxygen species mediate Na^+-induced SOS1 mRNA stability in *Arabidopsis*. *Plant J.*, 53: 554–565.

Clark, L.J., Whalley, W.R. and Barraclough, P.B. 2003. How do roots penetrate strong soil? *Plant and Soil*, 255: 93–104.

Clark, L.J., Ferraris, S., Price, A.H. and Whalley, W.R. 2008. A gradual rather than abrupt increase in soil strength gives better root penetration of strong layers. *Plant and Soil*, 307: 235–242.

Cobb, B.G., Malcolm C. D., David L., Andrews, J.J. MacAlpine, Tricia, D.L. Danielson, Turn Bough, M.A. and Davis, R. 1995. How Maize Seeds and Seedlings Cope with Oxygen Deficit. *Hort. Science*, 30(6): 1160-1164.

Cooksey, J.R., B.A. Kahn, and J.E. Motes. 1994. Plant morphology and yield of paprika pepper in response to method of stand establishment. *Hort Science*, 29: 1282–1284.

Cortes, P., Espelta, J.M., Savé, R. and Biel, C. 2004. Effects of a nursery CO_2 enriched atmosphere on the germination and seedling morphology of two Mediterranean oaks with contrasting leaf habit. *New Forest*, 28: 79–88.

Costa, R.N. T., De Vasconselos, J.P., Da Silva, L.A. and Ness, R.L.L. 2008. Effect of excess of water on sugar beet yield components. *Horticultura Brasileira*, 26: 74–77.

Creelman, R.A., H.S. Mason, R.J. Bensen, J.S. Boyer, and J.E. Mullet. 1990. Water deficit and abscisic acid cause differential inhibition of shoot versus root growth in soybean seedlings. *Plant Physiol.*, 92: 205–214.

Croser, C., Renault, S., Franklin, J. and Zwiazek, J. 2001. The effect of salinity on the emergence and seedling growth of *Picea mariana*, *Picea glauca*, and *Pinus banksiana*. *Environmental Pollution*, 115: 9–16.

Cuartero J., Bolarin M.C., Asins M.J, and Moreno, V. 2006. Increasing salt tolerance in the tomato. *J. Exp. Bot.*, 57(5): 1045-1058.

Cui, S., Sadayoshi, K., Ogawa, Y. and Nii, N. 2004. Effects of water stress on sorbitol content in leaves and roots, anatomical changes in cell nuclei, and starch accumulation in leaves of young peach trees. *Journal of the Japanese Society for Horticultural Science*, 73: 25–30.

Davis, W.J., J. Metcalfe, T.A. Costa, and A.R. da Costa. 1986. Plant growth substances and the regulation of growth under drought. *Aust. J. Plant Physiol.*, 13: 105–125.

Deak, K.I., and Malamy, J. 2005. Osmotic regulation of root system architecture. *The Plant Journal: Cell and Mo. Biol.,* 43(1): 17-28.

Debeaujon, I., and Koornneef, M. 2000. Gibberellin requirement for *Arabidopsis* seed germination is determined both by testa characteristics and embryonic abscisic acid. *Plant Physiol.*, 122: 415–424.

Deborah, V., Malone, R.P. and Dix, P.J. 2011. Increased tolerance to abiotic stress in tobacco plants expressing a barley cell wall peroxidase. *J Plant Sci.,*6(1): 1-13.

Demidchik, V., Shabala, S.N., Coutts, K.B., Tester, M.A. and Davies, J.M. 2003. Free oxygen radicals regulate plasma membrane Ca^{2+} and K^+ permeable channels in plant root cells. *J Cell Sci.*, 116: 81-88.

De Smet, I., Zhang, H., Inzé, D. and Beeckman, T. 2006. A novel role for abscisic acid emerges from underground. *Trends in Plant Science*, 11(9): 434-439.

De Sousa, M.A. and Lima, M.D.B. 2010. Influence of suppression of the irrigation in stages of growth of bean cv. Carioca comum. *Bioscience Journal*, 26: 550–557.

Devaiah, B.N., Karthikeyan, A.S., and Raghothama, K.G. 2007a. RKY75 Transcription factor is a modulator of phosphate acquisition and root development in *Arabidopsis*. *Plant Physiology*, 143(4): 1789-1801.

Devaiah, B.N., Nagarajan, V.K., and Raghothama, K.G. 2007b. Phosphate homeostasis and root development in *Arabidopsis* are synchronized by the zinc finger transcription factor ZAT6. *Plant Physiology*, 145(1): 147-159.

Devaiah, B.N., Madhuvanthi, R., Karthikeyan, A.S., and Raghothama, K.G. 2009. Phosphate starvation responses and gibberellic acid biosynthesis are regulated by the myb62 transcription factor in *Arabidopsis. Molecular Plant,* 2(1):, 43-58.

Dickin, E. and Wright, D. 2008.The effect of winter water logging and summer drought on the growth and yield of winter wheat (*Triticum aestivum* L.). *European J. Agron.*, 28: 234–244.

Disante, K.B., Fuentes, D. and Cortina, J. 2011. Response to drought on Zn-stressed *Quercus suber* L. seedlings. *Environmental and Experimental Botany*, 70: 96–103.

Do Amarante, L., Lima, J.D. and Sodek, L. 2006. Growth and stress conditions cause similar changes in xylem amino acids for different legume species. *Environmental and Experimental Botany*, 58: 123–129.

Drew, M.C. 1992. Soil aeration and plant root metabolism. *Soil Sci.*, 154: 259– 268.

Dubrovsky, J.G., Sauer, M., Napsucialy-mendivil, S., Ivanchenko, M.G., Friml, J., Shishkova, S., Celenza, J. and Benkova, E. 2008. Auxin acts as a local morphogenetic trigger to specify lateral root founder cells. *Proc Natl Acad Sci U S A.*, 105(25): 8790-8794.

Dufault, R.J. 1986. Influence of nutritional conditioning on muskmelon transplant quality and early yield. *J. Amer. Soc. Hort. Sci.*, 111: 698–703.

Dufault, R.J. 1994. Long-term consequences and significance of short-term pretransplant nutritional conditioning. *Hort Technology*, 4: 41–42.

Dufault, R.J. and R.R. Melton. 1990. Cyclic cold stresses before transplanting influence tomato seedling growth, but not fruit earliness, fresh market yield, or quality. *J. Amer. Soc. Hort. Sci.*, 115: 559–563.

Eissenstat, D.M. and Yanai, R.D. 2002. Root life span, efficiency, and turnover. In "*Plant Roots: The Hidden Half.* 3rd Edition", Eds. Y. Waisel, A. Eshel, and U. Kafkafi, Marcel

Eglinton, G. and Hamilton, R.J. 1967. Leaf epicuticular waxes. *Science*, 156(3780): 1322-1335.

Ennos, A.R. 1990. The anchorage of leek seedlings: The effect of root length and soil strength. *Ann. Bot.*, 65: 409–416.

El-Keblawy, A. 2004. Salinity Effects on Seed Germination of the Common Desert Range Grass, *Panicum Turgidum. Seed Sci. Technol.*, 32: 873-878.

EL-Melegi, El-sayed, A., Mahdia, F.G., Fouad, H.M. and Mona, A.I. 2004. Responses to NaCl salinity of tomato cultivated and breeding lines differing in salt tolerance in callus cultures. *Int. J. Agric. Biol.*, 6(1): 19- 26.

Ercoli, L., Lulli, L., Mariotti, M., Mosani, A. and Arduini, I. 2008. Postanthesis dry matter and nitrogen dynamics in durum wheat as aff ected by nitrogen supply and soil water availability. *Eur J Agron.*, 28: 138-147.

Evlagon, D., I. Ravina, and P.M. Neumann. 1992. Effects of salinity stress and calcium on hydraulic conductivity and growth in maize seedling roots. *J. Plant Nutr.*, 15: 795–803.

Feng, S., Martinez, C., Gusmaroli, G., Wang, Y., Zhou, J., Wang, F., Chen, L., Yu, L., Iglesias-Predraz, J.M., Juan, M., Kircher, S., Schafer, E., Fu, X., Fan, L.M. and Deng, X.W. 2008. Coordinatedregulation of *Arabidopsisthaliana* development by light and gibberellins. *Nature*, 451: 475–479.

Fernández, J.A., Balenzategui, L., Bañón, S. and Franco, J.A. 2006. Induction of drought tolerance bypaclobutrazol and irrigation deficit in *Phillyrea angustifolia* during the nursery period. *Scientia Horticulturae*, 107: 277–283.

Feser, C., St. Hilaire, R. and Van Leeuwen, D. 2005. Development of in-ground container plants of Mexican elders exposed to drought. *Hort. Science*, 40: 446–450.

Finch-Savage,W.E., and Leubner-Metzger, G. 2006. Seed dormancy and the control of germination. *New Phytol.*, 171: 501–523.

Finkelstein, R.R. 1994. Mutations at two new *Arabidopsis* ABA response loci are similar to the abi3 mutations. *Plant J.*, 5: 765– 771.

Finkelstein, R.R., Wang, M.L., Lynch, T.J., Rao, S., and Goodman, H.M. 1998. The *ARABIDOPSIS* abscisic acid response locus AB14 encodes an AP*ETALA*2 domain protein. *Plant Cell*, 10: 1043–1054.

Finkelstein, R.R., and Lynch, T. 2000. The *Arabidopsis* abscisic acid response gene ABI5 encodes a basic leucine zipper transcription factor. *Plant Cell*, 12: 599–609.

Fitter, A.H. 1991. Characteristics and functions of root systems, In "Plant roots, the hidden half", Eds. Y. Waisel, A. Eshel, and U. Kafkafi, Marcel Dekker, New York. pp. 3–24.

Foolad, M.R. 1996. Response to selection for salt tolerance during germination in tomato seed derived from P.I. 174263. *J. Am. Soc. Hort. Sci.*, 121: 1006-1011.

Fornes, F., Belda, R.M., Carrión, C., Noguera, V., Garcíaagustín, P. and Abad, M. 2007. Pre-conditioningornamental plants to drought by means of saline water irrigation as related to salinity tolerance. *Scientia Horticulturae*, 113: 52–59.

Fraire-velázquez, S., Sánchez-calderón, L. and Guzmán-gonzález, S. 2012. Abiotic stress response in plants: integrative genetic pathways and overlapping reactions between abiotic and biotic stress responses. In "Abiotic stress: new research Croatia" Eds. N. Haryana, and S. Punj, Nova Science.

Francini, A. and Sebastiani, L. 2010. Copper effects on Prunus persica in two different grafting combinations (*P. persica, P. amygdalus* and *P. cerasifera*). *Journal of Plant Nutrition*, 33: 1338–1352.

Franco, J.A., Abrisqueta, J.M. and Hernansaez ,A. 1995. Root development of almond rootstocks in a young almond orchard under trickle irrigation as affected by almond scion cultivar. *Journal of Horticultural Science*, 70: 597–607.

Franco, J.A., Bañón, S., Fernández, J.A. and Leskovar, D.I. 2001. Effect of nursery regimes and establishment irrigation on root development of *Lotus creticus* seedlings following transplanting. *Journal of Horticultural Science & Biotechnology*, 76: 174–179.

Franco, J.A., Cros, V., Bañón, S., González, A. and Abrisqueta, J.M. 2002a. Effects of nursery irrigation on post planting root dynamics of *Lotus creticus* in semiarid field conditions. *Hort. Science*, 37: 525–528

Franco, J.A., Cros, V., Bañón, S. and Martínez-Sánchez, J.J. 2002b. Nursery irrigation regimes and establishment irrigation affect the post planting growth of *Limonium cossonianum* in semiarid conditions. *Israel Journal of Plant Sciences*, 50: 25–32.

Franco, J.A., Martínez-Sánchez, J.J., Fernández, J.A. and Bañón, S. 2006. Selection and nursery production of ornamental plants for landscaping and xero gardening in semiarid environments. *Journal of Horticultural Science & Biotechnology*, 81: 3–17.

Franco, J.A., Arreola, J., Vicente, M.J. and Martínezsánchez, J.J. 2008. Nursery irrigation regimes affect the seedling characteristics of *Silene vulgaris* as they relate to potential performance following transplanting into semi-arid conditions. *Journal of Horticultural Science & Biotechnology*, 83: 15–22.

Franco, J.A., Cros,V., Vicente, M.J. and Martínez-Sánchez, J.J. 2011. Effects of salinity on the germination, growth, and nitrate contents of purslane (*Portulaca oleracea*) cultivated dunder different climatic conditions. *Journal of Horticultural Science & Biotechnology*, 86: 1–6.

François, B. and Bahizire. 2007. Effect of salinity on germination and seedling growth of canola (*Brassica napus* L.). *The degree of Master of Agricultural Sciences at the University of Stellenbosch.*

Fujii, H., Verslues, P.E., and Zhu, J.K. 2007. Identification of two protein kinases required for abscisic acid regulation of seed germination, root growth, and gene expression in *Arabidopsis. Plant Cell.* 19: 485–494.

Fujii, H., Chinnusamy, V., Rodrigues, A., Rubio, S., Antoni, R., Park, S-Y., Sean, R.C., Sheen, J., Rodriguez, P.L. and Zhu, J.K. 2009. *In vitro* reconstitution of an abscisic acid signalling pathway. *Nature*, 462: 660–664.

Gabbrielli, R., Pandolfini, T., Espen, L. and Palandri, M.R. 1999. Growth, peroxidase activity and cytological modifications in *Pisum sativum* seedlings exposed to Ni^{2+} toxicity. *Journal of Plant Physiology*, 155: 639–645.

Galvan-ampudia, C.S, and Testerink, C. 2011. Salt stress signals shape the plant root. *Curr. Opi. Pl.Biol.,* 14(3): 296-302.

Gao, X,, Zou, C., Wang, L. and Zhang, F. 2004. Silicon improves water use efficiency in maize plants. *J Plant Nutr.*, 27(8): 1457-1470.

Gao, X,, Zou, C., Wang, L. and Zhang, F. 2006. Silicon decreases transpiration rate and conductance from stornata of maize plants. *J Plant Nutr.*, 29: 1637-1647.

Garg, B.K., Vyas, S.P., Kathju, S. and Lahiri, A.N. 1998. Influence of water deficit stress at various growth stage on some enzymes of nitrogen metabolism and yield in clusterbean genotypes. *Ind. J. Plant Physiol.*, 3: 214 -218.

Garg,, B.K., Kathju, S. and Burman, U. 2001. Influence of water stress on water relations, photosynthetic parameters and nitrogen metabolism of moth bean genotypes. *Biol. Plant*, 44: 289-292.

Garg, N., and Manchanda, G. 2009. ROS generation in plants: Boon or bane?, *Plant Biosystems - An International Journal Dealing with all Aspects of Plant Biology*, 143(1): 81-96.

Garg, G. 2010. Response in germination and seedling growth in *Phaseolus mungo* under salt and drought stress. *J. Environ. Biol.*, 31: 261-264.

Garham, J., Hughes, L.Y. and Wynjanes, R.G. 1981. Low molecular weight carbohydrates in some salt stressed plants. *Physiol. Plant.*, 53: 27–33.

Garcia, M.E., Lynch, T., Peeters, J., Snowden, C., and Finkelstein, R. 2008. A small plant-specific protein family of ABI five binding proteins (AFPs) regulates stress response in germinating *Arabidopsis* seeds and seedlings. *Plant Mol Biol.*, 67: 643–658.

Garton, R.W. and I.E. Widders. 1990. N and P preconditioning of small plug seedlings influence growth and yield of processing tomatoes. *Hort Science*, 25: 655–657.

Gaul, D., Hertel, D., Borken, W., Matzner, E. and Leuschner, C. 2008. Effects of experimental drought on the fine root system of mature Norway spruce. *Forest Ecology and Management*, 256: 1151–1159.

Ghannadha, M.R., Omidi, M., Shahi, R.A. and Poustini, K. 2005. A study of salt tolerance in genotypes of bread wheat using tissue culture and germination test. *Iranian J Agri Sci.*, 36(1): 75-85.

Ghoulam, C., Foursy, A. and Fares, K. 2002. Effects of salt stress on growth, inorganic ions and proline accumulation in relation to osmotic adjustment in five sugar beet cultivars. *Environ Exp Bot.*, 47:39–50.

Gill, P.K., Sharma, A.D., Sing, P. and Bhullar, S.S. 2003. Changes in germination, growth and soluble sugar contents of Sorghum bicolor L. Moench seeds under various abiotic stresses. *Plant Growth Regul.*, 40: 157-162.

Gonzalez, A., Bermejo, V. and Gimeno, B.S. 2010. Effect of different physiological traits on grain yield in barley grown under irrigated and terminal water deficit conditions. *J Agric Sci.*, 148: 319–328.

Gonzalez, A., Martin, .I, Ayerbe, L. 2009. Yield and osmotic adjustment capacity of barley under terminal water-stress conditions. *J Agron Crop Sci.*, 194: 81–91.

Goto, T., Matsuno, T., Yoshida, Y. and Kageyama, Y. 2002. Photosynthetic, evapotranspiratory and leaf morphological properties of chrysanthemum grown under root restriction as affected by fertigation frequency. *Journal of the Japanese Society for Horticultural Science*, 71: 277–283.

Grabov, A. 2007. Plant KT/KUP/HAK potassium transporters: single family- multiple functions. *Annals of Botany*, 99(6): 1035-1041.

Grant, C.A., Flaten, D.N.,Tomasiewicz, D.J. and Sheppard, S.C. 2001. The importance of early season phosphorus nutrition. *Canadian Journal of Plant Science*, 81: 211–224.

Granato, T.C. and D.C. Raper, Jr. 1989. Proliferation of maize (Zea maysL.) roots in response to localized supply of nitrate. *J. Expt. Bot.*, 40: 263–275.

Groppa, M.D., Zawoznik, M.S.,Tomaro, M.L. and Benavides, M.P. 2008. Inhibition of root growth and polyamine metabolism in sunflower (*Helianthus annuus*) seedlings under cadmium and copper stress. *Biological Trace Elements Research*, 126: 246–256.

Grover, A., Kappor, A., Lakshmi, S.O., Agarwal, S., Agarwal, S.K., Agarwal, M. and Dubey, H. 2001. Understanding molecular alphabets of the plant abiotic stress responses. *Curr. Sci.*, 80: 206-216.

Guilfoyle, T.J., and Hagen, G. 2007. Auxin response factors. *Curr Opin Plant Biol.*, 10: 453–460.

Gulzar, S., Khan, M.A. and Ungar, L.A. 2003. Salt tolerance of a coastal salt marsh grass. *Comm. Soil Sci. Plant Anal.*, 34: 2595-2605.

Guo, J., Liang, J., and Chen, J.G. 2007. RACK1: a versatile scaffold protein in plants? *Int J Plant Dev Biol.,* 1: 95–105.

Guo, J., Wang, J., Xi, L., Huang, W.-D., Liang, J., and Chen, J.-G. 2009. RACK1 is a negative regulator of ABA responses in *Arabidopsis*. *J Exp Bot.*, 60: 3819–3833.

Gutierrez, L., Van Wuytswinkel, O., Castelain, M., and Bellini, C. 2007. Combined networks regulating seed maturation. *Trends Plant Sci.*, 12:, 294–300.

Guzmán, M. and Sánchez, A. 2003. Influence of nitrate end calcium increments on development, growth and early yield in sweet pepper plants. *Acta Horticulturae*, 609: 207–211.

Habash, D.Z., Kehel, Z. and Nachit, M. 2009. Genomic approaches for designing durum wheat ready for climate change with a focus on drought. *J Exper Bot.*, 60(10): 2805–2815.

Hadas, A. 1976. Water uptake and germination of leguminous seeds in soils of changing matrix and osmotic water potential. *J Exp Bot.,* 28: 977- 985.

Hagemeyer, J. and Breckle, S.W. 2002. Trace element stress in roots. In "*Plant Roots: The Hidden Half.* 3rd Edition", Eds. Y. Waisel, A. Eshel, and U. Kafkafi, Marcel Dekker, New York,NY, USA. pp. 763–785.

Hajibagheri, M.A., Yeo, A.R., Flowers, T.J. and Collins, J.C. 1989. Salinity resistance in *Zea mays* uses of potassium, sodium and chloride, cytoplasmic concentrations and microsomal membrane lipids. *Plant Cell Environ.*, 12: 753-757.

Hajibagheri, M.A., Yeo, A.R., Flowers, T.J. and Collins, J.C. 1989. Salinity resistance in *Zea mays* uses of potassium, sodium and chloride, cytoplasmic concentrations and microsomal membrane lipids. *Plant Cell Environ.*, 12: 753-757.

Hardenburg, R.E., A.E. Watada, and C.Y. Wang. 1986. The commercial storage of fruits, vegetables, and florist and nursery stocks. *U.S. Dept. Agr. Hdbk.* 66.

Hampson, C.R. and Simpson, G.M. 1990. Effect of temperature, salt and osmotic potential in early growth of wheat (*Triticum aestivum*) germination. *Can. J. Bot.*, 68: 524-528.

Harrison-Murray, R.S. and D.T. Clarkson. 1973. Relationships between structural development and the absorption of ions by the root system of *Cucurbita pepo*. *Planta*, 114: 1–16.

Hasegawa, P.M., Bressan, R.A., Zhu, J.K. and Bohnet, H.J., 2000. Plant cellular and molecular responses to high salinity. *Annu. Rev. Plant Phys.*, 51: 463-499.

He, X.Q., Du, C., Shao, Z. and Li, Q. 2009. Effect of salt and water stress on seed germination of *Dianthus Chinensis* L. *Academic Conference of Horticultural Science and Technology Proceedings*, pp. 60-62.

He, J., Tan, L.P. and Lee, S.K. 2009. Root-zone temperature effects on photosynthesis, C^{14-} photoassimilate partitioning and growth of temperate lettuce (*Lactuca sativa* cv. 'Panama') in the tropics. *Photosynthetica*, 47: 95–103.

Hilhorst, H.W.M. 2007. Definition and hypotheses of seed dormancy. In: Seed Development, Dormancy and Germination. *Annual Plant Reviews*, 27:, Chap 4.

Hirsch, R.E., Lewis, B.D., Spalding, E.P. and Sussman, M.R. 1998. A role for the AKT1 potassium channel in plant nutrition. *Science*, 280(5365): 918-921.

Hirschi, A, Cecchini, M, Steinhardt, R. C, Schamber, M. R, Dick, F. A, & Rubin, S. M. 2010. An overlapping kinase and phosphatase docking site regulates activity of the retinoblastoma protein. *Nature structural & molecular biology*, 17(9): 1051-1057.

Hodge, A., Berta, G., Doussan, C., Merchan, F. and Crespi, M. 2009. Plant root growth, architecture and function.*Plant Soil*, 321(1-2): 153-187.

Hole, D.J., B.G. Cobb, P.S. Hole, and M.C. Drew. 1992. Enhancement of anaerobic respiration in root tips of *Zea mays* following low-oxygen (hypoxic) acclimation. *Plant Physiol.*,99:213–218.

Horchani, F., Khayati, H., Raymond, P., Brouquisse, R. and Aschi-Smiti, S. 2009. Contrasted effects of prolonged root hypoxia on tomato root and fruit (*Solanum lycopersicum*) metabolism. *Journal of Agronomy and Crop Science*, 195: 313–318.

Hsiao, T.C. and Xu, L.K. 2000. Sensitivity of growth of roots versus leaves to water stress: biophysical analysis and relation to water transport. *J. Expt. Bot.,* 51: 1595–1616.

Huang, J. and Redmann, R.E. 1995. Salt tolerance of Hordeum and Brassica species during germination and early seedling growth. *Can. J. Plant Sci.*, 75: 815-819.

Huang, B.R. and NeSmith, D.S. 1999. Soil aeration effects on root growth and activity. *Acta Horticulturae*, 504: 41–49.

Huang, B.R. 2008. Mechanisms and strategies for improving drought resistance in turf grass. *Acta Horticulturae*, 783: 221–227.

Hund, A., Ruta, N. and Liedgens, M. 2009. Rooting depth and water use efficiency of tropical maize inbred lines differing in drought tolerance. *Plant and Soil*, 318: 311–325.

Hundertmark, M., Buitink, J., Leprince, O., and Hincha, D.K. 2011. The reduction of seed-specific dehydrins reduces seed longevity in *Arabidopsisthaliana. Seed Sci Res.*, 21: 165–173.

Irmak, S., Haman, D.Z., Irmak, A., Jones, J.W., Campbell, K. L. and Crisman, T.L. 2004. Measurement and analyses of growth and stress parameters of *Viburnum odoratissimum* (Ker-gawl) grown in a multi-pot box system. *Hort. Science*, 39: 1445–1455.

Irmak, S., Haman, D.Z., Irmak, A., Jones, J.W., Tonkinson, B., Burch, D., Yeager, T.H. and Larsen, C. 2005. Root zone temperatures of *Viburnum odoratissimum* grown in a multiport box system and conventional systems: Measurement and analyses of temperature profiles and predicting root zone temperatures. *Hort. Science*, 40: 808–818.

Iqbal, M., and Ashraf, M. 2007 Seed treatment with auxins modulates growth and ion partitioning in saltstressed wheat plants. *J. Integr. Plant Biol.*, 49: 1003-1015.

Jackson, M.B. 1979. Rapid injury to peas by soil water logging. *J. Sci. Food Agr.*, 30: 143–152.

Jackson, M.B. and M.C. Drew. 1984. Effects of flooding on growth and metabolism of erbaceous plants. In "Flooding and plant growth," Ed. T.T. Kozlowski, Academic, Orlando, Fla, pp. 47–128.

Jamil, M. and Rha, E.S. 2004. The effect of salinity (NaCl) on the germination and seedling of sugar beet (*Beta vulgaris* L.) and cabbage (*Brassica oleracea capitata* L.). *Kor. J. Plant Res.*, 7: 226-232.

Jamil, M., Lee, C.C., Rehman, S.U., Lee, D.B., Ashraf, M, and Rha, E.S. 2005. Salinity (NaCI) tolerance of brassica species at germination and early seedling growth. *Electro. J. Environ. Agric. Food Chem.*, 7: 116-121.

Jiang, W.S. and Liu, D.H. 2010. Pb-induced cellular defense system in the root meristematic cells of *Allium sativum* L.*BMC Plant Biology*, 10: Art. No. 40, 8 pp.

Jiang, C., Gao, X., Liao, L., Harberd, N.P. and Fu, X. 2007. Phosphate starvation root architecture and anthocyanin accumulation responses are modulated by the gibberellin-DEL LA signaling pathway in *Arabidopsis. Plant Physiol.*, 145(4): 1460-1470.

Johnson, J., B.G. Cobb, and M.C. Drew. 1989. Hypoxic induction of anoxia tolerance in roots in *Zea mays. Plant Physiol.*, 91: 837–841.

Jovanovic, M., Rielefebvre, V., Laporte, P., Gonzales-rizzo, S., Lelandais-briére, C., Frugier, F., Hartmann, C., and Crespi, M. 2007. How the Environment Regulates Root Architecture in Dicots. *Adv. Bot. Res.,* 46: 35-74.

Jung, H.J., and Kang, H. 2007. Expression and functional analyses of microRNA417 in *Arabidopsis thaliana* under stress conditions. *Plant Physiol Biochem.*, 45: 805–811.

Kachout, S.S., Mansoura, A.B., Leclerc, J.C., Mechergui, R., Rejeb, M.N. and Ouerghi, Z. 2009. Effects of heavy *metals* on antioxidant activities of *Atriplex hortensis* and *A. rosea. Journal of Food, Agriculture & Environment*, 7: 938–945.

Kachout, S.S., Mansoura, A.B., Hamza, K.J., Leclerc, J.C., Rejeb, M.N. and Ouerghi, Z. 2011. Effect of water stress on plant growth in *Atriplex hortensis* L. *Journal of Horticultural Science & Biotechnology*, 86: 101–106.

Kadir, S., Sidhu, G. and Al-Khatib, K. 2006. Strawberry (*Fragaria ananassa* Duch.) growth and productivity as affected by temperature. *Hort. Science*, 41: 1423-1430.

Kagale, S., Divi, U.K., Krochko, J.E., Keller, W.A., and Krishna, P. 2007. Brassinosteroid confers tolerance in *Arabidopsisthaliana* and *Brassica napus* to a range of abiotic stresses. *Planta,*225: 353–364.

Kahn, B.A., P.J. Stoffella, R.F. Sandsted, and R.W. Zobel. 1985. Influence of flooding on root morphological components of young black beans. *J. Amer. Soc. Hort. Sci.*, 110: 623–627.

Kahn, B.A. and P.J. Stoffella. 1991. Nodule distribution among root morphological components of field-grown cowpeas. *J. Amer. Soc. Hort. Sci.*, 116: 416–420.

Kakkar, R.K., Bhaduri, S., Rai, V.K. and Kumar, S. 2000. Amelioration of NaCl stress by arginine in rice seedling: Changes in endogenous polyamines. *Biologia Plantarum*, 43: 419–422.

Karantev, K., Yordanova, R. and Popova, L. 2006. Salicylic acid decrease cadmium toxicity in maize plants. *Plant Physiol Special Issue*: 45-52.

Karmoker, J.L. and R.F. Van Steveninck. 1979. The effect of abscisic acid on sugar levels in seedlings of *Phaseolus vulgaris* L. cv. Redland Pioneer. *Planta*, 146: 25–30.

Karni, L., Aktas, H., Deveturero, G. and Aloni, B. 2010. Involvement of root ethylene and oxidative stress-related activities in pre-conditioning of tomato transplants by increased salinity. *J. Hort. Sci. & Biotech.,* 85: 23–29.

Kant, S., Bi, Y-M. and Rothstein, S.J. 2011. Understanding plant response to nitrogen limitation for the improvement of crop nitrogen use efficiency. *Journal of Experimental Botany*, 62(4): 1499-1509.

Kaur, S. A., Gupta, K., and Kaur, N. (2002) Effect of osmo and hydropriming of chickpea seeds on seedling growth and carbohydrate metabolism water deficit stress. *Plant growth Regul.* 37: 17-22.

Kavikishore, P.B., Sangam, S., Amrutha, R.N., Srilakshmi, P., Naidu, K.R., Rao, K.R.S.S., Sreenath, R., Reddy, K.J., Theriappan, P. and Sreenivasalu, N. 2005 Regulation of proline biosynthesis, degradation, uptake and transport in higher plants. Its implications in plant growth and abiotic stress tolerance. *Current Sci.*, 88(3): 424-436.

Kaya, M.D., Kaya G., Çikili Y. and Çiftçi C.Y. 2006a. Effects of NaCl on the germination, seedling growth and water uptake of triticale. *Turk. J. Agric. For.*, 30: 39-47.

Kaya, M.D., Okçu G., Atak M., Çikili Y. Kolsarici, O. 2006b Seed treatments to overcome salt and drought stress during germination in sunflower (*Helianthus annuus* L.). *Eur. J. Agron.* 24: 291-295.

Kant, S., Bi, Y-M. and Rothstein, S.J. 2011. Understanding plant response to nitrogen limitation for the improvement of crop nitrogen use efficiency. *Journal of Experimental Botany*, 62(4): 1499-1509.

Kefu, Z., Munns, R. and King, R. 1991. Abscisic acid levels in Nacl treated barley, cotton and saltbush. *Func. Pl. Biol.,* 18(1): 17-24.

Keiffer, C.H. and Ungar, L,A. 1997. The effect of level extend exposure to hyper saline conditions on the germination of five inland halophyte species. *Am J Bot.*, 84: 104-111.

Keller, B. and C.J. Lamb. 1989. Specific expression of a novel cell wall hydroxyproline-rich glycoprotein gene in lateral root initiation. *Genes Dev.* 3: 1639–1646.

Kent, L.M. and A. Lauchli. 1985. Germination and seedling growth of cotton: Salinity–calcium interactions. *Plant Cell Environ.*, 8: 155–159.

Keyhani, E. and Keyhani, J. 2008. Effect of copper and glucose on root development and corm anti-oxidative stress enzymes activity in saffron (*Crocus sativus* L.). *Acta Horticulturae*,779: 677–687.

Khan M.A. and Rizvi, Y. 1994. Effect of salinity, temperature and growth regulators on the germination and early seedling growth of *Atriplex griffithii* var. Stocksii. *Can. J. Bot.* 72: 475–479.

Khan M.A. and Gulzar S. 2003. Germination Responses of *Sporobolus ioclados*: A Saline Desert Grass. *J. Arid Environ.*,53: 387-394.

Khan M.A., Qayyum, A. and Noor E 2007. Assessment of wheat genotypes for salinity tolerance. Proceeding of 8th African Crop Science Conference, Oct. 27-31, El-Minia, Egypt. pp. 75-78.

Kharkina, T.G., Ottosen, C.O. and Rosenqvist, E. 1999. Effects of root restriction on the growth and physiology of cucumber plants. *Physiologia Plantarum*, 105: 434–441.

Khayyat, M., Rajaee, S., Sajjadinia, A., Eshghi, S. and Tafazoli, E. 2009. Calcium effects on changes in chlorophyll contents, dry weight and micronutrients of strawberry (*Fragaria nanassa* Duch.) plants under salt-stress conditions. *Fruits*, 64: 53–59.

Khurana, N. and Chatterjee, C. 2007. Changes in metabolism, root quality and nutritive characters of carrot under copper stress. *Indian Journal of Horticulture*, 64: 335–339.

Kim, S.G., Lee, A.K., Yoon, H.K., and Park, C.M. 2008. A membrane-bound NAC transcription factor NTL8 regulates gibberellic acid-mediated salt signaling in *Arabidopsis* seed germination. *Plant J.*, 55: 77–88.

Klepper, B. 1990. Root growth and water uptake. In "Irrigation of agricultural crops" Eds. B.A. Stewart and D.R. Nielsen, Amer. Soc. Agron., Madison, Wis., pp. 281–322..

Klepper, B. 1991. Crop root system response to irrigation. *Irrigation Sci.*, 12: 105–108.

Klepper, B. and R.W. Rickman. 1991. Predicting root development of crop plants. In "Predicting crop phenology", Ed.T. Hodges, CRC Press, Boca Raton, Fla. pp. 85–99.

Ko, J.H., Yang, S.H., and Han, K.H. 2006. Upregulation of an *Arabidopsis* RING-H2 gene, XERICO, confers drought tolerance through increased abscisic acid biosynthesis. *Plant J.*, 47: 343–355.

Koike, T., Kitao, M., Quoreshi, A.M. and Matsuura, Y. 2003. Growth characteristics of root-shoot relations of three birch seedlings raised under different water regimes. *Plant and Soil*, 255: 303–310.

Koh, S., Lee, S.C., Kim, M.K., Koh, J.H., Lee, S., An, G., Choe, S. and Kim, S.R. 2007. T-DNA tagged knockout mutation of rice OsGSK1, an orthologue of *Arabidopsis* BIN2, with enhanced tolerance to various abiotic stresses. *Plant Mol Biol.*, 65: 453–466.

Kosova´, K., Pra´il, T.I., and Vý´ta´mva´s, P. 2011. Role of Dehydrins. In "*Plant Stress Response Handbook of Plant and Crop Stress*", 3rd ed. Mohammad Pessarakli, ed. Boca Raton, FL: CRC Press

Kulkarni, M. and Deshpande, U. 2007. Gradient *in vitro* testing of tomato (*Solanum lycopersicon* L.) cultivars by inducing water deficit – a new approach to screen germplasm for drought tolerance. *Asian Journal of Plant Science*, 6: 934–940.

Kulkarni, M., Borse, T. and Chaphalkar, S. 2008. Mining anatomical traits: A novel modelling approach for increased water use efficiency under drought condition in plants. *Czech Journal of Genetics and Plant Breeding*, 44: 11–21.

Kulkarni, M. and Phalke, S. 2009. Evaluating variability of root size system and its constitutive traits in hot pepper (*Capsicum annuum* L.) under water stress. *Scientia Horticulturae*,120: 159–166.

Kumar S.G., Reddy A.M. and Sudhakar, C. 2003. NaCl eff ects on proline metabolism in two high yielding genotypes of mulberry (*Morus alba* L.) with contrasting salt tolerance. *Plant Sci*, 165: 1245-1251.

Kutz, A., Müller, A., Hennig, P., Kaiser, W.M., Piotrowski, M. and Weiler, E.W. 2002. A role for nitrilase 3 in the regulation of root morphology in sulphur-starving *Arabidopsis thaliana. Pl. J.,* 30(1): 95-106.

Laby, R.J., Kincaid, M.S., Kim, D., and Gibson, S.I. 2000. The *Arabidopsis* sugar-insensitive mutants sis4 and sis5 are defective in abscisic acid synthesis and response. *Plant J.*, 23: 587–596.

Lallu, Dixit, R.K. 2005. Salt tolerance of Mustard genotype at seedling stage. *Indian J Plant Physiol.*, 14(2): 33-35.

Lammens, T., Boudolf, V., Kheibarshekan, L., Zalmas, L.P., Gaamouche, T., Maes, S., Vamstraelen, M., Kondorosi, E.La., Thangue, N.B., Govaerts, W., Inze, D. and De Vevlder, L. 2008. Atypical E2F activity restrains APC/CCCS52A2 function obligatory for endocycle onset. *Proc. Nat. Acad. Sci. USA,* 105(38): 14721-14726.

Latimer, J.G. 1991. Mechanical conditioning for control of growth and quality of vegetable transplants. *Hort Science*, 26: 1456–1461.

Latimer, J.G. and C.A. Mitchell. 1988. Effects of mechanical stress or abscisic acid on growth, water status, and leaf abscisic acid content of eggplant seedlings. *Scientia Hort.*, 36: 37–46.

Lauchli, A. and E. Epstein. 1990. Plant responses to saline and sodic conditions. In "Agricultural salinity assessment and management", Ed. K.K. Tanji, *Amer. Soc. Civil Eng. Manuals Rpts. Eng. Practice*, 71: 113–137.

Lecompte, F., Bressoud, F., Pares, L. and De Bruyne, F. 2008. Root and nitrate distribution as related to the critical plant N status of a fertigated tomato crop. *Journal of Horticultural Science & Biotechnology*, 83; 223–231.

Lee, S.C., Cheng, H., King, K.E., Wang, W., He, Y., Hussain, A., Lo, J., Harberd, N.P. and Peng, J. 2002. Gibberellin regulates *Arabidopsis* seed germination via RGL2, a GAI/RGA-like gene whose expression is upregulated following imbibition. *Genes Dev.*, 16: 646–658.

Lee, S.H., Singh, A.P., Chung, G.C., Ahn, S.J., Noh, E.K. and Steudle, E. 2004. Exposure of roots of cucumber (*Cucumis sativus*) to low temperature severely reduces root pressure,ydraulic conductivity and active transport of nutrients. *Physiologia Plantarum*, 120: 413–420.

Lee, S.H., Singh, A.P. and Chung, G.C. 2004. Rapid accumulation of hydrogen peroxide in cucumber roots due to exposure to low temperature appears to mediate decreases in water transport. *J. Expt. Bot.,* 55(403): 1733-1741.

Lee, S.H., Choi, J.H., Kim, W.S., Han, T.H., Park, Y.S. and Gemma, H. 2006. Effect of soil water stress on the development of stone cells in pear (*Pyrus pyrifolia* cv. 'Niitaka')lesh. *Scientia Horticulturae*, 110: 247–253.

Leskovar, D.I., D.J. Cantliffe, and P.J. Stoffella. 1989. Pepper (*Capsicum annuum* L.) root growth and its relation to shoot growth in response to nitrogen. *J. Hort. Sci.*, 64: 711–716.

Leskovar, D.I., D.J. Cantliffe, and P.J. Stoffella. 1990. Root growth and root– shoot interaction in transplants and direct seeded pepper plants. *Environ. Expt. Bot.*, 30: 349–354.

Leskovar, D.I. and D.J. Cantliffe. 1991. Tomato transplant morphology affected by handling and storage. *Hort Science*, 26: 1377–1379.

Leskovar, D.I. and D.J. Cantliffe. 1992. Pepper seedling growth response to exogenous abscisic acid. *J. Amer. Soc. Hort. Sci.*, 117: 389–393.

Leskovar, D.I. and D.J. Cantliffe. 1993. Comparison of plant establishment method, transplant or direct-seeding, on growth and yield of bell pepper. *J. Amer. Soc. Hort. Sci.*, 118: 17–22.

Leskovar, D.I. and R.R. Heineman. 1994. Growth of 'TAM-Mild Jalapeño-1' pepper seedlings as affected by greenhouse irrigation systems. *Hort Science*, 29: 1470–1474.

Leskovar, D.I., D.J. Cantliffe, and P.J. Stoffella. 1994. Transplant production systems influence growth and yield of fresh market tomatoes. *J. Amer. Soc. Hort. Sci.*, 119: 662–668.

Leskover, D. and Stoffella, P.J. 1995. Vegetable seedling root systems: morphology, development, and importance. *Hort. Science*, 30(6): 1153-1159.

Letey, J., L.H. Stolzy, and G.B. Blank. 1962. Effect of duration and timing of low soil oxygen content on shoot and root growth. *Agron. J.*, 54: 34–37.

Leubner-Metzger, G. 2003. Functions and regulation of b-1,3- glucanase during seed germination, dormancy release and after ripening. *Seed Sci. Res.*, 13: 17–34.

Levitt, J. 1980. In "Responses of Plants to Environmental Stresses, 11th Ed.", Academic Press, New York. Li, J., Nagpal, P., Vitart, V., McMorris, T.C., and Chory, J. 1996. A role for brassinosteroids light-dependent development of *Arabidopsis*. *Science,*272: 398–401.

Li, J., and Jin, H. 2006. Regulation of brassinosteroid signaling. *Trends Plant Sci.*, 12: 37–41.

Li, S., Chen, J. and Zuo, Q. 2007. Influences of optimizing fertilization on the growth and yield of rice variety Wandao68. *J Anhui Agric Sci.*, 35: 8571–8573.

Liang, Y., Wong, J.W.C. and Wei, L. 2005. Silicon-mediated enhancement of cadmium tolerance in maize (*Zea mays* L.) grown in contaminated soil. *Chemosphere,*58: 475-483.

Liao, W., Huang, G., Yu, J., Zhang, M. and Shi, X. 2011. Nitric oxide and hydrogen peroxide are involved in indole-3-butyric acid-induced adventitious root development in marigold. *Journal of Horticultural Science & Biotechnology*, 86: 159–165.

Liptay, A. 1987. Field survival and establishment of tomato transplants of various age and size. *Acta Hort.*, 220: 203–209.

Liu, P.P., Montgomery, T.A., Fahlgren, N., Kasschau, K.D., Nonogaki, H., and Carrington, J.C. 2007. Repression of AUXIN RESPONSE FACTOR10 by microRNA160 is critical for seed germination and post-germination stages. *Plant J.*, 52: 133–146.

Lin, C. C. and Kao, C.H. 1995. NaCl stress in rice seedlings. Stach mobilization and the influence of gibberelic acid on seedling growth. *Bot. Bull. Acad. Sin.*, 36:, 169–173.

Linkies, A., Graber, K., Knight, C., and Leubner-Metzger, G. 2010. The evolution of seeds. *New Phytol.*, 186: 817–831.

Liu, H., and Stone, S.L. 2010. Abscisic acid increases *Arabidopsis* ABI5 transcription factor levels by promoting KEG E3 ligase self-ubiquitination and proteasomal degradation. *Plant Cell*, 22: 2630–2641.

Ljung, K, Hull, A.K, Kowalczyk, M, Marchant, A, Celenza, J, Cohen, J.D. and Sandberg, G. 2002. Biosynthesis, conjugation, catabolism and homeostasis of indole-3-acetic acid in *Arabidopsis thaliana. Plant Mol Biol.*, 49(3-4): 249-272.

Llave, C., Xie, Z., Kasschau, K.D., and Carrington, J.C. 2002. Cleavage of Scarecrow-like mRNA targets directed by a class of *Arabidopsis* miRNA. *Science*, 297: 2053–2056.

Lopez-Molina, L., and Chua, N-H. 2000. A null mutation in a bZIP factor confers ABA-insensitivity in *Arabidopsisthaliana. Plant Cell Physiol.*, 41: 541–547.

Lopez-Molina, L., Mongrand, B., McLachlin, D.T., Chait, B.T., and Chua, N.H. 2002. ABI5 acts downstream of ABI3 to execute an ABA dependent growth arrest during germination. *Plant J.*, 32: 317–328.

Lopez-bucio, J., Hernandez-abreu, E., Sanchez-calderon, L., Nieto-jacobo, M.F., Simpson, J. and Herrera-estrella, L. 2002. Phosphate availability alters architecture and causes changes in hormone sensitivity in the *Arabidopsis* root system. *Plant Physiol.*, 129(1): 244-256.

López-bucio, J. Cruz-Ramýìrez A, Herrera-Estrella L. 2003. The role of nutrient availability in regulating root architecture. *Curr. Opinion Pl. Biol.,* 6(3): 280-287

López-bucio, J, Cruz-ramírez, A, Pérez-torres, A, Ramírez-pimentel, R, Sánchez-calderón, L, and Herrera-estrella, L. 2005. Root architecture. In "*Plant Architecture and its Manipulation*" Ed. C.G.N. Turnbull, Oxford UK: Blackwell, pp. 315.

López-Marín, H.D., Rao, I.M. and Blair, M.W. 2009. Quantitative trait loci for root morphology traits under aluminium stress in common bean (*Phaseolus vulgaris* L.). *Theoretical and Applied Genetics*, 119: 449–458.

Loveys, B.R. and Davies, W.J. 2004. Physiological approaches to enhance water use efficiency in agriculture: Exploiting plant signalling in novel irrigation practice. In "*Water Use Efficiency in Plant Biology*", Ed. M. Bacon, Blackwell Publishing, Oxford, UK. 113–141.

Lynch, J. and J.J. van Beem. 1993. Growth and architecture of seedling root of common bean genotypes. *Crop Sci.*, 33: 1253–1257.

Ma, Y., Szostkiewicz, I., Korte, A., Moes, D., Yang, Y., Christmann, A., *et al.* 2009. Regulators of PP2C phosphatase activity function as abscisic acid sensors. *Science*, 324: 1064–1068.

Macdonald, J.E. and Owens, J.N. 2006. Morphology, physiology, survival and field performance of containerized coastal Douglas fir seedlings given different dormancy induction regimes. *Hort. Science*, 41: 1416–1420.

Macdonald, M.T., Lada, R.R., Hoyle, J. and Robinson, A. R. 2009. Ambiol preconditioning can induce drought tolerance in abscisic acid-deficient tomato seedlings. *Hort Science*, 44: 1890–1894.

Macdonald , J.E. and Owens, J.N. 2010. Physiology and growth of containerized coastal Douglas fir seedlings given different durations of short days to induce dormancy. *Hort. Science*, 46: 342–346.

MacIsaac, S.A., V.K. Sawhney, and Y. Pohorecky. 1989. Regulation of lateral root formation in lettuce (*Lactuca sativa*) seedling roots: Interacting effects of α-naphthalene acetic acid and kinetin. *Physiol. Plant.*, 77: 287–293.

MacLeod, R.D. 1990. Lateral root primordium inception in *Zea mays* L. *Environ. Expt. Bot.*, 30: 225–234.

Malamy, J.E. 2005. Intrinsic and environmental response pathways that regulate root system architecture. *Plant, cell and Env.*, 28(1): 67-77.

Markesteijn, L. and Poorter, L. (2009). Seedling root morphology and biomass allocation of 62 tropical tree species in relation to drought- and shade-tolerance. *Journal of Ecology*, 97: 311–325.

Masle, J. 2002. High soil strength: Mechanical forces at play on root morphogenesis and in root:shoot signaling. In "*Plant Roots: The Hidden Half.* 3rd Edition", Eds. Y. Waisel, A. Eshel, and U. Kafkafi, Marcel Dekker, New York, NY, USA. pp. 807–819.

Martinez C.A., Maestri M. and Lani, E.G. 2003. In vitro salt tolerance and proline accumulation in Andean potato (*Solanum* spp.) differing in frost resistance. *Plant. Sci.*, 116: 117-184.

Matsumoto, H. 2002. Plant roots under aluminium stress: Toxicity and tolerance. In "*Plant Roots: The Hidden Half.* 3rd Edition", Y. Waisel, A. Eshel, and U. Kafkafi, Marcel Dekker, New York, NY, USA. 821–838.

Mathers, H.M., Lowe, S.B., Scagel, C., Struve, D.K. and Case, L.T. 2007. Abiotic factors influencing root growth of woody nursery plants in containers. *Hort. Technology*, 17: 151–162.

Matsuki, S., Ogawa, K., Tanaka, A. and Hara, T. 2003. Morphological and photosynthetic responses of *Quercus crispula* seedlings to high-light conditions. *Tree Physiology*, 23: 769–775.

Marschner, H. and V. Romheld. 1983. In vivo measurement of root induced pH changes at the soil–root surface: Effect of plant species and nitrogen source. *Z. Planzenphysiol. Bodenkd.*, 111: 241–51.

Marschner, H., V. Romheld, W.J. Horst, and P. Martin. 1986. Root-induced changes in the rhizosphere. Importance for mineral nutrition on plants. *Z. Planzenernaehr. Bodenkd.*, 119: 441–456.

Martinez C.A., Maestri M. and Lani, E.G. 2003. In vitro salt tolerance and proline accumulation in Andean potato (*Solanum* spp.) differing in frost resistance. *Plant. Sci.*, 116: 117-184.

Masoni, A., Ercoli, L., Mariotti, M. and Arduini, I. 2007. Postanthesis accumulation and remobilization of dry matter, nitrogen and phosphorus in durum wheat as affected by soil type. *Eur J Agron.*, 26:179-186.

Matilla, A., Gallardo, M., and Puga-Hermida, M.I. 2005. Structural, physiological and molecular aspects of heterogeneity in seeds: a review. *Seed Sci. Res.*, 15: 63–76.

Mayer, A. M. and Poljkoff-Mayber, A. 1975. The germination of seeds. 2nd edition. Pergamon press, New York, 167–193.

Maurel, C., Simonneau, T. and Sutka, M. 2010.The significance of roots as hydraulic rheostats. *Journal of Experimental Botany*, 61: 3191–3198.

Mauromicale, G. and Licandro, P. 2002. Salinity and temperature effects on germination, emergence and seedling growth of globe artichoke. *Agronomie*, 22: 443–450.

McCully, M.E. and A.J. Canny. 1988. Pathways and processes of water and nutrient movement in roots. *Plant Soil*, 111: 159–170.

McKee, J.M.T. 1981. Physiological aspects of transplanting vegetables and other crops. I. Factors which influence re-establishment. *Hort. Abstr.*, 51: 265–272.

Mcmichael, B.L. and Burke, J.J. 2002. Temperature effects on root growth. In "*Plant Roots: The Hidden Half*. 3rd Edition, Eds. Y. Waisel, A. Eshel, and U. Kafkafi, Marcel Dekker,New York, NY, USA. pp. 717–728.

Meier, H. and Reid, J.S.G. 1982. Reserve polysaccharides other than starch in higher plants. In *Encyclopedia of Plant Physiol*ogy", New series, Vol. 13a, Eds.

Melcher, K., Ng, L-M., Zhou, X.E., Soon, F.-F., Xu, Y., SuinoPowell, K.M., Park, S-Y., Joshua, J.W., Fujii, H., Chinnusamy, V., Kovach, A., Li, J., Wang, Y., Li, J., Peterson, F.C., Jensen, D.R., Yong, E-L., Volkman, B.F., Cutler, S.R., Zhu, J.K., and Xu, H.E. 2009. A gate–latch–lock mechanism for hormone signalling by abscisic acid receptors. *Nature*, 462: 602–608.

Melchior, W. and E. Steudle. 1993. Water transport in onion (*Allium cepa* L.) roots. *Plant Physiol.*,101:1301–1315.

Milchunas, D.G.,Morgan, J.A.,Mosier, A.R. and Lecain, R.D. 2005a. Root dynamics and demography in short grass steppe under elevated CO2, and comments on mini rhizotron methodology. *Global Change Biology*, 11: 1837–1855.

Milchunas, D.G., Mosier, A.R., Morgan, J.A., Lecain, D.R., King, J.Y. and Nelson, J.A. 2005b. Root production and tissue quality in a short grass steppe exposed to elevated CO_2: Using a new in growth method. *Plant and Soil*, 268: 111–122.

Miller, D.E. 1986. Root systems in relation to stress tolerance. *Hort Science*, 21: 963–970.

Miller, G., Suzuki, N., Ciftci-yilmaz, S and Mittler, R. 2010. Reactive oxygen species homeostasis and signalling during drought and salinity stresses. *Plant, Cell and Environment,* 33(4): 453-467.

Miralles, J., Nortes, P.A., Sánchez-Blanco, M.J., Martínezsánchez, J.J. and Bañón, S. 2009. Above-ground and pot-in pot production systems for *Myrtus communis* L. *Transactions Of the ASABE*, 52: 93–101.

Miralles, J., Martínezsánchez, J.J., Franco, J.A. and Bañón, S. 2011. *Rhamnus alaternus* growth under four simulated shade environments: Morphological, anatomical and physiological responses. *Scientia Horticulturae*, 127: 562–570.

Mitra, J. 2001. Genetics and genetic improvement of drought resistance in crop plants. *Current science.* 80(6): 758 – 763.

Miura, K., Rus, A., Sharkhuu, A., Yokoi, S., Karthikeyan, A.S., Raghothama, K.G. Baek, D. Koo, Y.D., Jin, J.B., Bressan, R.A., Yun, D.J. and Hasegawa, P.M. 2005. The *Arabidopsis* SUMO E3 ligase SIZ1 controls phosphate deficiency responses. *Proc. Nat. Acad. Sci., USA.,* 102(21): 7760-7765.

Miura, K., Lee, J., Jin, J.B., Yoo, C.Y., Miura, T., and Hasegawa, P.M. 2009. Sumoylation of ABI5 by the *Arabidopsis* SUMO E3 ligase SIZ1 negatively regulates abscisic acid signaling. *Proc Natl Acad Sci USA*, 106: 5418–5423.

Moinuddin, Khanna-Chopra, R. 2004. Osmotic adjustment in chickpea in relation to seed yield and yield parameters. *Crop Sci.*, 44: 449–455.

Monneveux, P., Sanchez, C., Beck, D. and Edmeades, G.O. 2006. Drought tolerance improvement in tropical maize source populations: evidence of progress. *Crop Sci.*, 46: 180–191

Moreno, J.L., Hernandez, T. and Garcia, C. 1999. Effects of a cadmium-containing swage sludge compost on dynamics of organic matter and microbial activity in an arid soils. *BiolFert Soils.*,28:230-237.

Morgan, J.A., Mosier, A. R., Milchunas, D.G., Lecain, D.R., Nelson, J.A. and Parton, W.J. 2004. CO_2enhances productivity but alters species composition and reduces forage quality in the Colorado short grass steppe. *Ecological Applications*, 14: 208–219.

Mortimer, J.C., Laohavisit, A., Miedema, H. and Davies, J.M. 2008. Voltage, reactive oxygen species and the influx of calcium. *Plant Signaling & Behavior*, 3(9): 698-699.

Mumford, C., Vickers, A.W. and James, I.T. 2004. A feasibility study into the use of reflected light to improve turf grass growth in shaded areas. *Acta Horticulturae*, 661: 565–569.

Munns, R. and Termaat, A. 1986. Whole plant responses to salinity. *Aust J Physiol.*, 13: 143-160.

Munns, R. 2002. Comparative physiology of salt and water stress. *Plant, Cell and Environment*, 25: 239–250.

Munns, R, and Tester, M. 2008. Mechanisms of salinity tolerance. *Annual Review of Plant Biology*, 59: 651-681.

Munoz-Arboleda, F., Mylavarapu, R.S., Hutchinson, C.M. and Portier, K.M. 2006. Root distribution under seepage irrigated potatoes in northeast Florida. *American Journal of Potato Research*, 83: 463–472.

Murungu, F.S., Nyamugafata, P., Chiduza, C., Clark, L.J., and Whalley,W.R. 2003 Effects of seed priming, aggregate size and soil matric potential on emergence of cotton (*Gossypium hirsutum* L.) and maize (*Zea mays* L). *Soil and Tillage Research*, 74: 161-168.

Mwale, S., Hamusimbi, C., Mwansa, K. 2003. Germination, emergence and growth of sunflower (*Helianthus annuus* L.) in response to osmotic seed priming. *Seed SciTechnol.,*31: 199-206.

Naaz, S. and Pandey, S.N. 2010. Effects of industrial water on heavy m*et al* accumulation, growth and biochemical responses of lettuce (*Lactuca sativa* L.). *Journal of Environmental Biology*, 31: 273–276.

Nachit, M.M. and Elouafi, I. 2004. Durum adaptation in the Mediterranean dryland: breeding, stress physiology, and molecular markers. In "*Challenges and strategies for dryland agriculture*" Eds. S.C. Rao, and J. Ryan, CSSA Special Publication 32. Madison, Wisconsin, USA: Crop Science Society of America Inc., American Society of Agronomy Inc. 203–218.

Nagajyoti, P.C., Lee, K.D. and Sreekanth, T.V.M. 2010. Heavy *metals*, occurrence and toxicity for plants: a review. *Environ Chem Lett.,*8: 199–216.

Nagarajan, V.K. and Smith, A. P. 2012. Ethylene's role in phosphate starvation signaling: more than just a root growth regulator. *Plant and Cell Physiology*, 53(2): 277-286.

Nakabayashi, K., Okamoto, M., Koshiba, T., Kamiya, Y., and Nambara, E. 2005. Genome-wide profiling of stored mRNA in *Arabidopsisthaliana* seed germination: epigenetic and genetic regulation of transcription in seed. *Plant J.*, 41: 697–709.

Nakashima, K., Fujita, Y., Kanamori, N., Katagiri, T., Umezawa, T., Kidokoro, S., Maruyama, K., Yoshida, T., Ishiyama, K., Kobayashi, M., Shinozaki, K. and Yamaguchi-Shinozaki, K. 2009. Three *Arabidopsis* SnRK2 protein kinases, SRK2D/SnRK2.2, SRK2E / SnRK2.6/ OST1 and SRK2I/SnRK2.3, involved in ABA signaling are essential for the control of seed development and dormancy. *Plant Cell Physiol.*, 50: 1345–1363.

Nakamura I., Murayama S. and Tobita S. 2002. Effect of NaCl on the photosynthesis, water relations, and free proline accumulation in the wild *Oryza* species. *Plant. Prod. Sci.*, 5: 305-310.

Nakano, Y. 2007. Response of tomato root systems to environmental stress under soilless culture. *Japan Agricultural Research Quarterly*, 41: 7–15.

Navarro, A., Vicente, M.J., Martínez-Sánchez, J.J., Franco, J.A., Fernández, J.A. and Bañón, S. 2008. Influence of deficit irrigation and paclobutrazol on plant growth and water status in *Lonicera implexa* seedlings. *Acta Horticulturae*, 782: 299–304.

Navarro, L., Bari, R., Achard, P., Lison, P., Nemri, A., Harberd, N.P. and Jones, J.D. 2008. DELLAs control plant immune responses by modulating the balance of jasmonic acid and salicylic acid signaling. *Curr Biol.*, 18: 650–655.

Natarajan, S. and Kuehny, J.S. 2008. Heat tolerance and heat preconditioning of *Salvia splendes*. *Acta Horticulturae*, 766: 175–182.

Neves, N.R., Oliva, M.A., Centeno, D.C., Costa, A.C., Ribas, R.F. and Pereira, E.G. 2009. Photosynthesis and oxidative stress in the restinga plant species *Eugenia uniflora* L. xposed to simulated acid rain and iron ore dust deposition: potential use in nvironmental risk assessment. *Sci Total Environ.*, 407: 3740–3745.

Neumann, P.M. 2003. Transient resistance to root growth inhibition by moderate water stress: A possible explanation. In "*Roots: The Dynamic Interface between Plants and the Earth*," Ed. J. Abe, Kluwer Academic Press, Dordrecht, The Netherlands, pp. 439–443.

Nicoll, B.C., Gardiner, B.A. and Peace, A.J. 2008. Improvements in anchorage provided by the acclimation of forest trees to wind stress. *Forestry*, 81: 389–398.

Nieuwland, J., Maughan, S., Dewitte, W., Scofield, S., Sanz, L. and Murray, J.A. 2009. The Dtype cyclin CYCD4;1 modulates lateral root density in *Arabidopsis* by affecting the basal meristem region. *Proc. Nat. Acad. Sci. USA,* 106(52): 22528-22533.

Nii, N., Pan, C.X., Ogawa, A,Y. and Cui, S. 2004. Anatomical features of the cell wall growth in the cortical cells outside the endodermis and the development of the *Casparian* strip in loquat trees. *Journal of the Japanese Society for Horticultural Science*, 73: 411–414.

Niu, G., Rodriguez, D.S., Rodriguez, L. and Mackay, W. 2007. Effect of water stress on growth and flower yield of big bend bluebonnet. *Hort. Technology*, 17: 557–560.

Niu, G., Rodriguez, D.S. and Mackay, W. 2008. Growth and physiological responses to drought stress in four oleander clones. *J. Am. Soc. Hort. Sci.,* 133: 188–196.

Niu, G. and Rodriguez, D.S. 2009. Growth and physiological responses of four rose rootstocks to drought stress. *J. Am. Soc. Hort. Sci.*, 134: 2002–209.

Nishimura, N., Kitahata, N., Seki, M., Narusaka, Y., Narusaka, M., Kuromori, T., Asami, T., Shinozaki, K. and Hirayama, T. 2005. Analysis of ABA Hypersensitive Germination2 revealed the pivotal functions of PARN in stress response in *Arabidopsis*. *Plant J.*, 44: 972–984.

Nishimura, N., Sarkeshi, A., Nito, K., Park, S.Y., Wang, A., Carvalho, P.C., Lee. S., Caddell, D.F., Cutler, bv S.R., Chory, J., Yates, J.R. and Schroeder, J.I. 2010. PYR/PYL/RACR family members are major in vivo ABI1 protein phosphatase 2C interacting proteins in *Arabidopsis*. *Plant J.,*. 61: 290–299.

Nonogaki, H. 2008a. Seed germination and reserve mobilization. In "*Encyclopedia of Life Sciences. Chichester*": John Wiley and Sons, Ltd.

Nonogaki, H. 2008b. Repression of transcription factors by microRNA during seed germination and post germination: another level of molecular repression in seeds? *Plant Signal Behav., 3:* 65–67

Ochoa, J., Bañón, S., Fernández, J. A., Franco, J.A. and González, A. 2003. Influence of cutting position and rooting media on rhizogenesis in oleander cuttings. *Acta Horticulturae*, 608: 101–106.

Ogawa, D, Abe, K, Miyao, A, Kojima, M, Sakakibara, H, Mizutani, M, Morita, M., Toda, Y., Hobo, T., Sato, Y., Hattori, T., Hirochika, H. and Takeda, S. 2011. RSS1 regulates the cell cycle and maintains meristematic activity under stress conditions in rice. *Nature Communications*, 2: 278

Oge, L., Bourdais, G., Bove, J., Collet, B., Godin, B., Granier, F., Boutin, J.P., Job, D., Jullian, M. and Grappin, P. 2008. Protein repair L-isoaspartyl methyltransferase1 is involved in both seed longevity and germination vigor in *Arabidopsis*. *Plant Cell*, 20: 3022–3037.

Okamoto, M., Tatematsu, K., Matsui, A., Morosawa, T, Ishida, J, Tanaka, M, *et al.* 2010. Genome-wide analysis of endogenous abscisic acid-mediated transcription in dry and imbibed seeds of *Arabidopsis* using tiling arrays. *Plant J.* 62: 39–51.

Okcu, G, .Kaya, M.D. and Atak, M. 2005. Effects of salt and drought stresses on germination nd seedling growth of pea (*Pisum sativum* L.). *Turk J Agric Forest.*,29: 237-242.

Omar, M.A., Omar, F.A. and Samarrai, S.M. 1982. Effect of different soil treatments on tomato plants grown in Wadi Fatima soil. B. Effect of salinity treatments. *Technical Report, Faculty of Meteorol-Environ and Arid Land Agric*, pp 26.

Ostonen, I., Puttsepp, U., Biel, C., Alberton, O., Bakker, M.R., Lohmus, K., Majdi, H., Metcalfe, D., Olsthoorn, A.F.M., Pronk, A.,Vanguelova, E.,Weih, M. and Brunner, I. 2007. Specific root length as an indicator of environmental change. *Plant Biosystems*, 141: 426–442.

Ozdener, Y. and Guray Kutbay, H. 2008. Effect of salinity and temperature on the germination of *Spergularia marina* seeds and ameliorating effect of ascorbic and salicylic acids. *J. Environ. Biol.*, 29: 959-964.

Pan, C.X., Nakao, Y. and Nii, N. 2006. Anatomical development of *phi* thickening and the Casparian strip in loquat roots. *Journal of the Japanese Society for Horticultural Science*, 75: 445–449.

Papadopoulos, A.P. and H. Tiessen. 1987. Root and air temperature effects on the elemental composi-tion of tomato. *J. Amer. Soc. Hort. Sci.*, 112: 988–993.

Park, S.Y., Fung, P., Nishimura, N., Jensen, D.R., Fujii, H., Zhao, Y., Lumba, Y., Sanitago, J., Rodriques, A., Chow, T.F., Alfred, S.E., Bonetta, D., Finkelstein, R., Provart, N.J., Desveaux, D., Rodriquez, P.L., McCourt, P., Zhu, J.K., Schroeder, J., Volkman, B.F. and Cutler, S.R. 2009. Abscisic acid inhibits type 2C protein phosphatases via the PYR/PYL family of START proteins. *Science*, 324:1068–1071.

Passardi, F., Tognolli, M., De Meyer, M., Penel, C. and Dunand, C. 2006. Two cell wall associated peroxidases from Arabidopsis influence root elongation. *Planta*,.223(5): 965-974.

Patade, V.Y., Maya, K., and Zakwan, A. 2011. Seed priming mediated germination improvement and tolerance to subsequent exposure to cold and salt stress in capsicum. *Res. J. Seed Sci.*, 4(3):125 -136.

Patade, V.Y., Bhargava, S., and Suprasanna, P. 2009. Halopriming imparts tolerance to salt and PEGinduced drought stress in sugarcane. *Agric. Ecosyst. Environ*. 134: 24-28.

Patade V.Y., Maya K. and Zakwan, A. 2011. Seed priming mediated germination improvement and tolerance to subsequent exposure to cold and salt stress in capsicum. *Res. J. Seed Sci.*, 4(3): 125 -136.

Peleg, Z., Reguera, M., Tumimbang, E., Walia, H. and Blumwald, E. 2011. Cytokinin-mediated source /sink modifications improve drought tolerance and increase grain yield in rice under water-stress. *Plant Biotechol J.*, 9(7): 747-758.

Peres, A., Churchman, M. L, Hariharan, S, Himanen, K, Verkest, A, Vandepoele, K,, Magyar, Z., Van Der Schueren, E., Beemster, G.T., Frankard, V., Larkin, J.C., Inze, D. and De Veylder, L. 2007. Novel plant-specific cyclin-dependent kinase inhibitors induced by biotic and abiotic stresses. *The Journal of biological chemistry*, 282(35): 25588-25596.

Phene, C.J., K.R. Davis, R.B. Hutmacher, B. Bar-Yosef, D.W. Meek, and J. Misaki. 1991. Effect of high frequency surface and subsurface drip irrigation on root distribution of sweet corn. *Irrigation Sci.*, 12: 135–140.

Planchais, S., Samland, A.K. and Murray, J.A. 2004. Differential stability of *Arabidopsis* Dtype cyclins: CYCD3;1 is a highly unstable protein degraded by a proteasome-dependent mechanism. *The Plant Journal*, 38(4): 616-625.

Popova, L., Mslenkova', L., Yordanova, R., Ivanova, A., Krantev, A., Szalai, G. and Janda, T. 2009. Exogenous treatment with salicylic acid attenuates toxicity in pea seedlings. *Plant Physiol Biochem.,*47: 224-231.

Postaire, O., Tournaire-Roux, C., Grondin, A., Boursiac, Y., Morillon, R., Schäffner, A.R. and Maurel, C. 2010. A PIP1 aquaporin contributes to hydrostatic pressure-induced water transport in both the root and rosette of *Arabidopsis*. *Plant Physiol.,* 152: 1418–1430.

Pradet, A. and J.L. Bomsel. 1978. Energy metabolism in plants under hypoxia and anoxia. In "*Plant life in anaerobic environments*", Eds. D.D. Hook and R.M.M. Crawford, Ann Arbor Science, Ann Arbor, Mich., pp. 89–118.

Prado, F. E., Boero, C., Gallarodo, M. and Gonzalez, J.A. 2000. Effect of NaCl on germination, growth and soluble sugar content in *Chenopodium quinoa* willd seeds. *Bot. Bull. Acad. Sin.*, 41:, 27–34.

Penfield, S., Li, Y., Gilday, A.D., Graham, S., and Graham, I.A. 2006. *Arabidopsis* ABA INSENSITIVE4 regulates lipid mobilization in the embryo and reveals repression of seed germination by the endosperm. *Plant Cell,*18: 1887–1899.

Perruc, E., Kinoshita, N., and Lopez-Molina, L. 2007. The role of chromatin-remodeling factor PKL in balancing osmotic stress responses during *Arabidopsis* seed germination. *Plant J.,* 52: 927–936.

Pires, N. D. and Dolan, L. (2012). Morphological evolution in land plants: new designs with old genes. *Philosophical Transactions of the Royal Society B: Biological Sciences,* 367(1588):, 508-18.

Poysa, V.W., C.S. Tan, and J.A. Stone. 1987. Flooding stress and the root development of several tomato genotypes. *Hort Science*, 22: 24–26.

Pratap, V. and Sharma, Y.K. 2010. Impact of osmotic stress on seed germination and seedling growth in black gram (*Phaseolus Mungo*). *J. Environ. Biol.*, 31: 721-726.

Puppala, N., Fowler, J.L., Poindexter, L. and Bhardwaj, H.L. 1999. Evaluation of salinity tolerance of canola germination. In "*Perspectives on new crops and new uses*", Ed. J. Janick, pp. 251-253.

Quesada, V., Garcý'a-Martý'nez, S., Piqueras, P., Ponce, M.R., and Micol, J.L. 2002. Genetic architecture of NaCl tolerance in *Arabidopsis*. *Plant Physiol.*, 130: 951–963.

Qi, X., Wu, Z., Li, J., Mo, X., Wu, S., Chu, J. and Wu, P. 2007. AtCYT-INV1, a neutral invertase, is involved in osmotic stress-induced inhibition on lateral root growth in *Arabidopsis. Plant Mol Biol.*, 64(5): 575-587.

Qin, L., He, J., Lee, S.K. and Dodd, I.C. 2007. An assessment of the role of ethylene in mediating lettuce (*Lactuca sativa*) root growth at high temperatures. *Journal of Experimental Botany*, 58: 3017–3024.

Rad, M. and Rad, J. 2013. Effects of Abiotic Stress Conditions on Seed Germination and Seedling Growth of Medical Plant, Hyssop (*Hyssopus officinalis* L.). *Int. J. Agric. Crop Sci.*, 5(21): 2593-2597.

Radziejwoski, A., Vlieghe, K., Lammens, T., Berckmans, B., Maes, S., Jansen, M.A., Knappe, C., Albert, A., Seiditz, H.K., Bahnweg, G., Inze, D. and De Vevlder, L. 2011. Atypical E2F activity coordinates PHR1 photolyase gene transcription with endore duplication onset. *The EMBO journal*, 30(2): 355-363.

Rajjou, L., Gallardo, K., Job, C., and Job, D. 2006. Proteome analysis for the study of developmental processes in plants. In: Plant Proteomics. C. Finnie, ed. (Blackwell Publishing, Oxford), pp. 151–184.

Rahnama, A., Munns, R., Poustini, K. and Watt, M. 2011. A screening method to identify genetic variation in root growth responses to a salinity gradient. *J. Expta. Bot.,* 62: 69–77.

Ramden, H.A., Niemi, S.A. and Hadathi, Y.K.A. 1986. Salinity and seed germination of corn and soyabean. *Iraqui J Agri Sci.*, 4(2): 97-102.

Ramagopal, S. 1990. Inhibition of seed germination by salt and its subsequent effect on embryonic protein synthesis in barley. *J. Plant Physiol.*, 136: 621-625.

Raskin, I. and Ensley, B.D. 2000. Phyto-remediation of Toxic*Metals*. Using Plants to Clean up the Environment. Wiley and Sons, NewYork. Raza SH.

Raven, J. A, and Edwards, D. (2001). Roots: evolutionary origins and biogeochemical signifi cance. *Journal of Experimental Botany,* 52 (Root Special Issue): 381-401.

Raven, Peter, H., Ray, F., Evert; S., and Eichhorn, E. 2005. In "Biology of Plants, 7th Edition", New York: W.H. Freeman and Company Publishers. pp. 504–508.

Reichheld, J.P., Lardon, T.V.F., Van Montagu, M. and Inzé, D. 1999. Specific checkpoints regulate plant cell cycle progression in response to oxidative stress. *The Plant Journal*, 17(6): 647-656.

Remans, T., Nacry, P., Pervent, M., Filleur, S., Diatloff, E., Mounier, E., Tillard, P., Forde, B.G. and Goion, A. 2006. The *Arabidopsis* NRT1.1 transporter participates in the signaling pathway triggering root colonization of nitrate-rich patches. *Proceedings of the Nat. Acad. Sci. USA*, 103(50): 19206-19211.

Reyes, E. and P.H. Jennings. 1994. Response of cucumber (*Cucumis sativus* L.) and squash (*Cucurbita pepo* L. var. *melopepo*) roots to chilling stress during early stages of seedling development. *J. Amer. Soc. Hort. Sci.*, 119: 964–970.

Reyes, J.L., and Chua, N.H. 2007. ABA induction of miR159 controls transcript levels of two MYB factors during *Arabidopsis* seed germination. *Plant J.*, 49: 592–606.

Richards, D.E., King, K.E., Ait-ali, T. and Harberd, N.P. 2001. How Gibberellin regulates plant growth and development: A molecular genetic analysis of gibberellin signaling. *Ann. Rrev. Pl. Physiol. Pl. Mol. Biol.,* 52: 67-88.

Riou-khamlichi, C, Menges, M, Healy, J. M, & Murray, J. A. 2000. Sugar control of the plant cell cycle: differential regulation of Arabidopsis D-type cyclin gene expression. *Mol Cell Biol.,* 20(13): 4513-4521.

Robert-Seilaniantz, A., Navarro, L., Bari, R., and Jones, J.D. 2007. Pathological hormone imbalances. *Curr Opin Plant Biol.*, 10: 372–379.

Romero, P., Fernández-Fernández, J.I. and Martinezcutillas, A. 2010. Physiological thresholds for efficient regulated deficit-irrigation management in wine grapes grown under semi-arid conditions. *American Journal of Enology and Viticulture*, 61: 300–312.

Ron, D., Chen, C.H., Caldwell, J., Jamieson, L., Orr, E., and Mochly-Rosen, D. 1994. Cloning of an intracellular receptor for protein kinase C: a homolog of the b-subunit of G-proteins. *Proc Natl Acad Sci USA*, 91: 839–843.

Rose, D.A., Ghamarnia, H.M. and Gowing, J.W. 2010. Development and performance of wheat roots above shallow saline groundwater. *Aus. J. Soil Res.,* 48: 659–667.

Rost, T.L. and H. Toriyama. 1991. A comparative analysis of the root systems of light- and dark-grown seedlings of pea (*Pisum sativum* cv. Alaska). *Plant Sci.*, 75: 117–121.

Rutto, K.L., Mizutani, F. and Kadoya, K. 2002. Effect of root-zone flooding on mycorrhizal and non-mycorrhizal peach (*Prunus persica* Batsch) seedlings. *Scientia Horticulturae*, 94: 285–295.

Ruzicka, K., Ljung, K., Vanneste, D., Podhorska, R., Beeckman, T., Frimi, J. and Benkova, E. 2007. Ethylene regulates root growth trough effects on auxin biosynthesis and transport dependent auxin distribution. *The Plant Cell*, 19: 2197–2212.

Rouhi, H.R., Aboutalebian, M.A., and Sharif-zadeh, F. 2011. Effects of hydro and osmopriming on drought stress tolerance during germination in four grass species. *Int. J. Agrisi.*, 1(2): 107-114.

Rowse, H.R. 1974. The effect of irrigation on the length, weight, and diameter of lettuce roots *Plant Soil*, 40: 381–391.

Russell, R.S. 1977. Plant root systems: Their function and interaction with the soil. McGraw-Hill, Berkshire.

Sadeghi, H., Khazaei, F., and land Sh, S. 2011 Effect of seed osmopriming on seed germination behavior and vigor of soyeBean (*Glycin max* L..). *APKN J. Agric. Biol. Sci.*, 6(1):

Saglio, P.H., M.C. Drew, and A. Pradet. 1988. Metabolic acclimation to anoxia induced by low (2-4 kPa partial pressure) oxygen pretreatment (hypoxia) in root tips of Zea mays. *Plant Physiol.*, 86: 61–66.

Saglam, S., Day, S., Kaya, G., and Gurbuz, A. 2010 Hydropriming increases germination of lentil (*Lens cutinaris* Medik) under water stress. *Notulae Scientia Biologicae.* 2(2): 103- 106.

Samson, B.K., Hasan, M. and Wade, L.J. 2002. Penetration of hardpans by rice lines in the rain-fed lowlands. *Field Crops Research*, 76: 175–188.

Sánchez-Blanco, M.J., Ferrández, T., Navarro, A., Bañón, S. and Alarcón, J.J. 2004. Effects of irrigation and air humidity preconditioning on water relations, growth and survival of *Rosmarinus officinalis* plants during and after transplanting. *Journal of Plant Physiology*, 161: 1133–1142.

Sanchez-calderon, L., Lopez-bucio, J., Chacon-lopez, A., Cruz-ramirez, A., Nieto-jacobo, F., Dubrovsky, J.G, and Herrera-Estrella, L. 2005. Phosphate starvation induces a determinate developmental program in the roots of *Arabidopsis thaliana. Plant & Ccell Physiol.*, 46(1): 174-84.

Sánchez-calderón, L., López-bucio, J., Chacón-lópez, A., Gutiérrez-ortega, A., Hernández-abreu, E., and Herrera-estrella, L. 2006. Characterization of low phosphorus insensitive mutants reveals a crosstalk between low phosphorus-induced determinate root development and the activation of genes involved in the adaptation of *Arabidopsis* to phosphorus deficiency. *Pl. Physiol.*, 140(3): 879-889.

Sanders, D.C., T.A. Howell, M.M.S. Hile, L. Hodges, and C.J. Phene. 1989. Tomato root development affected by traveling trickle irrigation rate. *Hort Science*, 24: 930–933.

Sangakkara, U.R., Amarasekera, P. and Stamp, P. 2010. Irrigation regimes affect early root development, shoot growth and yields of maize (*Zea mays* L.) in tropical minor seasons. *Plant, Soil and Environment*, 56: 228–234.

Santiago, J., Dupeux, F., Round, A., Antoni, R., Park, S.-Y., Jamin, M., Sean, R.C., Rodriguez, P.L., Marquez, J.A. 2009. The abscisic acid receptor PYR1 in complex with abscisic acid. *Nature*, 462: 665–668.

Santos,T.P., Lopes, C.M., Rodrigues, M.L., Souza, C.R., Silva, J.R., Maroco, J.P., Pereira, .S. and Chaves, M.M. 2007. Effects of deficit irrigation strategies on cluster microclimate for improving fruit composition of Moscatel field-grown grapevines. *Scientia Horticulturae*, 112: 321–330.

Satti, S.M. and Al-Yahyai, R.A. 1995. Salinity tolerance in tomato: Implications of potassium, calcium and phosphorus. *Soil Sci. Plant Anal.*, 26: 2749-2760.

Schaller, G. 1987. pH changes in the rhizosphere in relation to the pH-buffering of soils. *Plant Soil*,97:439–444.

Schlossberg, M.J., Karnok, K.J. and Lanfry, G. 2002. Estimation of viable root-length density of heat-tolerant 'Crenshaw' and 'L93' creeping bentgrass by an accumulative degree-day model. *Journal of the American Society for Horticultural Science*, 127: 224–229.

Schmidt, W. and Schikora, A. 2001. Different pathways are involved in phosphate and iron stress-induced alterations of root epidermal cell development. *Plant Physiol.*, 125(4): 2078-2084.

Schopfer, P. 2006. Biomechanics of plant growth. *Am. J. Bot.*, 93: 1415–1425.

Shakirova, F.M., Sakhabutdinova, A.R., Bezrukova, M.V., Fatkhutdinova, R.A., and Fatkhutdinova, D.R. 2003. Changes in the hormonal status of wheat seedlings induced by salicylic acid and salinity. *Plant Sci.*, 164: 317–322.

Sharma, P. and Dubey, R.S. 2007. Involvement of oxidative stress and role of anti-oxidative defense system in growing rice seedlings exposed to toxic concentrations of aluminum. *Plant cell reports*, 26(11): 2027-2038.

Shao, G.C., Liu, N., Zhang, Z.Y., Yu, S.E. and Chen, C.R. 2010. Growth, yield and water use efficiency response of greenhouse grown hot pepper under time-space deficit irrigation. *Scientia Horticulturae*, 126: 172–179.

Sharp, R.E., Poroyko, V., Hejlek, J.G., Spollen, W.G., Springer, G.K., Bohnert, H.J. and Nguyen, H.T. 2004. Root growth maintenance during water deficits: Physiology to functional genomics. *Journal of Experimental Botany*, 55: 2343–2351.

Shi, H., Bressan, R., Hasegawa, P.M., and Zhu, J.-K. 2005. Sodium. In "*Plant Nutritional Genomics*". Eds. M. Broadlay and P. White, Blackwell Publishing, London, pp. 127–149.

Shi, K., Ding X.T., Dong, D.K., Zhou, Y.H. and Yu, J.Q. 2008. Root restriction-induced limitations to photosynthesis in tomato (*Lycopersicon esculentum* Mill.) leaves. *Scientia Horticulturae*, 117: 197–202.

Shober, A.L., Moore, K.A., Wiese, C., Scheiber, S.M., Gilman, E.F., Paz, M., Brennan, M.M. and Vyapari, S. 2009. Post transplant irrigation frequency affects growth of container grown sweet viburnum in three hardiness zones. *Hort. Science*, 44: 1683–1687.

Shokohifard, G., Sakagam, K.H. and Matsumoto, S. 1989. Effect of amending materials on growth of radish plant in salinised soil. *J. Plant Nutr.*, 12: 1195-1294.

Silber, A., Levi, M., Cohen, M., David, N., Shataynmetz, Y. and Assouline, S. 2006. Response of *Leucadendron* 'Safari Sunset' to irrigation and fertilisation levels. *Journal of Horticultural Science & Biotechnology*, 81: 355–364.

Sidari, M., Muscolo, A., Anastasi, U., Pretti, G. and Santonoceto, C. 2007. Response of four genotypes of lentil to salt stress conditions. *Seed Sci. Technol.*, 35: 497-503.

Sindhoj, E., Hansson, A.C., Andren, O., Katterer, T., Marissink, N. and Pettersson, R. 2000. Root dynamics in a semi-natural grassland in relation to atmospheric carbon dioxide enrichment, soil water and shoot biomass. *Plant and Soil*, 223: 253–263.

Skopelitis D.S., Paranychianakis N.V. and Paschalidis K.A. 2006. Abiotic stress generates ROS that signal expression of anionic glutamate dehydrogenases to form glutamate for proline synthesis in tobacco and grapevine. *Plant. Cell*. 18: 2767- 2781.

Sliwinska, E., Bassel, G.W., and Bewley, J.D. 2009. Germination of *Arabidopsisthaliana* seeds is not completed as a result of elongation of the radicle but of the adjacent transition zone and lower hypocotyl. *J. Expt. Bot.*, 60: 3587–3594.

Smith, D., 1967. Carbohydrates in grasses. II. Sugar and fructosan composition of the stem bases of bromegrass and timothy at several growth stages and in different plant parts at anthesis. *Crop Sci.*, 7: 62–67.

Smith, P.T. and Comb, B.G. 1991. Physiological and enzymatic activity of pepper seeds (*Capsicum annuum*) during priming. *Physiol. Plant*, 82: 71-78.

Snapp, S.S. and C. Shennan. 1992. Effects of salinity on root growth and death dynamics of tomato, *Lycopersicon esculentum* Mill. *New Phytol.*, 121: 71–79.

Snyman, H.A. 2004. Effects of various water application strategies on root development of *Opuntia ficus-indica* and *Opuntia robusta* under greenhouse growth conditions. *Journal of the Professional Association for Cactus Development*, 6: 35–61.

Soltani A., Gholipoor M. and Zeinali, E. 2006. Seed reserve utilization and seedling growth of wheat as affected by drought and salinity. *Environmental and Experimental Botany*. 55: 195-200.

Sperry, J.S., Stiller, V. and Hacke, U.G. 2002. Soil water uptake and water transport through root systems. In "*Plant Roots: The Hidden Half.* 3rd Edition", Eds. Y. Waisel, A. Eshel, and U. Kafkafi, Marcel Dekker, New York, NY, USA. pp. 663–681.

Souza, V.L., De Almeida, A.A.F., Lima, S.G.C., Cascardo, J.C.D., Silva, D.C., Mangabeira, P.A. O. and Gomes, F.P. 2011. Morpho-physiological responses and programmed cell death induced by cadmium in *Genipa americana* L. *Biometals*, 24: 59–71.

Spollen, W.G, and Sharp, R.E. 1991. Spatial distribution of turgor and root growth at low water potentials. *Plant Physiol.,*96(2): 438-443.

Sriroth, K., Piyachomkwan, K., Santisopasri, V. and Oates, C.G. 2001. Environmental conditions during root development: Drought constraint on cassava starch quality. *Euphytica*, 120: 95–101.

Steber, C.M., and McCourt, P. 2001. A role for brassinosteroids in germination in *Arabidopsis*. *Plant Physiol.*, 125: 763–769.

Stoffella, P.J., R.F. Sandsted, R.W. Zobel, and W.L. Hymes. 1979a. Root characteristics of black beans: I. Relationship of root size to lodging and seed yield. *Crop Sci.*, 19: 823–826.

Stoffella, P.J., R.F. Sandsted, R.W. Zobel, and W.L. Hymes. 1979b. Root characteristics of black beans: II. Morphological differences among genotypes. *Crop Sci.*, 19: 826–830.

Stoffella, P.J., R.F. Sandsted, R.W. Zobel, and W.L. Hymes. 1981. Root morphological characteristics of kidney beans as influenced by within-row spacing. *Hort Science*, 16: 543–548.

Stoffella, P.J. 1983. Root morphological characteristics of field-grown tomatoes. *Hort Science*, 18: 70–72.

Stoffella, P.J. and H.H. Bryan. 1988. Plant population influence growth and yield of bell peppers. *J. Amer. Soc. Hort. Sci.*, 113: 835–839.

Stoffella, P.J. and B.A. Kahn. 1986. Root system effects on lodging of vegetable crops. *Hort Science*, 21: 960–963.

Stoffella, P.J., M. Lipucci DiPaola, A. Pardossi, and F. Tognoni. 1988a. Root morphology and development of bell peppers. *Hort Science*, 23: 1074–1077.

Stoffella, P.J., A. Pardossi, and F. Tognoni. 1988b. Temperature and seed treatment effects on taproot growth of young bell peppers. *Adv. Hort. Sci.*, 2: 8–10.

Stoffella, P.J., M. Lippucci DiPaola, A. Pardossi, and F. Tognoni. 1991. Rhizosphere pH influences early root morphology and development of bell peppers. *Hort Science*, 26: 112–114.

Stoffella, P.J., M. Lipucci DiPaola, A. Pardossi, and F. Tognoni. 1992. Seedling root morphology and development and shoot growth after seed priming or pregermination of bell peppers. *Hort Science*, 27: 214–215.

Stone, S.L., Williams, L.A., Farmer, L.M., Vierstra, R.D., and Callis, J. 2006. KEEP ON GOING, a RING E3 ligase essential for *Arabidopsis* growth and development, is involved in abscisic acid signaling. *Plant Cell*, 18: 3415–3428.

Spollen, W.G, and Sharp, R.E. 1991. Spatial distribution of turgor and root growth at low water potentials. *Plant Physiol.,*96(2): 438-443.

Sun, F., Zhang, W., Hu, H., Li, B., Wang, Y., Zhao, Y., Li, K., Liu, M. and Li, X. 2008. Salt modulates gravity signaling pathway to regulate growth direction of primary roots in *Arabidopsis. Plant Physiology*, 146(1):,178-188.

Suralta, R.R., Inukai, Y. and Yamauchi, A. 2010. Dry matter production in relation to root plastic development, oxygen transport, and water uptake of rice under transient soil moisture stresses. *Plant and Soil*, 332: 87–104.

Subbarao, G.V., Nam, N.H., Chauhan, Y.S. and Johansen, C. 2000. Osmotic adjustment, water relations and carbohydrate remobilization in pigeonpea under water deficits. *J Plant Physiol.*, 157: 651–659.

Spollen, W.G, and Sharp, R.E. 1991. Spatial distribution of turgor and root growth at low water potentials. *Plant Physiol.,*96(2): 438-443.

Sun, F., Zhang, W., Hu, H., Li, B., Wang, Y., Zhao, Y., Li, K., Liu, M. and Li, X. 2008. Salt modulates gravity signaling pathway to regulate growth direction of primary roots in *Arabidopsis. Plant Physiology*, 146(1):,178-188.

Svistoonoff, S., Creff, A., Reymond, M., Sigoillot-claude, C., Ricaud, L, Blanchet, A., Blanchet, A., Nussaume, L. and Desnos, T. 2007. Root tip contact with low-phosphate media reprograms plant root architecture. *Nat Genet.*, 39(6): 792-796.

Tahir, I.S.A. and Nakata, N. 2005. Remobilization of nitrogen and carbohydrate from stems of bread wheat in response to heat stress during grain filling. *J Agron Crop Sci.*, 191: 106-115.

Taiz, L. and Zeiger, E. 2006. Plant Physiology, 4rd edn.Sinauer Associates.

Takahashi, N., Yamazaki, Y., Kobayashi, A., Higashitani, A., and Takahashi, H. 2003. Hydrotropism interacts with gravitropism by degrading amyloplasts in seedling roots of *Arabidopsis* and radish. *Plant Physiol.,* 132(2): 805-10.

Talanova, V.V. and Titov, A.F. 1994. Endogenous abscisic acid content in cucumber leaves under the influence of unfavourable temperatures and salinity. *J. Expt. Bot.,* 45(7): 1031-1033.

Tani, T., Kudoh, H. and Kachi, N. 2003. Responses of root length/leaf area ratio and specific root length of an understory herb, *Pteridophyllum racemosum*, to increases in irradiance. *Plant and Soil*, 255: 227–237.

Tangpremsri, T., Fukai, S. and Fischer, K.S. 1995. Growth and yield of Sorghum lines extracted from a population for differences in osmotic adjustments. *Aust J Agric Res.*, 46: 61–74.

Thiagarajan, V., Byrdin, M., Eker, A.P., Muller, P. and Brettel, K. 2011. Kinetics of cyclobutane thymine dimer splitting by DNA photolyase directly monitored in the UV. *Proc. Nat. Acad. Sci. USA*, 108(23): 9402-9407.

Thomson, A.M., Brown, R.A., Rosenberg, M.J., Izaurralde, R.C. and Benson, V. 2005. Climate change impacts for the conterminous USA: An integrated assessment. Part 3. Dry land production of grain and forage crops. *Clim. Change*, 69: 43-65.

Tremblay, N. and A. Gosselin. 1989. Growth and nutrient status of celery seedlings in response to nitrogen fertilization and NO_3 : NH_4 ratio. *Hort Science*, 24: 284–288.

Turhan, E. and Eris, A. 2005. Changes of micronutrients, dry weight, and chlorophyll contents in strawberry plants under salt stress conditions. *Communications in Soil Science and Plant Analyses*, 36: 1012–1028.

Tyburski, J.B., Dunajska, K. and Tretyn, A. 2009. Reactive oxygen species localization in roots of *Arabidopsis* thaliana seedlings grown under phosphate deficiency. *Pl. Growth Regul.,* 59: 27-36.

Tyburski, J.B., Dunajska, K. and Tretyn, A. 2010. A role for redox factors in shaping root architecture under phosphorus deficiency. *Plant Signal Behav.*, 5: 64-66.

Tyler, L., Thomas, S.G., Hu, J., Dill, A., Alonso, J.M., Ecker, J.R. and Sun, T-P. 2004. DELLA proteins and gibberellin-regulated seed germination and floral development in *Arabidopsis. Plant Physiol.,* 135: 1008–1019.

Tyree, M.T. 2003. Hydraulic properties of roots. In "*Root Ecology*",.Eds. H. De Kroon, and E.J.W. Visser, Springer-Verlag, Berlin, Germany. Pp. 125–150.

Ubeda-tomas, S, Federici, F, Casimiro, I, Beemster, G. T, Bhalerao, R, Swarup, R, Doemer, P., Haseloff, J. and Bennett, M.J. 2009. Gibberellin signaling in the endodermis controls *Arabidopsis* root meristem size. *Curr. Biol.,* 19(14):,1194-1199.

Umezawa, T., Nakashima, K., Miyakawa, T., Kuromori, T., Tanokura, M., Shinozaki, K. and Yamaguchi- Shinozaki, K. 2010. Molecular basis of the core regulatory network in ABA responses: sensing, signaling and transport. *Plant Cell Physiol.*, 51: 1821–1839.

Vadez, V., Kashiwagi, J., Krishnamurthy, L., Serraj, R., Sharma, K.K., Devi, J., Bhatnagar-Mathur, P., Hoisington, D., Chandra, S., Gaur, P.M., Nigam, S.N., Rupakula, A., Upadhyaya, H.D., Hash, C.T. and Rizvi, S.M.H. 2005. Recent advances in drought research at ICRISAT: Using root traits and rd29a::DREB1A to increase water use and water use efficiency in drought-prone areas. Poster presented at the Inter drought II conference, Rome 24-28 September

Vamerali, T., Saccomani, M., Bona, S., Mosca, G., Guarise, M. and Ganis, A. 2003. A comparison of root characteristics in relation to nutrient and water stress in two maize hybrids. *Plant and Soil*, 255: 157–167.

Varney, G.T., M.J. Canny, X.L. Wang, and M.E. McCully. 1991. The branch roots of *Zea*. I. First order branches, their numbers, sizes and division into classes. *Ann. Bot.*, 67: 357–364.

Vicente-agullo, F., Rigas, S., Desbrosses, G., Dolan, L., Hatzopoulos, P. and Grabov, A. 2004. Potassium carrier TRH1 is required for auxin transport in *Arabidopsis* roots. *The Pl. Journal*, 40(4): 523-535.

Vidoz, M.L., Loreti, E., Mensuali, A., Alpi, A. and Perata, P. 2010. Hormonal interplay during adventitiousroot formation in flooded tomato plants. *Plant Journal*, 63: 551–562.

Villar-Salvador, P., Planelles, R., Enríquez, E., Peñuelas Rubira, J. 2004. Nursery cultivation regimes, plant functional attributes, and field performance relationships in the Mediterranean oak *Quercus ilex* L. *Forest Ecology and Management*, 196: 257–266.

Villora, G., Moreno, D.A., Pulgar, G. and Romero, L. 2000. Yield improvement in zucchini under salt stress: Determining micronutrient balance. *Scientia Horticulturae*, 86: 175–183.

Vlot, A.C., Dempsey, D.A., and Klessig, D.F. 2009. Salicylic acid, a multifaceted hormone to combat disease. *Annu Rev Phytopathol.*, 47: 177–206.

Vyas, S.P., Kathju, S., Garg, B.K. and Lahiri, A.N. 1996. Activities of nitrate reductase and ammonia assimilating enzymes of moth bean under water stress. *Sci. Cult.*, 62: 213-214.

Vysotskaya, L.B., Korobova, A.V. and Kudoyarova, G.R. 2008. Abscisic acid accumulation in the roots of nutrient limited plants: Its impact on the differential growth of roots and shoots. *Journal of Plant Physiology*, 165: 1274–1279.

Waisel, Y. and Eshel, A. 2002. Functional diversity of various constituents of a single root system. In "*Plant Roots: The Hidden Half*. 3rd Edition". Eds. Y. Waisel, A. Eshel, and U. Kafkafi, Marcel Dekker, New York, NY, USA. pp. 157–174.

Walter, A., Spies, H., Terjung, S., Kusters, R., Kirchgessener, N. and Schurr, U. (2002). Spatio-temporal dynamics of expansion growth in roots: Automatic quantification of diurnal course and temperature response by digital image sequence processing. *Journal of Experimental Botany*, 53: 689–698.

Wample, R.L. and R.M. Reid. 1978. Control of adventitious root production and hypocotyl hypertrophy of sunflower (*Helianthus annum*) in response to flooding. *Physiol. Plant.*,44: 351–358.

Wang, E.L. and Smith, C.J. 2004. Modelling the growth and water uptake function of plant root systems: a review. *Australian Journal of Agricultural Research*, 55: 501–523.

Wang, Z., Cao, G., Wang, X., Miao, J., Liu, X., Chen, Z., Qu, L.J. and Gu, H. 2008a. Identification and characterization of COI1-dependent transcription factor genes involved in JA-mediated response to wounding in *Arabidopsis* plants. *Plant Cell Rep.*, 27: 125–135.

Wang X.S. and Han, J.G. 2009. Changes in proline content, activity, and active isoforms of antioxidative enzymes in two alfalfa cultivars under salt stress. *Agric. Sci. China*, 8: 431-440.

Wang, K., Liu, Y., Dong, K., Dong, J., Kang, J., Yang, Q., Zhou, H. and Sun, Y. 2011. The effect of NaCl on proline metabolism in *Saussurea amara* seedlings. *Afr. J. Biotechnol.*,10: 2886-2893.

Ward, J.T., Lahner, B., Yakubova, E., Salt, D.E, and Raghothama, K.G. 2008. The effect of iron on the primary root elongation of *Arabidopsis* during phosphate deficiency. *Plant Physiol.*, 147(3): 1181-1191.

Wasilewska, A., Vlad, F., Sirichandra, C., Redko, Y., Jammes, F., Valon, C., Frei dit Frey, N. and Leung, J. 2008. An update on abscisic acid signaling in plants and more. *Mol Plant*, 1: 198–217.

Weaver, J.E. 1926. Root development of field crops. McGraw-Hill, New York.

Weaver, J.E. and W.E. Bruner. 1927. Root development of vegetable crops. McGraw-Hill, New York.

Weitbrecht, K., Mu¨ller, K. and Leubner-Metzger, G. 2011. First off the mark: early seed germ ination.*J. Expt. Bot.*, 62(10): 3289-3330.

Welbaum, G.S., Tissaoui, T. and Bradford, K.J. 1991. Water relations of seed development and germination in muskmelon (*Cucumis melo* L.). III. Sensitivity of germination to water potential and abscisic acid during development. *Plant Physiol.*, 92: 1029-1037.

Westgate, M.E. and J.S. Boyer. 1985. Osmotic adjustment and the inhibition of leaf, root, stem, and silk growth at low water potentials in maize. *Planta*, 164: 540–549.

Weston, L.A. and B.H. Zandstra. 1989. Transplant age and N and P nutrition effects on growth and yield of tomatoes. *Hort Science*, 24: 88–90.

Wolters, H., and Jurgens, G. 2009. Survival of the flexible: hormonal growth control and adaptation in plant development. *Nat Rev Genet.*, 10: 305–317.

Widders, I.E. and O.A. Lorenz. 1982. Potassium nutrition during tomato plant development. *J. Amer. Soc. Hort. Sci.*, 107: 960–964.

Wien, H.C., P.L. Minotti, and V.P. Grubinger. 1993. Polyethylene mulch stimulates early root growth and nutrient uptake of transplanted tomatoes. *J. Amer. Soc. Hort. Sci.*, 118: 207–211.

Wilcox, G.E. and C.L. Pfeiffer. 1990. Temperature effect on seed germination, seedling root development and growth of several vegetables. *J. Plant Nutr.*, 13: 1393–1403.

Wu, Q.S. and Zou, Y.N. 2010. Beneficial roles of *Arbuscular mycorrhizas* in citrus seedlings at temperature stress. *Scientia Horticulturae*, 125: 289–293.

Xiong, L. and Zhu, J.K. 2002. Molecular and genetic aspects of plant responses to osmotic stress. *Plant Cell Environ.*, 25: 131–139.

Xiong, Y.C., Li, F.M, and Zhang, T. 2006. Performance of wheat crops with different chromosome ploidy: root-sourced signals, drought tolerance, and yield performance. *Planta,* 224(3): 710-718.

Xiong, J., Fu, G.F., Tao, L.X. and Zhu, C. 2010. Roles of nitric oxide in alleviating heavy m*et al* toxicity in plants. *Archives of Biochemistry and Biophysics*, 497: 13–20.

Yancey, P. H., Clark, M.E., Hand, S.C., Bowlis, R.D. and Somero, G.N. 1982. Living with water stress: evolution of osmolyte system. *Science,*217: 1214–1222.

Yang, Y., Han, C., Liu, Q., Lin, B. and Wang, J.W. 2008. Effect of drought and low light on growth and enzymatic antioxidant system of *Picea asperata* seedlings. *Acta Physiologiae Plantarum*, 30: 433–440.

Yang, F., Hu, J., Li, J., Wu, X. and Qian, Y. 2009. Chitosan enhances leaf membrane stability and antioxidant enzyme activities in apple seedlings under drought stress. *Plant Growth Regul.,*58:131–136.

Yazici I., Turkan I., Sekmen A.H. and Demiral, T. 2007. Salinity tolerance of purslane (*Portulaca oleracea* L.) is achieved b enhanced antioxidative system, lower level of lipid peroxidation and proline accumulation. *Environ. Exp. Bot.*, 61: 49-57.

Yeh, D.M., Lin, L. and Wright, C.J. 2000. Effects of mineral nutrient deficiencies on leaf development, visual symptoms and shoot-root ratio of *Spathiphyllum. Scientia Horticulturae*, 86: 223–233.

Yi, K., Wu, Z., Zhou, J., Du, L., Guo, L., Wu, Y. and Wu, P. 2005. OsPTF1, a novel transcription factor involved in tolerance to phosphate starvation in rice. *Plant Physiol.,* 138(4): 2087-2096.

Yin, P., Fan, H., Hao, Q., Yuan, X., Wu, D., Pang, Y., Yan, C., Li, W., Wang, J. and Yan, N. 2009. structural insights into the mechanism of abscisic acid signalling by PYL proteins. *Nat Struct Mol Biol.*, 16: 1230–1236.

Yousifi, N., Slama, I., Ghnaya, T., Savoure, A. and Abdelly, C. 2010. Effects of water deficit stress on growth, water relations and osmolyte accumulation in *Medicago truncatula* and *M. laciniata* populations. *Comptes Rendus Biologies.*, 333: 205–213.

Younis, S.A., Shahatha, H.A., Hagop, P.G. and Al-Rawi, F.I. 1991. Effect of salinity on the viability of rice seeds. *Plant growth, drought and salinity in the Arab region*, pp. 235-244.

Youssef, R.A. and M. Chino. 1989. Root-induced changes in the rhizosphere of plants. I. pH changes in relation to the bulk soil. *Soil Sci. Plant Nutr.*, 35: 461–468.

Yi, K., Wu, Z., Zhou, J., Du, L., Guo, L., Wu, Y. and Wu, P. 2005. OsPTF1, a novel transcription factor involved in tolerance to phosphate starvation in rice. *Plant Physiol.,* 138(4): 2087-2096.

Yu, H., Luo, N., Sun, L. and Liu, D. 2012. HPS4/SABRE regulates plant responses to phosphate starvation through antagonistic interaction with ethylene signalling. *J. Expt. Bot.,* 63(12): 4527-4538.

Yuan, K., Rashotte, A.M., and Wysocka-Diller, J.W. 2011. ABA and GA signaling pathways interact and regulate seed germination and seedling development undersalt stress. *Acta Physiol Plant,*3:261–271.

Yupsanis, T., Moustakas, M., Eleftherrious, P. and Damalanidou, K 1994. Protein Phosporylation dephosphorylation in alfalfa seeds germinating under salt stress. *J. Plant Physiol.*, 143: 234-240.

Zalba, P. and Peinemann, N. 1998. Salinity – fertility interactions on early growth of maize and nutrient uptake. *Edafologia*, 5: 29-39.

Zdeneck, P. 1999. Influence of nickel contaminated soils on lettuce and tomatoes. *ScientiaHort.*,81: 243-250.

Zentella, R., Zhang, Z.L., Park, M., Thomas, S.G., Endo, A., Murase, K., Fleet, C.M., Jikumaru, Y., Nambara, E., Kamiya, Y. and Sun, T.P. 2007. Global analysis of della direct targets in early gibberellin signaling in *Arabidopsis. Plant Cell*, 19: 3037–3057.

Zhang, H. and Forde, B.G. 2000. Regulation of Arabidopsis root development by nitrate availability. *Journal of Experimental Botany*, 51(342): 51-59.

Zhang, X., Garreton, V., and Chua, N.H. 2005. The AIP2 E3 ligase acts as a novel negative regulator of ABA signaling by promoting ABI3 degradation. *Genes Dev.,* 19: 1532–1543.

Zhang, S., Cai, Z., and Wang, X. 2009a. The primary signaling outputs of brassinosteroids are regulated by abscisic acid signaling. *Proc Natl Acad Sci USA*, 106: 4543–4548.

Zhang, H.M., Zhang, S.S., Meng, Q.M., Zou, J., Jiang, W.S. and Liu, D.H. 2009b. Effects of aluminium on nucleoli in root tip cells, root growth and the antioxidant defense system in *Vicia faba* L. *Acta Biologica Cracoviensia Series Botanica*, 51: 99–106.

Zheng, S.J. and Yang, J.L. 2005. Target sites of aluminium phytotoxicity. *Biologia Plantarum*, 49: 321–331.

Zeng, H., and Hua, Q.T.X. 2010. *Arabidopsis* brassinosteroid mutants det2-1 and bin2-1 display altered salt tolerance. *J Plant Growth Regul.*, 29: 44–52.

Zhou, Y. Q., Huang, S.Z., Yu, S.L., Gu, J.G., Zhao, J.Z., Han, Y.L. and Fu, J.J. 2010. The physiological response and subcellular localization of lead and cadmium in *Iris pseudacorus* L. *Ecotoxicology*, 19: 69–76.

Zhu, J.K. 2001. Cell signaling under salt, water and cold stress. *Curr. Opin. Plant Biol.*, 4: 401–406.

Zhu, J.K. 2002. Salt and drought stress signal transduction in plants. *Annu Rev Plant Biol.*, 53: 247–273.

Zhu, J.K. 2003. Regulation of ion homeostasis under salt stress. *Curr Opin Plant Biol.*, 6: 441–445.

Zhu, Q., Zhang, J., Gao, X., Tong, J., Xiao, L., Li, W. and Zhang, H. 2010. The *Arabidopsis* AP2/ERF transcription factor RAP2.6 participates in ABA, salt and osmotic stress responses. *Gene*, 457:, 1–12.

Zivack, M., Repkava, J., Olsavska, K. and Brestic, M. 2009. Osmotic adjustment in winter wheat varieties and its importance as a mechanism of drought tolerance. *Cereal Res Commun.*, 37: 569–572.

Zobel, R.W. 1975. The genetics of root development, In "*The development and function of roots*", Eds. J.G Torrey and D.F. Clarkson, Academic, London, pp. 261–275.

Zobel, R.W. 1986. Rhizogenetics (root genetics) of vegetable crops. *Hort Science*, 21: 956–959.

Zobel, R.W. 1989. Steady-state control and investigations of root system morphology, In "*Applications of continuous and steady-state methods to root biology*" Eds. J.G. Torrey and L.J. Winchip, Klewer Academics, Rosdrecht, The Netherlands, pp. 165–182.

Zoghlami, L.B., Djebali, W., Abbes, Z., Hediji, H., Maucourt, M., Moing, A., Brouquisse, R. and Chaibi, W. 2011. Metabolite modifications in *Solanum lycopersicum* roots and leaves under cadmium stress. *African Journal of Biotechnology*, 10: 567–579.

3

Nitrogen Fixation Under Salinity, Low and High Temperature and Moisture

Raj Bahadur*

Even though our atmosphere consists of nearly 80% of N_2, it is the limiting element in most ecosystems. Bio-available nitrogen can enter an ecosystem via atmospheric nitrogen deposition, mostly in the form of ammonium and nitrogen oxides. The source of this nitrogen is mostly anthropogenic for these compounds, but nitrogen oxides are also formed by lightning (about 10% of nitrogen deposition on the land). However, the largest input of nitrogen comes from nitrogen fixed by micro-organisms; called diazotrophs.These can be either free living or associated to plants from, mainly, the family of the Fabaceae. Nitrogen input from legumes can be a sustainable source of nitrogen in agricultural systems. The symbiotic micro-organisms in the root nodules, the rhizobia, can take up gaseous di-nitrogen from the air and 'fix' the nitrogen into molecules that can be assimilated by the plant. In return, the plant provides the rhizobia with a carbon source in the form of di-carboxylic acids. The enzyme responsible for the nitrogen fixation, nitrogenase, is irreversibly damaged when exposed to oxygen. The plant produces leg hemoglobin, a protein related to human hemoglobin to provide the rhizobia in the nodules with oxygen, often giving functional nodules a pink color. The symbiosis first appeared around 58 million years ago when the Papilionoideae (a subfamily of the Fabaceae) underwent genome duplication. Interestingly, the genes involved in the signaling seem to be originally involved in the symbiotic relationship with mycorrhiza. Thus, it seems that only through this whole genome duplication, genes became available that were free to fulfill this new function in the communication with rhizobia, thereby enabling the legumes to start this almost unique symbiotic relationship.

**Scientist (Plant Physiology) Narendra Dev University of Agriculture & Technology, Kumarganj Faizabad-224229 (U.P.) India*

General circulation models (GCM) predict that ambient temperature in the Mediterranean basin is going to increase 1–3^{0}C by the middle of this century (Iker *et al.*, 2007). Rising temperature will increase evapotranspiration rates and exacerbate low water availability problems commonly observed in Mediterranean environments, in which current annual potential evapotranspiration is often nearly twice the amount of rainfall (Sabate *et al.*, 2002). It has long been known that drought during the warmest period of the year is the major stress factor limiting plant species distribution and growth (Mooney, 1983). Most studies analyzing the effect of environments on plant growth have considered temperature and water availability separately. However, it is well known that the effect of combined stresses on plant growth causes alterations that cannot be predicted if they are analyzed alone, such as those resulting from synergistic and antagonistic phenomena (Valladares and Pearcy, 1997). For example, Chaves and Pereira (2004) observed that although photochemical processes are very resistant to low water availability, a down regulation of the photosynthetic apparatus occurs when plants are exposed simultaneously to drought and elevated temperature conditions. In addition, Kaiser (1987) suggested that the particular way in which water stress is imposed might be of special importance not only in understanding the field response to drought, but also in evaluating the plant's capacity to acclimate to stress. Withholding water is the most common method for short-term experiments, but sustained or cyclic water stress is needed to simulate more realistic responses to drought (Pennypacker *et al.*, 1990).

The inhibitory effects of elevated temperatures target photosynthesis, water status, and N_2 fixation in the case of N_2-fixing plants (Bushby, 1982). Many during summer drought decrease due to stomatal closure (Penuelas *et al.*, 1998; Llusi'a and Penuelas, 2000) and metabolic limitation (rubisco activity, RuBP regeneration, *etc*., Tretiach *et al.*, 1997; Larcher, 2000). CO_2 fixation may be affected by an altered excitation energy distribution associated with changes of thylakoid membrane fluidity and by damaged Calvin cycle enzyme activities. Enhanced CO_2/O_2 solubility coupled with diminished specificity of rubisco for CO_2 associated with plants grown under high temperature is reflected in enhanced photorespiration and decreased photosynthesis rates (Jordan and Ogren, 1984; Brooks and Farquhar, 1985). Besides, Stitt *et al.* (1991) explained reduced CO_2 fixation rates as a consequence of lower photo assimilate requirements, which lead to carbohydrate accumulation and the inhibition of Calvin cycle enzyme activities.

The sensitivity of legume-*Sinorhizobium* symbiosis to ambient stresses like temperature and low water availability is also well known (Hungria and Vargas, 2000; Streeter, 2003; Curtis *et al.*, 2004; Thomas *et al.*, 2004). Furthermore, some authors (Castellanos *et al.*, 1996; Thomas *et al.*, 2004) suggest that the effect of water deficit on N accumulation and N_2 fixation is larger than that on biomass accumulation. Alfalfa is temperate forage frequently exposed to low water availability, N-deficient soils and high temperature conditions. Photosynthesis supplies organic carbon to nodules, where it is used by the nitrogenase enzyme in the bacteroid inside nodules as a source of energy and reducing power to fix N_2 (Azcon-Bieto *et al.*, 2000). The products of N_2 fixation, either amides or ureids, are exported to the plant via the xylem (Schubert *et al.*, 1995; Walsh, 1995). Factors that increase photosynthesis increase N_2 fixation, while factors decreasing photosynthesis tend to decrease N_2 fixation. This coupling results in the regulation of nitrogenase activity in plants by photosynthesis (carbon supply), nitrogen availability (N source strength) and N demand (N sink strength).

Serraj *et al.* (1999) suggest that N_2 fixation is also regulated by nodule permeability to O_2 and feedback regulation by N compounds under conditions of low water availability. A lower rate of water movement out of the nodule during drought stress may restrict export of N_2 fixation products, thus inhibiting nitrogenase activity via a feedback mechanism (Serraj *et al.*, 1999). The optimum temperature range for root-nodule symbiosis is between 25 and 33^{0}C (Pankhurst and Sprent, 1976), although reductions in N_2 fixation activity have been observed above 28^{0}C (Hungria and Franco, 1993). Elevated temperature might affect N_2 fixation directly or indirectly. Direct inhibition by temperature is a consequence of decreased nodule development (Piha and Munns, 1987), functionality (Piha and Munns, 1987; Hern´andez-Armenta *et al.*, 1989) and accelerated nodule senescence. Indirect inhibition is related to temperature effects on root hair formation depression, reduction of nodulation sites and modified adherence of bacteria to root hairs (Frings, 1976).

3.1. Nitrogen Fixation in Saline Soil

Salinity leads to several physiological stresses in plants and consequently few plants can tolerate significant salinity levels in their root medium for any length of time. Of all the worlds' species, about 1% are considered halophytes (Rozema and Flowers, 2008), which are defined by Flowers and Colmer (2008) as able to complete their lifecycles under saline conditions corresponding to at least 200mM of NaCl in the root medium. Non-halophytes are called glycophytes. The negative effects of salt on plant growth and development has been shown by numerous experiments, reviewed several times and has been discussed in length in this special issue. The negative effects of salt on plant growth are a considerable problem for agriculture and thus the world food production.

Salinization has been an issue in agriculture for thousands of years. Currently it is causing problems in many parts of the world (especially in arid and semi arid regions, Manchanda and Garg, 2008); and it is predicted to get worse under climate change conditions with increasing weather extremes and rising seawater levels (Rozema and Flowers, 2008). Worldwide, up to 20% of arable land surface is salt affected (FAO, 2002). Soil salinity relates to the build-up of salts in soil and can be both natural and anthropogenic. A soil is considered saline if the electric conductivity of a saturated paste (equivalent to the available salts in the soil pore water) of that soil is over 4dS/m (equivalent to ±40 mM NaCl, which is roughly equivalent to $40/550\times100 = 7\%$ of seawater salinity).

Notwithstanding the sensitivity of many plants to salinity, some plants can survive and grow vigorously under saline conditions (Rozema and Schat, 2012). The use of such plants would be extremely helpful not only to reclaim salinized areas, but also because it would allow us to use brackish or slat water for irrigation in agriculture. The use of brackish and saline water for irrigation is related to the scarcity of freshwater in the world, especially in areas that receive little annual rainfall. About 1% of the world's water is fresh (>0.05% dissolved salts), about97% of the water is seawater (<3% of dissolved salts; Rozema and Flowers, 2008) and the remainder is of intermediate salinity. Humans use large quantities of fresh water (and the demand increases faster than the human population grows) for a variety of activities including industry (20%), domestic use (10%) and, most notably, agriculture which accounts for around 70% of global fresh water consumption.

The salinity problems and the scarcity of fresh water point to a solution via the use of salt-tolerant plants in agricultural production. Few crop plants can tolerate even moderate levels of salinity however. Attempts to improve the salinity tolerance of conventional crops have not been successful so far. This is because tolerance to salinity is a complex trait (Delgado *et al.*, 1994), involving multiple genes and having evolved multiple times independently among different lineages (Flowers *et al.*, 2010a), leading to different mechanisms of salinity tolerance. A different promising strategy is to focus on the de novo domestication of halophytes.

Sea water contains many elements required for plant growth. This would reduce the need for fertilization of arable lands under saline irrigation for many (micro) nutrients. However, the element plants require in the highest quantities, nitrogen (N), is present in sea water in very small quantities, and would therefore have to be supplemented. Conventional farming uses large quantities of fertilizer to supply crops with nitrogen but the process of artificial nitrogen fixation, the Haber–Bosch process, occurs under high temperatures and pressures and thus requires high energy inputs, rendering it unsustainable in the long term (Peoples *et al.*, 1995).

3.2. Legumes and Nitrogen Fixation

Legumes (Fabaceae) are a family of dicotyledonous plants, 88% of which (Defaria *et al.*, 1989) form a symbiosis in their root systems (sometimes on their shoots; Loureiro *et al.*, 1995) with nitrogen fixing microbes (diazotrophs) in specialized organs called nodules. Legumes are important food crops all over the world (*e.g.* Dita *et al.*, 2006), in part because of their high nitrogen content which in turn is a result of the symbiosis with the rhizobia. Green manure adds nitrogen to the soil and improves soil quality by increasing the organic matter content of the soil. Of the plants' total nitrogen content, on average 70–80% is fixed by the symbiotic rhizobia (Peoples *et al.*,1995).The remaining plant nitrogen is taken up by the root system as in other plants. The increase of yield after growing green manure is most pronounced in the first season after growing the legumes (Sullivan, 2003).

The symbiotic microorganisms in the root nodules, the rhizobia, can take up gaseous di-nitrogen (N_2) from the air and 'fix' the nitrogen into molecules (ammonia or amino acids) that can be assimilated by the plant. In return, the plant provides the rhizobia with a carbon source in the form of di-carboxylic acids (*i.e.* Soussi *et al.*, 1999). The enzyme responsible for the nitrogen fixation, nitrogenase, is irreversibly damaged when exposed to oxygen. The plant produces leg hemoglobin, a protein related to human hemoglobin to provide the rhizobia in the nodules with oxygen, often giving functional nodules a pink color.

The symbiosis first appeared around 58 million years ago when the Papilionoideae (a subfamily of the Fabaceae) underwent genome duplication (Young *et al.*, 2011). Interestingly, the genes involved in the signaling seem to be originally involved in the symbiotic relationship with mycorrhiza. Thus, it seems that only through this whole genome duplication, genes became available that were free to fulfill this new function in the communication with rhizobia, thereby enabling the legumes to start this almost unique symbiotic relationship (Young *et al.*, 2011).

3.2.1. Legumes in Saline Agriculture; Nitrogen Fixation Capabilities

In many African countries, fertilizer use reflects the economic status of the respective countries more than the actual N requirement of the crop. As a result fertilizer use varies greatly among countries, from over 700 kg N/ha/yr in, for example, in Egypt, Costa Rica and Malaysia to less than1kg in many African countries, for example Niger, Namibia and Mozambique (FAO, 2008).

However, for the purpose of comparison it is assumed 100 kg N/ha/yr as a reasonable approximation of fertilizer needs for most crops. For example, average nitrogen fertilizer use in the United States of America has been a little under 95kgN/ha/yr (USDA ERS) between1995 and 2005. Table1summarizes the amount of nitrogen some frequently used legumes can fix (after Phillips, 1980) which allows us to compare the addition of nitrogen via green manure with fertilizer use. Most legumes in Table1 are easily able to fix100 kg N/ha/yr, and this figure fits well with findings in other studies; reported values of 200–300 kg N/ha/yr are no exceptions (Peoples *et al.*, 1995). However, about 40–60% (Sullivan, 2003) of the nitrogen fixed is available for the subsequent crop.

Table 1: Estimates of the amount of nitrogen legumes can add to agricultural fields through nitrogen fixation (after Phillips,1980).The values are in kilograms of nitrogen per hectare per year, which can be compared with and are in the same range as the addition of synthetic fertilizer at most conventional farming systems around the world. Methods used for measuring nitrogen fixation differed within and between species for the subsequent crop.

S. No.	Species	kg N/ha/yr
1.	*Medicago sativa*	270–300
2.	*Glycine max*	75–311
3.	*Phaseolus vulgaris*	25–64
4.	*Vicia faba*	326–648
5.	*Pisum sativum*	52–69
6.	*Trifolium subterraneum*	58–138
7.	*Trifolium pratense*	163

The rest of the nitrogen is incorporated in the soil microbial cycle and eventually lost by denitrification and leaching as ammonium. Considering a loss of around 50% of N and that 75% of total plant nitrogen comes from atmospheric N_2, if we aim for the addition of 100 kg N/ha/yr available for a subsequent crop, the legume used as green manure should fix around 265 kg N/ha/yr. Based on Table 1, two legumes (*Medicago sativa* and *Vicia faba*) are capable of fixing this amount of nitrogen with the maximum rate of N fixation more than double this requirement. The use of legumes as an alternative to the use of fertilizer, *i.e.* as green manure, is thus a viable strategy based on these calculations. The rates reported here coincide broadly data from the literature (Peoples *et al.*, 1995).

3.2.2. Legumes in Saline Agriculture; Salinity Tolerance of Legumes

In the four categories defined by Maas and Hoffman (1977), the majority of legumes are sensitive or moderately sensitive (*i.e.*, show 80% of biomass production as compared to non-saline conditions at 3–6 dS/m, respectively). Legumes not only suffer directly from salt stress like all plants, the salinity tolerance of their symbionts also potentially plays a role. The functional symbiosis is even more sensitive to salinity than either partner is. Salinity interferes with various phases of the initiation of the symbiosis, nodule formation and functionality (*i.e.*, nitrogen fixation). Fig.1 shows four steps at which salinity can interfere with root hair nodulation. These steps were chosen not because they necessarily reflect biologically important, discrete phases in the nodulation process, but are nevertheless useful to structure our understanding of potential and realized disruptions caused by salinity. The steps are: A) root hair formation, which focuses on plant performance in general without symbiosis, B) communication, which considers the signal exchange between the plant and rhizobia and the sensitivity of the rhizobia to salinity, C) curling, which consists of the physical attachment and subsequent root hair deformations and infection thread formation, and finally, D) the intact, functional nodule, which includes nitrogen fixation rates under salinity, number of nodules, nodule weight etcetera.

Most legumes under saline conditions show reduced growth, have lower numbers of root hairs and the root hairs are more often deformed (Fig.1A). Normal root hair development is necessary for nodule formation and this is therefore the first inhibitory consequence of salinity. A high number of rhizobia is also necessary for successful nodulation (Fig.1B), and there is evidence that rhizobia numbers are reduced under salinity (Tu, 1981). However, most studies find that rhizobia are more tolerant than the associated legume and there is a significant interaction between rhizobium identity and salt tolerance of the symbiosis (*i.e.*, nitrogen fixation ability under salt depends on identity of rhizobium; Bala *et al.*, 1990; Hashem*et al.*,1998; Saxena and Rewari,1992). Studies that compare mineral N-fed chickpeas (*Cicer arietinum*) with symbiotically fixing legumes, generally find that the functional symbiosis is most sensitive to salinity (Zurayk *et al.*, 1998). Chickpea is one of the most salt sensitive legumes, with some varieties of chickpea (*Cicer arietinum*) failing to survive at even 25mM of salt in hydroponics (Flowers *et al.*, 2010b).

To initiate nodulation, both symbiotic partners (host plant and microbe) excrete molecules to communicate with each other. Host plants excrete flavonoids to attract rhizobia. This signal exchange can be negatively affected by salinity (Miransari and Smith, 2009; Oldroyd and Downie, 2004). Via chemo taxis, the microorganisms approach the root while also excreting molecules, so-called Nod factors. This leads to a physical attachment of the microbes to the root hairs (Fig.1C) (Oldroyd and Downie, 2004).These will start to deform (curl and expand).The root hair in vaginate sand the diazotrophs travelin wards (infection thread formation) until they are released inside cortical cells where they are encapsulated by a membrane of the host plant, forming a symbiosome (Oldroyd and Downie, 2004). Root hair curling, expansion and infection thread formation canal be negatively affected by salinity (Fig.1D) (Miransari and Smith, 2009; Zahranand Sprent, 1986), although this is not always so (Borucki and Sujkowska, 2008) and depends on the salinity level. Finally, as a result of salinity many studies find that less nodules are formed, they weighless, contain more Reactive Oxygen Scavengers (ROS), less leg hemoglobin (Lb, involved in the oxygen supply to the diazotrophs), have reduced nitrogenase activity and often have a generally altered morphological appearance and sometimes show necrosis (Bouhmouch *et al.*, 2005; Delgado *et al.*,1994; Soussi *et al.*,1999; Yousef and Sprent,1983; Zahran and Sprent, 1986).

In short, all steps in nodulation are at least potentially negatively affected by salinity with the result that a reduction in nodule number, weight and functioning (nitrogen fixation capacity) is often observed. This is also reflected by the more severe influence of salt on root growth than shoot growth (Cordovilla *et al.*, 1999; Borucki and Sujkowska, 2008). There are also some noticeable exceptions however. Under low to moderate salinity levels, an increase in total nodule dry weight has been found in *Pisum sativum* at 25 mM NaCl (Borucki and Sujkowska, 2008) and *C. arietinum* ILC1910 at100 mM NaCl (Soussi*et al.*, 1999). It has been proposed that this increase in total nodule number (Borucki and Sujkowska, 2008) or biomass (Soussi *et al.*, 1999) can offset the harmful effects of salinity (Borucki and Sujkowska, 2008). Borucki and Sujkowska (2008) found that the actelyene reduction capacity decreased at concentrations above 50 mM NaCl, but given the higher number of nodules, total nitrogen fixation was higher as compared to control treatments at 50 mM. However, Rao *et al.* (2002) report that total plant nitrogen content was significantly correlated to nodule number but not to nodule biomass per plant, serving as indirect evidence that salt exposed legumes cannot simply compensate lower fixation rates with a higher number of nodules. Fig. 2 shows the response curves of nitrogen fixation to salinity of different species and varieties of species of legumes expressed as a reduction in nitrogen fixation as compared to the control (0mM NaCl) treatment. In most cases, the rate of nitrogen fixation decreases with increasing salinity. In two cases (*C. arietinum* ILC 1919 and *Prosopis juliflora*) there seems to be a stimulation of the rate of nitrogen fixation upon low salinity levels. The legumes and their associated rhizobia differ in their ability to maintain adequate levels of nitrogen fixation under increasing salinity. Many species drop to below 50% of nitrogen fixation at only 50 mM of NaCl. There is a striking difference between different varieties of the same species. *Phaseolus vulgaris* contender for example appears to be the most tolerant species (still fixing about 50% of the amount of nitrogen at100 mM expressed as amount of nitrogen fixed at 0mM), whereas *P. vulgaris* COCOT stops fixing nitrogen all together at 50 mM of NaCl.

A regression line through the points (except where there were only two points, *i.e.* the two *Medicago ciliaris* cultivars) and displayed the regressions lope for broad comparison between the species and varieties (at both *M. ciliaris* the slope of the line connecting the two points). However, there is no a priori reason to assume that the relationship between the rate of nitrogen fixation and salt concentration is linear. Some of the relations in Fig.2 can be considered linear, others show apparent stimulation of salinity on nitrogen fixation rate under salt (*C. arietinum* ILC1919 and *P. juliflora*) and some species seem to level off at intermediate salinity levels (*P. sativum, M. sativa* and *P. juliflora*). Three tree species are more salt tolerant than all the other legumes in Fig.2: *P. juliflora, Leucaena leucophela* and *Acacia nilotica* and have multiple uses and are being grown as fodder. Nitrogen fixation capacity is reduced slightly at 9 dS/m (<90 mM NaCl). Nitrogen fixation in the Acacia and Prosop is species has been measured upto12 dS/m (<120 mM NaCl) at which nitrogen fixation rate can still be considered high (more than 80% in *P. juliflora*). *P. juliflora* also shows no sign of a reduction in the rate of nitrogen fixation and has the lowest regressions lope (0.17). Of the remaining species in Fig.2, *Glycine max* is the most salt tolerant. Medicago truncatula also shows little reduction in nitrogen fixation rate, but has only been measured upto50 mM NaCl. The other Medicago species and varieties tested show comparable sensitivity to salinity with regression slopes between 0.45 and 0.86.

Data from *Pisum sativum, Vicia faba, Glycine max* and *Phaseolus vulgaris* cv. Contender from Delgado *et al.*(1994); *Medicago truncatula* and *Lotus japonicus* from López *et al.*, 2008; *Cicer arietinum* Pedrosalino and *C. arietinum* ILC1919 from Soussi *et al.* (1999); *Phaseolus vulgaris* BAT 477(BAT) and *P. vulgaris* COCOT from Jebara *et al.* (2010); *Acacia nilotica, Leucaena leucophela* and *Prospis juliflora* from Bala *et al* .(1990): data from the most tolerant *Rhizobium* strain used in their experiments); Medicago ciliaris 1.8 and *M. ciliaris* 11.9 from Ben Salah *et al.* (2010); *M. sativa* from Serray and Drevon (1998).

To see whether nitrogen fixation is more sensitive to salinity than growth (*i.e.*, the functional symbiosis more sensitive to salinity than either partner is), as often is suggested and reported (i.e., Delgado *et al.*,1994;) both growth and nitrogen fixation relative to the performance at 0 mM NaCl of the same legumes were plotted as Fig.2 (Fig.3). Generally nitrogen fixation decreases faster than biomass production. There are some exceptions however, for example both *Medicago* species. Most tellingly, most salt tolerant legumes in these graphs, *L. leucophela, A. nilotica* and *P. juliflora*, show a pattern of higher salt tolerance of nitrogen fixation than biomass production. Even though these graphs do not constitute formal statistical testing of this observation, the lines suggest

Schematic drawing Non-saline Saline	Phase in nodulation	Problems under salinity (references)	References and Species
 A	**Roothair formation**	↓plant growth ↓number of root hairs (2) Roothairs deformed (2, 7)	2, 7: *Glycine max*
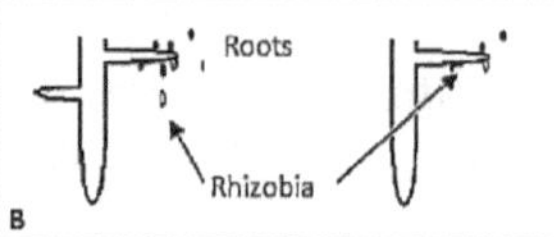 Roots Rhizobia B	**Communication, Rhizobia growth** and **physical contact**	↓ possibilities for signal exchange between partners (2) ↓ growth of Rhizobia (7, but 6 no effect)	2, 6, 7: *Glycine max*
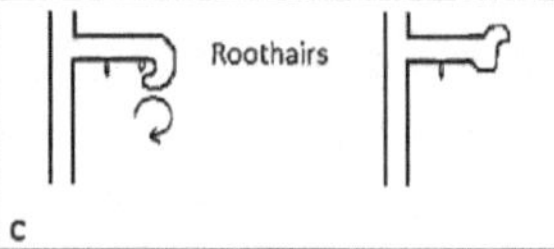 Roothairs C	**Roothair curling** and **Infection thread formation**	↓curling (2, 5, 7) ↓root hair expansion (5) ↓infection thread formation (2) bacterial release from infection thread into cells disturbed	2, 7: *Glycine max* 4, 5: *Vicia faba*
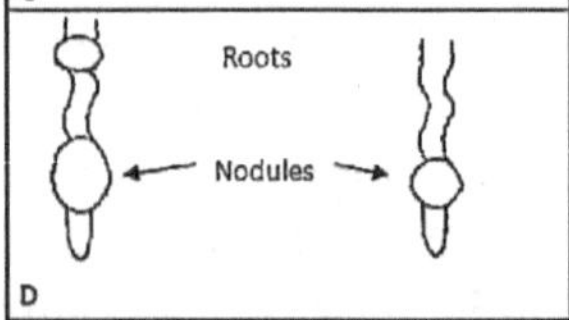 Roots Nodules Nodules D	**Nodule formation** and **functioning** (i.e. **nitrogen fixation**)	↓nodule number (1, 4, 5, 6, 7, 9). Nodule morphology altered; necrosis in nodule (3) ↑ nodule weight (1, 4, 6) ↓ nodule weight (4, 8, 9) ↑sugars and proteins in nodules (1, 8) ↓Leghemoglobin (Lb) content (8; high salt concentrations) ↓Acetylene reduction activity (1, 6, 8)	1: *Cicer arietinum* (Two cultivars) 4, 5, 8: *Vicia faba* 6, 7, 8: *Glycine max* 8: *Pisum sativum* 8, 9: *Phaseolus vulgaris*

Fig. 1: Four different phases of nodule formation and nodule functioning and the (potential influence of NaCl stress on those phases. 1. Soussi *et al.* (1999); 2. Miransari and Smith (2009); 3. Borucki and Sujkowska (2008); 4.Yousef and Sprent (1983); 5. Zahran and Sprent (1986); 6. Delgado *et al.* (1994); 7. Bouhmouch *et al.* (2005).

that the response of nitrogen fixation is a good indicator for general tolerance to salinity in legumes because the most tolerant species and genotypes do not seem to show this increased sensitivity of nitrogen fixation over growth under salinity (Soussi *et al.*, 1999; Rao *et al.*, 2002; Ashraf and Bashir, 2003). Under higher salt concentrations, the early steps in nodulation are most likely responsible for the decrease in nodule number observed (Rao *et al.*, 2002). The number of root hairs and therefore potential infection sites is negatively infected by salt (Miransari and Smith, 2009). The influence of calcium (Ca^{2+}) seems to be critical in the decrease of root hair number and extension. In *Hordeum vulgare*, root hair density decreases upon salinity (Shabala *et al.*, 2003) and root hair extension was decreased in *Arabidopsis thaliana* (Halperin *et al.*, 2003). In both these plants, the negative effect of salt on the root hairs could be alleviated or even completely reversed by increasing the extracellular concentration of calcium. Salinity seems to have a direct effect on calcium activity (as shown by model susing GEOCHEM software Shabala *et al.* (2003)) but also causes root hairs to excrete calcium from organelles which is subsequently lost from the cells in *Zea mays* (Lynch *et al.*, 1989). Bibikova *et al.* (1997) showed that the direction of root hair growth is determined by an intra cellular calcium gradient in *A. thaliana*.

This suggests a key mechanism for the explanation of the reduction in nodule number upon salinity, since directed root hair extension is often needed for successful nodule formation (Oldroyd and Downie, 2004). The distorted morphology observed in some nodules under salinity may also be a consequence of altered calcium and balance. The further study of the exact mechanisms of the influence of increased salinity in the root zone on calcium and the subsequent consequences for the altered calcium balance on root hair curling deserved future studies to elucidate further the negative effects of NaCl on nodulation. Delgado *et al.* (1994) suggests that under low salinity (slightly reduced growth of the host) the supply of oxygen to the nodules is the limiting factor. Ashraf and Bashir (2003) suggest that the ability to keep nodules hydrated is a key factor in the ability of legumes to maintain nitrogen fixation under salinity.

Data from *Pisum sativum*, *Vicia faba*, *Glycine max* and *Phaseolus vulgaris* cv. Contender from Delgado *et al.*(1994); *Medicago truncatula* and *Lotus japonicus* from López *et al.*, 2008; *Cicer arietinum* Pedrosalino and C. arietinum ILC1919 from Soussi *et al.* (1999); *Phaseolus vulgaris* BAT 477 (BAT) and COCOT from Jebara *et al.*(2010); *Acacia nilotica*, *Leucaena leucophela* and *P. vulgaris Prospis juliflora* from Bala *et al.* (1990:data from the most tolerant Rhizobium strain used in their experiments); *Medicago ciliaris* 1.8 and *M. ciliaris* 11.9 from Ben Salah *et al.*(2010); *M. sativa* from Serray and Drevon (1998).

The ability to counteract the osmotic stress caused by salinity by the production and accumulation of osmolytes such asproline and glycine betaine further increases nodule function under salinity (Ashraf and Bashir, 2003) and is a general sign of the ability of the plant to respond to increased levels of NaCl in its environment.

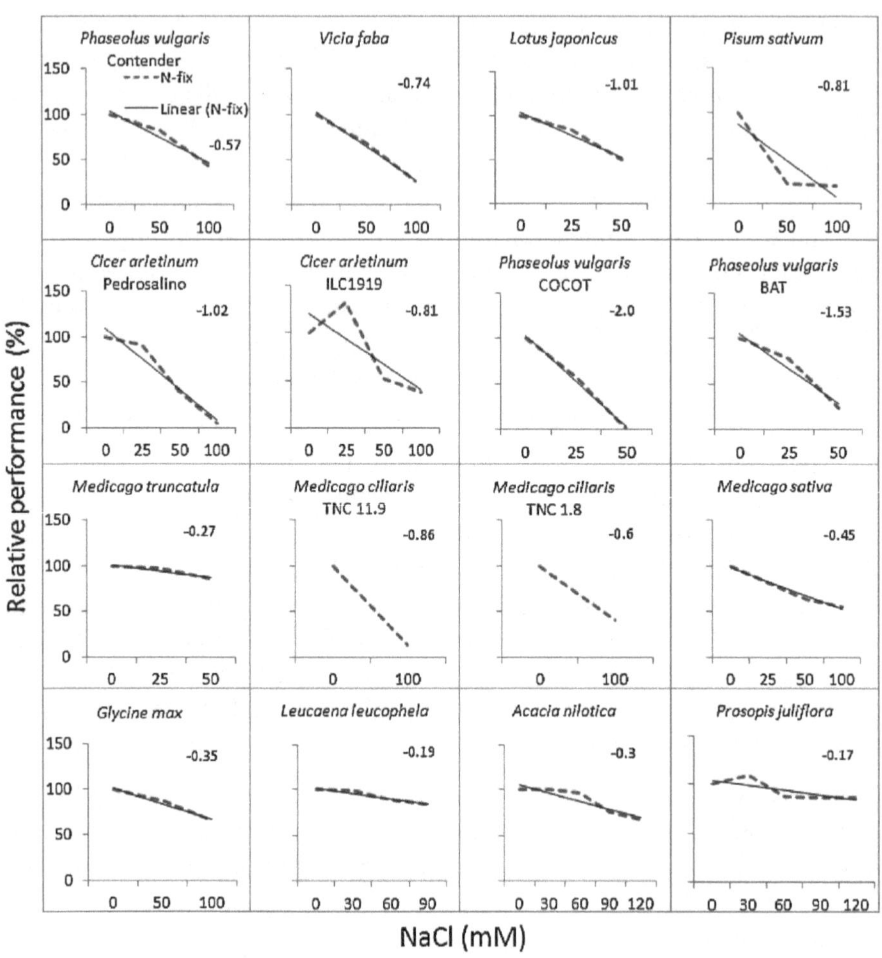

Fig. 2: The rate of nitrogen fixation (dotted line) as a function of salinity (legend in the upper left panel). Solid line is the regression line and the number in the upper right corner of each panel is the slope of the regression line. Nitrogen fixation rate at 0mM NaCl in the root medium was taken as control (i.e.100%) and nitrogen fixation at higher salinity levels is expressed as a reduction compared to the control to compensate for differences in methodology and units. The salt concentration expressed as NaCl concentration is an approximation in the last three panels (Leucaena leucophela, Acacia nilotica and Prosopis juliflora) is an approximation since the salinity levels in the original publication is expressed in electrical conductivity (EC) units. The conversion used was1ECH"10mMNaCl.

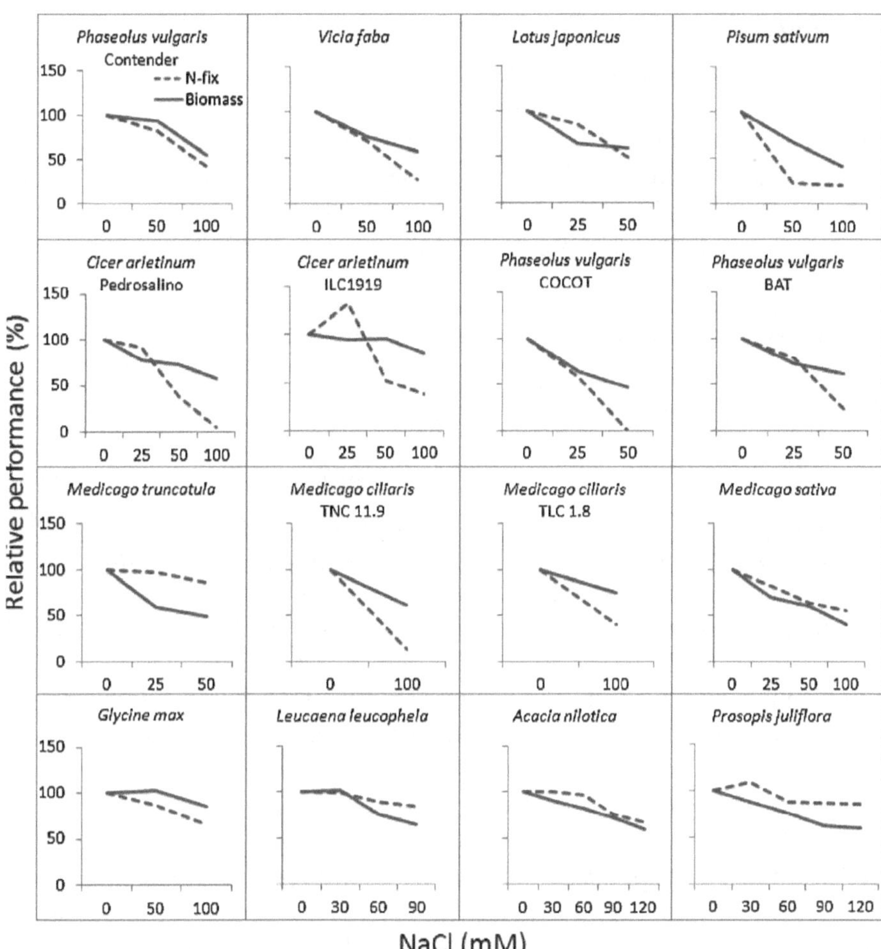

Fig.3: The rate of nitrogen fixation (dotted line) and biomass production (solid line) with increased salinity expressed as a percentage of the control treatment (0mM).The NaCl concentrations in *Leucaena leucophela*, *Acacia nilotica* and *Prosopis juliflora* represent an approximation. EC levels in the original publications were converted to mM NaCl using1 ECH≈10 mM NaCl.

3.2.3. Improvement of Salinity Tolerance of Legumes

The use of modern molecular techniques and the introduction of two model legume species (*M. truncatula* and *Lotus japonicus*; Cook, 1999; Dita *et al.*, 2006) has contributed to a more profound understanding of the molecular mechanisms of salinity stress and tolerance in legumes and the functional symbiosis (Manchanda and Garg, 2008), and in the process opened up possibilities for improving salinity tolerance. In general, approaches to improvement can be broadly divided into-

- Looking for tolerant populations in the wild (Ben Salah *et al.*, 2010),
- Research into variation in salinity tolerance among varieties (*i.e.*, Soussi *et al.*, 1999; Teakle *et al.*, 2006) and
- Improving salinity tolerance by altering the genetic makeup of the plant.

Verdoy *et al.*, (2006) were the first to increase the salinity tolerance of a legume by genetically altering the plant. By transforming *M. truncatula* plants with a gene for proline synthesis from another legume (*Vigna aconitifolia*), the authors were able to increase proline content in the nodules of the transformed plants. This led to plants with increased salinity tolerance as shown by a smaller reduction in nitrogen fixation rate under salt stress than the decrease observed in control plants. Increased resistance to salinity of the nitrogen fixation process was also achieved by de la Pena *et al.* (2010). They transformed *M. truncatula* with a gene coding for a cyanobacterial flavodoxin, an electron transfer protein known to generally increase resistance to abiotic stress. Under 100 mM NaCl, growth reduction and nitrogen fixing reduction still took place in the transgenic plants as well as in the control plants, but this was less so for nitrogen fixation (de La Pena *et al.*,2010). Genetic tools are capable to increase salinity tolerance in legumes (Dita *et al.*, 2006).

Yet other approaches to improving yields of legumes under saline conditions include the following (a) selection on salt tolerant rhizobia (Bala *et al.*, 1990; Elsheikh and Wood, 1995; Hashem *et al.*, 1998). For example, Bala *et al.* (1990) showed that *A. nilotica* inoculated with a tolerant Rhizobia strain was still able to maintain a nitrogen fixation rate of nearly 70% at 12 dS/m as compared to the maximum fixation rate (at 2 dS/m). (b) Adding minerals or other substances to the soil. For example, Rao and Sharma (1995) were able to improve whole plant nitrogen content and shoot biomass in *C. arietinum* by adding small quantities of nitrogen to the sand used in their experiment (most pronounced effects at salinity levels of 5.6 and 7.4 dS/m). The beneficial effects of adding nitrogen can probably beat tribute to a better performance under salinity of fertilized legumes, as has also been found by Yousef and Sprent (1983). El-Hamdaoui *et al.* (2003) were able to counteract the negative effects of

75 mM NaCl on *P. sativum* growth, nodule formation and nitrogen fixation rate by adding calcium and boron to the experimental field. Also, there are some indications that inoculation with Vesicular Mycorrhizal fungi help to improve salt tolerance of legumes (Asghari, 2008; Garg and Chandel, 2011) as well. Finally, adding gen is teinto the rhizobia inoculums resulted in better nitrogen nfixation rates in the functional symbiosis in soybean (*G. max*, Miransari and Smith, 2009).

3.2.4. Legumes of the HALOPH Database

Salt tolerant legumes

There are a number of legumes that are naturally moderately salt tolerant. Table.2 lists all the Papilionaceae of the HALOPH database of salt tolerant plants which represents the most complete list available of salt tolerant legumes. Some genera, for example *Melilotus* do not appear on this list but contain some tolerant species and will therefore be discussed separately. There are forty six legumes that are tolerant to salinity levels ranging from 8 dS/m to sea water salinity. This includes annuals and perennials, and legume species from tropical, subtropical and temperate regions. However, most of the species are trees and herbaceous perennials. Some species in the list like *Hedysarum carnosum* (tolerant without yield loss and loss of nitrogen fixation capacity upto100 mM NaCl; Kouas *et al.*, 2010) are already being used as fodder species in northern Africa. Most of the species on the list have relatively small native ranges and areas yet not broadly agronomically or economically exploited. Rogers *et al.* (2006) advocate a local approach to develop legumes for saline agriculture. First, suitable legume candidates should be identified. The selected species will then have to be screened for their salinity tolerance. Next, tolerant legumes have to be screened for secondary metabolites. The efficiency with which legumes maintain the ability to effectively fix nitrogen under saline and nonsaline conditions varies between species (Fig.2), and both for the use as green manure as well as the use as fodder and forage, high nitrogen contents are to be preferred.

Once screenings of species have led to the identification of promising candidates, the secondary metabolites of the species have to be evaluated. Secondary metabolites are known to increase in many plant species under salinity-and other stresses. As the reaction of a plant to cope with stress, this plant response is often an unwanted side effect of salinity stress. However, it may be worth exploring the potential of exploiting this response for our benefit. Some HALOPH legumes produce commercially and pharmaceutically useful secondary metabolites. For example, *Lathyrus littoralis* and *Canavalia obtusifolia* produce canavanine, a non protein amino acid produced to protect the plant from herbivory. Another HALOPH legume that produces a secondary metabolite that protects it from predation is *Lotus tenuis*. This species produces roteone, a broad pesticide. These examples hint at the possibility that saline agriculture may reduces the need for pesticides. *L. tenuis* is a promising species for several reasons, as it is also used as forage and additionally to its salinity tolerance (over 80% of biomass production of control treatments at 200 mM of NaCl) shows good tolerance to water logging (Teakle *et al.*, 2006). Another species that produces a useful molecule is *Alhagi maurorum*. It produces lupeol, a pharmaceutically active substance (Laghari *et al.*, 2011).

Other salt tolerant legumes

Other legumes are perhaps not as salt tolerant as the species on the HALOPH list but may still be of use. Ashraf and Bashir (2003) describe the effects of NaCl on nodules and other plant parts of two legume species, *P. vulgaris* and *Sesbanea aculeata*. *Sesbanea aculeata* is frequently used as a green manure in salinized areas (Ashraf and Bashir, 2003).The authors chose as salt treatments those values that were previously shown to cause a 50% reduction in growth. *Sesbanea acueate* was exposed to a salt concentration of 13 dS/m (~130 mM NaCl) which means it has moderate salt tolerance. Additionally, nodule number in *Sesbanea aculeate* was less adversely affected by salinity (albeit they were showed a stronger decrease in size) and nodules accumulated more proline and glycine betaine, and less sodium and chloride than *P. vulgaris*. Finally respiration and stomatal conductance was reduced less upon salinity in *Sesbania aculeate* than in *P. vulgaris*.

Another legume sometimes used in salinized areas is *Trifolium alexandrinum*. For example, Agarwal *et al.*, (2010) show that there is genetic variation in *T. alexandrinum* with the more tolerant genotypes only showing significant reductions in biomass productions at salinity levels of 7 and 14 dS/m (~70 and140 mM NaCl). Another example is *Trifolium fragiferum* (strawberry clover), occurring in the upper parts of salt marshes (Rumbaugh *et al.*, 1993) is considered moderately tolerant (*i.e.*, 80% of biomass production as compared to control at 11 dS/m; Maas and Hoffman (1977) and is sometimes used in Australia as a pasture legume in salt affected areas (Rogers *et al.*, 2008). Perhaps the most frequently mentioned moderately salt tolerant legume is lucerne, *M. sativa*. The species is also considered moderately tolerant (~11 dS/m) and in Fig.2 it appears intermediate in its reduction in nitrogen fixation under increasing salinity levels, compared to the other species in Fig. 2. Moreover, it shows genetic variation in its salinity tolerance (Rogers, 2001). The best performing plants under salinity described in Rogers (2001) were not originally selected for salinity tolerance but for general vigorous growth under non-saline conditions. Thus, surprisingly, the cultivar that was selected for salt tolerance

Table 2: List of salt tolerant nitrogen fixing plants (from the HALOPH database): The authors do not claim this list to be exhaustive; however, it should represent some of the potential tolerant legumes that may be exploited in a saline agricultural system.

S.No.	Species	Life form (height in m)	Plant type	NaCl	Economic uses
1.	*Aganope heptaphylla*	Shrub	Hydro halophyte	SW	
2.	*Alhagi maurorum Medik*	Weedy	Chaemaephyte	n. a.	Medical
3.	*Canavalia maritima*	Herbaceous perennial	Psammophyte	n. a.	Land use (sand stabilization); ornamental
4.	*Canavalia obtusifolia*	Herbaceous perennial	Psammophyte	n. a.	Land use (sand stabilization); ornamental
5.	*Canavalia rosea*	Herbaceous perennial	Psammophyte	n. a	Land use (sand stabilization); ornamental
6.	*Dalbergia amerimnion*	Tree (2)	Psammophyte	SW	
7.	*D. caudenatensis*	Vine (perennial)	Hydrohalophyte	n. a.	
8.	*D. ecastophyllum*	Shrub	Hyhal	n. a.	Ornamental; fish poisons
9.	*D. menoides*	Vine (perennial)	Hyhal	n. a.	
10.	*D. sissoo*	Tree (8)	Xerohalophyte (dry environment)	n. a.	Timber (fuel); fodder; ornamental
11.	*Daviesia hakeoides*	Shrub	Xerohalophyte (inland salt dessert)	n. a.	
12.	*Derris trifoliata*	Shrub	Hydrohalophyte	SW	Fodder; fibers (cordage); fish poisons
13.	*Drepanocarpus lunatus*	Tree (6)	Hydrohalophyte	n. a.	Timber; medical
14.	*Erythrina herbacea*	Shrub	Hydrohalophyte	n. a.	Domestic (cosmetics); medical
15.	*Erythrina variegata*	Tree (8)	Hydrohalophyte	n. a.	Ornamental
16.	*Geoffroea decorticans*	Tree (4)	Phreatophyte	n. a.	Timber (fuel); forage
17.	*Hedysarum carnosum*	Annual	Xerohalophyte (dry environment)	20	Fodder
18.	*Indigofera spinosa*	Herbaceous perennial	Xerohalophyte (inland salt dessert))	n. a.	Timber (fuel)
19.	*Inocarpus fagifer*	Tree (20)	Hydrohalophyte	SW	Timber (fuel); timber (construction)
20.	*Lathyrus littoralis*	Herbaceous perennial	Psammophyte	n. a.	Ornamental
21.	*Lathyrus palustrus*	Herbaceous perennial	Hydrohalophyte	n. a.	
22.	*Lotus creticus*	Herbaceous perennial	Psammophyte	8	Land use (sand stabilization); ornamental; forage

Contd.

S.No.	Species	Life form (height in m)	Plant type	NaCl	Economic uses
23.	*L. cytisoides*	Herbaceous perennial	Chaemaephyte asmophyte	8	Land use (sand stabilization); ornamental
24.	*L. halophilus*	Herbaceous perennial	Psammophyte	n. a.	Forage
25.	*L. jolyigi*	Herbaceous perennial	Psammophyte	n. a.	
26.	*L. nuttalianus*	Herbaceous perennial	Psammophyte	n. a.	
27.	*L. preslii*	Herbaceous perennial	Psammophyte	n. a.	Forage
28.	*L. tenuis*	Herbaceous perennial	Hydrohalophyte	n. a.	
29.	*Medicago litoralis*	Annual	Psammophyte	n. a.	
30.	*M. marina*	Herbaceous perennial	Psammophyte	n. a.	
31.	*Milletia hemsleyana*	Tree (4)	Hydrohalophyte	n. a.	
32.	*Muellera frutescens*	Shrub	Hydrohalophyte	SW	
33.	*Muellera monoliformis*	Tree (4)	Hydrohalophyte	SW	
34.	*Ormocarpum verrucosum*	Tree (4)	Hydrohalophyte	SW	Timber (fuel); chemicals (tannins)
35.	*Pongamia pinnata*	Tree (20)	Hydrohalophyte	SW	Timber (construction)
36.	*P. velutina*	Tree (20)	Hydrohalophyte	SW	Timber (construction)
37.	*Pterocarpus draco*	Tree (25)	Xerohalophyte (dry environment)	n. a.	Timber; medical
38.	*P. indicus*	Tree (87)	Hydrohalophyte	n. a.	Timber (construction)
39.	*P. officinalis*	Tree (12)	Hydrohalophyte	SW	Timber (carpentery, carvings)
40.	*Sophora tomentosa*	Shrub	Hydrohalophyte	n. a.	
41.	*Tephrosia purpurea*	Herbaceous perennial	Xerohalophyte (inland salt dessert)	n .a.	Green manure
42.	*Trifolium maritimum*	Herbaceous perennial	Psammophyte	25	Forage
43.	*T. resupinatum*	Herbaceous perennial	Psammophyte	n. a.	Green manure
44.	*T. tomentosum*	Herbaceous perennial	Psammophyte	n. a.	
45.	*T. wormskjoldii*	Herbaceous perennial	Psammophyte	n. a.	
46.	*Vigna marina*	Herbaceous perennial	Psammophyte	n. a.	

(Sirosal) was not the most salt tolerant (Rogers, 2001).Varieties also differ in their tolerance to waterlogging (Rogers *et al.*, 2008).These facts make this species a valuable crop to continue investigating.

Rogers *et al.* (2008), screening on salinity and water logging tolerance of 19 species of the genus *Melilotus* that could potentially be developed as fodder species in Australia, resulted in the identification of three species with good biomass production under the two stresses. The combination of tolerance to waterlogging and salinity stress is important, since many forage areas in Australia are situated in flood plains and hence prone to flooding. Since the physiological consequences of salinity stress sometimes coincide with those of other stresses such as drought (Munns and Tester, 2008), it may be fruitful lto use knowledge on tolerance to other stresses in the identification of potential legumes to be used in saline areas or agriculture.

3.3. Perspective for Saline Agriculture

Of four of some of the most commonly grown legumes in agriculture, *i.e.*, pea (*P. sativum*), chickpea (*C. arietinum*), common bean (*P. vulgaris*) and soybean (*Glycine max*), soybean is the most salt tolerant (Delgado *et al.*,1994). Additionally, the rate of nitrogen fixation is only moderately reduced under salinity (Fig.2). In soybean there is a large variation in salinity tolerance (Wang and Shannon, 1999) and progress is being made in identifying the regions in the DNA responsible for increased salinity (Hamwieh *et al.*, 2011). Using natural variation in salinity tolerance in legumes, combined with insight in the physiological mechanisms of the tolerance and the underlying genetics, open sup possibilities for integrating legumes in more sustainable forms of saline agriculture. This can be reached by the development of tolerant genotypes (de la Pena *et al.*, 2010; Hamwieh *et al.*, 2011; Verdoy *et al.*, 2006), by adding elements to the soil such as small quantities of nitrogen (Rao and Sharma, 1995) or genistein to the inocula used by farmers (Miransari and Smith, 2009) or inoculate the fields with suitable mycorrhizal fungi (Garg and Chandel, 2011). Even though Rhizobia are generally more salt tolerant than their host plants, they can sometimes make a difference in the salt tolerance of the functional symbiosis and this line of research should not be abandoned because there may be progress left to be made (Bala *et al.*, 1990). Together with the recent promising results in increasing N_2 fixation under salt stress by up-regulation of a single membrane ion co transporter involved in salinity tolerance (Verdoy *et al.*, 2006) and in the increased productivity of crop plants grown under a tripartite symbiosis (rhizobia and arbuscular mycorrhizal fungi), we can adopt a positive view on the future regarding the development of salt tolerant, nitrogen fixing legumes, both for forage crops and as green manure.

3.4. Impacts of Agriculture and Global Climate Change Terrestrial Biogeochemical Cycle of Nitrogen

Carbohydrates, lipids, proteins, and nucleic acids are the major groups of these organic compounds that are required by life. Carbon, hydrogen, oxygen, nitrogen and phosphorus are the five principal elements used to different extents to construct these molecules-of-life. In these compounds, C atoms are bonded together in various arrangements to form the basic structural skeleton of the molecule. Hydrogen and oxygen, in turn, are associated with these atoms. The proportions of carbon, hydrogen and oxygen and their relative geometric positions contribute variation to the chemical properties of the organic molecules. While carbon, hydrogen and oxygen are important constituents of each of the four groups of organic molecules that are required by life, nitrogen is used only in proteins and nucleic acids, and phosphorus is principally a component of nucleic acids (and phospholipids). The latter two elements, nitrogen and phosphorus, because of their unique characteristics, impart additional specific chemical properties to the molecules enabling them, in turn, to fulfill different cellular functions.

Of these five major nutrient elements and additional macro nutrients (potassium, sulphur, calcium, iron, magnesium) and micro nutrients (manganese, boron, copper, irin, chlorine, cobalt, molybdenum and zinc), elements that are required to lesser degrees by life, nitrogen is typically the limiting nutrient of plant growth. That is, it is the nutrient that is available in the least amount relative to its need by the plant—growth is determined by its abundance. Consequently in terrestrial ecosystems, N availability exerts a strong control on plant productivity.

It should thus be apparent that C, H, O, N, and P (along with all other required nutrients) are linked proportionally (stoichiometrically) for the proper health of organisms. Therefore, even if C, H, O, P, Ca, Fe, and all of the other required nutrients are in abundance, if N is proportionately scarce in the soil, the plant's ability to produce N-containing organic compounds for continued growth will be limited. Consequently, growth will be restricted. The C:N stoichiometric linkage is very important to the processes of decomposition of plant litter by soil microbes and will be emphasized in this computational science module. Interestingly unlike most of the other plant nutrient elements that are derived from the soil (*e.g.*, K, Ca, and P), N has multiple valence states and consequently multiple chemical forms—including gaseous and nongaseous forms. Within soils, bacteria typically mediate chemical transformations among these forms of N. Such changes can result in the incorporation of N into, or loss from, the soil as well as in the production of soil N forms that differ in their availability to plants. Environmental conditions exert a significant control over the microbial-mediated processes that transform N. In particular, soil temperature and the amounts of soil water and oxygen, along with the C:N ratio of plant litter are of a principal importance. As a consequence, plant growth is not only affected directly by climate, but also indirectly by it though climate's effects on the soil microbial community.

Modern intensive agriculture relies on practices that not only influence the microclimate (temperature and moisture) of the soil, but also its chemistry—particularly that of N, P, and K. Not only do crop plants have a high demand for these major nutrients, but the nutrients also are removed during harvest rather than recycled as in natural terrestrial ecosystems. While fertilizer supplements do return these nutrients to the soil, additions of irrigation water, manipulation during plowing, sowing, and treatment of weeds and parasites and herbivores produce alterations to the N biogeochemistry of the agricultural field; alterations which, in turn, impact crop productivity. Consequently, for optimal long-term success in providing food for the World's hungry, it is necessary to have a sound understanding of the linkages between agricultural practices, N biogeochemistry, and crop productivity. The importance of this understanding will only be magnified over the remainder of the 21st century should predicted global climate change occur.

3.4.1. *Nitrogen Chemistry and Terrestrial Life*

While N_2 is the most common constituent of the Earth's atmosphere, (constituting 79% of the total gases by molar content), the majority (98%) of the Earth's total N is found in rocks and minerals—typically in very low concentrations. Of the remaining 2% of total nitrogen, atmospheric N_2 represents 1.2%, leaving organic N in living and dead organic matter contributing only 0.8% of the total (Paul and Clark, 1989). Although N exists in such a high concentration in the atmosphere, to plants this represents the equivalent of the ocean-full of salt water to which the Ancient Mariner in the poem *The Rime of the Ancient Mariner* by Samuel Taylor Coleridge commented—"Water, water, every where, nor any drop to drink." For plants, the refrain could be "Nitrogen gas, nitrogen gas everywhere, but not a molecule to take up." This is because N_2 is unusable by plants, and of the other principal inorganic forms of nitrogen that occur in the environment, NH_3, NH_4^+, NO_2^- and NO_3^-, only the second and the fourth are the forms that can be taken up by plants. Bacteria can use all four forms. Once taken up by plants or microbes, these forms of N can be transformed into the N-containing organic forms that are used by life: proteins, nucleic acids, and microbial cell wall constituents (chitin and peptidoglycans).

While plants capture and transform inorganic N into N-containing organic molecules that can then be transferred to, and among, animal members of food webs, bacteria mediate important N transformations in soils and sediments. Nitrogen fixation is a bacterial process that transforms N_2 into NH_4^+, thus providing an input of N that is usable as a nutrient within ecosystems. Nitrification transforms NH_4^+ to NO_2^- and this then to NO_3^-. As mentioned previously, NO_3^- can be taken up by plants, and it also has the potential to be leached readily from soils. Also, microbial denitrification can transform NO_3^- to NO_2^-, NH_4^+, NH_3 and N_2. While NH_4^+, can be taken up from the soil by plants or immobilized by clay minerals, diffusion of the gaseous NH_3 and N_2 molecules can cause N loss from soils.

3.5. Nitrogen Fixation

With the exception of water-associated H and O, all soil-derived nutrients (N, P, K, S, Ca, Fe, Mg, *etc.*) are introduced initially into terrestrial food webs through the weathering of soil minerals. While this does occur with N, however, soil mineral weathering contributes typically a minor addition that by itself would be insufficient to produce the productivity found in terrestrial plant communities. The principal N input to terrestrial ecosystems stems from the ability of certain bacteria to transform atmospheric N_2 into NH_4^+ which the bacteria can then incorporate into organic-N compounds. This nitrogen fixation occurs asymbiotically or symbiotically with plants and, in the case of lichens, with fungi. Free-living asymbiotic, N-fixing heterotrophs, (including *Azotobacter and Clostridium*) use organic compounds as their energy source. The majority of these species require aerobic conditions, although some are facultative, surviving under limited oxygen levels. While the rate of asymbiotic N fixation is low per hectare, it is significant on a global basis. In addition, there exist free-living N-fixing photoautotrophs; cyanobacteria, that are found in soil and aquatic systems. Other N-fixers include those that are associated with grasses where they occur in the plant rhizosphere (the area of soil surrounding the plant root) and on leaf surfaces.

In symbiotic relationships between bacteria and plants, energy- and nutrient-containing organic compounds are supplied by the plant to the bacteria. This increases the efficiency of symbiotic bacterial N fixation and NH_4^+production. Any surplus $NH_4^+$is excreted by the bacterium and acquired by the host plant. Leguminous plants, including those used as food crops (*e.g.*, soybeans, beans, and peas) and cover crops (*e.g.*, clovers and alfalfa) are the plants best known for their symbiotic relationship with N-fixing bacteria. In these plants, species of *Rhizobium* infect the roots, triggering the formation of nodules that the bacteria then occupy. Symbiotic relationships also exist between nitrogen-fixing microbes and non-leguminous plant species. The nitrogen-fixing actinomycete, *Frankia*, forms symbiotic relationships with 170 known species within eight plant families. Globally, symbiotic N_2 fixation represents 8.5% of the N that is taken up by plants. The addition of N through fertilizer applications is about 50% of this amount. Due to a low efficiency in the transfer of fertilizer N to plants, however, this represents approximately 2% of the total plant N uptake (Paul and Clark, 1989).

While the plant is surrounded by an atmosphere that is rich in N_2 plant uptake of N occurs through the roots, and both NH_4^+ and NO_3^- are forms that are taken up. Within the plant, NH_4^+ is used to form proteins, with energy being required for the necessary chemical transformations. Nitrate is used to form proteins as well, but first it must be reduced to NH_4^+ in a process that requires additional energy to be used by the plant. Consequently, plant uptake of NH_4^+ is preferable to that of NO_3^-, because it requires less energy expenditure by the plant to transform it into organic-N forms. However, Nitrate is, more mobile than NH_4^+ within the soil which produces a greater delivery of it to the root. Within the plant, this N can be incorporated into proteins and nucleic acids becoming part of cellular constituents. Parasitism and herbivory can then transfer these organic-N compounds from the plant to the consumers of the food chain. Predation can transfer the N to higher trophic levels. At all levels within the food chain, through death, litter formation, or production of animal waste (urine or feces), N-containing organic compounds along with other organic molecules are transferred to the soil. Once there, the decomposer food web uses this material for its energy and nutrient (including N) needs.

3.5.1. Decomposition

Following the introduction of litter to the soil, amino sugars and nucleic acids in the litter can be used as energy sources or building blocks for other organic molecules by the soil decomposers. Consequently, N in these molecules may become part of the soil decomposer food web. Alternatively, N that is contained in these molecules may be transformed by the microbes to NH_4^+ that can be released to the soil through a general process that is termed mineralization. This inorganic form of N can be taken up by plants and other microbes or immobilized by clay minerals. While the majority of soil organic-N is present in decomposing plant and microbial matter, up to 40% occurs in by-products of decomposition that are somewhat resistant to further decomposition. Having been altered greatly by the complexities of decomposition, the primary biological source of this mixture of principally phenolic compounds is unrecognizable.

3.5.2. Nitrification

Nitrification consists of the step-wise transformation of NH_4^+ to NO_2^- and then to NO_{3-} by a group of chemoautotrophic bacteria. These bacteria, which are obligate aerobes, obtain their energy through the oxidation of NH_4^+ or NO_2^- and their C from CO_2 or soil carbonates. Two groups of oxidizers are involved in nitrification: the NO_2^- formers include members of the *Nitrosoccus* and *Nitrobacter* genera and the NO_3^- formers are *Nitrobacter*. In this latter oxidative transformation, oxygen is taken from water and not O_2. Heterotrophic nitrification can also occur in soils, but is considered insignificant in agricultural settings. A variety of environmental factors influences the rate of nitrification through effects on the activity of nitrifying bacteria. The optimum soil acidity range for nitrifying organisms is pH 6.6 - 8.0 with nitrifying rates dropping off on either side of this range. With nitrifying bacteria being obligate aerobes, O_2 availability in the soil is of great importance to their functioning. Consequently, soil moisture content exerts a strong influence on nitrification.

Water logging of soils decreases aeration through slowing the rate of diffusion of O_2 into, and within, the soil. As a result, aerobic cellular respiration by soil microbes will consume O_2 at a rate greater than it is replaced leading to anaerobic soil conditions. Alternatively, soil dryness will limit bacterial activity due to their requirement for moisture. Thus, there exists an intermediate range in soil moisture that is conducive to nitrification. Temperature also influences nitrification. Nitrification rates are typically low below 5^{0}C and above 4^{0}C, the optimum temperature range being 30 – 35^{0}C. In temperate regions, through a combination of the influences of moisture and temperature, soil nitrification is greatest in the spring and the fall (Paul and Clark, 1989). Following the production and release of NO_3^- into the soil through nitrification, there are several things that can happen. Nitrate can be taken up by organisms to fill their nutritional need for N. Within organisms, multiple biochemical reactions of the assimilatory reduction biological pathway transform NO_3^- to NH_4^+ that then can be processed into proteins. Unlike NH_4^+, NO_3^- is not readily adsorbed by clay minerals. As a consequence, it can be leached from the rooting zone in soils. This leaching reduces soil N availability to plants and can lead to the export of soil N to the ground water and, subsequently, to surface waters.

3.5.3. Denitrification

Conversely under anaerobic soil conditions, NO_3^- can be used as an electron acceptor during cellular respiration of certain bacteria. These denitrifying bacteria are principally heterotrophic; common forms include *Pseudomonas*, *Bacillus*, and *Alcaligenes*. In a series of reactions, NH_4^+ and other forms of reduced-N molecules can be produced. In enyzmatic denitrification, NO_3^- is first transformed to NO_2^- and then to N_2O and N_2. In dissimilatory reduction, NO_3^- is also used as an electron acceptor, and NH_4^+, but not gaseous N forms, is produced (Paul and Clark, 1989).

As with both nitrogen fixation and nitrification, environmental conditions exert a significant influence on the rate of denitrification. Sufficient soil water is required to produce the necessary anaerobic soil conditions. The optimal range of soil pH for denitrifying bacteria is pH 6-8, and it can occur down to pH 4. As with other microbial reactions, the rate of denitrification increases exponentially with temperature increase and occurs from 5^{0}C to 7.5^{0}C. Lastly, a simple statistic can be used to emphasize the importance of this microbial process in agricultural soils. On average, 15% of N in fertilizer is lost to the atmosphere through denitrification (Paul and Clark, 1989).

3.5.4. Bio-geochemical Linkage of Nitrogen and Carbon

While the above-mentioned bacterial processes exert a great influence on the soil chemistry of N, an understanding of the N cycle is incomplete if its relationship with C is not considered. The stoichiometric relationship between N and C in the organic molecules of living tissue is of great importance as a factor that influences the processes of decomposition and, consequently, the biological cycling of soil N and C. Although highly variable, the C:N ratio in trees, crops, and grasses averages 50:1 (Paul and Clark, 1989). In plants, while this ratio for the cytoplasm is substantially lower, high amounts of C-rich, N-poor organic compounds of the cell wall dramatically increase the whole-plant C:N ratio to the high value just stated. This ratio in plant tissue has an important influence on food chain transfers. For proper health, herbivores require a minimum amount of N in their food. As a consequence, it benefits herbivores to be selective in their consumption of plant species or specific tissues which do not contain overly high C:N ratio compounds.

As mentioned, the C:N ratio of dead biomass also influences soil decomposition and the resultant mineralization of N. Agricultural soils have approximately 45% C and 6-9% N which gives a C:N ratio of 10:1. Typical crop residues have 40% C and 1.6% N to produce a C:N ratio of 25:1. The C:N ratio of soil organisms is lower, ranging from 5:1 to 8:1. For the different groups of soil microorganisms, the C:N ratio also varies due to the relative proportions of C and N in cytoplasm and cell wall components (peptidoglycan in bacteria and chitin in fungi). Fungi contain approximately 45% C and 3-10% N to yield a C:N ratio with a range of 4.5:1 to 15:1. For bacteria, the ratio ranges from 3:1 to 5:1 (Paul and Clark, 1989).

In agricultural soils having the typical crop residue C:N ratio of 25:1, the N needs of the decomposer microbes can be met solely through the N that is available in the plant litter. Consequently, this produces no net mineralization or immobilization of N by the microbial food web. When the C:N ratio in litter is greater than 25:1, however, there is insufficient N to maintain optimal microbial growth. To meet this N need of the microbes, additional N may be acquired from soil inorganic-N or from the use of soil organic matter (C:N of approximately 10:1). If this is the case, microbial growth and decomposition rates will not be limited by N availability. If this supplemental N is not available, the rate of decomposition will be reduced (Paul and Clark, 1989).

3.5.5. Soil Organic Matter

Given the great importance of its influence on soil N dynamics and its role as a storage compartment of C in the global C cycle, it is desirable to explore further the organic matter component of soils. As a by-product of terrestrial food chains, dead organic matter accumulates in soils. This organic matter serves as a food source for the decomposer food web of the soil, as well as, stores a substantial amount of the world's C.

Soil organic matter (SOM) also is important for its capacity to store soil nutrients (including N) and to increase the water holding capacity of soils. In addition, it influences soil structure. It helps form soil mineral grains into various sized aggregates that, in turn, produce open pathways which facilitate soil aeration and water movement. Decomposition processes produce a wide variety of chemical forms of soil C. Some of these are fairly ephemeral while others are relatively resistant to decomposition and consequently enhance the accumulation of C in the soils of most terrestrial ecosystems. This amount of C stored in soil varies geographically due to the relative influences of climate on the processes of photosynthesis (production of organic matter) and cellular respiration (destruction of organic matter). The source of soil organic C is atmospheric CO_2 that enters terrestrial food chains through photosynthesis. This C can then be transferred to soils as dead organic matter in the form of plant litter, organic matter solubilized from plant tissues, food web wastes, and dead consumers within food webs. Organic matter is contributed to soils primarily as plant material in either dissolved or particulate forms. Particulate forms consist principally of carbohydrates (including cellulose, hemicellulose, and lignin) and proteins. The former constitutes 85% or more of a plant's biomass, and the latter constitutes a lesser amount (typically less than 15%) (Paul and Clark, 1989). Once on or within the soil, this dissolved and particulate organic matter is modified by the decomposer food web. This produces a mixture of SOM that includes these raw materials for the decomposer food web plus their decomposing residues, decomposition by-products, highly resistant humic matter, and even the microorganisms, themselves.

Chemically, SOM consists of a wide range of organic compounds within each of four categories—carbohydrates (of microbial origin, which constitute 10-20% of the SOM); N-containing compounds (20%); aliphatic fatty acids, alkanes, and other related compounds (10-20%); and aromatic compounds which make up the remainder of the material (Paul and Clark, 1989). The latter group is quite resistant to decomposition and is the predominant constituent of soil humic matter. Humic matter is dark and spongy in texture and is composed of a mixture of degradation residues that are fairly resistant to degradation by decomposition—somewhat the end-product of decomposition. Inputs of dead plant matter to the soil occur at the soil surface from the above-ground plant parts and below the surface from plant roots. Decomposition occurs in both above- and below-ground locations, and there is a tendency for dissolved and fine particulate SOM to be transferred from the soil surface downward into the soil through the action of percolating soil water. As a consequence, SOM accumulates both on the soil's mineral surface and in the upper portion of the mineral soil.

The amount of organic matter in a soil is influenced by the climatic conditions of moisture and temperature and various soil conditions including nutrient availability, water holding capacity, and aeration. Moisture influences the rate of photosynthesis, and therefore the production of organic matter in terrestrial ecosystems. Seasonality in moisture availability can produce a seasonality in photosynthesis. Water availability and, consequently, photosynthesis are decreased during dry seasons and during periods when soil water is frozen. Both water and temperature influence decomposition rates, and thus the accumulation of SOM and the generation of CO_2. Being ectotherms, the metabolic rates of decomposers (and therefore the decomposition rate) increase with increasing temperatures to a point, then fall off when temperatures are hotter than optimal for the decomposers. Likewise, the amount of water in a soil can limit decomposition if the soil is too dry or too wet. Waterlogged soils produce anaerobic conditions that slow decomposition rates. Under these moisture conditions, facultative and obligate anaerobic microorganisms can be active and can produce CH_4.

Globally, total annual plant production is estimated at 40×10^9 mg (Paul and Clark, 1989). Decomposition almost equals this amount, and consequently there is a net accumulation of organic matter within soils. Geographic variations in climatic factors (water and temperature) have produced a geographic variation in the accumulation of SOM, with some ecosystems having large quantities while other contain small amounts. For example in grasslands, the annual dry season limits decomposition producing the accumulation of SOM. Tropical wet forests have a greater production of biomass, but lower soil C than temperate forests. Greater accumulation of SOM in the latter forests is due to colder winter temperatures that limit decomposition. The highest accumulation of SOM occurs in highly productive wetlands where the rate of decomposition is slowed due to O_2 limitation in the waterlogged soils.

3.5.6. Perturbations and N and C Biogeochemistry

Human perturbations of ecosystems alter environmental conditions that influence the microbial processes of the N and C biogeochemical cycles. Land use changes that decrease plant cover (*e.g.*, deforestation for development or agriculture) can reduce C storage in the soil through decreasing litter inputs to the soil and increasing the rate of decomposition in the soil through increased soil warming. Globally, deforestation and other land use changes contribute approximately 1.5 Gt of C to the atmosphere each year (Paul and Clark, 1989). To compensate for such losses in agricultural systems, animal manure and green manures (cover crops that are plowed under) provide supplemental additions of organic matter to the soil. Soil N is also impacted through such land use changes. In agricultural soils, N is removed with harvested plant material and lost through soil erosion, leaching, and volatilization. To counter the losses, supplemental additions of inorganic N fertilizers (e.g., urea, nitrate, and ammonium) and organic N in the form of animal wastes can be added. When these additions are excessive, runoff and leaching can contaminate water bodies. Crop rotation, which incorporates the use of N fixing plants (*e.g.,* alfalfa), adds N to the soil.

Other farming practices alter the soil's aeration, water, and temperature regimes, which in turn can alter the rates of microbial processes leading to losses through N leaching and emissions of N_2, CO_2, and CH_4. Irrigation can lead to soil water logging and associated changes in aeration status. The use of heavy mechanical farm equipment can compact the soil. The traditionally used plow, the moldboard plow, turns over the soil to the depth of the plow. This disrupts soil structure altering aeration and water retention, as well as exposes bare soil making it susceptible to erosion. No-till approaches better maintain soil structure and decrease potential erosion, but require the use of herbicides. While agricultural practices produce local alterations in the N and C biogeochemical cycles, cumulative land use change over the Earth translates to global-scale disruptions of the cycles. With deforestation and other land-use change contributing significantly to increases in atmospheric CO_2, the potential for climate alteration at the global-scale increases. Global climate change has the potential to impact agricultural systems in different manners. Predicted increased temperature would likely increase the metabolic rates of soil microbes, thereby accelerating the rates of decomposition and N-transformations. For some geographic regions drier conditions are predicted, while for others precipitation may increase. Such variation would produce quite different geographic effects on microbial activity and the biogeochemical cycles of N and C. This, in turn, would contribute to significant variability in effects of climate change on agricultural productivity among different geographical regions.

3.5.7. Organic Nitrogen

Soil organic matter is the major storehouse of many plant nutrients in soils, including nitrogen, phosphorus, sulfur, calcium and magnesium. Soil organic matter is composed of a stable material called **humus**, an easily decomposed material (litter), soil microbes and some other organic molecules. Typically, humus will contain 45 to 55 percent carbon and about 5 percent nitrogen. In other words, soil humus typically has a carbon to nitrogen ratio (C:N ratio) of approximately 12:1 for surface soils and 8-10:1 for subsurface soils.

3.5.8. Mineralization

Nitrogen that is present in soil organic matter, crop residues and manures is converted to the inorganicform by the process of **mineralization**. Initially, larger organic matter molecules are broken down into smaller ones, with soil microorganisms attacking these remaining materials by producing specific enzymes. The transformation of organic nitrogen to the ammonia and ammonium forms is referred to as **ammonification**. The resultant ammonium can then be transformed to the nitrate form by a process called **nitrification**. Since the decomposition process is carried out by living organisms, it is affected by several environmental variables, including soil moisture, temperature, pH, the C:N ratio and the type of organic materials in the residue. Cotton and grain sorghum stalks and small grain residues are relatively high in carbon and low in nitrogen (having C:N ratios greater than 30:1). Soil organisms may need additional nitrogen to decompose these residues. When the nitrogen supply is limited, soil microbes compete with plants for fertilizer N in a process called **immobilization**. When these microbes die, nitrogen tied up in the decomposition process again becomes available for crop use.

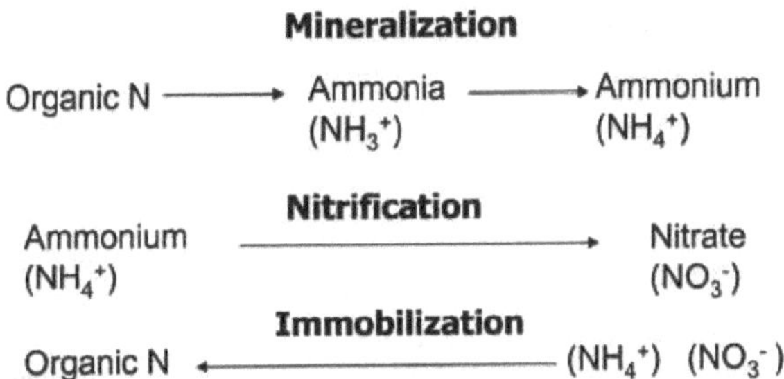

3.6. Factors Influencing Biological Nitrogen Fixation in Legumes

The complex process of legume BNF is affected by environmental conditions such as temperature, water content, N concentration, root zone pH, plant nutrient status including C and N substrates in roots, and genetic variation in potential N fixation capacity. It is also affected by plant nutritional status such as phosphorus (P) and potassium (K) levels that control nodule growth and nitrogenase activity directly or indirectly (Havelka *et al.*, 1982).

3.6.1. Temperature

Generally, soil temperature inhibits legume BNF through its control on nodulation, nodule establishment, and nitrogenase activity when it is either too high or too low (Roughley and Dart, 1970; Whitehead, 1995). Therefore, minimum and maximum soil temperatures, and the range of temperature between these which are favourable for N fixation, could be used to define the response of N fixation to soil temperature. Soil temperature in the root zone is one of controlling factors for nodulation and nodule establishment. For example, the nodulation of arrow leaf clover (*Trifolium vesiculosum* Savi.) is accelerated at a root temperature of 25⁰C compared with that growing at both 18⁰C and 32⁰C (Schomberg and Weaver, 1992). However, the response of nodule establishment to soil temperature differs between species and varieties. In soybean (*Glycine max* (L.) Merr.), more nodules are produced in the early growth stage at 25⁰C, while 20⁰C is optimal for nodule size after nodule generation is completed compared to 15⁰C and 30⁰C (Lindemann and Ham, 1979). In contrast, nodule establishment is enhanced with increasing temperature in the range of 10–35⁰C for white clover (*Trifolium repens* L.) regardless of the varieties and the rhizobia strains (Whitehead, 1995). Nitrogenase activity responds slightly differently to soil temperature between species. There are a large number of studies on the response of N fixation to temperature in legume crops, some of which are summarized in Table I. Minimum temperatures for N fixation differ among species from 2⁰C to 10⁰C, and normally tropical and subtropical legumes have higher minimum temperatures than temperate species. Nitrogenase activity is high around 12–35⁰C and reachesmaximum at 20–25⁰C in most legumes. N fixation in subterranean clover (*Trifolium subterraneum* L.) is very active at a wide range of temperatures, from 5⁰C to 30⁰C, but declines dramatically with low temperature and almost stops at 2⁰C. Generally, 35⁰C or 40⁰C is the upper limit of temperature for legume BNF.

3.6.2. Soil Water Status

In a similar manner to soil temperature, soil water content in the root zone controls N fixation through nodule establishment and nodule activity, plus gas permeability (Weisz *et al.*, 1985; Weisz and Sinclair, 1987; Sinclair *et al.*, 1987). Soil water deficit inhibits N fixation (Whitehead, 1995; Goh and Bruce, 2005), and the inhibition is reinforced as drought stress becomes more intense (Albrecht *et al.*, 1984). N fixation by peanut (*Arachis hypogaea* L.) grown in soil at two-thirds of field water capacity is reduced by 18–40% compared to that in well-watered soil, and by 44–69% when the soil water is one-third of field capacity

(Pimratch *et al.*, 2008). In addition, waterlogging can seriously reduce N fixation through depression of the establishment and activity of nodules (Havelka *et al.*, 1982). It is not easy to quantify the relationship between N fixation and soil water status precisely, due to the limitations of experimental controls, stress periods and plant recovery (Ledgard and Steele, 1992). Pimratch *et al.* (2008) tried to determine N fixation under drought stress with multiple-linear regression but the correlation coefficients between the reduction of N fixation under drought compared to under field capacity and the soil water deficit was not statistically significant. A sigmoid function has been used to describe the response of N fixation by common bean (*Phaseolus vulgaris* L.), black gram (*Vigna mungo* (L.) Hepper) and cowpea (*Vigna unguiculata* (L.) Walp) to soil water stress, showing a sharper decline in N fixation as soil water stress becomes more severe (Serraj and Sinclair, 1998; Sinclair *et al.*, 1987).

3.6.3. Nitrogen Concentration in the Root Zone

It has been widely reported that soil mineral N in the root zone inhibits legume nodulation (Abdel Wahab *et al.*, 1996; Herridge *et al.*, 1984), nodule establishment (Imsande, 1986) and nitrogenase activity (Purcell and Sinclair, 1990; Eaglesham, 1989) as it costs less energy for legumes to take up N from soil than fix N biologically from the atmosphere (Cannell and Thornley, 2000; Phillips, 1980; Wood, 1996). Normally, the severity of inhibition of N fixation by soil mineral, N increases with soil mineral N content (Macduff *et al.*, 1996; Lamb *et al.*, 1995; Waterer and Vessey, 1993a; Chalifour and Nelson, 1987). It has been reported that a certain concentration of mineral N in the root zone, defined as "starter N", stimulates nodule establishment and N fixation compared to non-mineral N in some circumstances. And the concentrations of "starter N" that stimulate legume BNF vary widely with cultivar and growth conditions but are normally less than 4mM for ammonium (NH_4^+) and less than 2 mM for nitrate (NO_3^-) (Schomberg and Weaver, 1992; Gulden and Vessey, 1997; Gan *et al.*, 2004). However, the time of external N application in relation to legume growth stage affects nodule growth and N fixation, and the later N is applied, the less nodule growth and N fixation is inhibited in field pea (*Pisum sativum* L.) (Jensen, 1986; Waterer and Vessey, 1993b). This is probably due to nodules being well established in the early stages before N is applied. Moreover, the inhibition of N fixation by NO_0^- was more severe than that of NH_4^+ in white clover, field pea and soybean (Svenning *et al.*, 1996; Bollman and Vessey, 2006; Gan *et al.*, 2004) even though high amounts of either NH_4^+ or NO_3^- retarded N fixation.

Table 3: Response of N fixation to soil temperature (^0C) for legume species

	Species	Minimum	Optimum range	Maximum	Reference
1.	Alfalfa (*Medicago sativa* L.)	2	20-25	-	Waughman (1977)
			35	40	Dart and Day (1971)
2.	Arrowleaf clover (*Trifolium vesiculosum*)	-	25	-	Schomberg and Weaver (1992)
3.	Barrel medic (*Medicago truncatula* Gaertn.)	2	20	35	Dart and Day (1971)
4.	Big-leaved lupin (*Lupinus polyphyllus* Lindl.)	1.5	25		Waughman (1977)
5.	Birdsfoot trefoil (*Lotus corniculatus* L.)	10	25-27	35	Rao (1977)
6.	Common bean (*Phaseolus vulgaris* L.)	-	-	35	Piha and Munns (1987)
				40	Michiels *et al.* (1994)
7.	Common vetch (*Vicia sativa* L.)	2	20	40	Dart and Day (1971)
8.	Cowpea (*Vigna unguiculata* (L.) Walp)	52	40	-	Dart and Day(1971)
			20-30	40	Dart and Day (1971)
9.	Faba bean (*Vicia faba* L.)	55	20	-	Waughman (1977)
			15-25		Halliday (1975)
10.	Field/garden pea (*Pisum sativum* L.)	0.5	25	-	Waughman (1977)
			20-26		Lie (1971)
11.	Guar (*Cyamopsis tetragonoloba* Guar)	-	-	37-40	Pate (1961)
12.	Narrowleaf lupin (*Lupinus angustifolius* L.)	10	20-30	-	Halliday (1975)
13.	Peanut (*Arachis hypogaea* L.)	-	-	40	Kishinevsky *et al.* (1992)
14.	Purple vetch (*Vicia atropurpurea* Desf.)	-2	2425	-35	Pate (1961)
					Dart and Day (1971)
15.	Red clover (*Trifolium pratense* L.)	—	12-26	-	Small and Joffe (1968)
			27		Kuo and Boersma (1971)
16.	Soybean (*Glycine max* L.)	-52	20-25	-	Lindemann and Ham (1979)
			20-35	40	Dart and Day (1971)
			5-30	-	Dart and Day (1971)
17.	Subterranean clover (*Trifolium subterraneum* L.)	5-5-	12-32	-	Gibson (1971)
			15		Roughley and Dart (1969)
			13-26		Halliday and Pate (1976)
			26		Small and Joffe (1968)
18.	White clover (*Trifolium repens* L.)	-9	21	-	Masterson and Murphy (1976)
			26		Frame and Newbould (1986)

3.6.4. Carbon Demand for Fixation

Photosynthate partitioned to roots supports nodule growth, provides energy for N fixation, maintains a functional population of rhizobia, and allows the synthesis of amino compounds produced from N fixation (Layzell *et al.*, 1979; King *et al.*, 1986). Even though it is difficult to distinguish the proportion of CO_2 generated by N fixation from that generated by respiration for nodule growth and maintenance (Warembourg and Roumet, 1989), the correlation between the rate of CO_2 produced from either nodulated roots or nodules and N fixation rate may be used to evaluate C consumption by N fixation (Lawrie and Wheeler, 1973; Mahon, 1977a, b; Warembourg and Roumet, 1989). The C cost per unit fixed N (g C per g N fixed) varies widely with species, growth stage and environmental conditions, and ranges from 1.4 to 8.5 (Phillips, 1980; Minchin *et al.*, 1981; Sheehy, 1987; Schulze *et al.*, 1999). For example, it is reported that the C cost is 1.54 g C g N fixed"1 in cowpea and 3.64 g C g N fixed"1 in white lupin (*Lupinus albus* L.) from nodules during early vegetative stage (Layzell *et al.*, 1979; Layzell *et al.*, 1981) while it is 6.3–6.8 g C g N fixed"1 for soybean, cowpea and white clover from nodulated roots at periods of intense N fixation. The C cost determined by CO_2 released from nodulated roots is generally higher than that from nodules as the former includes root respiration. The C cost of N fixation also varies with growth stages (Twary and Heichel, 1991), but it is a matter of debate that the C cost increases (Warembourg and Roumet, 1989; Voisin *et al.*, 2003) or decreases (Adgo and Schulze, 2002) with the course of legume life cycle. In addition, the strain of *Rhizobium* may affect the C cost significantly. For example, the C cost in alfalfa (*Medicago sativa* L.) nodules formed by strain P207 is an average 9.4 molC molN"[1] which is 59% higher than that of nodules formed by strain I02F51, 5.9 molC molN"[1] (Twary and Heichel, 1991).

3.6.5. Seasonal Regulation of Legume Biological Nitrogen Fixation

The rate of legume BNF changes with physiological growth stages. It is low in the early growth stages while nodules are establishing (Lawrie and Wheeler, 1973) and reaches a maximum value between early flowering and early seed-filling, depending on the species and growing conditions (Lawn and Brun, 1974; Klucas, 1974; Nelson *et al.*, 1984; Jensen, 1987). After the peak, N fixation decreases dramatically or even ceases during seed-filling (Herridge and Pate, 1977; Beverly and Jarrell, 1984; Sinclair *et al.*, 1987), due to nodule senescence (Lawrie and Wheeler, 1973) and poor C supply as a result of the strong demand for seed dry matter accumulation (Herridge and Pate, 1977; Voisin *et al.*, 2003, 2007).

3.7. Quantification of Biological Nitrogen Fixation

Legume biological nitrogen fixation may be quantified by direct measurement, estimation based on yield or with empirical models, or simulation by crop models. The methods used to measure N fixation directly so far, such as the acetylene reduction/hydrogen increment assay, N difference, 15N-labelling and ureide, have been thoroughly reviewed (Herridge *et al.*, 2008; Carlsson and Huss-Danell, 2008).

3.7.1. Estimation with Empirical Models

A static estimation of N fixation during the whole growing season may use either economic yield or above-ground dry matter. The equation is:

$$Nfix = \alpha * DM * fleg * Ncon * \%Ndfa * (1 + Rroot)* \quad (1)$$

where DM is the yield or aboveground dry matter, fleg is the proportion of legume if it is intercropped, Ncon is the N concentration in the legume, %Nd f a is the proportion of total plant N derived from N_2 fixation, and Rroot is the ratio of the fixed N belowground to the fixed N aboveground. Values for the parameters fleg, Ncon and Rroot for different species at a range of sites are summarized in Table II. α is a parameter that has different definitions according to author. For example, α may represent the decline in *%Nd f a* under high soil mineral N conditions, in terms of net mineral N input (g N m^{-2}), and is calculated as:

$$\alpha = 1 - \beta * Nnet.inorg \quad (2)$$

where, β evaluates the sensitivity of legume BNF to soil mineral N (Korsaeth and Eltun, 2000). This is set to 0.028 for white clover (*Trifolium repens* L.) and red clover (*Trifolium pratense* L.), 0.043 for grey peas (*Pisum arvense* L.) and common vetch (*Vicia sativa* L.). Alternatively, α is used as a modifier to compensate for the proportion of fixed N transferred to a companion crop, consumed by grazing animals, or lost by immobilization (Høgh-Jensen *et al.*, 2004).

The second method to estimate N fixation is by empirical models based on the correlation of fixed N in the final yield against variables, such as harvested dry matter or the proportion of legume in mixed leys. A linear equation has been fitted to the measured data of mixed white clover and grass swards at different sites from four countries. This showed a significant correlation between fixed N and the extra dry matter of mixed leys, compared with corresponding pure grass either for cut swards or grazed swards (Watson and Goss, 1997). Kristensen *et al.* (1995) found fixed N at harvest increased linearly with clover

Parameter values used to estimate N fixation in equation

Species	Condition	N_{con} (%)	% *Ndf*	R_{root}	Reference
1. White clover	Stock camp	4.9	61	-	Ledgard *et al.* (1987)
2. White clover	Gentle slope	4.9	82	-	Ledgard *et al.* (1987)
3. Subterranean clover	Gentle slope	4.9	82	-	Ledgard *et al.* (1987)
4. Subterranean clover	Steep slope	4.9	82	-	Ledgard *et al.* (1987)
5. Slender birds-foot trefoil	Steep slope	4.9	82	-	Ledgard *et al.* (1987)
6. Lucerne	Cut 1–2 years	2.7	74	0.25	Høgh-Jensen *et al.* (2004)
7. Red clover	Cut 1–2 years	3.0	74	0.25	Høgh-Jensen *et al.* (2004)
8. White clover	Intercropping with grass, cut 1–2 years	4.3	95	0.25	Høgh-Jensen *et al.* (2004)
9. White clover	Intercropping with grass, grazed 1–2 years	4.3	80	0.25	Høgh-Jensen *et al.* (2004)
10. Red clover	Intercropping with grass, cut 1–2 years	3.0	95	0.25	Høgh-Jensen *et al.* (2004)
11. Red clover	Intercropping with grass, grazed 1–2 years	3.0	80	0.25	Høgh-Jensen *et al.* (2004)
12. White clover	Intercropping with grass,cut >2 years	4.3	95	0.25	Høgh-Jensen *et al.* (2004)
13. White clover	Intercropping with grass,grazed >2 years	4.3	75	0.25	Høgh-Jensen *et al.* (2004)
14. White clover	1st cut 2nd cut	3.94	92.3	For all cuts	Korsaeth and Eltun (2000)
		3.49	92.3	(year 1, 2 and 3)	
15. Red clover	1st cut 2nd cut	3.14	92.3	-	Korsaeth and Eltun (2000)
		2.91	92.3		
16. Grey peas	-	2.80	65.4	0.045	Korsaeth and Eltun (2000)
17. Common vetch	-	3.00	65.4	0.27	Korsaeth and Eltun (2000)

dry matter content in mixed swards through statistically analyzing the experimental data from different sites with distinct soil types and irrigation schemes.

Values of parameters, *c* and *d*, in linear empirical models expressed in equation.

Condition	c	d	Model reference
Cut swards	6.8	0.067	Watson and Goss (1997)
Grazed swards	–168.1	0.067	Watson and Goss (1997)
1–2 cropping years	18	4.47	Kristensen *et al.* (1995)
3–5 cropping years	19	2.77	Kristensen *et al.* (1995)

The equation for the estimation of N fixation (kg N ha^{-1}) is summarized as,

$$Nfix = c + d * Leg \quad (3)$$

Where, *Leg* is the extra dry matter increment (kg ha^{-1}) in white clover mixed leys compared with a pure grass ley in Watson's estimation, and clover dry matter content (%clover) in clover mixed leys for Kristensen's study (1995), respectively; and *c* and *d* are parameters the values of which are summarized in Table 3.

The first method described above is apparently a direct estimation of N fixation, and the parameter values can be easily measured on site or estimated from the literature. It does not strictly require an adequate dataset for multiple years to determine the parameters, so it is easy to use. However, when determining the parameter values, data from years of abnormal weather should be avoided, and the properties of the soil should be relatively stable year-on-year.As these equations are independent from environmental factors such as soil properties and weather conditions, they are only applicable and accurate for similar sites and average weather conditions. In addition, the parameter values need to be adjusted if the equations are used for different sites or legumes.

In contrast, the second method is based on statistical correlation and assumes that N fixation has a strong linear relationship to the variables. It is more flexible to use and can be applied to one specific site or multiple sites with different soil types, depending on how the empirical relationship is developed and which sites the data were obtained from. This approach has a higher data requirement compared to the first method, and the data should be representative and adequate to guarantee the correlation and determine the parameter values. However, as with the first method, these approaches are restricted to specific sites because the equation is not able to represent the interaction between plant and environment mechanistically.

In simulation models of biological nitrogen fixation in legumes, the most popular method to estimate the rate of legume biological nitrogen fixation is a potential or maximum fixation rate modified by the influence of environmental factors. The potential fixation rate is estimated based on either a demand-uptake mechanism or on the dry matter of plant tissues, and is varied with plant growth stages. The environmental factors normally include soil temperature, soil or plant water content, soil mineral N or substrate N concentration in plant tissues and substrate C concentration in the plant. Other factors, such as soil pH, salinity and the supply of other nutrients, have not been included in models to date. The most-used recent simulation models where a legume N fixation function has been implemented (Tab. IV). As crop models may be used under different circumstances, the estimation of N fixation may have various versions within the same model. The general expression of the calculation in the majority of the reviewed models can be written as:

$$N_{fix} = N_{fixpot} f_T f_W f_N f_C f_{gro} \quad (4)$$

While, in the EPIC and STICS models it is:

$$N_{fix} = N_{fixpot} f_T \min(f_W, f_N) f_{gro} \quad (5)$$

where, $Nfixpot$ is the potential N fixation rate (g N fixed day^{-1}), fT is the influence function of soil temperature, fW is a soil water deficit or flooding function, fN is the function of soil mineral N or root substrate N concentration, fC is the function of substrate C concentration in plant or root, $fgro$ is the influence factor of growth stage and min is the mathematical function to take a minimum value between fW and fN. There is an extra function, fa, representing the limitation by anoxia in the STICS model.

3.7.2. Potential N Fixation Rate

There are two definitions on a potential legume biological nitrogen fixation rate used in the models based either on the difference between N demand and uptake by a legume, or on the N fixation capacity of legume nodules. The EPIC and APSIM models use variations of the first definition to estimate potential N fixation rate. The EPIC model assumes that the total plant N demand is equal to the potential N fixation (Bouniols *et al.*, 1991; Cabelguenne *et al.*, 1999). APSIM defines critical N concentrations for plant tissues and uses these to estimate N demand by maintaining non-stressed N levels in plant tissues and supporting the N demand of new tissues. This N demand is met by either N uptake from soil and/or N fixation. The former has a higher priority because the process is less energy consuming than N fixation (Macduff *et al.*, 1996); N

Simulation models that include legume biological nitrogen fixation, and the factors considered in each model.* indicates model named here after the first author's name. *fT* , *fW*, *fN*, *fC* and *fgro* are the factor of soil temperature, soil/plant water, soil/plant nitrogen, plant carbon and plant growth stage, respectively, used in equations (4, 5)

Model	Factors					Simulated legume species	Reference
	f_T	f_W	f_N	f_C	f_{gro}		
Sinclair Model*		√					
						soybean	Sinclair (1986)
		√			√	soybean, cowpea, black gram	Sinclair *et al.* (1987)
EPIC		√	√		√	soybean	Sharpley and William (1990); Bouniols *et al.* (1991); Cabelguenne *et al.* (1999)
Hurley Pasturte Model	√	√		√	√	White clover	Thornley (1998); Thornley and Cannell (2000); Thornley (2001)
			√	√		Field pea	Eckersten *et al.* (1996)
Schwinning Model*			√				White clover Schwinning and Parsons (1996); Schmid *et al.* (2001)
CROPGRO	√	√		√	√	Soybean, peanut, drybean, velvet bean, faba bean, cowpea	Boote *et al.* (1998); Sau *et al* (1999); Hartkamp *et al.* (2002); Boote *et al.* (2002, 2008)
SOILN		√	√	√		White clover	Wu and McGechan (1999)
APSIM		√		√	√	Soybean, chickpea, peanut, mungbean, lucerne	Herridge *et al.* (2001); Robertson *et al.* (2002)
Soussana Model*			√			White clover	Soussana *et al.* (2002)
STICS	√	√	√		√	Field pea and other legumes	Brisson *et al.* (2009); Crre-Hellou *et al.* (2007, 2009)

fixation is only calculated if N uptake can not meet the plant N demand. Thus the potential N fixation is assumed to be the difference between plant N demand and N uptake (Herridge *et al.*, 2001; Robertson *et al.*, 2002).The second definition is based on the strong relationship between N fixation and either nodule size/biomass (Weisz *et al.*, 1985; Voisin *et al.*, 2003) or root biomass (Voisin *et al.*, 2007). As the biomass of both nodules and roots are difficult to measure in the field, some studies have used above-ground biomass to replace nodule/root biomass, based on the relationship between these two variables (Bell *et al.*, 1994; Yu *et al.*, 2002). The potential fixation rates used in the models are shown in Table 5.

3.7.3. Impact of Soil Temperature

The impact of soil temperature on N fixation rate is assumed to follow a four-threshold-temperature function in most of the reviewed models:

$$fr = \begin{cases} 0 & (T < Tmin \ or \ T > Tmax) \\ \dfrac{T - Tmin}{ToptL - Tmin} & (Tmin \le T \le ToptL) \\ 1 & (ToptL \le T \le ToptH) \\ \dfrac{Tmax - T}{Tmax - ToptH} & (ToptH < T < Tmax) \end{cases} \tag{6}$$

Where *T* is soil temperature (æ%C), *Tmin* is the minimum temperature below which N fixation ceases, *ToptL* and *ToptH* define the optimal temperature range within which the response function to soil temperature is the unit, and *Tmax* is the maximum temperature above which fixation stops. The values of these four temperatures vary among models and are shown in Table 6. A slightly different function is used in the Hurley Pasture model to simulate the influence of temperature on N fixation and most plant rate responses of white clover (Thornley, 1998):

$$fT = T2(45 - T) \times 10{-}4. \tag{7}$$

There are variations in the response functions adopted in the reviewed models. For example, the functions for white clover in Hurley Pasture and SOILN differ in curve shapes and threshold temperatures. There is a much wider range of temperature for N fixation in Hurley Pasture than SOILN, perhaps because the function in the Hurley Pasture model is not specially assigned to the N fixation module, but is used to simulate most plant processes. The function reaches unit at 20^{0}C for both models and the maximum value of the function occurs at 30°C in Hurley Pasture, while the function is zero in SOILN at the same temperature. The value of the function in Hurley Pasture is higher than SOILN when the soil temperature is below 10°C, but this reverses when soil temperature is in the range 10 to 20°C. Over 20°C there is a large difference between Hurley Pasture and SOILN. This may be because Hurley Pasture was developed in southern Britain, where the temperature is often between 4 and 16°C (Thornley and Cannell, 2000) and thus it might not perform well at much higher temperatures.

3.7.4. Impact of Soil Water Status

Soil water stress in the form of either a deficit or excess in the root zone could inhibit nodule nitrogenase activity. The effect of water deficit is considered in some models, but only few models take account for the influence of excess water. An exponential or linear equation derived from experimental data to represent the response of legume biological nitrogen fixation rate to soil water deficit is incorporated into most of the reviewed models. In STICS, the soil is divided into depth layers, and the proportion of these soil layers where water contents are above permanent wilting point is used as the water deficit factor. In Sinclair's model, an empirical sigmoid relationship between relative nodule nitrogenase activity, expressed as C_2H_2, and transpirable soil water was developed based on experimental data (Sinclair, 1986; Sinclair *et al.*, 1987):

$$f_W = -1 + \frac{2}{1 + e\,(-m^*f_{TSW} + n)} \tag{8}$$

Where *fTSW* is the fraction of transpirable soil water, and the total transpirable soil water content is defined as the difference between field capacity and the soil water content when the transpiration rate from drought-stressed plants decreases to less than 10% of that from well-watered plants (Sinclair, 1986); *m* and *n* are constants which defines the sensitivity of legume biological nitrogen fixation to low soil water content. Values of *m* and *n* are 6 and 0 for soybean, 9 and 0.03 for both cowpea and black gram, respectively.

A linear function is incorporated into APSIM, EPIC (Sharpley and Williams, 1990; Bouniols *et al.*, 1991; Cabelguenne *et al.*, 1999) and SOILN (Wu and McGechan, 1999):

$$f_w = \begin{cases} 0 & (W_f \le Wa) \\ \varphi_1 + \varphi_2 \cdot W_f & (W_a < W_f < W_b) \\ 1 & (W_f \ge W_b) \end{cases} \tag{9}$$

where W_f is the relative available soil water, expressed as the ratio of available soil water content to that at field capacity, φ_1 and φ_2 are coefficients, and

Potential fixed N rate used in the models. Value in APSIM varies with growth stage.

Model	Species	Maximum specific N fixation rate	Unit	Based	Reference
Sinclair Model	soybean	$0.55 – 0.7x10^{-3}$	g N g shoot DM^{-1} d^{-1}	Shoot biomass	Sinclair (1986); Sinclair *et al.* (1987)
	cowpea	$0.7 – 0.8 x10^{-3}$	g N g shoot DM^{-1} d^{-1}	Shoot biomass	
	Black gram	$0.7 – 0.75x10^{-3}$	g N g shoot DM^{-1} d^{-1}	Shoot biomass	
APSIM	legumes	1.0	$– 6.0x10^{-3}$	g N gshoot DM^{-1} d^{-1}	Shoot biomass (APSIM source code: http:// apsrunet. apsim.info/svn/ development/trunk/apsim/)
	Soybean, chick pea, Mungbean, peanut, lucerne	N demand-uptake	g N d^{-1}	N demand after N uptake	Herridge *et al.* (2001); Robertson *et al.* (2002)
EPIC	soybean	total N demand	g N d^{-1}	total N demand	Bouniols *et al.* (1991); Cabelguenne *et al.* (1999)
Hurley Pasture Model	clover	0.175	g N g nodule DM^{-1} d^{-1}	nodule biomass	Brugge and Thornley (1984)
	whiteclover	$0.05- 1.0x10^{-3}$	g N g nodule DM^{-1} d^{-1}	root biomass	Thornley, 1998; Thornley and Cannell (2000); Thornley (2001)
	pea	0.012-0.027	g N g nodule DM^{-1} d^{-1}	root biomass	Eckersten *et al.* (2006)
CROPGRO	faba bean	0.05	g N g nodule DM^{-1} d^{-1}	nodule biomass	Boote *et al.* (2002)
	soybean	0.045	g N g nodule DM^{-1} d^{-1}	nodule biomass	
SOILN	white clover	0.1106	g N g nodule DM^{-1} d^{-1}	nodule biomass	Wu and McGechan (1999)
Soussana Model	white clover	0.04	g N g nodule DM^{-1} d^{-1}	root biomass	Soussana *et al.* (2002)
STICS	legumes	-	g N g nodule DM^{-1} d^{-1}	above ground growth rate	Brisson *et al.* (2002)
	field pea	0.028	g N g nodule DM^{-1} d^{-1}	above ground growth rate	Corre-Hellou *et al.* (2007); Corre-Hellou *et al.* (2009)

Values of threshold temperatures (æ °C) used in three models that use equation (6) to simulate the effect of temperature on legume biological nitrogen fixation.

Model	Species	T_{min}	T_{optL}	T_{optH}	T_{max}	Reference
CROPGRO	soybean	5	20	35	44	Boote *et al.* (2008)
	cowpea	5	20	35	44	
	velvet bean	5	23	35	44	
	dry bean	4	19	35	44	
	peanut	7	23	34	44	
	faba bean	1	16	25	40	
SOILN	white clover	9	13	26	30	Wu and McGechan (1999)
STICS	legume	0	15	25	35	Brisson *et al.* (2009)

Parameter values in the response function to soil water deficit.

Model	APSIM											EPIC	SOILN
Parameter value	soybean	peanut	Navy bean	mungbean	lupin	Lucern	Faba bean	Cowpea	chickpea	Butterfly pea	Field pea	soybean	White clover
φ_1	-0.33	0	0	0	0	0	0	0	0	0	0	-0.82	0.216
φ_2	1.67	2.5	2	2	2	1.25	2	2	2	2	2	1.82	0.789
W_b	0.8	0.4	0.5	0.5	0.5	0.8	0.5	0.5	0.5	0.5	0.5	1	1
W_a	0.2	0	0	0	0	0	0	0	0	0	0	0.45	-

W_b is a critical value of W_f above which N fixation is not inhibited by soil water content and W_a is the threshold below which N fixation is totally restricted by soil water deficit.

Although a linear expression similar to equation (9) is used in CROPGRO to simulate the impact of water stress on N fixation, the variable *Wf* is defined as the ratio of root water uptake to transpiration demand (Boote *et al.*, 2008). Furthermore, the value of *Wf* on the current day is kept to compare with the average of *Wf* for the last eight days, to account for the prolonged impact of any previous severe drought on N fixation. The final value of *Wf* is equal to the lesser of these two values. The upper soil layer above 30 cm depth is prone to be dry or wet when drought or irrigation occurs, and this causes a lag between water potential in the root nodules and that in the soil (Albrecht *et al.*, 1984). Therefore, water potential within roots could be a more stable indicator to quantify the relationship between water stress and N fixation. In the Hurley Pasture model, it is assumed that N fixation is controlled through chemical activity in roots, which is further influenced by root water potential and soil temperature. Thus the effect of water on N fixation is combined together with temperature (Thornley, 1998):

$$f_w = \frac{20 * [18 * \Delta_{rt}]}{e\ 8314 * (Ts + 273.15)} \qquad (10)$$

Where ö*rt* is root water potential (J kg-1) and *Ts* is soil water temperature (ÚC). Compared with water deficit, the influence of excessive water on the simulation of N fixation has received very little attention in published studies. In Sinclair's model, a simple assumption is adopted that N fixation is set to zero once flooding happens (Sinclair, 1986; Sinclair *et al.*, 1987). In STICS, the restraint of legume biological nitrogen fixation by excessive water is considered as a stress from anoxia, which is calculated as the proportion of soil layers that are in anaerobic conditions in the nodulation zone (Brisson *et al.*, 2009). The evaluation of N fixation inhibition by water excess is incorporated into CROPGRO (Boote *et al.*, 2008) by computing the fraction of pore space filled with water. When pore space is fully filled with water, N fixation is inhibited, but the validation of this rule to date is insufficient.

3.7.5. Effect of Soil Mineral N or Internal Substrate N

A natural logarithmic function to represent the effect of soil mineral N content on N fixation by white clover is incorporated in SOILN model:

$$f_N = \begin{cases} 1 - 0.0784\, lnN_s & (N_s \geq 1) \\ 1 & (N_s < 1) \end{cases} \qquad (11)$$

where *Ns* is soil mineral N concentration (mg N m^{-3}). As N uptake and N fixation are the main N source for legumes, the influence of soil mineral N on legume biological nitrogen fixation rate could be presented indirectly by its influence on N uptake from soil. In Schwinning's model, such a scheme is used:

$$f_N = \varepsilon\, x\, \left(1.0 - f_{Nup}\right) = \varepsilon\, x \left[1.0 - \text{fmax}\frac{1}{1 + K_N / N_s}\right] \qquad (12)$$

where ε is the efficiency of legume biological nitrogen fixation and a value of 0.6 was used for white clover, *fmax* is a maximal fraction of N derived from soil mineral N uptake, which is assumed to be 0.85 for white clover to make sure that N fixation still retains even in high soil mineral N concentrations in the root zone; *KN* is the nitrate content (g N m^{-2}) at which an N uptake rate reaches half its maximal rate and *NS* is the actual soil nitrate content (g N m^{-2}). Even though the schemes in SOILN and Schwinning's model are different and *fN* in Schwinning's model is never greater than 0.6 due to the limitation by legume biological nitrogen fixation efficiency, they describe a similar trend of N fixation response to soil mineral N by white clover.

The values of *fN* from these two functions are very close while soil mineral N is higher than 0.05 g N m^{-2} (Fig. 2). The impact of soil mineral N is assessed as a linear function when soil nitrate concentration (*NsNitra*) is between 10 and 30 g N m^{-3} within 0.3 m top soil in EPIC (Sharpley and Williams, 1990; Cabelguenne *et al.*, 1999):

$$f_N = \begin{cases} 1 & (N_{sNitra} \leq 10) \\ 1.5 - 0.05 N_{sNitra} & (10 < N_{sNitra} < 30) \end{cases} \qquad (13)$$

A similar function is incorporated in STICS as in EPIC, with a different value of *NsNitra*. Moreover, in STICS, nodulation progress is also inhibited by high soil nitrate concentration, which is represented by a reduction of potential N fixation rate. Once the soil nitrate concentration in the nodulation zone is greater than a critical value, *Nfixpot* equals a baseline value; otherwise, *Nf ixpot* is kept at the normal value (Brisson *et al.*, 2009). Both the Hurley Pasture Model and Soussana's model use plant substrate N concentration to simulate the influence of N nutritional status on N fixation rate:

$$f_N = \frac{1}{1 + N_{inter} / K_r} \qquad (14)$$

where *Ninter* (g N g^{-1}r.wt) is root substrate N concentration in the Hurley Pasture Model or plant substrate N concentration in Soussanna's model, respectively; and *Kr* is a coefficient inhibiting N fixation at high internal N concentration and it is set to 0.02 for white clover and 0.01 for field pea in the Hurley Pasture Model (Thornley, 1998, 2001; Eckersten *et al.*, 2006), 0.004 for both a normal cultivar and a low-soil-N-uptake breeding cultivar, and 0.0004 for a low-N-fixation breeding cultivar in Soussana's model for white clover (Soussana *et al.*, 2002).

3.7.6. Influence of Plant Substrate C or C supply

Biological N fixation has a high energy demand and the rate of legume biological nitrogen fixation depends on the C supply, which is the energy source for N fixation. Experimental observations demonstrate that the enhancement of photosynthetic C allocated to roots which is available for N fixation and nodule establishment increases nodule numbers and N fixation rate (Voisin *et al.*, 2003; Haase *et al.*, 2007; Kirizii *et al.*, 2007). However, among the reviewed models only the Hurley Pasture and CROPGRO models implemented this mechanism. In the Hurley Pasture Model, a Michaelis-Menten function is used to demonstrate the effect of root substrate C on the N fixation of white clover and field pea:

$$f_c = \frac{1}{1 + K_c / C_r} \tag{15}$$

where *Cr* is legume root C substrate concentration (g C g^{-1} r.wt), and *Kc* is a Michaelis-Menten constant, set to be 0.01 for white clover (Thornley, 2001) and 0.035 for field peas (Eckersten *et al.*, 2006).

In CROPGRO, photosynthetic carbohydrate supports not only the energy consumption of N fixation but is also the substrate supply for nodule biomass accumulation. Carbohydrate allocated to nodulated roots is divided into three parts with priorities: requirement for minimum nodule growth, the cost for N fixation and the requirement for nodule normal growth (Fig. 4.). First, it needs to guarantee the minimum daily nodule growth (step 1). Then the carbohydrate left over after the reduction for the minimum nodule growth is assumed to be potentially available for N fixation to determine potential N fixation rate (step 2), and in this step carbohydrate is not truly used to fix N until the actual N fixation rate is determined (step 3). If there is any carbohydrate left after the third step, the remainder will be used to produce more nodules (step 4) (Boote *et al.*, 2008).

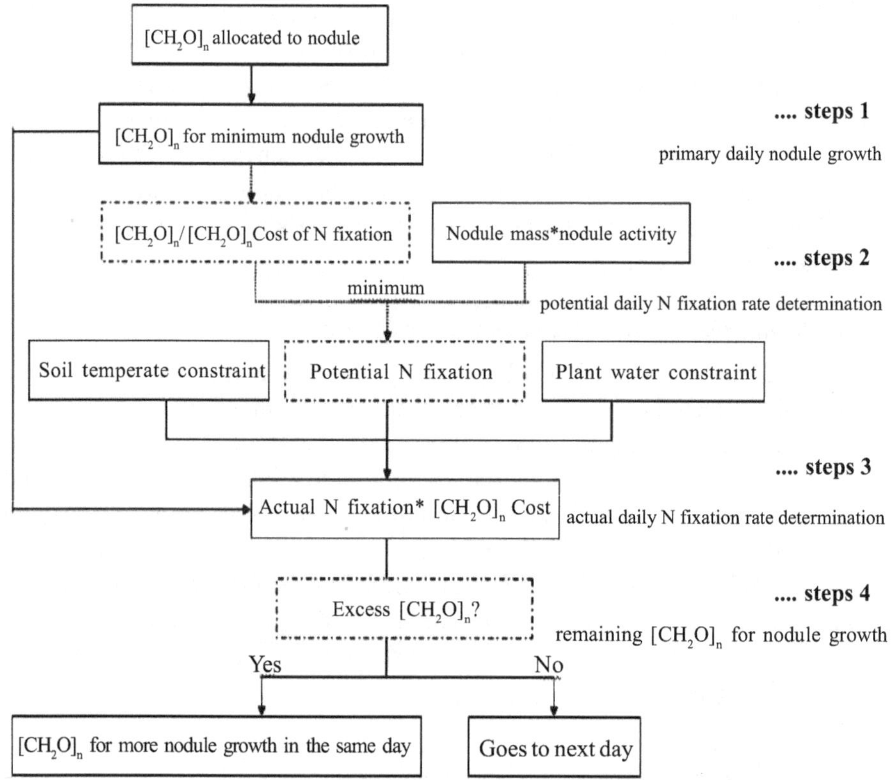

Fig. 4: The representation of daily carbohydrate routes in nodules used in CROPGRO (Boote *et al.*, 2008). $[CH_2O]n$ in dotted line means that the amount of $[CH_2O]n$ here is only used for calculation, neither for fixation nor nodule growth.

3.7.7. Changes in N Fixation with Plant Growth Stage

The quantitative effect of plant growth stage on legume biological nitrogen fixation rate is considered in very few models and in general the process is stopped forcibly after the legume attains a certain growth stage. For example, N fixation ceases at the beginning of seed growth for cowpea and black gram, but continues until the end of seed-filling for soybean in Sinclair's model (Sinclair *et al.*, 1987), whereas it stops at physiological maturity in CROPGRO (Boote *et al.*, 2008). A more specific function, similar to the temperature response function, is incorporated into EPIC and STICS to simulate the seasonal change of N fixation (Sharpley andWilliams, 1990; Bouniols *et al.*, 1991; Cabelguenne *et al.*, 1999):

$$f_{gro} = \begin{cases} 0 & (g < g_{min} or g > g_{max}) \\ \dfrac{g - g_{min}}{g_{optL} - g_{min}} & (g_{min} \leq g \leq g_{optL}) \\ 1 & (g_{optL} \leq g \leq g_{optH}) \\ \dfrac{g_{max} - g}{g_{max} - g_{optH}} & (g_{optH} < g < g_{max}) \end{cases} \quad (16)$$

where *gmin* is a period before which no fixation happens because of inadequate nodulation, expressed as the percentage of total growing period required; g*optL* and g*optH* are the beginning and end time within which legume BNF rate is not limited by growth stage; and g*max* is the time after which N fixation ceases because of nodule senescence. The values of the parameters for soybean are set to 10%, 30%, 60%, and 80% of the life cycle by Bouniols *et al.* (1991), and 15%, 30%, 55% and 75% by Cabelguenne *et al.* (1999). In the STICS model, g*min* and g*max* are the times of nodule initiation and senescence respectively, and g*optL* and g*optH* correspond to 27% and 80% of nodule life (Brisson *et al.*, 2009).

References

Abdel Wahab A.M., Zahran H.H. and Abd-AllaM.H. 1996. Root-hair infection and nodulation of four grain legumes as affected by the form and the application time of nitrogen fertilizer, *Folia Microbiol*, 41: 303–308.

Adgo E. and Schulze J. 2002. Nitrogen fixation and assimilation efficiency in Ethiopian and German pea varieties, *Plant Soil*, 239: 291–299.

Agarwal S., Singh B.K, and Alka, P. 2010. Salinity tolerance of Berseem (*Trifolium alexandrinum* L.) during germination and early seedling growth. *Vegetos*, 23: 63–82.

Aguirreolea J.,and Sanchez-Diaz M. 1989. CO_2 evolution by nodulated roots in *Medicago sativa* L. under water stress. *J. Plant Physiol.*, 134: 598–602.

Albrecht S.L., Bennett J.M. and Boote K.J. 1984. Relationship of nitrogenase activity to plant water stress in field-grown soybeans, *Field Crops Res.*,8: 61–71.

Andersen M.K., Hauggaard-Nielsen H., Ambus P. and Jensen E.S. 2004. Biomass production, symbiotic nitrogen fixation and inorganic N use in dual and tri-component annual intercrops, *Plant Soil*, 266: 273–287.

Asghari H.R. 2008.Vesicular-arbuscular (VA) mycorrhizae improve salinity tolerance in pre-inoculation subterranean clover (*Trifolium subterraneum*) seedlings. *International Journal of Plant Production*, 2: 243–256.

Ashraf M. and Bashir A. 2003. Salt stress induced changes in some organic metabolites and ionic relations in nodules and other plant parts of two crop legumes differing in salt tolerance. *Journal of Flora*, 198: 486–498.

Azcon Bieto J., Fleck I., Aranda X. and Xambo A. 2000. Fotos´ýntesis en unambiente cambiante. In, *"Fundamentos de Fisiolog´ýa Veget al,"* Eds. J. Azcon-Bieto, M. Talon, McGraw-Hill Interamericana, Madrid, Spain, pp.203–217.

Baroli I. and Melis A. 1998. Photoinhibitory damage is modulated by the rate of photosynthesis and by photosystem II light-harvesting chlorophyll antenna size. *Planta*, 205: 288–296.

Bala N., Sharma P.K. and Lakshminarayana K. 1990. Nodulation and nitrogen-fixation by salinity-tolerant rhizobia in symbiosis with treelegumes. *Agriculture, Ecosystems & Environment,* 33: 33–46.

Bell M.J., Wright G.C., Suryantini, and Peoples M.B. 1994. The N_2-fixing capacity of peanut cultivars with differing assimilate partitioning characteristics, *Aust. J. Agric. Res.*, 45: 1455–1468.

Ben Salah I., Slatni T., Albacete A., Gandour M., Andujar C.M., Houmani H., Ben Hamed K., Martinez V., Perez-Alfocea F. and Abdelly C. 2010. Salt tolerance of nitrogen fixation in *Medicago ciliaris* is related to nodule sucrose metabolism performance rather than antioxidant system. *Symbiosis*, 51: 187–195.

Bergmeyer H.V. 1974. Enzymes I: oxidoreductase, transferase. *Methods of Enzymatic Analysis,*, vol. III, third ed. Weinheim Deerfield Beach, Florida.

Berry J., and BjÖrkmam O. 1980. Photosynthetic response and adaptation to temperature in higher plants. *Ann. Rev. Plant Physiol.* 31: 491–543.

Bethlenfalvay G.J. and Phillips D.A. 1978. Interactions between symbiotic nitrogen fixation, combined-Napplication, and photosynthesis in *Pisum sativum*, *Physiol. Plant*, 42: 119–123.

Beverly R.B. and Jarrell W.M. 1984. Cowpea response to N form, rate, and timing of application, *Agron. J.*, 76: 663–668.

Bibikova T.N., Zhigilei A. and Gilroy S. 1997. Root hair growth in Arabidopsis thaliana is directed by calcium and an endogenous polarity. *Planta*, 203: 495–505.

Blumenthal J.M. and Russelle M.P. 1996. Subsoil nitrate uptake and symbiotic dinitrogen fixation by alfalfa, *Agron. J.* 88: 909–915.

Bolanos L., Esteban E., de Lorenzo C., Fernandez-Pascual M., de Felipe M.R., Garate A., and Bonilla I. 1994. Essentiality of boron for symbiotic dinitrogen fixation in pea (*Pisum-sativum*) rhizobium nodules, *Plant Physiol.*, 104: 85–90.

Bollman M.I. and Vessey J.K. 2006. Differential effects of nitrate and ammonium supply on nodule initiation, development, and distribution on roots of pea (*Pisum sativum*), *Can. J. Bot.*, 84: 893–903.

Boote K.J., Hoogenboom G., Jones J.W. and Ingram K.T. 2008. Modeling nitrogen fixation and its relationship to nitrogen uptake in the CROPGRO model. In, "*Modeling*," Eds. L. Ma, L.R. Ahuja, T.W. Bruulsem, CRC Press, Florence, USA, pp. 13–46.

Boote K.J., Jones J.W., Hoogenboom G. and Pickering N.B. 1998. The CROPGRO model for grain legumes. In, *"Understanding Options for Agricultural Production"* Eds. G.Y. Tsuji, G. Hoogenboom, P.K. Thornton, pp. 99–128.

Boote K.J.,MinguezM.I. and Sau F. 2002. Adapting the CROPGRO legume model to simulate growth of faba bean, *Agron. J.*, 94: 743–756.

Borucki W. and Sujkowska M. 2008. The effects of sodium chloride-salinity upon growth, nodulation,and root nodule structure of pea (*Pisum sativum* L.) plants. *Acta Physiologiae Plantarum*, 30: 293–301.

Bouhmouch I., Souad-Mouhsine B., Brhada F. and Aurag J. 2005. Influence of host cultivars and rhizobium species on the growth and symbiotic performance of *Phaseolus vulgaris* under salt stress. *Journal of Plant Physiology*, 162: 1103–1113.

Bouniols A., Cabelguenne M., Jones C.A., Chalamet A., Charpenteau J.L. and Marty J.R. 1991. Simulation of soybean nitrogen nutrition for a silty clay soil in southern France, *Field Crops Res..*, 26: 19–34.

Boyer J.S., Cavalieri A.J. and Schulze E.D. 1985. Control of the rate of cell enlargement: excision,wall relaxation, and growth-induced water potentials. *Planta*, 163: 527–543.

Bradford M.M. 1976. A rapid and sensitive method for the quantitation of microgram quantities of protein utilizing the principle of protein-dye binding. *Ann. Biochem.*, 72: 248–254.

Brioua A.H. and Wheeler C.T. 1994. Growth and nitrogen fixation in *Alnus glutinosa* (L.) Gaertn under carbon dioxide enrichment of the root atmosphere. *Plant Soil*, 162: 183–191.

Brisson N., Launay M., Mary B. and Beaudoin N. 2009. Nitrogen transformations. In, "*Conceptual basis,formalisations and parameterization of the STICS crop model,*" Eds. N. Brisson, M. Launay,

B. Mary, N. Beaudoin, Quæ, Versailles Cedex, France, pp. 141–165.

Brooks A. and Farquhar, G.D. 1985. Effect of temperature on the CO_2/O_2 specificity of ribulose-1, 5-bisphosphate carboxylase/oxygenase and the rate of respiration in the light. *Planta,* 165: 397–406.

Brugge R. and Thornley J.H.M. 1984. Shoot root nodule partitioning in a vegetative legume – A model, *Ann. Bot.*, 54: 653–671.

Bushby H.V.A. 1982. Ecology. In, "*Nitrogen Fixation, Vol. 2: Rhizobium*" Ed. W.J. Broughton,. Clarendon Press, Oxford, UK, pp. 35–75.

Cabelguenne M., Debaeke P. and Bouniols A. 1999. EPIC phase, a version of the EPIC modelsimulating the effects of water and nitrogen stress on biomass and yield, taking account of developmental stages: validation on maize, sunflower, sorghum, soybean and winter wheat, *Agric. Syst.*, 60:175–196.

Cannell M.G.R. and Thornley J.H.M. 2000. Modelling the components of plant respiration: some guiding principles, *Ann. Bot.*, 85: 45–54.

Carlsson G. and Huss-Danell K. 2008. How to quantify biological nitrogen fixation in forage legumes in the field. In, "*Biological Nitrogen Fixation: Towards Poverty Alleviation through Sustainable Agriculture*," Eds. F.D. Dakora, S.B.M. Chimphango, A.J. Valentine, C. Elmerich, and W.E. Newton, Springer Netherlands, Dordrecht, pp. 47–48.

Castellanos J.Z., Pena-Cabriales J.J. and Costa-Gallegos J.A.A. 1996. ^{15}N determined dinitrogen fixation capacity of common bean (*Phaseolus vulgaris*) cultivars under water stress. *J. Agric. Sci. Camb.*, 126: 327–333.

Chalifour F.P. and Nelson L.M. 1987. Effects of continuous combined nitrogen supply on symbiotic dinitrogen fixation of faba bean and pea inoculated with different rhizobial isolates, *Can. J. Bot.*, 65: 2542–2548.

Chaudhary M., Adu-Gyamfi J., Saneoka H., Nguyen N., Suwa R., Kanai S., El Shemy H., Lightfoot D. and Fujita K. 2008. The effect of phosphorus deficiency on nutrient uptake, nitrogen fixation and photosynthetic rate in mashbean, mungbean and soybean, *Acta Physiol. Plant.,* 30: 537–544.

Chaves M.M. and Pereira, J.S. 2004. Respuestas de la estabilidad de las plantas al estr´es m´ultiple y la habilidad de enfrentarse a un ambiente cambiante. In, "*La Ecofisiolog´ýa Veget al Una Ciencia de S´ýntesis,*" Eds. M.J. Reigosa, N. Pedrol and A. S´anchez, Thomson Editores Spain Paraninfo, S.A., Madrid, Spain, pp. 577–602.

Christensen, J.H.B. Hewitson, A. Busuloc, A. Chen, X. Gao, I. Held, R. Jones, R. Kolli, W. Kwon, R. Laprise, V. Magaña Rueda, L. Mearns, C. Menendez, J Raisanen, A. Rinke, A. Sarr, and P. Whetton 2007. Regional climate projections. In: *Climate Change 2007: The Physical Science* Basis. Contribution of Working Group I to the Fourth Assessment Report of Intergovernmental *Panel on Climate Change.* [Solomon, S., D. Qin, M. Manning, Z. Chen, M marquis, K. Averyt, M. Tignor, and H. Miller (eds.)]. Cambridge University Press, Cambridge, United Kingdom.

Collins M., Lang D.J. and Kelling K.A. 1986. Effects of phosphorus, potassium, and sulfur on alfalfa nitrogen-fixation under field conditions, *Agron. J.*, 78:, 959–963.

Cook D. 1999. Medicago truncatula: A model in the making! *Curr. Opin. Plant Biol.*, 2: 301–304.

Cordovilla M.D.P., Ligero F. and Lluch C. 1999. Effects of NaCl on growth and nitrogen fixation and assimilation of inoculated and KNO3 fertilized *Vicia faba* L. and *Pisum sativum* L. plants. *Plant Science*, 140:127–136.

Corre-Hellou G., Fustec J., Crozat Y. (2006). Interspecific competition for soil N and its interaction with N_2 fixation, leaf expansion and crop growth in pea–barley intercrops, *Plant Soil*, 282: 195–208.

Corre-Hellou, G., Brisson, N., Launay, M., Fustec, J. and Crozat Y. 2007. Effect of root depth penetration on soil nitrogen competitive interactions and dry matter production in pea–barley intercrops given different soil nitrogen supplies, *Field Crops Res.*, 103: 76–85.

Corre-Hellou, G., Faure, M., Launay, M., Brisson, N. and Crozat, Y. 2009. Adaptation of the STICS intercrop model to simulate crop growth and N accumulation in pea-barley intercrops, *Field Crops Res.*, 113: 72–81.

Cowan, M. 1978. The influence of nitrate, water potential and oxygen tension on nitrogen fixation in detached pea roots. *Plant Foods Hum. Nutr*, 28: 65–69.

Craufurd, P.Q., Wheeler, T.R., Ellis, R.H., Summerfield, R.J. and Williams, J.H. 1999. Effect of temperature and water deficit on water-use efficiency, carbon isotope discrimination, and specific leaf area in peanut. *Crop Sci.*, 39: 136–142.

Curtis, J., Shearer, G. and Kohl, D.H. 2004. Bacteroid proline catabolism affects N_2 fixation rate of drought-stressed soybeans. *Plant Physiol.*, 136: 3313–3318.

Dart, P. and Day, J. 1971. Effects of incubation temperature and oxygen tension on nitrogenase activity of legume root nodules. *Plant Soil*, 35: 167–184.

De Luis I. 2000. Efectos del aumento de la concentraci´on de CO_2 atmosf´erico en plantas de alfalfa fijadoras de nitr´ogeno bajo condiciones de estr´es. Thesis. Universidad de Navarra, Pamplona, Spain.

Defaria, S.M., Lewis, G.P., Sprent, J.I. and Sutherland, J.M. 1989. Occurrence of nodulation in the Fabaceae. *New Phytologist*, 111: 607–619.

De Luis, I., Irigoyen J.J. and S´anchez-D´ýaz, M. 1999. Elevated CO_2 enhances plant growth in droughted N_2-fixing alfalfa without improving water status. *Physiol. Plant*, 107: 84–89.

Dela Pena, T.C., Redondo, F.J., Manrique, E., Lucas, M.M. and Pueyo, J.J. 2010. Nitrogen fixation persists under conditions of salt stress in transgenic *Medicago truncatula* plants expressing a cyanobacterial flavodoxin. *Plant Biotechnology Journal*, 8: 954–965.

Delgado, M.J., Ligero, F. and Lluch, C. 1994. Effects of salt stress on growth and nitrogen-fixation by pea, faba-bean, commonbean and soybean plants. *Soil Biology and Biochemistry*, 26: 371–376.

Deroche, M.E., Carrayol, E. and Jolivet, E. 1983. Phosphoenolpyruvate carboxylase in legume nodules. *Physiol. V´eg.*, 21: 1075–1081.

Dita, M.A., Rispail, N., Prats, E., Rubiales, D. and Singh, K.B. 2006. Biotechnology approaches to overcome biotic and abiotic stress constraints in legumes. *Euphytica*, 147: 1–24.

Duffy, J., Chung, C., Boast, C. and Franklin, M. 1975. A simulation model of biophysiochemical transformations of nitrogen in tile-drained corn belt soil, *J. Environ. Qual.*, 4: 477–486.

Eaglesham, A.R.J. 1989. Nitrate inhibition of root-nodule symbiosis in doubly rooted soybean Plants. *Crop Sci.*, 29: 115–119.

Eckersten, H., Af Geijersstam, L. and Torssell, B. 2006. Modelling nitrogen fixation of pea (*Pisum sativum* L.). *Acta Agric. Scand. B* , 56: 129–137.

El-Hamdaoui, A., Redondo-Nieto, M., Torralba, B., Rivilla, R., Bonilla, I. and Bolanos, L. 2003. Influence of boron and calcium on the tolerance to salinity of nitrogen- fixing pea plants. *Plant and Soil*, 251: 93–103.

Elsheikh, E.A.E. and Wood, M. 1995. Nodulation and N-2fixation by soybean inoculated with salt-tolerant rhizobia or salt-sensitive brady rhizobia in saline soil. *Soil Biology and Biochemistry*, 27: 657–661.

Escalona J.M., Flexas J. and Medrano, H. 1999. Stomatal and nonstomatal limitations of photosynthesis under water stress in field-grown grapevines. *Aust. J. Plant Physiol.*, 26: 421–433.

Etheridge, D.M., Steele, L.P., Langenfelds, R.L., Francey, R.J., Barnola, J.-M. and Morgan, V.I. 1998. Historical CO_2 record derived from a spline fit (20 year cutoff) of the Law Dome DE08 and DE08-2 ice cores. http://cdiac.ornl.goftp/trends/co2/lawdome.smoothed.yr20

Evans, H.J. 1974. Symbiotic nitrogen fixation in legume nodules. In, "*Research Experiences in Plant Physiology*," Ed. M.J. Moore, Springer–Verlag, NY, USA, pp. 417–426.

Evans J., McNeill, A.M., Unkovich, M.J., Fettell, N.A. and Heenan, D.P. 2001. Net nitrogen balances for cool-season grain legume crops and contributions to wheat nitrogen uptake: a review. *Aust. J.Exp. Agric.*, 41: 347–359.

Evans, J., O'Connor, G.E., Turner, G.L., Coventry, D.R., Fettell, N., Mahoney, J., Armstrong, E.L. and Walsgott, D.N. 1989. N_2 fixation and its value to soil N increase in lupin, field pea and other legumes in south-eastern Australia, *Aust. J. Agric. Res.*, 40: 791–805.

Evans, J., Turner, G.L., O'Connor, G.E. and Bergersen, F.J. 1987. Nitrogen fixation and accretion of soil nitrogen by field-grown lupins (*Lupinus angustifolius*), *Field Crops Res.*, 16: 309–322.

FAO. 2002. World agriculture towards 2015/2030–summary report. Rome, Italy. Available at:http://www.fao.org/docrep/004/y3557e/y3557e08.htm#m

FAO. 2008. The State of Food and Agriculture. Rome, Italy. Available at: http://www.fao.org/docrep/011/i0100e/i0100e00.htm

Farquhar, G.D. and Sharkey T.D. 1982. Stomatal conductance and photosynthesis. *Ann. Rev.Plant Physiol.*, 33: 317–345.

Faurie, O. and Soussana, J.F. 1993 Oxygen-induced recovery from shortterm nitrate inhibition of N_2 fixation in white clover plants from spaced and dense stands, *Physiol. Plant.*, 89: 467–475.

Fellers, U., Crafts-Brandner, S.J. and Salvucci, M.E. 1998. Moderately high temperatures inhibit ribulose-1, 5-bisphosphate carboxylase/oxygenase (rubisco) activase-mediated activation of rubisco. *Plant Physiol.*, 116: 539–546.

Flechard, C.R., Ambus, P., Skiba, U., Rees, R.M., Hensen, A., van Amstel, A., van den Pol-vanDasselaar, A., Soussana, J.-F., Jones, M., Clifton-Brown, J., Raschi, A., Horvath, L., Neftel, A.,Jocher, M., Ammann, C., Leifeld, J., Fuhrer, J., Calanca, P., Thalman, E., Pilegaard, K., Di Marco, C., Campbell, C., Nemitz, E., Hargreaves, K.J., Levy, P.E., Ball, B.C., Jones, S.K., van de Bulk, W.C.M., Groot, T., Blom, M., Domingues, R., Kasper, G., Allard, V., Ceschia, E., Cellier, P., Laville P., Henault C., Bizouard F., Abdalla M., Williams, M., Baronti S., Berretti, F. and Grosz, B. 2007. Effects of climate and management intensity on nitrous oxide emissions in grassland systems across Europe, *Agric. Ecosyst. Environ.*, 121: 135–152.

Flowers, T.J. and Colmer T.D. 2008. Salinity tolerance inhalophytes. *New Phytologist*, 179: 945–963.

Flowers, T.J., Galal, H.K. and Bromham, L. 2010a. Evolution of halophytes: multiple origins of salt tolerance in land plants. *Functional Plant Biology*, 37: 604–612.

Flowers, T.J., Gaur, P.M., Gowda, C.L.L., Krishnamurthy, L., Samineni, S., Siddique, K.H.M., Turner, N.C., Vadez, V., Varshney, R.K. and Colmer T.D. 2010b. Salt sensitivity in chickpea. *Plant,Cell & Environment*, 33: 490–509.

Frame, J. and Newbould, P. 1986. Agronomy of white clover, *Adv. Agron.* 40: 1–88.

Gan, Y., Stulen, I., van Keulen, H. and Kuiper, P.J.C. 2004. Low concentrations of nitrate and ammonium stimulate nodulation and N2 fixation while inhibiting specific nodulation (nodule DW g^{-1} root dry weight) and specific N_2 fixation (N_2 fixed g^{-1} root dry weight) in Soybean. *Plant Soil*, 258: 281–292.

Frings, J.F.J. 1976. The rhizobium-pea symbiosis as affected by high temperatures. Ph.D. Thesis, Wageningen Agricultural University, Germany.

Garg, N., Geetanjali. 2007. Symbiotic nitrogen fixation in legume nodules: process and signaling. A review, *Agron. Sustain. Dev.*, 27: 59–68.

Garg, N. and Chandel, S. 2011. Effect of mycorrhizal inoculation on growth, nitrogen fixation, and nutrient uptake in *Cicer arietinum* (L.) under salt stress. *Turkish Journal of Agriculture and Forestry*, 35: 205–214.

Gibson, A. 1971. Factors in the physical and biological environment affecting nodulation and nitrogen fixation by legumes. *Plant Soil*, 35: 139–152.

Goh, K.M. and Bruce, G.E. 2005. Comparison of biomass production and biological nitrogen fixation of multi-species pastures (mixed herb leys) with perennial ryegrass-white clover pasture with and without irrigation in Canterbury, New Zealand. *Agric. Ecosyst. Environ.* 110: 230–240.

Goh, K.M., Edmeades, D.C. and Robinson, B.W. 1978. Field measurements of symbiotic nitrogen fixation in an established pasture using acetylene reduction and a ^{15}N method, *Soil Biol. Biochem.*, 10: 13–20.

Gulden, R. and Vessey, J.K. 1997. The stimulating effect of ammonium on nodulation in *Pisum sativum* L. is not long lived once ammonium supply is discontinued. *Plant Soil*, 195: 195–205.

Haase, S., Neumann, G., Kania, A., Kuzyakov, Y., Romheld, V. and Kandeler, E. 2007. Elevation of atmospheric CO_2 and N-nutritional status modify nodulation, nodule-carbon supply, and root exudation of *Phaseolus vulgaris* L., *Soil Biol. Biochem.*, 39: 2208–2221.

Halliday, J. 1975. An interpretation of seasonal and short term fluctuations in nitrogen fixation, PhD thesis, University of Western Australia, Perth.

Halliday, J. and Pate, J.S. 1976. The acetylene reduction assay as a means of studying nitrogen fixation in white clover under sward and laboratory conditions, *Grass Forage Sci.*, 31: 29–35.

Halperin, S.J., Gilroy, S. and Lynch, J.P. 2003. Sodium chloride reduces growth and cytosolic clacium, but does not affect cytosolic pH, in root hairs of *Arabidopsis thaliana* L. *Journal of Experimental Botany,* 54: 1269–1280.

Hamwieh, A., Tuyen, D.D., Cong, H., Benitez, E.R., Takahashi, R. and Xu, D.H. 2011. Identification and validation of a major QTL for salt tolerance in soybean. *Euphytica*, 179: 451–459.

Hartkamp, A.D., Hoogenboom, G. and White, J.W. 2002. Adaptation of the CROPGRO growth model to velvet bean (*Mucuna pruriens*): I. Model development. *Field Crops Res.*, 78: 9–25.

Hashem, F.M., Swelim, D.M., Kuykendall, L.D., Mohamed, A.I., Abdel-Wahab, S.M. and Hegazi, N.I. 1998. Identification and characterization of salt and thermo-tolerant Leucaena nodulating rhizobium strains. *Biology and Fertility of Soils*, 27: 335–341.

Hauggaard-Nielsen, H., Ambus, P. and Jensen, E.S. 2001. Interspecific competition, N use and interference with weeds in pea-barley intercropping, *Field Crops Res.*, 70: 101–109.

Hauggaard-Nielsen, H., Ambus, P. and Jensen, E.S. 2003. The comparison of nitrogen use and leaching in sole cropped versus intercropped pea and barley. *Nutr. Cycl. Agroecosyst.*, 65: 289–300.

Hauggaard-Nielsen, H., Andersen, M.K., Jørnsgaard, B. and Jensen, E.S. 2006. Density and relative frequency effects on competitive interactions and resource use in pea–barley intercrops, *Field Crops Res.*, 95: 256–267.

Havelka, U.D., Boyle, M.G. and Hardy, R.W.F. 1982. Biological nitrogen fixation. In, "*Nitrogen in Agricultural Soils*", Ed. F.J. Stevenson, ASA, Madison, Wisconsin, USA, pp. 365–422.

Hernandez-Armenta, R., Wien, H. and Eagleshman, A.R.J. 1989. Carbohydrate partitioning and nodule functioning in common bean after heat stress. *Crop Sci.*, 29: 1292–1297.

Herridge, D.F. and Pate J.S. 1977. Utilization of net photosynthate for nitrogen fixation and protein production in an annual legume. *Plant Physiol.*, 60: 759–764.

Herridge, D.F., Turpin, J.E. and Robertson, M.J. 2001. Improving nitrogen fixation of crop legumes through breeding and agronomic management: analysis with simulation modeling. *Aust. J. Exp. Agric.*, 41: 391–401.

Herridge, D.F., Peoples, M.B. and Boddey, R.M. 2008. Global inputs of biological nitrogen fixation in agricultural systems, *Plant Soil*, 311: 1–18.

Hilbert, D.W. and Canadell, J. 1995. Biomass partitioning and resource allocation of plants from Mediterranean-type ecosystems: possible responses to elevated atmospheric CO_2. In, "*Global Change and Mediterranean-Type Ecosystems*" Eds. J.M. Moreno, W.C. Oechel, Springer Verlag, NY, USA, pp. 76–101.

Hungria, M. and Franco, A.A. 1993. Effects of high temperature on nodulation and nitrogen fixation by *Phaseolus vulgaris* L. *Plant Soil*, 149: 95–102.

Hungria, M. and Vargas, M.A.T. 2000. Environmental factors affecting N_2 fixation in grain legumes in the tropics, with an emphasis on Brazil. *Field Crop Res.*, 65: 151–164.

Hunt, S. and Layzell, D.B. 1993. Gas exchange of legume nodules and the regulation of nitrogenase activity. *Ann. Rev. Plant Physiol. Plant Mol. Biol.*, 44: 483–511.

Høgh-Jensen, H., Loges, R., Jørgensen, F.V., Vinther, F.P. and Jensen, E.S. 2004. An empirical model for quantification of symbiotic nitrogen fixation in grass-clover mixtures. *Agric. Syst.*, 82: 181–194.

Hossain, S.A., Strong, W.M., Waring, S.A., Dalal, R.C. and Weston, E.J. 1996. Comparison of legume-based cropping systems at Warra, Queensland. II. Mineral nitrogen accumulation and availability to the subsequent wheat crop. *Aust. J. Soil Res.*, 34: 289–297.

Howard, J.B. and Rees, D.C. 1996. Structural basis of biological nitrogen fixation, *Chem. Rev.*, 96: 2965–2982.

Huang, C.Y., Boyer, J.S. and Vanderhoef, L.N. 1975. Limitation of acetylene reduction (nitrogen fixation) by photosynthesis in soybean having low water potentials. *Plant Physiol.*, 56: 228–232.

Ibekwe, A.M., Angle, J.S., Chaney, R.L. and van Berkum, P. 1997. Enumeration and N_2 fixation potential of *Rhizobium leguminosarum* biovar *trifolii* grown in soil with varying pH values and heavy m*et al* concentrations. *Agric. Ecosyst. Environ.*, 61: 103–111.

Imsande, J. 1986. Inhibition of nodule development in soybean by nitrate or reduced nitrogen. *J. Exp. Bot.*, 37: 348–355.

Irigoyen, J.J., Emerich, D.W. and Sanchez-Dýaz, M. 1992.Water stress induced changes in oncentrations of proline and total soluble sugars in nodulated alfalfa (*Medicago sativa*) plants. *Physiol. Plant*, 84: 55–60.

ISEOS. (2009a). The DNDC Model (Version 9.3) 2009. www.dndc.sr.un.edu

ISEOS. (2009b). *User's Guide for the DNDC Model.* (Version 9.3). www.dndc.sr.un.edu

Jacobsen, T. and Adams, R.M. 1958. Salt and silt in ancient Mesopotamian agriculture. *Science,* 128: 1251–1258.

Jarvis, C.E. and Walker, J.R.L. 1993. Simultaneous, rapid, spectrophotometric determination of total starch, amylose and amylopectin. *J. Sci. Food Agric.*, 63: 53–57.

Jebara, S., Drevon, J.J. and Jebara, M. 2010. Modulation of symbiotic efficiency and nodular antioxidant enzyme activities in two *Phaseolus vulgaris* genotypes under salinity. *Acta Physiologiae Plantarum,* 32(925):, 932.

Jensen, E.S. 1986. The influence of rate and time of nitrate supply on nitrogen fixation and yield in pea (*Pisum sativum* L.). *Fert. Res.* 10: 193–202.

Jensen, E.S. 1987. Seasonal patterns of growth and nitrogen fixation in field-grown pea. *Plant Soil,* 101: 29–37.

Jensen, E.S. 1997. The role of grain legume N2 fixation in the nitrogen cycling of temperate cropping systems, DSc Thesis, Riso-R-885, Risø National Laboratory, Denmark.Jones, F.R. and Tisdale, W.B. 1921. Effect of soil temperature upon the development of nodules on the roots of certain legumes. *J. Agric. Res.*, 22: 17–37.

Jordan, D.B. and Ogren, W.L. 1984. The CO_2/O_2 specificity of ribulose 1, 5-bisphosphate carboxylase / oxygenase. *Planta*, 161: 308–313.

Kaiser, W.M. 1987. Methods for studying the mechanism of water stress effects on photosynthesis. In, "*Plant Response to Stress*", Eds. J.D. Tenhunen, F.M. Catarino, O.L. Lange, W.C. Oechel, NATO ASI Series, vol. G15. Springer–Verlag, Germany, pp. 77–93.

King, B.J., Layzell, D.B. and Canvin, D.T. 1986. The role of dark carbon dioxide fixation in root nodules of soybean. *Plant Physiol.*, 81: 200–205.

Kirizii, D.A., Vorobei, N.A. and Kots, S.Ya. 2007). Relationships between nitrogen fixation and photosyn-thesis as the main components of the productivity in alfalfa. *Russ. J. Plant. Physiol.,* 54: 589–594.

Kishinevsky, B.D., Sen, D. and Weaver, R.W. 1992. Effect of high root temperature on *Bradyrhizobium*-peanut symbiosis. *Plant Soil*, 143: 275– 282.

Klucas, R.V. 1974. Studies on soybean nodule senescence. *Plant Physiol.*, 54: 612–616.

Korsaeth, A. and Eltun, R. 2000. Nitrogen mass balances in conventional, integrated and ecologicalcropping systems and the relationship between balance calculations and nitrogen runoff in an 8-year field experiment in Norway. *Agric. Ecosyst. Environ.*, 79: 199–214.

Kouas, S., Slatni T., Ben Salah I. and Abdelly, C. 2010. Eco-physiological responses and symbiotic nitrogen fixation capacity of salt exposed *Hedysarum carnosum* plants. *African Journal of Biotechnology*, 9: 7462–7469.

Kramer, P.J. and Boyer, J.S. 1995. Growth In, "*Water Relations of Plants and Soils*", Academic Press Inc., San Diego, USA, pp. 313–342.

Kristensen, E.S., Høgh-Jensen, H. and Kristensen, I.S. 1995. A simple model for estimation of atmospherically-derived nitrogen in grass-clover systems. *Biol. Agric. Hortic.*, 12: 263–276.

Kuo, T. and Boersma, L. 1971. Soil water suction and root temperature effects on nitrogen fixation in soybeans. *Agron. J.*, 63: 901–904.

Laghari, A.H., Memon, S., Nelofar, A. and Khan, K.M. 2011. Alhagi maurorum: aconvenient source of lupeol. *Industrial Crops and Products*, 34: 1141–1145.

Lamb, J.F.S., Barnes, D.K., Russelle, M.P., Vance, C.P., Heichel, G.H. and Henjum, K.I. 1995. Ineffectively and effectively nodulated alfalfas demonstrate biology nitrogen fixation continues with high nitrogen fertilization. *Crop Sci.*, 35: 153–157.

Larcher, W. 1995. Carbon utilization and dry matter production. In, "*Physiological Plant Ecology*", Springer–Verlag, Germany, pp. 57–166.

Larcher, W. 2000. Temperature stress and survival ability of Mediterranean schlerophyll plants. *Plant Biosyst.*, 134: 279–295.

Lawn, R.J. and Brun, W.A. 1974. Symbiotic nitrogen fixation in soybeans. I. effect of photosynthetic source-sink manipulations. *Crop Sci.*, 14: 11–16.

Lawrie, A.C. and Wheeler, C.T. 1973. The supply of photosynthetic assimilates to nodules of *Pisum Sativum* L. in relation to the fixation of nitrogen. *New Phytol.*, 72: 1341–1348.

Layzell, D.B., Pate, J.S., Atkins, C.A. and Canvin, D.T. 1981. Partitioning of carbon and nitrogen and the nutrition of root and shoot apex in a nodulated legume. *Plant Physiol.*, 67: 30–36.

Layzell, D.B., Rainbird, R.M., Atkins, C.A. and Pate, J.S. 1979. Economy of photosynthate use in nitrogen-fixing legume nodules: observations on two contrasting symbioses, *Plant Physiol.*, 64: 888–891.

Le Roux, M.R., Khan, S. and Valentine, A.J. 2008. Organic acid accumulation may inhibit N_2 fixation in phosphorus-stressed lupin nodules. *New Phytol.* 177: 956–964.

Ledgard, S.F. and Steele, K.W. 1992. Biological nitrogen fixation in mixed legume/grass Pastures. *Plant Soil*, 141: 137–153.

Ledgard, S.F., Brier, G.J. and Littler, R.A. 1987. Legume production and nitrogen fixation in hill pasture communities. *New Zeal. J. Agric. Res.*, 30: 413–421.

Leegood, R.C. and Edwards, G.E. 1996. Carbon metabolism and photorespiration: temperature dependence in relation to other environmental factors. In, "*Photosynthesis and the Environment*", Ed. N.R. Baker, Kluwer, UK, pp. 192–221.

Legg, B.J., Day, W., Lawlor, D.W. and Parkinson, K.J. 1975. The effects of drought on barley growth: models and measurements showing the relative importance of leaf area and photosynthetic rate. *J. Agric. Sci.*, 92: 703–716.

Lie, T.A. 1971. Symbiotic nitrogen fixation under stress conditions, *Plant Soil*, 35: 117–127.

Liechenthaler, H.K. 1987. Chlorophyll and carotenoids: pigments of photosynthetic biomembranes. In, "*Methods in Enzymology. Plant Cell Membranes*", Eds. L. Packer, R. Douce, Academic Press, USA, pp. 350–382.

Lilley, R.Mc. and Walker, D.A. 1974. An improved spectrophotometric assay of ribulose bisphosphate carboxylase. *Biochem. Biophys. Acta*, 358: 226–229.

Lindemann, W.C. and Ham, G.E. 1979. Soybean plant growth, nodulation, and nitrogen fixation as affected by root temperature. *Soil Sci. Soc. Am. J.*, 43: 1134–1137.

Lloret, F., Casanovas, C. and Penuelas, J. 1999. Seedling survival of Mediterranean shrubland species in relation to root:shoot ratio, seed size and water and nitrogen use. *Funct. Ecol.*, 13: 210–216.

Llusia, J. and Penuelas, J. 2000. Seasonal patterns of terpene content and emission from seven Mediterranean woody species in field conditions. *Am. J. Bot.*, 87: 133–140.

López, M., Herrera Cervera, J.A., Iribarne, C., Tejera, N.A. and Lluch, C. 2008. Growth and nitrogen fixation in *Lotus japonicus* and *Medicago truncatula* under NaCl stress: nodule carbon metabolism. *Journal of Plant Physiology*, 165: 641–650.

Loureiro, M.F., James, E.K., Sprent, J.I. and Franco, A.A. 1995. Stem and root nodules on the tropical wet land legume *Aeschynomene fluminensis*. *New Phytologist*, 130: 531–544.

Lynch, J., Polito, V.S. and Läuchi, A. 1989. Salinity stress increases cytoplasmic Ca activity in maize root protoplasts. *Plant Physiology*, 90: 1271–1274.

Lynd, J.Q., Hanlon, E.A., Jr. and Odell, G.V., Jr. 1984. Nodulation and nitrogen fixation by arrowleaf clover: effects of phosphorus and potassium. *Soil Biol.Biochem.*, 16: 589–594.

Maas, E.V. and Hoffman, G.J. 1977.Crop salt tolerance, current assessment. *Journal of Irrigation and Drainage Division*, ASCE, 103: 115–134.

Macduff, J.H., Jarvis, S.C. and Davidson, I.A. 1996. Inhibition of N_2 fixation by white clover (*Trifolium repens* L.) at low concentrations of NO_3 in flowing solution culture. *Plant Soil*, 180: 287–295.

Mahon, J.D. 1977a. Respiration and the energy requirement for nitrogen fixation in nodulated pea roots. *Plant Physiol.*, 60: 817–821.

Mahon, J.D. 1977b. Root and nodule respiration in relation to acetylene reduction in intact nodulated peas. *Plant Physiol.*, 60: 812–816.

Malezieux, E., Crozat, Y., Dupraz, C., Laurans, M., Makowski, D., Ozier-Lafontaine, H., Rapidel, B., de Tourdonnet, S. and Valantin-Morisonm M. 2009. Mixing plant species in cropping systems: concepts, tools and models. A review. *Agron.Sustain. Dev.*, 29: 43–62.

Manchanda, G. and Garg, N. 2008. Salinity and its effects on the functional biology of legumes. *Acta Physiologiae Plantarum*, 30: 595–618.

Masterson, C.L. and Murphy, P.M. 1976. Application of the acetylene reduction technique to the study of nitrogen fixation by white clover in the field. In, "*Symbiotic Nitrogen Fixation in Plants*", Ed. P.S. Nutman, Cambridge University Press, London, UK, pp. 299–316.

Mc Callum, M.H., Peoples, M.B. and Connor, D.J. 1999. Contributions of nitrogen by field pea (*Pisum sativum* L.) in a continuous cropping sequence compared with a lucerne (*Medicago sativa* L.)-based pasture ley in the Victorian Wimmera. *Aust. J. Agric. Res.*, 51: 13–22.

Michiels, J., Verreth, C. and Vanderleyden, J. 1994. Effects of temperature stress on bean-nodulating *Rhizobium* strains. *Appl. Environ. Microbiol.*, 60: 1206–1212.

Miransar,i M. and Smith, D.L. 2009. Alleviating salt stress on soybean (*Glycine max* (L.) Merr.) – Brady rhizobium japonicum symbiosis, using signal molecule genistein. *EuropeanJournal of Soil Biology*, 45:146–152.

Minchin, F.R., Summerfield, R.J., Hadley, P., Roberts, E.H. and Rawsthorne, S. 1981. Carbon and nitrogen nutrition of nodulated roots of grain legumes. *Plant Cell Environ.*, 4: 5–26.

Mooney, H.A. 1983. Carbon-gaining capacity and allocation patterns of Mediterranean climate plants. In, "*Mediterranean Type Ecosystems: The Role of Nutrients*", Eds. F.J. Kruger, D.T. Mitchel, J.U.M. Jarvis, Springer, Berlin, Germany, pp. 103–119.

Munns, D. 1970. Nodulation of *Medicago sativa* in solution culture V.calcium and pH requirements during infection. *Plant Soil*, 32: 90–102.

Munns, R. and Tester, M. 2008. Mechanisms of salinity tolerance. *Annual Review of Plant Biology*, 59: 651–681.

Nelson, L.M. and Child, J.J. 1981. Nitrogen-fixation and hydrogen metabolism by *Rhizobium-leguminosarum* isolates in pea root nodules. *Can. J. Microbiol.*, 27: 1028–1034.

Nelson, D.R., Bellville, R.J. and Porter, C.A. 1984. Role of nitrogen assimilation in seed development of soybean. *Plant Physiol.*, 74: 128–133.

Ogaya, R. and Penuelas, J. 2003. Comparative field study of *Quercus ilex* and *Phillyrea latifolia*: photosynthetic response to experimental drought conditions. *Environ. Exp. Bot.*, 50: 137–148.

Oldroyd, G.E.D. and Downie, J.A. 2004. Calcium, kinases and nodulation signalling in legumes. *Nature Reviews Molecular Cell Biology*, 5: 566–576.

Oliveira, G. and Penuelas, J. 2001. Allocation of observed light energy into photochemistry and dissipation in a semideciduous and an evergreen Mediterrenean woody species during winter. *Aust. J. Plant Physiol.*, 28: 1–10.

Pankhurst, C.E. and Sprent, J.I. 1976. Effects of temperature and oxygen tension on the nitrogenase and respiratory activities of turgid and water-stressed soybeans and French bean root nodules. *J.Exp. Bot.*, 27: 1–9.

Pate, J.S. 1961. Temperature characteristics of bacterial variation in legume symbiosis, *Nature*, 192: 637–639.

Paul, E.A. and Clark, F.E. 1989. In, "*Soil Microbiology and Biochemistry*", Academic Press. San Diego, CA. pp. 275.

Pennypacker, B.W., Leath, K.T., Stout, W.L. and Hill, Jr. R.R. 1990. Technique for simulating field drought stress in the greenhouse. *Agron. J.*, 82(5): 951–957.

Penuelas, J., Filella, I., Llusia, J., Siscart, D. and Pinol, J. 1998. Comparative field study of spring and summer leaf gas exchange and photobiology of the Mediterranean trees *Quercus ilex* and *Phyllyrea latifolia*. *J. Exp. Bot.*, 49: 229–238.

Peoples, M.B., Herridge, D.F. and Ladha, J.K. 1995. Biological nitrogen fixation an efficient source of nitrogen for sustainable agricultural production. *Plant and Soil*, 174: 3–28.

Peoples, M.B., Lilley, D.M., Burnett, V.F., Ridley, A.M. and Garden, D.L. 1995. Effects of surface application of lime and superphosphate to acid soils on growth and N_2 fixation by subterranean clover in mixed pasture swards. *Soil Biol. Biochem.*, 27: 663–671.

Piha, M.I. and Munns, D.N. 1987. Sensitivity of the common bean (*Phaseolus vulgaris*) symbiosis to high soil temperature. *Plant Soil*, 98: 183–194.

Pimratch, S., Jogloy, S., Vorasoot, N., Toomsan, B., Patanothai, A., Holbrook, C.C. 2008. Relationship between biomass production and nitrogen fixation under drought-stress conditions in peanut genotypes with different levels of drought resistance, *J. Agron. Crop Sci.*, 194: 15–25.

Postgate, J.R. 1982. Biological nitrogen fixation: fundamentals, *Philos. Trans. R. Soc. B*, 296:, 387–375.

Purcell, L.C. and Sinclair, T.R. 1990. Nitrogenase activity and nodule gas permeability response to rhizospheric NH_3 in soybean. *Plant Physiol.*, 92: 268–272.

Purcell, L.C., de Silva, M., King, C.A. and Kim, W.H. 1997. Biomass accumulation and allocation in soybean associated with genotypic differences in tolerance of nitrogen fixation to water deficits. *Plant Soil*, 196: 101–113.

Rao, D.L.N., Giller, K.E., Yeo, A.R. and Flowers, T.J. 2002.The effects of salinity and sodicity upon nodulation and nitrogen fixation in chickpea (*Cicer arietinum*). *Annals of Botany*, 89: 563–570.

Rao, D.L.N. and Sharma, P.C. 1995. Alleviation of salinity stress in chickpea by rhizobium inoculation or nitrate supply. *Biologia Plantarum*, 37: 405–410.

Rao, V.R. 1977. Effect of root temperature on the infection processes and nodulation in *Lotus* and *Stylosanthes*. *J. Exp. Bot.*, 28: 241–259.

Rice, W.A., Clayton, G.W., Olsen, P.E. and Lupwayi, N.Z. 2000. Rhizobial inoculant formulations and soil pH influence field pea nodulation and nitrogen fixation. *Can. J. Soil Sci.*, 80: 395–400.

Robertson, M.J., Carberry, P.S., Huth, N.I., Turpin, J.E., Probert, M.E., Poulton, P.L., Bell, M., Wright, G.C., Yeates, S.J. and Brinsmead, R.B. 2002. Simulation of growth and development of diverse legume species in APSIM. *Aust. J. Agric. Res.*, 53: 429–446.

Rogers, M.E. 2001. The effect of saline irrigation on lucerne production: shoot and root growth, ion relations and flowering in *Cidence* in six cultivars grownin northern Victoria, Australia. *Irrigation Science*, 20:, 55–64.

Rogers, M.E., Colmer, T.D., Frost, K., Henry, D., Cornwall, D., Hulm, E., Deretic, J., Hughes, S.R. and Craig, A.D. 2008. Diversity in the genus Melilotus for tolerance to salinity and water logging. *Plant and Soil*, 304: 89–101.

Rogers, M.E., Craig, A.D., Munns, R., Colmer, T.D., Nichols, P.G.H., Malcolm, C.V., Barrett-Lennard, E.G, Brown, A.J., Semple, W.S., Evans, P.M., Cowley, K., Hughes, S.J., Snowball, R., Bennett, S.J., Sweeney, G.C., Dear, B.S. and Ewing, M. 2006.The potential for developing fodder plants for the salt-affected areas of southern and eastern Australia: an overview. *Australian Journal of Experimental Agriculture*, 45: 301–329.

Roughley, R.J. and Dart, P.J. 1969. Reduction of acetylene by nodules of *Trifolium subterraneum* as affected by root temperature, *Rhizobium* strain and host cultivar. *Arch. Microbiol.*, 69: 171–179.

Roughley R. and Dart, P. 1970. Root temperature and root-hair infection of *Trifolium subterraneum* L. cv. Cranmore. *Plant Soil*, 32: 518–520.

Rozema J. and Schat, H. 2012. Salt tolerance of halophytes, research questions reviewed in the perspective of saline agriculture. *Environmental and Experimental Botany,* 92: 83-95.

Rumbaugh, M.D., Pendery, B.M. and James, D.W. 1993.Variation in the salinity tolerance of strawberry clover (*Trifolium fragiferum* L.). *Plant and Soil*, 153: 265–271.

Ryle, G.J.A., Powell, C.E. and Gordon, A.J. 1979. The respiratory costs of nitrogen fixation in soyabean, cowpea, and white clover I. Nitrogen fixation and the respiration of the nodulated root. *J. Exp. Bot.*, 30: 135–144.

Sabate, S., Gracia, C.A. and Sanchez, A. 2002. Likely effects of climate change on growth of *Quercus ilex*, *Pinus halepensis*, *Pinus pinaster*, *Pinus sylvestris* and *Fagus sylvatica* forests in the Mediterranean region. *Forest Ecol. Manag.*, 162: 23–37.

Sall, K. and Sinclair, T.R. 1991. Soybean genotypic differences in sensitivity of symbiotic nitrogen fixation to soil dehydration. *Plant Soil*, 133: 31–37.

Sau, F., Boote, K.J. and Ruiz-Nogueira, B. 1999. Evaluation and improvement of CROPGRO-soybean model for a cool environment in Galicia, northwest Spain. *Field Crops Res.*, 61: 273–291.

Saxena, A.K. and Rewari, R.B. 1992. Differential responses of chickpea (*Cicer arietinum* L.) – rhizobium combinations to saline soil-conditions. *Biology and Fertility of Soils*, 13: 31–34.

Schiltz, S., Munier-Jolain, N., Jeudy, C., Burstin, J. and Salon, C. 2005. Dynamics of exogenous nitrogen partitioning and nitrogen remobilization from vegetative organs in pea revealed by ^{15}N in vivo labelling throughout seed filling. *Plant Physiol.*, 137: 1463–1473.

Schmid, M., Neftel, A., Riedo, M. and Fuhrer, J. 2001. Process-based modelling of nitrous oxide emissions from different nitrogen sources in mown grassland. *Nutrient Cycling in Agroecosystem*, 60: 177–187.

Schomberg, H.H. and Weaver, R.W. 1992. Nodulation, nitrogen fixation, and early growth of arrowleaf clover in response to root temperature and starter nitrogen. *Agron. J.*, 84: 1046–1050.

Schubert, S., Serraj, R., Plies-Balzer, E. and Mengel, K. 1995. Effect of drought stress on growth, sugar concentrations and amino acid accumulation in N_2-fixing alfalfa. *J. Plant Physiol.*, 146: 541–546.

Schulze, J., Adgo, E. and Merbach, W. 1999. Carbon costs associated with N_2 fixation in *Vicia faba* L. and *Pisum sativum* 1. over a 14-day period. *Plant Biol.*, 1: 625–631.

Schwinning, S. and Parsons, A.J. 1996. Analysis of the coexistence mechanisms for grasses and legumes in grazing systems, *J. Ecol.*, 84: 799–813.

Serraj, R. and Drevon, J.J. 1998. Effects of salinity and nitrogen source on growth and nitrogen fixation in alfalfa. *J. Plant Nutr.*, 21: 1805–1818.

Serraj R. and Sinclair, T.R. 1998. N_2 fixation response to drought in common bean (*Phaseolus vulgaris* L.). *Ann. Bot.*, 82: 229–234.

Serraj, R., Bona, S., Purcell, L.C. and Sinclair, T.R. 2001. Role of soil-oxygen enrichment in the response of nitrogen fixation to drought stress in field-grown soybean, *Soil Crop Sci. Soc. Fla. Proc.*, 60: 88–93.

Serraj, R. and Sinclair, T.R. 1998. Soybean cultivar for nodule formation and growth under drought. *Plant Soil*, 202: 159–166.

Serraj, R., Sinclair, T.R. and Purcell, L.A. 1999. Symbiotic N2 fixation response to drought. *J. Exp. Bot.,* 331: 143–155.

Shabala, S., Shabala, L. and vanVolkenburgh, E. 2003. Effect of calcium on root development and root ion fluxes in salinised barley seedlings. *Functional Plant Biology*, 30: 507–514.

Sharpley, A.N. and Williams, J.R. 1990. EPIC-erosion/productivity impact calculator 1. Model Documentation. *USDA Tech. Bull. No. 1768*, pp. 1–235.

Sheehy, J.E. 1987. Photosynthesis and nitrogen fixation in legume plants, *Crit. Rev. Plant Sci.,* 5: 121–159.

Sheehy, J.E., Woodward, F.I. and Gosse, G. 1991. Measurements of nitrogen fixation (C_2H_2), photosynthesis and respiration using an open system in the natural environment, *Ann. Bot.*, 67: 131–136.

Sinclair, T.R. 1986. Water and nitrogen limitations in soybean grain production I. model Development. *Field Crops Res.*, 15: 125–141.

Sinclair, T.R., Muchow, R.C., Ludlow, M.M., Leach, G.J., Lawn, R.J. and Foale, M.A. 1987. Field and model analysis of the effect of water deficits on carbon and nitrogen accumulation by soybean, cowpea and black gram. *Field Crops Res.*, 17: 121–140.

Small, J.G.C. and Joffe, A. 1968. Physiological studies on the genus *Trifolium* with special reference to the South African species. II. Influence of root temperature on growth,

nodulation and symbiotic nitrogen fixation. *S. African J. Agric. Sci.*, 11: 41–56.
Sokal R.R. and Rohlf, F.J. 1986. Introducci´on a la Bioestad´ýstica. Revert´e, Barcelona, Spain.
Soussana, J.F., Minchin, F.R., Macduff, J.H., Raistrick, N., Abberton, M.T. and Michaelson-Yeates, T.P.T. 2002. A simple model of feedback regulation for nitrate uptake and N_2 fixation in contrasting phenotypes of white clover. *Ann. Bot.*, 90: 139–147.
Soussi, M., Lluch, C. and Ocana, A. 1999. Comparative study of nitrogen fixation and carbon metabolism in two chick-pea (*Cicer arietinum* L.) cultivars under salt stress. *Journal of Experimental Botany*, 50: 1701–1708.
Stout W.L., Fales S.L., Muller L.D., Schnabel R.R. and Weaver S.R. 2000. Water quality implications of nitrate leaching from intensively grazed pasture swards in the northeast US, *Agric. Ecosyst. Environ.*, 77: 203–210.
Streeter J. 1988. Inhibition of legume nodule formation and N_2 fixation by nitrate, *Crit. Rev. Plant Sci.*, 7: 1–23.
Streeter J.G. (2003). Effects of drought on nitrogen fixation in soybean root nodules. *Plant Cell Environ*,. 26: 1199–1204.
Stitt M., von Shaewen A. and Willmitzer L. 1991. Sink regulation of photosynthetic metabolism in transgenic tobacco plants expressing yeast invertase in their cell wall involves a decrease of the Calvin cycle enzymes and an increase of glycolytic enzymes. *Planta*, 183: 40–50.
Sullivan P. 2003. Overview of cover crops and green manures. ATTRA. Available at https://attra.ncat.org/attra-pub/summaries/summary.php?pub=288
Sutcliffe J. 1980. Las PlantasylaTemperatura. Editorial Omega, Barcelona, Spain.
Svenning M.M., Junttila O. and Macduff J.H. 1996. Differential rates of inhibition of N_2 fixation by sustained low concentrations of NH^+_4 and NO^-_3 in northern ecotypes of white clover (*Trifolium repens* L.), *J. Exp. Bot.*, 47: 729–738.
Tans, P. 2010. Trends in carbon dioxide. www.esrl.noaa.gov/gmd/ccgg/trends
Teakle N.L., Real D. and Colmer T.D. 2006. Growth and ion relations in response to combined salinity and water logging in the perennial forage legumes Lotus cor- niculatus and Lotus tenuis. *Plant and Soil*, 289: 369–383.
Thomas Robertson M.J., Fukai S. and Peoples M.B. 2004. The effect of timing and severity of water deficit on growth, development, yield accumulation and nitrogen fixation of mungbean. *Field Crop Res.*, 86: 67–80.
Thornley J.H.M. 1998. Plant Submodel, In, "*Grassland Dynamics: An Ecosystem Simulation Model* ," Ed. J.H.M. Thornley, CAB International, Wallingford, Oxon, UK, pp. 53–57.
Thornley J.H.M. 2001. Simulating grass-legume dynamics: a phenomenological submodel, *Ann. Bot*, 88: 905–913.
Thornley J.H.M. and Cannell M.G.R. 2000. Dynamics of mineral N availability in grassland ecosystems under increased [CO_2]: hypotheses evaluated using the Hurley Pasture Model, *Plant Soil*, 224: 153–170.
Thornley J.H.M. Bergelson J. and Parsons A.J. 1995. Complex dynamics in a carbon-nitrogen model of a grass-legume pastures, *Ann. Bot.*,75: 79–94.
Tretiach, M., Bolognini, G. and Rondi A. 1997. Photosynthetic activity of *Quercus ilex* at the extremes of a transect between Mediterranean and submediterranean vegetation. *Trieste, NE, Italy. Flora*, 192:369–378.
Trindade, H., Coutinho, J., Jarvis, S. and Moreira, N. 2001. Nitrogen mineralization in sandy loam soils under an intensive double-cropping forage system with dairy-cattle slurry applications. *Eur. J.Agron.*, 15: 281–293.
Twary, S.N. and Heichel, G.H. Carbon costs of dinitrogen fixation associated with dry matter accumulation in alfalfa, *Crop Sci.*, 31: 985–992.

U.S. E.P.A. 2009. Future atmosphere changes in greenhouse gas and aerosol concentrations. www.epa.gov/climatechange/science/futureac.html

U.S. E.P.A. 2010. Global scale: forestry and agriculture in the global carbon cycle. www.epa.gov/sequestration/ccyle.html

Valladares F. and Pearcy R.W. 1997. Interactions between water stresses, sunshade acclimation, heat tolerance and photoinhibition in the sclerophyll *Heteromeles arbutifoliar*. *Plant Cell Environ.*, 20: 25–36.

Vance C.P. and Stade S. 1984. Alfalfa root nodule carbon dioxide fixation. II. Partial purification and characterization of root nodule phosphoenolpyruvate carboxylase. *Plant Physiol.*, 75: 261–264.

Vance C.P. and Heichel G.H. 1991. Carbon in N_2 fixation: limitation or exquisite adaptation.*Ann. Rev. Plant Physiol. Plant Mol. Biol.*, 42: 373–392.

Venkateswarlu B., Saharan N. and Maheswari M. 1990. Nodulation and N_2 (C_2H_2) fixation in cowpea and groundnut during water stress and recovery, *Field Crops Res.*, 25: 223–232.

Verdoy D., DelaPena T.C., Redondo F.J., Lucas M.M. and Pueyo J.J. 2006.Trans-genic Medicago truncatula plants that accumulate proline display nitrogen-fixing activity with enhanced tolerance to osmotic stress. *Plant, Cell & Environment*, 29: 1913–1923.

Voisin A.-S., Bourion V., Duc G. and Salon C. 2007. Using an ecophysiological analysis to dissect genetic variability and to propose an ideotype for nitrogen nutrition in pea *Ann. Bot.*, 100: 1525–1536.

Voisin A.-S., Salon C., Munier-Jolain N.G. and Ney B. 2002. Effect of mineral nitrogen on nitrogen nutrition and biomass partitioning between the shoot and roots of pea (*Pisum sativum L.*), *Plant Soil*, 242: 251–262.

Voisin A.-S., Salon C., Jeudy C. and Warembourg F.R. 2003. Symbiotic N2 fixation activity in relation to C economy of *Pisum sativum* L. as a function of plant phenology, *J. Exp. Bot.*, 54: 2733–2744.

Von Caemmerer S. and Farquhar G.D. 1981. Some relationships between the biochemistry of photosyn-thesis and the gas exchange of leaves. *Planta*, 153: 376–387.

Walsh K.B. 1995. Physiology of the legume nodule and its response to stress. *Soil Biol. Biochem*, 27: 637–655.

Wang D. and Shannon M.C. 1999. Emergence and seedling growth of soybean cultivars and maturity groups under salinity. *Plant and Soil*, 214: 117–124.

Warembourg F.R. and Roumet C. 1989. Why and how to estimate the cost of symbiotic N_2 fixation? A progressive approach based on the use of ^{14}C and ^{15}N isotopes, *Plant Soil*, 115: 167–177.

Waterer J.G. and Vessey J.K. 1993a. Effect of low static nitrate concentrations on mineral nitrogen uptake, nodulation, and nitrogen fixation in field pea, *J. Plant Nutr.*, 16:, 1775–1789.

Waterer J.G., and Vessey J.K. 1993b. Nodulation response of autoregulated or NH_4^+ –inhibited pea *(Pisum sativum)* after transfer to stimulatory (low) concentrations of NH_4^+. *Physiol. Plant*, 88: 460–466.

Waterer J.G., Vessey J.K. and David Raper J.C. 1992. Stimulation of nodulation in field peas (*Pisum sativum*) by low concentrations of ammonium in hydroponic culture, *Physiol. Plant*,. 86: 215–220.

Watson C.A. and Goss M.J. 1997. Estimation of N2-fixation by grass-white clover mixtures in cut or grazed swards, *Soil Use Manage.*, 13: 165–167.

Waughman G.J. 1977. The effect of temperature on nitrogenase activity, *J. Exp. Bot.* 28: 949–960.

Weatherley P.E. 1950. Studies in the water relations of the cotton plant. I. The field measurement of water deficits in leaves. *New Phytol.*, 49: 81–87.

Weisz P.R. and Sinclair T.R. 1987. Regulation of soybean nitrogen fixation in response to rhizosphere oxygen. II. Quantification of nodule gas permeability, *Plant Physiol.*, 84: 906–910.

Weis E. 1981. Reversible heat-inactivation of the Calvin cycle: a possible mechanism of the temperature regulation of photosynthesis. *Planta*, 151: 33–39.

Weisbach C., Hartwig U.A., Heim I. and N¨osberger J. 1996. Whole-nodule carbon metabolites are not involved in the regulation of the oxygen permeability and nitrogenase activity in white clovernodules. *Plant Physiol.*, 110: 539–545.

Weisz P.R., Denison R.F. and Sinclair T.R. 1985. Response to drought stress of nitrogen fixation (acetylene reduction) rates by field-grown soybeans, *Plant Physiol.*, 78: 525–530.

Whitehead D.C. 1995. Legumes: Biological nitrogen fixation and interaction with grasses, In, "Grassland Nitrogen," Ed. D.C. Whitehead D.C. CAB International, Wallingford, UK, pp. 36–57.

Wood, M. 1996. Nitrogen Fixation: How much and at what cost? In, "*Legumes in Sustainable Farming Systems*," Ed. D. Younie, Occasional Symposium No. 30, British Grassland Society, Reading, UK, pp. 26–35.

Wu L., Mc Gechan M.B. (1999). Simulation of nitrogen uptake, fixation and leaching in a grass/ white clover mixture, *Grass Forage Sci.*, 54: 30–41.

Yemm E.W. and Willis A.J. 1954. The estimation of carbohydrates in plant extracts by anthrone. *Biochem. J.*, 57: 508–514.

Young N.D., Debelle F., Oldroyd G.E.D., Geurts R., Cannon S.B., Udvardi M.K., Benedito V.A., Mayer K.F.X., GouzyJ., SchoofH., VandePeer Y., Proost S., Cook D.R., Meyers B.C., Spannagl M., Cheung F., DeMita S., Krishnakumar V., Gundlach H., ZhouS.G., MudgeJ., Bharti A.K., Murray J.D., Naoumkina M.A., Rosen B., Silverstein K.A.T., Tang, H.B., Rombauts S., Zhao P.X., Zhou P., Barbe V., Bardou P., Bechner M., Bellec A., Berger, A., Berges H., Bidwell S., Bissel- ing T., Choisne N., Couloux A., Denny R., Deshpande S., Dai X.B., Doyle J.J., Dudez A.M., Farmer A.D., Fouteau S., Franken C., Gibelin C., Gish J., Goldstein S., Gonzalez A.J., Green P.J., Hallab A., Hartog M., Hua A., Humphray S.J., Jeong D.H., Jing Y., Jocker A., Kenton S.M., Kim D.J., Klee K., Lai H.S., Lang C.T., Lin S.P., Macmil S.L., Magdelenat G., Matthews L.C., O'Bleness M.,Paule C.R., Poulain J., Prion F., Qin B.F., Qu C.M., Retzel E.F., Riddle C., Sallet E., Samain S., Samson N., Sanders I., Saurat O., Scarpelli C., Schiex T., Segurens B., Severin A.J., Sher- rier D.J., Shi R.H., Sims S., Singer S.R., Sinharoy S., Sterck L., Viollet A., Wang B.B., Wang K.Q., Wang M.Y., Wang X.H., Warfsmann J., Weissenbach J., White D.D., White J.D., Wiley G.B., Wincker P., Xing Y.B., Yang L.M., Yao Z.Y., Ying F., Zhai J.X., Zhou L.P., Zuber A., Denarie J., Dixon R.A., May G.D., Schwartz D.C., Rogers J., Quetier F., Town C.D., Roe B.A. 2011. The Medicago genome provides insight into the evolution of Rhizobial symbioses. *Nature*, 480: 520– 5247

Yousef A.N. and Sprent J.I. 1983. Effects of NaCl on growth nitrogen incorporation and chemical-composition of inoculated and NH_4NO_3 fertilized *Vicia faba* (L.) plants. *Journal of Experimental Botany*, 34: 941–950.

Yu M., Gao Q. and Shaffer M.J. 2002. Simulating interactive effects of symbiotic nitrogen fixation, carbon dioxide elevation, and climatic change on legume growth, *J. Environ. Qual.*, 31: 634–641.

Zahran H.H. and Sprent J.I. 1986. Effects of sodium-chloride and polyethylene-glycol on root-hair infection and nodulation of *Vicia faba* L. Plants by *Rhizobium-leguminosarum*. *Planta*, 167: 303–309.

Zurayk R., Adlan M., Baalbaki R. and Saxena M.C. 1998. Interactive effects of salinity and biological nitrogen fixation on chickpea (*Cicer arietinum* L.) growth. *Journal of Agronomy and Crop Science-Zeitschrift Fur Acker Und Pflanzenbau,* 180: 249–258.

4

Growth, Development and Phenology Under Abiotic Stress Condition

A. Bhattacharya

Great progress has been made in understanding the responses of plants to abiotic stress. Climatic models indicated that both average temperatures and extreme events frequency will increase in the Mediterranean basin. These changes are expected to have a great impact on agriculture in general and on crop phenology in particular. Consistent with the increasing temperatures, crop development is expected to be faster, thus phenological stages will be reached early and the length of the growth period of crops with determinate cycle (i.e. cereals, grapevine, etc.) will be shorter. These impacts, together with the higher risk to have climate extreme events during sensitive phonological phases, may have strong negative effects on final yield and on yield quality. However, these conclusions cannot be generalized, since the actual impact of phonological change needs to be assessed for specific cropenvironment combinations, providing the basis to formulate feasible adaptation options to climate change. In other terms, the simulated changes in phenology cannot be interpreted without considering the environmental context in which a species lives. For winter crops, the effect of predicted prolonged summer drought periods and heat weaves for the next decades may be smoothed or prevented due to the faster development that will allow escaping these and then avoiding reduction in final yields. In contrast, crops whose growing cycle takes place in summer time are likely to experience a severe reduction of final yield as the result of increased frequency of extreme climatic events and a reduced time for biomass accumulation to yield.

The world-wide population is predicted to increase from 7.2 billion to between 9.6 and 12.3 billion by 2100, increasing the demand on global food productivity

(Gerland *et al.*, 2014). However, nutrient deficiency, mineral toxicity and salinity severely limit the amount of land that is available for food production. In addition, world-wide crop losses due to extreme temperatures, drought, and flooding are estimated to rise about 50% by 2050 (Boyer, 1982). Water deficit was estimated to have caused Euro100 billion worth of damage in Europe between 1977 and 2007 (European Commission, 2007). The severe drought of 2012 affected 80% of the crop-growing territory in the USA (Hoerling *et al.*, 2013.), and such droughts are predicted to become more frequent as a result of global warming (Hansen *et al.*, 2010; Lawrimore *et al.*, 2011; Lobell and Gourdji, 2012; Cook *et al.*, 2015).

There is an urgent need for development of crop varieties with improved tolerance to abiotic stress, but the development of such improved varieties will require a thorough understanding of the mechanisms by which plants are affected by these conditions, and how they can tolerate them. Abiotic stress can affect cellular homeostasis via a variety of mechanisms. Freezing temperatures promote intra- and extra-cellular ice crystal formation, leading to cellular damage, dehydration and disruption of the plasma membrane (Pearce, 2011; Yamazaki *et al.*, 2008, 2009). High temperatures damage the plasma membrane by increasing its fluidity, leading to ion leakage. Heat stress can also result in inhibition of enzymatic activity and photosynthesis (Salvucci and Crafts-Brandner, 2004; Sharkev, 2005; Wahid *et al.*, 2007; Allakhverdiev *et al.*, 2008). High temperatures, salinity and drought stress are often accompanied by production of reactive oxygen species (ROS) which can lead to denaturation of functional and structural proteins (Gill and Tuteja, 2010; Suzuki *et al.*, 2012). Gradual exposure to these different stresses causes plants to become more tolerant, however, and acclimated plants can survive harsher conditions than non-acclimated plants. The acclimation process includes the expression of new transcripts and proteins, and the accumulation of metabolites that result in a new homeostatic state that is more compatible with the change in environmental conditions. This allows stabilization of proteins and membranes, osmotic adjustment and re-establishment of the redox balance (Krasenskv and Jonak, 2012); Mittler and Blumwald, 2015).

4.1. Factors of Abiotic Stress

It is definitively accepted that our climate is changing due to increased 'greenhouse gases' atmospheric concentrations and this change is expected to have important impact on different economic sectors (*e.g.* agriculture, forestry, energy consumptions, tourism, *etc.*) (Hanson *et al.*, 2006). In particular, for agriculture, such a change in climate may have significant impacts on crop growth and yield, since these are largely determined by the weather conditions during the growing season.

According to these premises, the assessment of cropping systems response to a warmer climate plays an important role for the evaluation of near future economic assets and the study of crop phenology response was indicated as a key stage for a better formulation of adaptation policies and options (Duchene and Schneider, 2005; Wolfe *et al.*, 2005; Sadras and Monzon, 2006). Phenology is in fact the most important attribute involved in the final yield assessment and consequently in the adaptation of crops to the changing environment. Both the timing of phenological stages and the relative duration of the pre and post-flowering phases are in fact critical determinant of yield (Sadras and Connor, 1991). The activity that is most demanding for a crop (*i.e.* reproductive phase) should take place at the time of optimal conditions (*i.e.* temperature and rainfall) (Visser and Both, 2005), whereas its duration should be as long as possible for optimal biomass partitioning to the fruit (Bindi *et al.*, 1992). Since crop development rate is highly temperature dependent, a warmer climate is expected to affect both these terms, by advancing phenological stages (shifting crop-growing period into a new climatic window) and by reducing the time for biomass accumulation (Bindi and Moriondo, 2005).

Additionally, a changing climate may exhibit increased climatic variability and this can produce relatively large changes in the frequency of extreme climatic events. Accordingly, the increase of extreme events centred at the time of sensitive growth stages are expected to have a great impact on final yield. Warmer and wetter future winters, as prospected in some areas, may cause advanced bud-burst leaving plants vulnerable to spring frosts (Moriondo and Bindi, 2007). On the other hand, increased dry spell may lead to a greater frequency of dry summers requiring irrigation for summer crops. Heat waves at anthesis should be also taken into account due to their effect on yield quality and quantity (Porter and Gawit, 1999). Accordingly, cultivated plants, as well as natural vegetation, will face environmental conditions never experienced before and the actual magnitude of climate change impact needs to be assessed for specific organism environment combinations.

The interactions between phenology-environment in a changing climate were highligheted in order to explain the possible impact on final yield and to formulate feasible adaptation strategies. In particular, papers dealing with climate change impact on the crop phenology of typical Mediterranean annual (sunflower, maize, potato and durum wheat) and perennial (grapevine and olive tree) crops were reviewed. The approach generally used in these papers (Osborne *et al.*, 2000, Moriondo and Bindi, 2006; Moriondo *et al*, 2011; Moriondo and Bindi, 2005, Bindi and Moriondo, 2005; Moriondo, 2003) lied on the use of crop growth and phenology simulation models coupled to the outputs of a General Circulation

Model (GCM). In particular, the phenology and yield of annual crops were simulated using the CropSyst model (Stockle *et al.* 2003); whereas the phenology of grapevine and olive was reproduced using the grapevine model developed by Bindi *et al.*, 1997 and growing degree-day accumulation models (Osborne *et al.*, 2000, Moriondo *et al.* 2001), respectively. The GCM data were directly used as model inputs or downscaled at higher resolution dynamically Regional Circulation Model (RCMs) or Empirically Downscaling (ED), for a better climate simulation at local scale.

Abiotic stresses are the main cause of extensive agricultural production losses worldwide (Barnabas *et al.* 2008; Athar and Ashraf 2009). Among abiotic stresses, drought, salinity and extreme temperatures are the major environmental constraints that modern agriculture has to cope with. It has been estimated that they may be responsible for over 50% yield reduction in major crop plants. However, severity of losses depends on the plant development stage at which the stress occurs, its intensity and duration (Bray *et al.* 2000a; Ashraf *et al.* 2008; Atteya 2003; Monneveux *et al.* 2006; Lafitte *et al.* 2007). According to the United Nations Food and Agriculture Organization (FAO), up to 26% of arable land is subjected to drought and over 20% of the irrigated land is salt-affected (Rehman *et al.* 2005; Pitman and Lauchli 2002). Thus, as the climatic conditions are getting worse, new resistant crop varieties are needed. It is possible to obtain them through the selection of cross-bred lines or by the means of genetic engineering. However, better understanding of the mechanisms involved in plant stress responses is necessary to reach that goal. Incorporation of modern knowledge into traditional breeding strategy is necessary for development of new varieties with improved adaptation to non-optimal environmental conditions. Plant responses towards abiotic stresses have been the subject of various researches since decades. Initially focused on the model plants, now moved to various crop plants like wheat, barley, rice, maize, and other economically important species. The development of the "omics" technologies (*e.g.* genomics, transcriptomics, proteomics, metabolomics) has revolutionized plant science research and has enabled more holistic study of interactions among biological components using models and/or networks to integrate genes, metabolites, proteins, regulatory elements, fluxes and other, this methodology is defined as systems biology (Yuan *et al.* 2008). Recent research permitted to explain functions of many key genes, proteins, metabolites and molecular networks involved in plant responses to salinity, drought, heat, cold, heavy metals and other abiotic stresses. It has been reported that under environmental stress during the flowering and fruiting period, abscission of flowers and flower buds may occur (Abdulmallick *et al.*, 2012) in sweet peeper (*Capsicum annum* L.) and this can result in serious yield decrease in pepper production (Minges and Warholic, 1973). The abscission can be caused by several factors such as extremes of temperature, lack of moisture or low light conditions (Wien, 1990). According to Dagdelen *et al.*, 2004, yield of pepper plant is seriously reduced when pepper is exposed to moisture stress during flowering and fruit formation stage.

Abiotic stresses such as drought, salinity, and high and low temperatures negatively affect plant growth and development, causing cellular water deficit, cell membrane injury, loss of enzyme activities, and other defects and resulting in severe reductions of crop yields (Zhu, 2002, Yamaguchi-Shinozaki and Shinozaki, 2006). Drought also enhances the damage caused by other stresses (Farooq *et al.*, 2009). Genes for acclimation to water deficit help to minimize loss of plant productivity during drought. Such genes have been identified and analyzed in many studies, especially during vegetative development (Kawasaki *et al.*, 2001; Zhu, 2001; Breshears *et al.*, 2005; Schröter *et al.*, 2005; Chaves *et al.,* 2009). For example, one of the plant responses to drought is to increase the expression levels of genes involved in osmolyte synthesis and metabolism, thereby enhancing drought tolerance (Kreps *et al.*, 2002; Yancey, 2005).

Plant hormone pathways and many gene families, including Pyrabactin resistance1 (PYR1), Protein phosphatase2C (PP2C), SNF1-related protein kinase (SnRK), Dehydration response element B (DREB), No apical meristem/arabidopsis transcription activation factor/cupshaped cotyledon (NAC) and Heat shock transcription factor (HSF) families, have important functions in response to different environmental stresses (Zhu, 2002; Ooka *et al.*, 2003; Busch *et al.*, 2005; Park *et aal.*, 2009: Kingler *et al.*, 2010). The hormone abscisic acid (ABA) plays key roles in response and acclimation to biotic and abiotic stresses during vegetative development (Fujita *et al.*,2006; Hirayama and Shinozaki, 2007). ABA is synthesized in response to drought and acts at least in part by binding to ABA receptors of the PYR/PYR1-LIKE (PYL)/REGULATORY COMPONENTS OF ABA RECEPTORS (RCAR) family in *Arabidopsis thaliana (*Fujii *et al., 2009;* Melcher *et al., 2009).* The ABA-PYR/PYL/RCAR complexes bind to and negatively regulate PP2Cs. The binding of PYR/PYL/RCAR to PP2Cs releases and facilitates the phosphorylation of SnRK2s, which then induce downstream responses, including the expression of transcription factors (Park *et al.*, 2009; Umezawa *et al.*, 2009; Cutler *et al.*, 2010).

Plant vegetative growth is greatly affected by environmental changes, including drought, especially during seed germination and seedling development (Yamaguchi-Shinozaki and Shinozaki, 2007). In contrast to the many studies on the effect of drought on vegetative development, there have been few reports about the effect of drought on reproductive growth. Crops such as wheat (*Triticum aestivum*) and rice (*Oryza sativa*) show partial male sterility under water deficit, leading to significant reduction of grain production (Sheoran and Saini, 1996; Lalonde *et al.*, 1997; Saini, 1997). Maize (*Zea mays*) exhibits embryo abortion soon after pollination in response to drought stress (Andersen *et al.*, 2002) and a transcriptomic analysis of fertilized maize ovaries under water deficit revealed that many events, such as ABA signaling and senescence, were activated by drought (Kakumanu *et al.*, 2012). Furthermore, *Arabidopsis* exhibits female reproductive organ abortion because of salt stress (Park *et al.*, 2004; Sun *et al.*, 2004). Another study uncovered altered expression of genes encoding 65 transcription factors in response to salt stress, including 6 ETHYLENE RESPONSE FACTOR (ERF)/AP2-family members and 11 NAC proteins (Sun *et al.*, 2005). However, these studies did not characterize reproductive development.

Plant vegetative growth is greatly affected by environmental changes, including drought, especially during seed germination and seedling development (Shinizaki and Yamaguchi-Shinozaki, 2007). In contrast to the many studies on the effect of drought on vegetative development, there have been few reports about the effect of drought on reproductive growth. Crops such as wheat (*Triticum aestivum*) and rice (*Oryza sativa*) show partial male sterility under water deficit, leading to significant reduction of grain production (Sheoran and Saini, 1996; Lalonde *et al.*, 1996; Saini, 1997). Maize (*Zea mays*) exhibits embryo abortion soon after pollination in response to drought stress (Anderson *et al.*, 2002). and a transcriptomic analysis of fertilized maize ovaries under water deficit revealed that many events, such as ABA signaling and senescence, were activated by drought (Kakumanu *et al.*, 2012). Furthermore, *Arabidopsis* exhibits female reproductive organ abortion because of salt stress (Park *et al.*, 2004; Sun *et al.*, 2004). Another study uncovered altered expression of genes encoding 65 transcription factors in response to salt stress, including 6 ETHYLENE RESPONSE FACTOR (ERF)/AP2-family members and 11 NAC proteins (Sun *et al.*, 2005). However, these studies did not characterize reproductive development.

Environmental factors — light, temperature, water, and soil — greatly influence plant growth and geographic distribution. These factors determine the suitability of a crop for a particular location, cropping pattern, management practices, and levels of inputs needed. A crop performs best and is least costly to produce if it is grown under the most favorable environmental conditions. To maximize the production of any crop, it is important to understand how these environmental factors affect plant growth and development.

4.2. Light

Sunlight is essential for any crop. Dry matter production often increases in direct proportion with increasing amounts of light. The amount of sunlight received by plants in a particular region is affected by the intensity of the incoming light and the day length. The light intensity changes with elevation, latitude, and season, as well as other factors such as clouds, dust, smoke, or fog. The total amount of light received by a crop plant is also affected by cropping systems and crop density. Different plants differ in their light requirements:

- **Full sun** plants thrive in full sun but grow poorly in shade.
- **Partial sun (or partial shade)** plants will produce an edible crop when grown in a shady location. However, these plants need at least 50-80% of full sun.
- **Full shade** plants thrive in 30-50% of full sun but weaken in full sun. Shading sometimes is used to inhibit pigment development in crops in which the lack of color is an important quality factor.

Among the main environmental factors, solar radiation is the most significant one that regulates the photosynthesis, and consequently, the plant survival, growth and adaptation. In any habitat the light intensity varies temporally (seasonally and diurnally) and spatially. Therefore, plants develop acclimation and plasticity to cope with the varying light regimes (Zhang *et al.* 2003). The majority of plant species have the ability to develop anatomical, morphological, physiological and biochemical alterations in response to different light intensities (Muraoka *et al.* 2002; Sousa Paiva *et al.* 2003; De Carvalho Gonçalves *et al.* 2005).

According to previous comparative studies, the biomass of roots, stems, leaves and whole plant as well as the photosynthetic rate, the transpiration and the stomatal conductance of water vapor decreased under low light (Zhang *et al.* 2003; An and Shangguan 2009; Wang *et al.* 2009; Mielke and Schaffer 2010). On the contrary, plant height increased at low light intensity (Yang *et al.* 2007; Wang *et al.* 2009). Besides, plant leaves expanded under high irradiance presented lower photosynthetic pigments contents than leaves expanded under low irradiance (De Carvalho Gonçalves *et al.* 2005; Yang *et al.* 2007; Mielke and Schaffer 2010).

Due to the tilt of the earth's axis and its travel around the sun, the day length (also called photoperiod) varies with season and latitude. Photoperiod controls flowering or the formation of storage organs in some species. Some plants flower when a specific day-length minimum has been passed:

- **Short day plants** flower when day length decreases.
- **Long day plants** flower when day length increases.

- **Day neutral plants** are not affected by day length, and can flower under any light period.

4.3. Temperature

Temperature influences photosynthesis, water and nutrient absorption, transpiration, respiration, and enzyme activity. These factors govern germination, flowering, pollen viability, fruit set, rates of maturation and senescence, yield, quality, harvest duration, and shelf life. Different plants have different temperature requirements. However, for most crop species, optimum temperatures usually range around 25°C. Temperature requirements (usually based on night temperature) of plants are given below by the cardinal values and derived range for "effective growth" (growth range) and "optimum growth" (optimum range) that Krug (1991) has used for major vegetables.

- **Hot:** growth range 18-35°C; optimum range 25-27°C.
- **Warm:** growth range (10) 12-35°C; optimum range 20-25°C.
- **Cool-hot:** growth range (5) 7-30°C; optimum range 20-25°C.
- **Cool-warm:** growth range (5) 7-25°C; optimum range 18-25°C.

Depending on the situation and the specific crop, ambient temperatures higher or lower than the effective growth range will reduce growth and delay development, and subsequently decrease yield and quality. The extremes may be considered killing frosts at about 0°C and death by heat and desiccation at about 40°C.

Exposure to excessive temperatures during development limits the yield of many of the world's major crops, especially in the tropics. Increasing global temperatures over the last three decades have resulted in significantly reduced yields in many crops. In addition to the general warming, a predicted increase in the occurrence of heat waves is likely to result in further yield losses (Long and Ort, 2010). Heat stress due to high ambient temperatures is a serious threat to crop production worldwide (Hall, 2001). Gaseous emissions due to human activities are substantially adding to the existing concentrations of greenhouse gases, particularly:

In daylight hours leaf temperatures are often higher than that of the surrounding air, as the canopy absorbs incident solar radiation. Overheating occurs when heat dissipation from the canopy is unable to keep pace with the thermal energy absorbed. This typically occurs when incident radiation is high and transpirational cooling is low. Even at temperate latitudes, such conditions often develop at midday when solar radiation peaks and soil water reserves are depleted. In warm, dry environments heat stress can persist for prolonged periods. As heat stress is frequently encountered in combination with water deficit and excess irradiance, it can be difficult to disentangle the effects of the three factors. Nonetheless, there is a distinct set of injuries and plant responses that are associated with heat stress. These are detailed in the following sections.

Of the major forms of abiotic stress plants are exposed to in nature, heat stress has an independent mode of action on the physiology and metabolism of plant cells. Although frequently, heat stress is compounded by additional abiotic stresses such as drought and salt stress, it is important to unravel the independent action and biological consequences of high temperature in order to ameliorate the effects of combined abiotic stress. The susceptibility to high temperatures in plants varies with the stage of plant development, heat stress affecting to a certain extent all vegetative and reproductive stages. The observed effects depend on species and genotype, with abundant inter- and intra-specific variations (Barnabás *et al.*, 2008; Sakata and Higashitani, 2008).

Various physiological injuries have been observed under elevated temperatures, such as scorching of leaves and stems, leaf abscission and senescence, shoot and root growth inhibition or fruit damage, which consequently lead to a decreased plant productivity (Vollenweider and Günthardt-Goerg, 2005). In many cases, plant architecture changes and hypocotyls and petioles elongate resembling the morphological responses of shade avoidance (Hua, 2009; Tian *et al.*, 2009). However, high temperatures reduce plant growth by affecting the shoot net assimilation rates and thus the total dry weight of the plant (Wahid *et al.*, 2007).

Higher plants exposed to excess heat, at least 5°C above their optimal growing conditions exhibit a characteristic set of cellular and metabolic responses required for the plants to survive under the high temperature conditions (Guy, 1999). These effects include changes in the organization of cellular structures, including organelles and the cytoskeleton, and membrane functions (Weis and Berry, 1988), accompanied by a decrease in the synthesis of normal proteins and the accelerated transcription and translation of heat shock proteins (HSPs; Bray *et al.*, 2000b), the production of phytohormones such as abscisic acid (ABA) and antioxidants and other protective molecules (Maestri *et al.*, 2002).

The changes in ambient temperature are sensed by plants with a complicated set of sensors positioned in various cellular compartments. The increased fluidity of the membrane leads to activation of lipid-based signaling cascades and to an increased Ca^{2+} influx and cytoskeletal reorganization. Signaling between these routes leads to the production of osmolytes and antioxidants in response to heat stress. For example, the *Arabidopsis* CNGC2 gene encodes a component of the membrane cyclic nucleotide gated Ca^{2+} channels that acts as the primary thermo-sensors of land plant cells. These channels in the plasma membrane respond to increments in the ambient temperature by triggering an optimal heat shock response (Saidi *et al.*, 2009). This data emphasizes the vital role of lipid membranes in response to heat stress in plants (Horváth *et al.*, 2012). Recently it was shown that differences in the tissue-specific activation of various signaling pathways may occur between vegetative and reproductive tissues (Mittler *et al.*, 2011).

Heat stress induces changes in respiration and photosynthesis and thus leads to a shortened life cycle and diminished plant productivity (Barnabás *et al.*, 2008). The early effects of thermal stress comprise of structural alterations in chloroplast protein complexes and reduced activity of enzymes (Ahmad *et al.*, 2010). In addition, by causing injuries to the cell membrane, organization of microtubules and ultimately to the cytoskeleton, heat stress changes membrane permeability and alters cell differentiation, elongation, and expansion (Potters *et al.*, 2008; Rasheed, 2009). For example, the photochemical modifications in the carbon flux of the chloroplast stroma and those of the thylakoid membrane system are considered the primary sites of heat injury (Wise et al., 2004), as photosynthesis and the enzymes of the Calvin–Benson cycle, including ribulose 1,5-bisphosphate carboxylase (Rubisco) and Rubisco activase are very sensitive to increased temperature and are severely inhibited even at low levels of heat stress (Maestri *et al.*, 2002; Morales *et al.*, 2003).

A specific effect of high temperatures on photosynthetic membranes includes the swelling of grana stacks and an aberrant stacking. Such structural changes are accompanied by ion-leakage from leaf cells exposed to heat and changes in energy allocation to the photosystems (Wahid and Shabbir, 2005; Allakhverdiev *et al.*, 2008). The maintenance of cellular membrane function under high temperature stress is thus essential for a sustained photosynthetic and respiratory performance (Chen *et al.*, 2010). The detrimental effects of heat on chlorophyll and the photosynthetic apparatus are also associated with the production of injurious reactive oxygen species (ROSs; Camejo *et al.*, 2006; Guo *et al.*, 2007). By increasing chlorophyllase activity and decreasing the amount of photosynthetic pigments, heat stress ultimately reduces the plant photosynthetic and respiratory activity (Todorov *et al.*, 2003; Sharkey and Zhang, 2010). With respect to reproductive success, a decline of photosynthesis will eventually result in limited resource availability for reproduction in parental and gametophytic tissues due to a reduction in energy reserves leading to plant starvation (Young *et al.*, 2004; Sumesh *et al.*, 2008).

A wide range of plant developmental and physiological processes is negatively affected by heat stress. When the stress occurs at key developmental stages such as reproduction, this becomes one of the major constraints of plant adaptation to a changing environment (Hall, 2001). For example, high temperature during wheat reproductive development hastened the decline in photosynthesis and leaf area, decreased shoot and grain mass as well as weight and sugar content of kernels, while also reducing water-use efficiency (Shah and Paulsen, 2003).

The effect of heat stress on staple crops like wheat can be severe. The impact varies depending on the developmental stage of the plants, with the most vulnerable stage being flowering. High temperatures shorten the duration of growth of both the leaves and the grains, accelerating their development and thus limiting the ability of the plant to accumulate the carbohydrate necessary for grain growth. In addition, heat stress before flowering can cause floret sterility, causing yield losses due to reduced grain number. This effect is most acute when heat occurs at or just after pollen meiosis, when carbohydrate supply to the developing pollen grains appears most critical. Grain size in heat stressed plants can be severely reduced, predominately from a reduction in starch, which makes up most of the mass of the grain.

Wheat (a temperate C_3 species) produces its most grain at temperatures below 26^{0}C; and its yield is reduced at higher temperatures. Yet in most grain-growing areas in the southern hemisphere, and in Mediterranean climates in the northern hemisphere, temperatures increase steadily during the growing season, and brief periods above 30^{0}C often occur during the grain-filling period. In many arid countries mean day temperatures can easily exceed wheat's high temperature threshold during much of the growing season and so heat stress can significantly reduce crop yield by accelerating plant senescence, diminishing seed number and final seed weight.

Where vegetation is sparse, maximum daytime temperatures occur at the soil surface where exposed soil absorbs solar radiation and quickly warms above ambient. Under such conditions soil temperatures can exceed 50^{0}C. Exposed soil is a particular problem when planting crops in warm regions where the dark, moist soil surface can reach high temperatures and severely inhibit germination and seedling emergence. A similar phenomenon has been seen in temperate climates when plastic mulch is used to artificially insulate the soil surface during planting (Farrell and Gilliland 2011).

At very high temperatures, severe cellular injury and even cell death may occur within minutes, which could be attributed to a catastrophic collapse of cellular organization (Schöffl *et al.*, 1999). At moderately high temperatures, injuries or death may occur only after long-term exposure. Direct injuries due to high temperatures include protein denaturation and aggregation, and increased fluidity of membrane lipids. Indirect or slower heat injuries include inactivation of enzymes in chloroplast and mitochondria, inhibition of protein synthesis, protein degradation and loss of membrane integrity (Howarth, 2005). Heat stress also affects the organization of microtubules by splitting and/or elongation of spindles, formation of microtubule asters in mitotic cells, and elongation of phragmoplast microtubules (Smertenko *et al.*, 1997). These injuries eventually lead to starvation, inhibition of growth, reduced ion flux, production of toxic compounds and reactive oxygen species (ROS) (Schöffl *et al.*, 1999; Howarth, 2005).

Immediately after exposure to high temperatures and perception of signals, changes occur at the molecular level altering the expression of genes and accumulation of transcripts, thereby leading to the synthesis of stress-related proteins as a stress tolerance strategy (Iba, 2002). Expression of heat shock proteins (HSPs) is known to be an important adaptive strategy in this regard (Feder and Hoffman, 1999). The HSPs, ranging in molecular mass from about 10 to 200 kDa, have chaperone like functions and are involved in signal transduction during heat stress (Schöffl *et al.*, 1999). The tolerance conferred by HSPs results in improved physiological phenomena such as photosynthesis, assimilate partitioning, water and nutrient use efficiency, and membrane stability (Camejo *et al.*, 2005; Ahn and Zimmerman, 2006; Momcilovic and Ristic, 2007). Such improvements make plant growth and development possible under heat stress. However, not all plant species or genotypes within species have similar capabilities in coping with the heat stress. There exists tremendous variation within and between species, providing opportunities to improve crop heat-stress tolerance through genetic means. Some attempts to develop heat-tolerant genotypes via conventional plant breeding protocols have been successful (Ehlers and Hall, 1998; Camejo *et al.*, 2005). Advanced techniques of molecular breeding and genetic engineering have provided additional tools, which could be employed to develop crops with improved heat tolerance and to combat this universal environmental adversary. However, to assure achievement of success in this strategy, concerted efforts of plant physiologist, molecular biologists and crop breeders are imperative (Wahid *et al.*, 2007).

4.3.1. Heat-stress Threshold

A threshold temperature refers to a value of daily mean temperature at which a detectable reduction in growth begins. Upper and lower developmental threshold temperatures have been determined for many plant species through controlled laboratory and field experiments. A lower developmental threshold or a base temperature is one below which plant growth and development stop. Similarly, an upper developmental threshold is the temperature above which growth and development cease. Knowledge of lower threshold temperatures is important in physiological research as well as for crop production. Base threshold temperatures vary with plant species, but for cool season crops 0⁰C is often the best-predicted base temperature (Miller *et al.*, 2001). Cool season and temperate crops often have lower threshold temperature values compared to tropical crops. Upper threshold temperatures also differ for different plant species and genotypes within species. However, determining a consistent upper threshold temperature is difficult because the plant behavior may differ depending on other environmental conditions (Miller *et al.*, 2001). In tomato, for example, when the ambient temperature exceeds 35⁰C, its seed germination, seedling and vegetative growth, flowering and fruit set, and fruit ripening are adversely affected. For other plant species, the higher threshold temperature may be lower or higher than 35⁰C. Upper threshold temperatures for some major crop species are displayed in Table. High temperature sensitivity is particularly important in tropical and subtropical climates as heat stress may become a major limiting factor for field crop production.

Threshold high temperatures for some crop plants

Crop Plants	Threshold temperature (⁰)	Growth Stage	Reference
Wheat	26	Post-anthesis	Stone and Nicolás, 1994
Corn	38	Grain filling	Thompson, 1986
Cotton	45	Reproductive	Rehman *et al.*, 2004
Pearl Millet	35	Seedling	Ashraf & Hafeez, 2004
Tomato	30	Emergence	Camp *et al.*, 2005
Brassica	29	Flowering	Morrison Stewart, 2002
Cool Season Pulses	25	Flowering	Siddique *et al.*, 1999
Groundnut	34	Pollen production	Vara Prasad *et al.*, 2000
Cowpea	41	Flowering	Patel Hall, 1990
Rice	34	Grain yield	Motita *et al.*, 2004

Brief exposure of plants to high temperatures during seed filling can accelerate senescence, diminish seed set and seed weight, and reduce yield (Siddique *et al.*, 1999). This is because under such conditions plants tend to divert resources to cope with the heat stress and thus limited photosynthates would be available for reproductive development. Another effect of heat stress in many plant species is induced sterility when heat is imposed immediately before or during anthesis. Pulse legumes are particularly sensitive to heat stress at the bloom stage; only a few days of exposure to high temperatures (30–35⁰C) can cause heavy yield losses through flower drop or pod abortion (Siddique *et al.*, 1999). In general, base and upper threshold temperatures vary in plant species belonging to different habitats. Thus, it is highly desirable to appraise threshold temperatures of new cultivars to prevent damages by unfavorable temperatures during the plant ontogeny.

4.3.2. Low Temperature Stress

Low temperature of the field environment restricted the vegetative growth and delayed all the phenological stages in comparison to control plants. It has been shown that chickpea (*Cicer arietinum* L.) lacks cold tolerance and is sensitive to chilling temperatures (>8°C), especially at its reproductive phase (Bakht *et al.* 2006). The reproductive structures can withstand temperature of 8°C minimum to 22°C maximum during the coldest period. The most advantageous temperature range for normal flowering, fertilization and seed set is 10 to 14°C average minimum temperature and 25 to 31°C average maximum temperature. Temperature within the chilling range can limit the growth and vigor of chickpea at all phenological stages but is considered most damaging to yield at reproductive stage. In India, chickpea is grown as a winter-season crop. The northern parts of the Indian subcontinent and southern Australia are the most affected regions due to chilling as the temperature is below 15°C at flowering (Clark, 2001; Bakht *et al.* 2006, Berger, 2007). During the reproductive phase, low temperature is detrimental to normal flowering and pod development, which causes prolonged reproductive phase, floral abortion, poor pollen germination, impaired ovule development, failure in pod set and reduction in seed filling that drastically affects the crop productivity (Nayyar *et al.* 2005). The reproductive phase, especially in early maturing chickpea genotypes, coincides with low temperature of the surroundings causing abortion of flowers and impairment in pod set (Kumar *et al.* 2005). A deep investigation of development of male and female components is required to prove the sensitivity of the developmental stage and actual cause of reproductive failure.

The effects of various abiotic stresses on structural and functional aspects of reproductive phase have been reported in soybean (*Glycine max*) (Gass *et al.* 1996, Kokubun *et al.* 2001), rice (*Oryza sativa*) (Imin *et al.* 2004), pea (*Pisum sativum*) (Guilioni *et al.* 1997) and chickpea (Kaur *et al.* 2008). In rice, low temperature during the reproductive stage has been found to cause degeneration of spikelets, incomplete panicle exsertion and increase in spikelet sterility (Terres, 1991). Subsequently, Takeoka *et al.* (1992) has identified that booting stage in rice is highly susceptible to chilling and the damage at this particular stage of reproductive development causes ~30–40% yield reduction in the temperate grown rice worldwide (Nishiyama, 1995, Andaya and Mackill, 2003). Cold stress induces flower abscission in soybean (Gass *et al.* 1996), pigeonpea (*Cajanus cajan*) (Sandhu *et al.* 2007) and chickpea (Nayyar *et al.* 2005), but the actual cause of reproductive failure is unclear.

Pollen sterility is a common cause of low fruit set under cold temperatures at flowering in rice, sorghum (*Sorghum bicolor*), strawberry (*Fragaria* sp) and tomato (*Lycopersicon esculantum*) (Lin and Peterson 1975, Toriyama and Hinata 1984). Low temperatures may cause deformation of flowers or floral parts (Rylski, 1979) leading to functional sterility or the formation of deformed fruit. It was observed that pod set failure and deformation in fruit could be associated with abnormal ovary development (Rylski, 1979). Fertilization process, subsequent embryo development and seed filling are potential targets of cold stress. It was observed that under cold conditions, the development of male gametophyte is impaired which leads to smaller pollen grains and an increase in non-viable pollen (Gudkova, 1980). Imin *et al.* (2004) while studying the effects of early cold stress on the maturation of rice anthers demonstrated for the first time that cold temperature stress at the young microspore stage enhances and induces partial degradation of proteins in the rice anthers at the trinucleate stage. Chickpea can experience yield losses up to 50% due to cold stress (Saxena and Johansen 1990) depending upon the severity of the stress and sensitivity of the genotype. Hence, incorporation of cold tolerance in chickpea cultivars is an important prerequisite for winter sowing of the crop (Singh *et al.* 1990, Bakht *et al.* 2006).

4.3.3. Plant Responses to Heat Stress

In tropical climates, excess of radiation and high temperatures are often the most limiting factors affecting plant growth and final crop yield. High temperatures can cause considerable pre- and post-harvest damages, including scorching of leaves and twigs, sunburns on leaves, branches and stems, leaf senescence and abscission, shoot and root growth inhibition, fruit discoloration and damage, and reduced yield (Guilioni *et al.*, 1997; Ismail and Hall, 1999; Vollenweider and Gunthardt-Goerg (2005). Similarly, in temperate regions, heat stress has been reported as one of the most important causes of reduction in yield and dry matter production in many crops, including maize (Giaveno and Ferrero, 2003).

High-temperature-induced modifications in plants may be direct as on existing physiological processes or indirect in altering the pattern of development. These responses may differ from one phenological stage to another. For example, long-term effects of heat stress on developing seeds may include delayed germination or loss of vigor, ultimately leading to reduced emergence and seedling establishment. Under diurnally varying temperatures, coleoptile growth in maize reduced at 40^{0}C and ceased at 45^{0}C (Weaich *et al.*, 1996). High temperatures caused significant declines in shoot dry mass, relative growth rate and net assimilation rate in maize, pearl millet and sugarcane, though leaf expansion was minimally affected (Ashraf and Hafeez, 2004; Wahid, 2007). Major impact of high temperatures on shoot growth is a severe reduction in the first internode length resulting in premature death of plants (Hall, 1992). For example, sugarcane plants grown under high temperatures exhibited smaller internodes, increased tillering, early senescence, and reduced total biomass (Ebrahim *et al.*, 1998).

Heat stress, singly or in combination with drought, is a common constraint during anthesis and grain filling stages in many cereal crops of temperate regions. For example, heat stress lengthened the duration of grain filling with reduction in kernel growth leading to losses in kernel density and weight by up to 7% in spring wheat (Guilioni *et al.*, 2003). Similar reductions occurred in starch, protein and oil contents of the maize kernel (Wilhelm *et al.*, 1999) and grain quality in other cereals under heat stress (Maestri *et al.*, 2002). In wheat, both grain weight and grain number appeared to be sensitive to heat stress, as the number of grains per ear at maturity declined with increasing temperature (Ferris *et al.*, 1998). In temperate and tropical lowlands, heat susceptibility is a cause of yield loss in common bean, *Phaseolus vulgaris* (Rainey and Griffiths, 2005) and groundnut, *Arachis hypogea* (Vara Prasad *et al.*, 1999). In tomato, reproductive processes were adversely affected by high temperature, which included meiosis in both male and female organs, pollen germination and pollen tube growth, ovule viability, stigmatic and style positions, number of pollen grains retained by the stigma, fertilization and post-fertilization processes, growth of the endosperm, proembryo and fertilized embryo (Foolad, 2005). Also, the most noticeable effect of high temperatures on reproductive processes in tomato is the production of an exserted style (*i.e.*, stigma is elongated beyond the anther cone), which may prevent self-pollination. Poor fruit set at high temperature has also been associated with lowlevels of carbohydrates and growth regulators released in plant sink tissues (Kinet and Peet, 1997). Growth chamber and greenhouse studies suggest that high temperature is most deleterious when flowers are first visible and sensitivity continues for 10–15 days. Reproductive phases most sensitive to high temperature are gametogenesis (8–9 days before anthesis) and fertilization (1–3 days after anthesis) in various plants (Foolad, 2005). Both male and female gametophytes are sensitive to high temperature and response varies with genotype; however, ovules are generally less heat sensitive than pollen (Peet and Willits, 1998). Overall, based on the available studies, it seems that plant responses to high temperature vary with plant species and phenological stages. Reproductive processes are markedly affected by high temperatures in most plants, which ultimately affect fertilization and post-fertilization processes leading to reduced crop yield.

4.4. Water

Water is absolutely essential for any plant species. Plants can be grouped according to their natural habitats with respect to water supply:

- **Hydrophytes** are plants that are adapted to living in water or in soil saturated with water. The hydrophytes usually have large interconnected intercellular gas-filled spaces in their root and shoot tissues (*aerenchyma*) to facilitate air l exchange.
- **Mesophytes a**re the most common terrestrial plants that are adapted to neither a long wet nor a long dry environment. Depending on the extension of their root systems and other plant features, however, their water requirement varies.
- **Xerophytes** are plants that can endure relatively long periods of drought. The xerophytes usually have special features such as reduced permeability to decrease water loss, swollen tissues to conserve water, or deep and extensive root systems to acquire water.

Water requirement Water is crucial for crop productivity and quality. However, crop water requirements differ according to plant and soil types. A plant's total sum of water requirement includes the water the crop uses by itself and also the losses due to evapotranspiration (which includes both plant transpiration and soil evaporation), water application, land preparation, and leaching during the crop growth period.

4.4.1. Drought

Drought is defined as a period without significant rainfall or soil moisture. Droughts may lead to plant water deficit (drought stress) and growth may be impacted. Drought stress usually occurs when soil water content is less than 50% of field capacity (*i.e.*, when the soil is full of water, hence 100%). Drought stress symptoms include wilting, droopy, curling or rolling of leaves; or browning of shoot tips. Among the mesophytes, the effect of drought stress varies with the species, variety, degree and duration of drought stress, and the growth stage. The yield formation stage is most sensitive for most vegetables. Periods of even short drought stress during this period can reduce yield.

Deficit water supply at any growth stage poses detrimental effects on crop growth and development in general but varies depending on the severity of stress and the crop growth stage. Effects of drought on morphological, physiological, and biochemical processes in plants are discussed below.

4.4.2. Flooding

Flooding occurs when water enters soil faster than it can drain away. Intense rainfall, river overflow, increased surface run-off, over-irrigation, and slow drainage through the soil profile all contribute to flooding, especially in lowland regions. Under waterlogged conditions all pores in the soil are filled with water, depriving the soil of oxygen. As a result, plant roots cannot obtain oxygen for respiration to maintain their activities for nutrient and water uptake. Weakend plants are susceptible to soil-borne diseases. Oxygen deficiency in the soil due to waterlogging also causes death of root hairs, and increases formation of compounds toxic to plant growth. All of these lead to retarded growth or the death of the plant. The extent of flooding damage depends upon the species or variety, stage of plant development, duration of flooding, water level in the soil, soil texture, temperature, and type of microorganisms present. High temperatures usually accelerate the damaging effects. Most mesophytes and xerophytes are sensitive to flooding. However, some species are able to tolerate flooding because of their abilities to increase porosity of the shoot base, or to replace damaged roots.

The extent of flooding damage depends upon the species or variety, stage of plant development, duration of flooding, water level in the soil, soil texture, temperature, and type of microorganisms present. High temperatures usually accelerate the damaging effects. Most mesophytes and xerophytes are sensitive to flooding. However, some species are able to tolerate flooding because of their abilities to increase porosity of the shoot base, or to replace damaged roots.

4.5. Soil

Soil is a natural medium that provides anchorage for the plant and supplies water and mineral nutrients for normal growth. Soil consists of mineral matter, organic matter, air, and water. The proportion of these four constituents and the types of mineral and organic material determines soil properties such as soil type, soil pH, and fertility.

4.5.1. Soil Type

Soil is made up in part of mineral particles grouped as sand (0.05 to 2 mm), silt (0.002 to 0.05 mm), and clay (<0.002 mm). The ratio of these determines soil types, such as **clay**, **clayey loam**, **loam**, **sandy loam**, and **sand**. Loam is composed of sand, silt, and clay in relatively even amounts, and exerts a greater influence on soil properties than does sand, silt, or clay. Soil type determines the soil's capacity to store water and nutrients, aeration, drainage, and ease of field operations. Sandy soils are easily tilled, well-drained and aerated but usually have low fertility and water-holding capacity. Clayey soils, on the other hand, are more fertile and have high water retention but are poorly drained and aerated.

4.5.2. Soil pH

Soil pH is a measure of the soil's acidity or alkalinity, and it affects the plant indirectly by influencing the availability of nutrients and the activity of microorganisms. Nutrients are most available at pH levels between 6.5 and 7.5. Nutrients in the soil may be chemically tied up or bound to soil particles and unavailable to plants if the pH is outside this range. Individual plants have pH preferences and grow best if planted in soils that satisfy their pH requirements.

4.5.3. Soil Fertility

Soil fertility is the inherent capacity of soil to provide plant nutrients in adequate amounts and in proper balance for the growth of specific plants. A fertile soil is usually rich in nitrogen, phosphorus, and potassium, and contains sufficient trace elements and soil organic matter that improves soil structure and soil moisture retention.

4.5.4. Soil Salinity

Soil salinity refers to the presence of excess salts in soil water, which often results from irrigated agriculture. After the plants take up the water, the dissolved salts from irrigated water start to accumulate in the soil. Soil salinity is usually measured as electrical conductivity (EC) of soil solution, and expressed in deci siemens per meter (dS/m). Excess salts generally affect plant growth by increasing osmotic tension in the soil, making it more difficult for the plants to take up water. Excessive uptake of salts from the soil by plants also may have a direct toxic effect on the plants. Soil salinity is most pronounced in arid areas.

4.5.4.1. Causes of Salinity

Natural cause: Most of the saline sodic soils are developed due to natural geological, hydrological and pedological processes. Some of the parent materials of those soils include intermediate igneous rocks such as phenolytes, basic igneous rocks such as basalt, undifferentiated volcanic rocks, sandstones, alluvium and lagoonal deposits (Wanjogu *et al.*, 2001). Climatic factors and water management may accelerate salinization. In arid and semi-arid lands evapo-transpiration plays a very important role in the pedogenesis of saline and sodic soils.

Another type of salinity occurs in coastal areas subjected to tides and the main cause is intrusion of saline water into rivers or aquifers. Coastal rice crops in Asia, for instance, are frequently affected by exposure to sea water brought in by cyclones around the Indian Ocean (Sultana *et al.*, 2001). Cyclic salts are ocean salts carried inland by wind and deposited by rainfall, and are mainly sodium chloride.

Depending on prevailing winds and distance from the sea-coast the rain water composition greatly varies. The composition of sea water is expressed as g kg^{-1} or ppt (parts per thousands) and is almost uniform around the globe. The electrical conductivity of sea water is 55 dS m^{-1} while that of rainwater is about 0.01 dS m^{-1}.

Anthropogenically induced salinity: Secondary salt affected soils are those that have been salinized by human caused factors, mainly as a consequence of improper methods of irrigation. Poor quality water is often used for irrigation, so that eventually salt builds up in the soil unless the management of the irrigation system is such that salts are leached from the soil profile. Szaboles (1992) estimated that 50% of all irrigated schemes are salt affected. Too few attempts have been made recently to access the degree of human-induced secondary salinization and, according to Flowers and Yeo (1995) this makes it difficult to evaluate the importance of salinity to future agricultural productivity. Anthropic salinization occurs in arid and semi arid areas due to waterlogging brought about by improper irrigation (Ponnamperuma, 1984). Secondary salt affected soils can also be caused by human activities other than irrigation and include, but are not limited to the following:

(a) **Deforestation**: It is recognized as a major cause of salinization and alkalinization of soils as a result of the effects of salt migration in both the upper and lower layers. Deforestation leads to the reduction in average rainfall and increased surface temperature (Hastenrath, 1991). Top thin soil rapidly gets eroded in the absence of soil green cover. Without the trees there to act as a buffer between the soil and the rain, erosion is practically inevitable. Soil erosion then leads to greater amounts of run-off and increased sedimentation in the rivers and streams. The combination of these factors leads to flooding and increased salinity of the soil (Hastenrath, 1991). The Indian plains formed by the rivers of north India increasingly getting salt affected as coastal areas of Ganges particularly lower Ganges plains and Sundarban estuarine areas. In southeast India, for example, vast areas of farmer forestland became increasingly saline and alkaline within a few years after the felling of the woods (Szaboles, 1994). In Australia, a country where one-third of the soils are sodic and 5% saline (Fitzpatrick, 1994), there is serious risk of salinization if land with shallow unconfined aquifers containing water with more than 0.25% total soluble salt is decreased of trees (Bui *et al.*, 1996).

(b) **Accumulation** of air-borne or water-borne salts in soils: Szaboles (1994) has reported that chemicals from industrial emissions may accumulate in the soil, and if the concentration is high enough, can result in salt accumulation in the upper layer of soil. Similarly, water with considerable salt concentration such as waste water from municipalities and sludge may contaminate the upper soil later causing salinization and/or alkalization (Bond, 1998).

(c) **Contamination** with chemicals: It often occurs in modern intensive agricultural systems, particularly in green houses and intensive farming systems.

(d) **Overgrazing:** This process occurs mainly in arid and semiarid regions, where the natural soil cover is poor and scarcely satisfies the fodder requirement of intensive animal husbandry (Szaboles, 1994). The natural vegetation becomes sparse and progressive salinization develops, and sometimes the process ends up in desertification as the pasture diminishes due to overgrazing.Not all plants respond to salinity in a similar manner; some crops can produce acceptable yields at much higher soil salinity than others. This is because some crops are better able to make the osmotic adjustments, enabling them to extract more water from a saline soil. For example, turnip and carrot are among the most sensitive vegetables and can tolerate soil salinities of only about 1 dS/m before yield declines. Zucchini, on the other hand, can tolerate soil salinity of up to 4.7 dS/m before yield reduces. The ability of a crop to adjust to salinity is extremely useful. In areas where a build-up of soil salinity cannot be controlled, an alternative crop can be selected that is both more tolerant of the expected soil salinity and able to produce economic yields.

4.6. Growth and Development

Establishment of an early and optimum crop stand is important for harvesting maximum productivity. However, if the crop experiences an early drought, thereby affecting germination, then the suboptimal plant population is the major cause of low grain yield. Early season drought severely reduces germination and stand establishment principally due to reduced water uptake during the imbibition phase of germination, reduced energy supply, and impaired enzyme activities (Okcu *et al.* 2005; Taiz and Zeiger 2010).

Growth is an irreversible increase in volume, size, or weight, which includes the phases of cell division, cell elongation, and differentiation. Both cell division and cell enlargement are affected under drought owing to impaired enzyme activities, loss of turgor, and decreased energy supply (Kiani *et al.* 2007; Farooq *et al.* 2009a; Taiz and Zeiger 2010). For example, drought decreases growth and productivity of sunflower (*Heliantus annuus* L.) owing to reductions in leaf water potential, rate of cell division, and enlargement primarily due to loss of turgor (Kiani *et al.* 2007; Hussain *et al.* 2009). Under drought, reduced dry matter accumulation occurs in all plant organs, although different organs manifest varying degrees of reduction. For instance, drought decreased shoot and flower fresh and dry weights of marigold (*Tagetes erecta* L.) plants (Asrar and Elhindi 2011). Likewise, drought considerably reduced shoot and root dry weights in Asian red sage (*Salvia miltiorrhiza* L.), although roots were less affected than shoots (Liu *et al.* 2011). Drought also decreased leaf area owing to loss of turgor and reduced leaf numbers (Farooq *et al.* 2010a).

Leaf area index (LAI) is the ratio of leaf area to ground area, which denotes the extent of assimilatory power of crops under field conditions. Drought decreases LAI in crop plants in general. For instance, Hussain *et al.* (2009) reported decline in LAI of sunflower exposed to drought at budding and flowering stages. Drought also suppresses leaf expansion and tillering (Kramer and Boyer 1995), and reduces leaf area due to early senescence (Nooden 1988). All these factors contribute to reduced dry matter accumulation and grain yield under drought. The study of different growth and developmental events in crop plants with respect to time is called crop phenology. Drought strongly affects crop phenology by shortening the crop growth cycle with a few exceptions. Limited water supply triggers a signal to cause an early switching of plant development from the vegetative to reproductive phase (Desclaux and Roumet 1996). For instance, total growth duration of both bread wheat (*Triticum aestivum* L.) and barley (*Hordeum vulgare* L.) decreased under drought (McMaster and Wilhelm 2003), which generally results in substantial yield reductions. The effect of drought is phase specific in most cases. For example, drought at pre-anthesis delayed flowering in quinoa (*Chenopodium quinoa* Wild.) and bread wheat plants (Majid *et al.* 2007; Geerts *et al.* 2008). Likewise, drought at anthesis commonly delays flowering in rice (*Oryza sativa* L.); interestingly, the longer the delay, the higher the yield penalty (Fukai 1999). In soybean (*Glycine max* L.), drought during grain filling hastened maturity but yield was down due to smaller grains (Desclaux and Roumet 1996). Different crops respond to drought differently. For instance, upon exposure to drought flowering is delayed in maize (*Zea mays* L.) (Abrecht and Carberry 1993), quinoa (Geerts *et al.* 2008), and rice (Fukai 1999), whereas in soybean (Desclaux and Roumet 1996), wheat, and barley (McMaster and Wilhelm 2003) drought hastened flowering and physiological maturity.

While drought occurs during the vegetative period of crop growth, it may substantially decrease economic yield. Drought stress during reproductive and grain filling phases is more devastating (Reddy *et al.* 2003; Vijay 2004; Yadav *et al.* 2004; Lafitte *et al.* 2007). Drought at flowering is critical as it can increase pollen sterility resulting in hampered grain set. In sunflower, for example, under drought at flowering, achene yield declined primarily due to less achenes (Hussain *et al.* 2008). In pearl millet (*Pennisetum glaucum* L. Leeke), drought at flowering increased the rate of ear abortion due to a decline in assimilate supply to developing ears (Yadav *et al.* 2004). In drought-stressed maize, kernel set was lost leading to low grain yield (Schussler and Westgate 1995). Likewise, water deficit at anthesis increased pod abortion which reduced yield in soybean (Liu *et al.* 2003).

Traditionally, 4 levels of soil salinity based on saline irrigation water have been distinguished, low salinity defined by electrical conductivity of less than 0.25 mmhos cm^{-1} (in current terminology equal to 0.25 dS m^{-1}); medium salinity (0.25 to 0.75 dS m^{-1}); high salinity (0.75 to 2.25 dS m^{-1}), and very high salinity with an electrical conductivity exceeding 2.25 dS m^{-1}. The common cations associated with salinity are Na^+, Ca^{2+} and Mg^{2+}, while the common anions are Cl^-, SO_4^{2-} and HCO^-_3. Since Na^+ in particular causes deterioration of the physical structure of soil and Na^+ and Cl^- both are toxic to plants are therefore considered the most important ions (Hasegawa *et al.*, 2000). Historically soils were classified as saline, sodic or salinesodic based on the total concentration of salt and the ratio of Na^+ to Ca^{2+} and Mg^{2+} in the saturated extract of the soil. Salinity occurs through natural or human induced processes that result in the accumulation of dissolved salts in the soil water to an extent that inhibits plant growth. Sodicity is a secondary result of salinity in clay soils, where leaching through either natural or human induced processes has washed soluble salts into the subsoil and left sodium bound to the negative charges of the clay due to an increase in its concentration. There is competition for fresh water among the municipal, industrial and agricultural sectors in several regions.

The consequence has been a decreased allocation of fresh water to agriculture (Tilman *et al.*, 2002). This phenomenon is expected to continue and to intensify in less developed, arid region countries that already have high population growth rates and suffer from serious environmental problems. For this reason there is increasing pressure to irrigate with water of certain salt content like ground water, drainage water and treated waste water.

According to Yeo (1998) and Grattan and Grieve (1999) the direct effect of salts on plant growth may be divided into three broad categories: (i) a reduction in the osmotic potential of the soil solution that reduces plant available water, (ii) a deterioration in the physical structure of the soil such that water permeability and soil aeration are diminished, and (iii) increase in the concentration of certain ions that have an inhibitory effect on plant metabolism (specific in toxicity and mineral nutrient deficiencies). The relative contribution of osmotic effects and specific in toxicities on yield is difficult to quantify. However, with most crops, Dasberg *et al.* (1991) reported that yield losses from osmotic stress could be significant before foliar injury is apparent.

Abiotic stress has main impact on the crop productivity worldwide, reducing average yields for major crop plants. These abiotic stresses are interconnected as osmotic stress, resulting in the disruption of ion distribution and homeostasis in the cell. It is mainly due to changes in the expression patterns of a group of genes, leads to responses that affect growth rates, productivity. Global climate changes are leading to increases in temperature and atmospheric CO_2 levels as well as disturbance in rainfall patterns. Prolonged periods of in-appropriate rainfall leads to drought. Severe drought conditions bring about a steadily decrease in soil water availability to plants and cause premature plant death, while intermittent drought conditions affect the plant growth and development but are not usually lethal. The ability to survive longer and maintain normal function under intermittent drought conditions leads to yields, which are much lower than those observed under hydrated conditions. Drought tolerance enables plants to grow and maintain high yields in spite of drought conditions and is a result of the plant's efforts to withstand from stress. If the tolerance is restricted to that particular generation, the plant is said to be acclimated to drought. If it persists over generations, the plant genotype is said to be adapted to drought conditions. First response of plants subjected to drought stress is growth arrest. Shoot growth reduction under drought lowers metabolic demands of the plant and mobilizes metabolites for the synthesis of protective compounds required for osmotic adjustment.

Cold stress is one of the main abiotic stresses that limit agricultural crop productivity by affecting their quality and post-harvest life. Because plants are immobile, they must modify their metabolism to survive such stress. Most temperate plants acquire chilling and freezing tolerance upon their exposure to sub-lethal cold stress, a process called cold acclimation. Still, many agronomically important crops are incompetent of cold acclimation. Cold stress affects virtually every aspects of cellular function in plants.

Soil salinity is a worldwide problem and poses a serious threat to entire world agriculture, because it reduces the crop yield in the affected areas. It is possibly the most imperative ecological restriction that causes extensive crop yield losses all over the world, and its threat is escalating day by day (Schwabele *et al.*, 2006). Increasing salinity reduces the average yield of major crops by more than 50% (Bray *et al.*, 2000b), and these losses are of great concern mainly for the agriculture based countries. Soil contains soluble salts of multifarious nature. High concentrations of salt impose both osmotic and ionic stresses on the plants which lead to several morphological and physiological changes (Jampeetong and Brix, 2009). Under salt stress the osmotic pressure in the soil solution exceeds the osmotic pressure in plant cells due to the presence of more salt, and thus, limits the ability of plants to take up water and minerals like K^+ and Ca^{2+} meanwhile Na^+ and Cl^- ions can enter into the cells and have direct toxic effects on cell membranes, as well as on metabolic activities in the cytosol. These primary effects of salinity stress causes some secondary effects like assimilate production, reduced cell expansion, and membrane function, as well as decreased cytosolic metabolism and production of ROS. Salinity stress limits crop growth and yield in different ways. Salt puts two primary effects on plants: osmotic stress and ionic toxicity. Under normal condition the osmotic pressure in plant cells is higher than that in soil solution. A clear stunting of plants is noticed due to salinity stress (Takemura *et al.*, 2002). Parida and Das (2005) reported that the detrimental effects of high salinity on plants can be observed at the whole-plant level as the death of plants and/or decreases in productivity. High salt content, especially chloride and sodium sulphates, affects plant growth by modifying their morphological, anatomical (Huang and Redmann, 1995), and physiological traits (Muscolo *et al.*, 2003). It is evident that there are big changes in morphology and anatomy of plants growing in saline soils (Dolatabadian *et al.*, 2011). Physiological responses to salinity include growth suppression and lowered osmotic potential (Marcum, 2006). Many researchers reported that with an increase in salinity there was a decrease in the development of the xylem. Pimmongkol *et al.* (2002) stated that the width of vascular bundles and diameters of rice stems decreased in NaCl medium.

The soil matrix and osmotic potentials are additive in lowering the free energy of water in the soil (Shalhevet, 1993), and a primary physiological response of plants to both conditions is to lower the cell water potential through the accumulation of organic and inorganic solutes so that the roots can continue to extract water from the soil solution (Flowers and Yeo, 1986). Hence, it is logical to think the two stress factors could be additive in affecting plant performance (Shalhevet, 1993). However, studies in which plants are grown in drying soils at different salinities show a more complicated response, in which soil salts actually mitigate some of the negative effects of water stress. For example, plants in drying soils usually survive longer in saline than in nonsaline soils, because salt-stressed plants grow less and therefore deplete soil moisture more slowly than nonsalt-stressed plants (Richards, 1992; Shalhevet, 1993). Furthermore, salt stress can increase leaf water use efficiency by reducing stomatal conductance to a greater extent than photosynthesis (Ayala and O'Leary, 1995), thereby allowing plants under salt stress to produce more dry matter than plants in nonsaline soil on the same quantity of water (Richards,

1992). Finally, salt stress can precondition plants to low soil water potential by allowing them to osmotically adjust, enhancing their ability to survive as the soil dries (Shalhevet, 1993). These generalizations appear to hold true for both C3 and C4 crop plants (McCree and Richardson, 1987) and halophytes (Richards, 1992). They have practical implications with regard to irrigation strategies for crop plants in salt-affected soils (Richards, 1992; Shalhevet, 1993), and they may also be relevant to the growth strategies of adapted native plants in saline, desert soils (McNaughton, 1991). Edward and Brown (1998) from a study of salt stress on the growth and water use efficiency of the xerohalophyte *Atriplex canescens* (Pursh.) Nutt. in drying soil It was conclude that tolerances to water and salt stress are linked through a common mechanism of Na uptake for osmotic adjustment in this species.

Many plants develop mechanisms either to exclude salt from their cells or to tolerate its presence within the cells. During the onset and development of salt stress within a plant, all the major processes such as photosynthesis, protein synthesis, and energy and lipid metabolism are affected. The earliest response is a reduction in the rate of leaf surface expansion, followed by a cessation of expansion as the stress intensifies (Parida and Das, 2005). Munns and Tester (2008) stated that salt-sensitive plants have reduced survival, growth, and development when exposed to even low to moderate salinities, whereas salt-tolerant species are able to grow and reproduce even at oceanic salinities. The only way to control the salinization process and to maintain the sustainability of landscape and agricultural fields is to combat the salinization problems by environmentally safe and clean techniques, as follows: use of salt-tolerant species (Beltrão *et al.*, 2009; Hamidov *et al.*, 2007).

4.6.1. Phases of Development Sensitive to Temperature or Photoperiod

The rate of development of plants is generally responsive to photoperiod for only part of their life cycle (*i.e.,* between emergence and flowering), though some post-flowering processes such as the rate of flower initiation are affected by photoperiod and can therefore influence the duration of the seed-filling period (Summerfield *et al.*, 1998). In contrast, temperature affects the rate of development throughout the life cycle (Roberts and Summerfield, 1985). Three distinct stages of pre-flowering development can be identified in plants, namely the pre-inductive or juvenile, inductive, and post-inductive phases. Photoperiod only affects the duration of the inductive phase (Yin *et al.*, 1997*b*), and in photoperiod-sensitive cultivars the effects of photoperiod are usually substantial and are the major determinant of flowering time. In most annual crop species examined, the duration of the pre-inductive or juvenile phase is short [days not weeks: soybean (Collinson *et al.*, 1992); lentil (Roberts *et al.*, 1986)], the most important exception being in rice (*O. sativa*) where the duration of the pre-inductive phase (sometimes called the basic vegetative phase or BVP, though this includes part of the inductive phase) may exceed 50 d (Vergara and Chang, 1976). A long juvenile trait has also been identified in soybean (Hinson, 1989).

Typically the response to temperature of many plant processes, including the rate of development (inverse of days to flower), is described in terms of cardinal temperatures (base or minimum, T_b; optimum, T_o; and maximum or lethal, T_m temperature) and the thermal sum (è) or rate (inverse of the duration). This response can be quantified by simple bi-linear or broken stick models with a sharply defined T_o or by linear models with a flat-top/plateau response that have a maximum rate of development over a range of temperatures. Curvilinear quadratic or beta-function models (Yin *et al.*, 1997*a*) that have a near optimal rate of development over a range of temperatures have also been used. This basic response to temperature is affected by photoperiod in photoperiod-sensitive genotypes and therefore must be determined under short-day or inductive photoperiod (Roberts and Summerfield, 1987; Yan and Wallace, 1998). These differences in model/interpretation are not particularly important in most natural growing seasons in current climates, where mean daily temperatures are mostly close to but below the T_o; the choice of model under these circumstances makes little difference to the predicted number of days to flowering (Summerfield *et al.*, 1993; Wallace and Yan, 1998). However, in future climates where temperature is expected to exceed T_o more frequently, the choice of temperature response function will be much more significant, and this is discussed further below.

Ambient or Tissue Temperature?

Ambient air temperature is the usual 'temperature' used to quantify responses to temperature, although of course for plant processes it is the temperature nearest/approximating that of the growing point or meristem that matters (Jamieson *et al.*, 1995). In many temperate crops, such as wheat, both soil and air temperature influence development; soil temperature while the apex is close to the ground, air temperature thereafter (Jamieson *et al.*, 1995). Similarly, in irrigated or flooded rice systems, water temperature, not air temperature, controls development until the apex is above the water (Collinson *et al.*, 1995). It is therefore important in quantifying and modeling responses to climate change to use the appropriate 'temperature' driver, and not simply ambient air temperature. However, it is also necessary to remember that climate model output used for climate change impacts studies will provide values of surface temperature that can be viewed as similar to the 2m height temperature recorded in weather stations. Therefore, quantitative relationships between the rate of development and temperature will often include a degree of uncertainty due to differences between where temperature is measured and where it is perceived by the plant.

Perhaps less obvious is that significant differences between air and tissue temperature are often found in controlled environments, and these differences may be critical for interpreting interactions where temperatures are close to the optimum. For example, Vara Prasad *et al.* (2000) recorded peanut (*Arachis hypogeae*) flower bud temperatures for plants raised in growth cabinets to be 0.4–4.3 °C below ambient air temperature over the range 28–48 °C. This has obvious implications for the quantification of responses to temperature for application to natural environments. Large differences between air and tissue temperatures can also arise where environmental conditions decrease transpirational cooling, for example with severe water deficits or at elevated [CO_2] (Ainsworth *et al.*, 2008).

4.7. Phenology

Circadian clocks are 24-h biological oscillators, which generally enable organisms to anticipate predictable, daily changes that are linked to the rotation of the earth. Such clocks have been found in a wide range of organisms, including cyanobacteria, fungi, insects, and mammals (Young and Kay, 2001). Having a clock is not essential for plant viability, and are rhythmic mutants are still capable of developing (Green *et al.*, 2002). However, there is convincing evidence that having a clock that is well tuned to diurnal cycles of light and darkness makes an important contribution to plant fitness. For example, plants whose clocks were a good match to their rhythmic environment performed best in competition experiments (Dodd *et al.*, 2005). This may be explained, at least in part, by the role of the clock to ensure the optimum use of starch at night while avoiding starvation before dawn (Graf *et al.*, 2010; Scialdone *et al.*, 2013; Mugford *et al.*, 2014; Webb and Satake, 2015). However, recent data suggest that the circadian clock may also contribute to plant fitness by enhancing their ability to tolerate abiotic stress.

The effects of the clock on plant physiology and development are largely mediated through the regulation of gene expression. Approximately 80% of transcripts cycle in *Arabidopsis* plants under diurnal light–dark or temperature cycles, and 30–40% of these oscillations persist upon transfer to constant conditions (Edwards *et al.*, 2006; Covington *et al.*, 2008; Michael *et al.*, 2008). Similar observations have been made in a number of crop plants and tree species, including rice, poplar, maize, tomato, and soybean (Facella *et al.*, 2008; Khan *et al.*, 2010; Filichkin *et al.*, 2011; Marcolino-Gomes *et al.*, 2014). Remarkably, the phasing of expression for the majority of genes was found to be conserved across monocots and dicots (Filichkin *et al.*, 2011). One particularly interesting outcome from transcriptomic analyses was that many genes that are responsive to abiotic stress were found to be under the control of the circadian clock. Thus, in *Arabidopsis*, approximately 50% of genes that are responsive to heat, osmoticum, salinity or water deprivation were found to be expressed rhythmically in constant light, and 40% of cold-regulated genes (Covington *et al.*, 2008). Rhythmic expression of abiotic stress-responsive genes was also reported for soybean and barley (Habte *et al.*, 2014; Marcolino-Gomes *et al.*, 2014).

Phenology, or the relationship between climate and the sequence and timing of developmental events or stages, provides a foundation for understanding crop development and growth (McMaster *et al.*, 2011). Flowering is a life history trait determined by plant genotype, the environment, and the interaction between them (Rathcke and Lacey, 1985). The timing of flower production can be critical for the evolutionary ecology of plant populations (Fox, 1990b). Synchronized flowering can increase the effective population size, reduce genetic structure, and maximize outcrossing (Primack, 1985). In contrast, asynchronous flowering can impede gene flow (Primack, 1985; Cruzan *et al.*, 1994) and reduce reproductive success (Rathcke and Lacey, 1985; Fox, 1990a). In addition to genetic mechanisms, flowering is also affected by environmental factors such as photoperiod (Shitaka and Hirose, 1998), temperature (Sandvik and Totland, 2000), herbivory (Pilson, 2000), and water stress (Fox, 1990a, 2001; Bram and Quinn, 2000; Stanton *et al.*, 2000). The potential impacts of climate change on natural and managed ecosystems are of concern and have been extensively evaluated by various simulation models (*e.g.* Tao *et al.*, 2000; Cramer *et al.*, 2001; Parry *et al.*, 2004) Observation of changes in plant phenology in response to abiotic stress can reveal a better understanding of interactions between stress atmosphere and the plant. Different phonological stages differ in their sensitivity to high temperature; however, this depends on species and genotype as there are great inter and intra-specific variations (Wollenweber *et al.*, 2003; Howarth, 2005). Heat stress is a major factor affecting the rate of plant development, which may be increasing to a certain limit and decreasing afterwards (Hall, 1992; Marcum, 1998; Howarth, 2005).

Since plant phenology is largely temperature dependent, there was an overall trend for earlier occurrence of key phenological events, and consequent shortening of crop growing cycle for winter and summer annual crops, grapevine and olive tree (Osborne *et al.*, 2000; Bindi and Moriondo, 2005; Moriondo *et al.*, 2007). More interestingly, when the effect of higher temperatures on the single phenological phase duration was investigated, an asymmetric impact on the length of vegetative and reproductive phases was evident both in annual and perennial crops, resulting in large fitness consequences. In winter crops, for instances, higher temperatures as simulated for the future scenarios, shortened the pre-flowering phase and this allowed grain filling periods to take place in cooler periods that favorites a longer time period for biomass partitioning to grains (Sadras and Monzon, 2006). Additionally, a shorter vegetative phase allow the crop to escape the summer drought resulting finally in an increased final yield.

This trend, confirmed by phenological observations over a long period reported by Duchêne and Schneider (2004) in Alsace, is of a major importance for grapevine yield that was simulated as progressively decreasing due to the combined effect of a shorter period for berry ripening and increased drought stress in summer (Moriondo and Bindi, 2005). In addition to the increase in mean temperature, also the change frequency of extreme events should be considered for a reliable climate change impact assessment on agriculture (Challinor *et al.*, 2005). A warmer climate may in fact exhibit an increased climatic variability and this can produce relatively large changes in the frequency of extreme climatic events. These, when occurring during key development stages of the crop may have a dramatic impact on final production even in case of general favourable weather conditions for the rest of the growing season (Porter and Gawit, 1999). Temperatures higher than 31^0 C at anthesis stage, inducing a reduced pollen and floret fertility, was demonstrated to be detrimental for both wheat and sunflower yield (Porter and Gawit, 1999). For grapevine, heat stresses ($T_{max} > 35^0$ C) during berry ripening are associated to a lower photosynthesis and yield through poor fruit set and small berries (Kliever, 1977), inhibition of colour development and partial or total failure of flavour ripening (Mullins *et al*, 1992). Frost events ($T_{min} < -2^0$ C) (Narciso *et al.*, 1992) at bud break are related to shoot loss and lower yield (Mullins et al, 1992). Temperature lower than -8^0 C may be lethal for olive tree whereas severe water stress during fruit ripening may cause severe yield loss. Accordingly, the shifting of phenological stages due to climate change may result in other important consequences when also change in extreme events frequency is taken into account.

Sexual reproduction and flowering in particular have been long recognized as extremely sensitive to heat stress, which often results in reduced crop plant productivity (Hedhly *et al.*, 2009; Thakur *et al.*, 2010). Studies carried out under glass and climate chambers suggest that high temperature is most deleterious at the stage of flower bud initiation, and that this sensitivity is maintained for 10–15 days (Hedhly *et al.*, 2009; Nava *et al.*, 2009). Many legumes and cereals show a high sensitivity to heat stress during flowering and severe reductions in fruit set have also been shown for several temperate and even tropical fruit trees species (Frank *et al.*, 2009; Saha *et al.*, 2010), most probably as result of reduced water and nutrient transport during reproductive development (Young *et al.*, 2004).

The male gametophyte is particularly sensitive to high temperatures at all stages of development, while the pistil and the female gametophyte are considered to be more tolerant (Hedhly, 2011). Heat stress often accelerates rather than delays the onset of anthesis, which means that the reproductive phase of development will be initiated prior to the accumulation of sufficient resources (Zinn *et al.*, 2010). The reproductive development in angiosperms takes place within two floral organs, the male stamen and the female pistil (Pachauri, 2007; Ahmad *et al.*, 2010). The developmental pathway for the male gametophyte (the pollen grain) starts with the separation of the reproductive tissues of the anther continues with meiosis of the pollen mother cell, followed by mitosis and microspore maturation that results in the mature pollen grain. After initiation, the highly specialized anther tissues will acquire non-reproductive (*e.g.*, the tapetum for support, the stomium for dehiscence) or reproductive functions (the pollen mother cell for pollen formation). Both tapetum and microspore development are essential for male fertility, as documented by numerous studies on male sterile mutants (Prasad *et al.*, 2003; Young *et al.*, 2004; Christensen and Christensen, 2007; Tauber *et al.*, 2007; Ainsworth *et al.*, 2008a; Jaggard *et al.*, 2010; Zinn *et al.*, 2010). Major variations in gene expression are observed under high temperature stress that are possibly linked to tapetum degeneration and pollen sterility in several plant species (Endo *et al.*, 2009).

Male sterility as a consequence of heat stress can be widely observed among many sensitive crop plants and the impairment of pollen development has been the main factor involved in reduced yield under heat stress (Sakata and Higashitani, 2008; Wassmann *et al.*, 2009). For example, in barley and *Arabidopsis* anthers developing under high temperature (30–35°C), cell-proliferation is arrested, vacuoles are distended, chloroplast development is altered, and mitochondrial abnormalities occur (Sakata *et al.*, 2010). Heat stress reduces carbohydrate accumulation in pollen grains and in the stigmatic tissue by altering assimilates partitioning and changing the balance between symplastic and apoplastic loading of the phloem (Taiz and Zeiger, 2006). Heat stress down-regulates sucrose synthase and several cell wall and vacuolar invertases in the developing pollen grains; as consequence, sucrose and starch turnover are disrupted and thus soluble carbohydrates accumulate in reduced levels (Sato *et al.*, 2006). In cowpea plants experiencing heat stress, drops in the concentration of soluble sugars in the anther walls, developing pollen grains, and in the locular fluid result in decreased sugar concentration in the mature pollen grains and decreased pollen viability (Ismail and Hall, 1999). In tomato, the reduction of sink- and source-strength even under moderately elevated temperatures leads to a depletion in available carbohydrates at critical stages of plant development, leading to reduced fruit set and other yield related parameters (Sato *et al.*, 2006). In sorghum, heat stress reduces the accumulation of carbohydrate in pollen grains and ATP in the stigmatic tissue (Jain *et al.*, 2007).

Heat stress also induces early abortion of tapetal cells, which cause the pollen mother cells to rapidly progress toward meiotic prophase and finally undergo programmed cell death (PCD), thus leading to pollen sterility (Oshino *et al.*, 2007; Sakata and Higashitani, 2008; Parish *et al.*, 2012). For example, under high temperature conditions, the structural abnormalities observed in developing microspores of snap bean anthers have been associated with tapetal degeneration due to malformations in the endoplasmic reticulum (Suzuki *et al.*, 2001). Reduced dehiscence of tomato anthers under heat stress is also accompanied by closure of the locules and thus reduced pollen dispersal in several crop plants (Peet *et al.*, 2002).

Continuing heat stress beyond a successful fertilization can also halt further development of the embryo (Barnabás *et al.*, 2008). Heat stress during seed development may result in reduced germination and loss of vigor, leading to reduced emergence and seedling establishment as has been shown for several crop plants (Akman, 2009; Ren *et al.*, 2009). In many temperate cereal crops, both grain weight and grain number appear to be impacted by heat stress, with a decline in grain number directly proportional with increasing temperatures during flowering and grain filling (Mahmood *et al.*, 2010). Quality reductions in starch, protein, and total oil yield in several crop species have been also associated with heat stress during seed development (Banowetz *et al.*, 1999). For example, high temperature during wheat reproductive development hastens the decline in photosynthesis and leaf area, decreases shoot and grain mass as well as weight and sugar content of kernels, and also reduces water-use efficiency (Shah and Paulsen, 2003). As a consequence, heat stress results in an altered nutritional flour quality (Hedhly *et al.*, 2009).

The responses of natural vegetation to climate change have been investigated by analysing satellite data, including changes in vegetation greenness (Zhou *et al.*, 2003), phenology (Zhang *et al.*, 2004), and net primary production (NPP) (Nemani *et al.*, 2003). The phenological seasons of natural vegetation have also been shown to change spatially and temporally in response to trends in climate change by using the observed data from phenological networks (e.g. Menzel *et al.*, 2001; Zheng *et al.*, 2002). However, most studies focus on changes in the natural vegetation; only a few deal with trends in agricultural and horticultural varieties despite their potential economic importance (Chmielewski *et al.*, 2004). In Germany, a shift in phenology of fruit trees and field crops due to increased temperature from 1961 to 2000 has been observed, but the changes in plant development are still moderate so no strong impacts on yield formation processes have been observed so far (Chmielewski *et al.*, 2004). Gradual temperature changes from 1982 to 1998 have caused a measurable impact on the yields of corn and soybeans in the United States (Lobell and Asner, 2003). Also, in the Philippines, rice grain yield was found to decline by 10% for each 1°C increase in growing-season minimum temperature in the dry season (January–April) from 1992 to 2003 (Peng *et al.*, 2004). Obviously, ongoing warming trend has had measurable impacts on the development and production of field crops, but the size and extent of the impacts have differed spatially and temporally. There is a clear and present need to synthesize crop yield and climate data from different areas to provide critically needed observational constraints on projections of the impacts of both climate change and management practices on future food production (Lobell and Asner, 2003). In China, mean temperature has increased in the last several decades, especially since the 1980s (Tao *et al.*, 2003). During 1951–1990, annual mean minimum temperature generally tended to increase all over China, with the largest increase in the north and smaller increases in the south. Annual mean minimum temperature increased significantly by 0.175 °C/decade for all of China. The largest trend was found in winter, with a warming rate of 0.417 °C/decade (Tao *et al.*, 2006). The annual mean maximum temperature showed a slight, but not statistically significant, increase (Zhai *et al.*, 1999). The explicit spatial and temporal changes in temperature, characterized by a marked asymmetry between maxima and minima, are presumed to have caused significant changes in crop development and production in China. Studies on the responses of field crops to such gradual climate changes on a decadal scale are scarce, however, although the impacts of seasonal and inter-annual climate variability on crop production have been investigated (Tao *et al.*, 2004).

The developmental stage at which the plant is exposed to the stress may determine the severity of possible damages experienced by the crop. Vulnerability of species and cultivars to high temperatures may vary with the stage of plant development, but all vegetative and reproductive stages are affected by heat stress to some extent. During vegetative stage, for example, high day temperature can damage leaf gas exchange properties. During reproduction, a short period of heat stress can cause significant increases in floral buds and opened flowers abortion; however, there are great variations in sensitivity within and among plant species (Guilioni *et al.*, 1997; Young *et al.*, 2004). Impairment of pollen and anther development by elevated temperatures is another important factor contributing to decreased fruit set in many crops at moderate-to-high temperatures (Sato *et al.*, 2006). The staple cereal crops can tolerate only narrow temperature ranges, which if exceeded during the flowering phase can damage fertilization and seed production, resulting in reduced yield (Porter, 2005). Under high temperature conditions, earlier heading is advantageous in the retention of more green leaves at anthesis, leading to a smaller reduction in yield (Tewolde *et al.*, 2006). Furthermore, high temperatures during grain filling can modify flour and bread quality and other physico-chemical properties of grain crops such as wheat (Perrotta *et al.*, 1998), including changes in protein content of the flour (Wardlaw *et al.*, 2002). Thus, for crop production under high temperatures, it is important to know the developmental stages and plant processes that are most sensitive to heat stress, as well as whether high day or high night temperatures are more injurious. Such insights are important in determining heat-tolerance potential of crop plants.

Agriculture flexibility in response to climate change, water shortages in the West Australian was evaluated and results showed that the decision to land use change leads to water shortages and reduced in agricultural profit. Temperature has a strong influence on rice growth and grain yield production warming tends to change rice phenology (Lu *et al.*, 2008) and magnitude the length of the growing season, permitting for earlier planting and delayed in harvesting date (Huang *et al.*, 2007). Maximum temperature are damage distractive to rice growth and essential development of rice yield (Yoshida, 1981). Using the simulation environmental models to analyze the effects of climate change on winter wheat and corn crops in the Switzerland was investigated (Finger and Schmid, 2008). Results showed that the yield was sensitive to climate change. Increasing temperatures and decreasing solar radiation during the growing season was identified as the main reasons for the decline in yield. The effect of minimum and maximum temperature on rice yields in tropical and subtropical areas was studied (Welch *et al.*, 2010). Results of this study showed that temperature and radiation have significant effect during both vegetative and ripening steps of the rice plant also minimum temperature reduced rice yield. Seedling, booting and heading-flowering are the essential of rice growth stages and they are the most vulnerable period of maximum temperature throughout the life cycle of rice plants (Nishiyama, 1976). Booting and heading- flowering stages are the most sensitive stages to low temperature (Yoshida, 1981). High temperature in the heading flowering stage effect on rice growth and rice productivity, especially in tropical areas where the temperature already near or above in the best possible range of rice (Sun and Huang, 2011). Climate variability is identify to increase the frequency of maximum temperature events that leads to decreasing crop yields (Rosenzweig *et al.*, 2001). Rice phenology is strongly influenced by the environment conditions in a certain range, an increase in temperature leading to decrease in vegetative phase. Low thermal stress in seedling and heading-flowering stages occurred, whereas high thermal stress during the end of growing season caused the sterile of seeds. on the other hand, the risk of cold damage during the rice growing make reduce yield (Shimono, 2011). By increasing in temperature, maturity stage was decline (Baker *et al.*, 1990). By increasing in seasonal temperatures, it will causes increase the risk of drought and the speed photosynthesis was limited (Tubiello *et al.*, 2007).

The timing of flowering, a critical stage of development in the life cycle of most plants when seed number is determined, is important for adaptation both to the abiotic stresses of temperature and water deficit constraints (Curtis, 1968) within the growing season. For example, in many annual crops, brief episodes of hot temperatures (>32–36 °C) can greatly reduce seed set, and hence crop yield, if they coincide with a brief critical period of only 1–3 d around the time of flowering (Vara Prasad *et al.*, 2000; Wheeler *et al.*, 2000; Jagadish *et al.*, 2008). Therefore, the moderation of crop development will be critical to the impacts of climate change on yield in two ways: through determining the season length, and hence the availability of radiation, water, and nutrient resources for growth; and by affecting the exposure of the crop to climate extremes. Adaptation to moderate changes in climate that influence temperature, season length, and planting dates, as well as the occurrence of abiotic stress, can be achieved by selecting varieties with appropriate flowering times and crop durations (Richards, 2006). Farmers and plant breeders have very successfully selected / manipulated life cycle duration and phenology to maximize the range of environments in which crops grow as well as their yield (Roberts *et al.*, 1996) at least for current climates. A major challenge for crop improvement is how to plan for future climate change. The genetic and environmental moderation of the timing of flowering is therefore central to the responses described above. The timing of flowering within a season is largely determined by responses to temperature and photoperiod, and in whole plants at suboptimal temperatures these quantitative responses are reasonably well understood (Wallace and Yan, 1998). However, responses to temperature and photoperiod at supra-optimal temperatures are poorly understood—though clearly these will become more important as the frequency of high temperature events increases under projected climate change. In recent years molecular biology has also greatly contributed to our understanding of flowering gene pathways (Baurle and Dean, 2006; Tsuji *et al.*, 2008), although the effects of temperature and temperature x photoperiod interactions on these pathways have not been studied.

4.7.1. Changes in Phenology of Cropping Systems

Earlier flowering and maturity have been observed and documented in crop plants (Williams *et al.*, 2004; Hu *et al.*, 2005; Menzel *et al.*, 2006; Tao *et al.*, 2006; Estrella *et al.*, 2007), as well as in natural communities (Fitter and Fitter, 2002), over the last 50 years from phenology networks and individual records. Menzel *et al.* (2006), for example, report that 78% of all observations in 21 European countries showed earlier flowering, with an advance in phenological events of 2.5 d per decade on average. In Germany, the phenology of 78 agricultural and horticultural events between 1951 and 2004 were, on average, 1.1–1.3 d earlier per decade (Estrella *et al.*, 2007). Likewise, winter wheat cv. Kharkof grown in the USA Great Plains has flowered 0.8–1.8 d earlier per decade (depending on location) since 1950 (Hu *et al.*, 2005).

In addition to phenology observations, Menzel *et al.* (2006) also reported that farmers' activities, such as sowing and harvesting, also occurred earlier, indicating a change in crop season length. Other studies have also shown that season length has increased, at least in mid to northern latitudes (White *et al.*, 1999; Menzel *et al.*, 2003), and this is associated with warmer temperatures in winter and spring. However, many of these changes in the timing of farming activities are driven by changes in farm management practices and the introduction of new cultivars. So, although several studies have associated these changes in phenology with warmer seasonal or winter/spring temperatures, earlier flowering in crop species may be related more to the earlier onset of farming activities than to temperature and hence past climate change *per se*. In addition, changes in crop management may also counter direct effects of temperature warming and the timing of farm operations, for example through a change to a longer duration variety. Therefore, studies that robustly attribute observed changes in phenology in ecosystems to changes in climate are rare for natural ecosystems (Root *et al.*, 2005) and not found for managed ecosystems.

4.7.2. Effect of CO_2 on Flowering

The effect of CO_2 on growth and development has been studied in many crop species. Springer and Ward (2007) recently summarized the effect of CO_2 on flowering time in 23 crop species in 33 papers that included experiments in growth cabinets/glasshouses, open-topped chambers, and field-based FACE (free air carbon enrichment) facilities. The majority of papers compared current ambient with a doubling of CO_2, *i.e.*, the expected CO_2 beyond 2070 depending on future greenhouse gas emissions. Effects of intermediate CO_2 representing short- and medium-term changes in CO_2 are not commonly reported, although changes to flowering time would be hard to detect at these intermediate CO_2.

Approximately half the studies cited by Springer and Ward (2007) reported earlier flowering time, and only four out of 33 (Ellis *et al.*, 1995, two species) reported delayed flowering in response to increased CO_2. Earlier flowering was reported in most crop species studied, including short-day soybean (*Glycine max*), rice (*Oryza sativa*), and cowpea (*Vigna unguiculata*) and long-day barley (*Hordeum vulgare*), pea (*Pisum sativum*), and faba bean (*Vicia faba*) species. The four papers reporting a delay in flowering were on maize, sorghum, and soybean, all short-day species. It has been suggested that short-day and long-day species respond differently to CO_2 (Reekie *et al.*, 1994). However, the four reports of delayed flowering were all from controlled-environment experiments, and there are no reports of delayed flowering from field-based experiments of short-day species.

Given that some authors have questioned the influence of artifacts associated with controlled environments on plant responses to climate (Long *et al.*, 2005), the most reliable guide to CO_2 effects on flowering should be those reported from the FACE experiments. In FACE experiments, CO_2 treatments are imposed on crops growing in large fields under well-managed farm conditions, *i.e.* as near 'natural' conditions as possible (Ainsworth *et al.*, 2008b). In general, the FACE experiments reported in Springer and Ward (2007), along with more recent FACE papers, suggest that CO_2 has little or no effect on flowering time in either C4 species (*e.g.*, maize; Leakey *et al.*, 2006) or C3 species (*e.g.*, rice, Shimono *et al.*, 2009). One FACE experiment on potato (*Solanum tubersosum*) by Miglietta *et al.* (1998) does show flowering occurring 5–7 d earlier at 460–660 μ mol mol^{-1} than at ambient CO_2, but this was associated with increased canopy temperature according to the paper. Although reductions in stomatal conductance (g_s) are commonly reported in FACE experiments at high CO_2 (Ainsworth and Long 2005; Ainsworth *et al.*, 2008), and this can increase canopy temperature (Long *et al.*, 2006), canopy temperature is not usually given.

At high CO_2, tissue temperatures are usually increased due to lower conductance, and care is therefore required when interpreting such data (Ainsworth *et al.*, 2008). For example, Vara Prasad *et al.* (2006) observed tissue temperature in sorghum to be 1.2–2.7° C warmer at high than at ambient CO_2 at ambient daytime air temperatures of 32–44° C. In this particular experiment, high CO_2 delayed flowering slightly at 36°C compared with 32 °C, suggesting a temperature x CO_2 interaction. However, this delay can be explained by higher tissue temperatures that were reported, resulting in mean temperature exceeding the optimum temperature (T_o) and hence causing a delay in flowering.

4.7.3. Temperature and Effects on Phenology

Given the apparent lack of a direct effect of CO_2 on rate of development, then temperature, and interactions with temperature, will be the most important aspect of human-induced climate change for crop development. The duration from sowing to flowering and maturity in plants without a vernalization requirement, or where that requirement has been met, is largely determined by responses to temperature and photoperiod (Wallace and Yan, 1998). While photoperiod-insensitive or day-neutral types are important in modern agriculture, especially in warm short-season environments, photoperiod sensitivity is the norm and is a very powerful adaptive mechanism (*e.g.*, soybean: Evans 1993; Roberts *et al.* 1996).

Global circulation models have projected that the global temperature will increase by 1.4-5.8 °C due to increase in concentrations of all greenhouse gases by the end of 21st *et al*century (Houghton, 2001). Heat stress during the reproductive phase can cause pollen sterility, tissue dehydration, lower CO_2 assimilation and increased photorespiration. Although high temperature accelerates growth, it also reduces the phenology, which is not compensated by the increased growth rate (Wardlaw and Moncur, 1995; Zahedi and Jenner 2003). However, temperatures more than 30 ºC during floret formation, may cause complete sterility (Saini, 1982). The reproductive stage is relatively more sensitive than the vegetative stage to heat stress in many crop species (Hall, 1992).

Photoperiod and Temperature Interactions

Most crop plants respond to photoperiod, and in general short- and long-day plants respond in a similar manner with photoperiods longer or shorter, respectively, than the critical or base photoperiod delaying flowering (*e.g.*, maize, a short-day species). In quantitative types, flowering is delayed but not prevented in the non-inductive photoperiod, whereas in qualitative types, if the photoperiod transgresses a critical threshold, flowering will not occur. While qualitative responses have been observed in some crop plants (*e.g.*, pigeonpea, (Carberry *et al.*, 2001); soybean (Roberts *et al.*, 1996), the photoperiod in most growing seasons does not transgress the ceiling or maximum photoperiod, or does so only for a short period, and hence most crop plants are effectively quantitative short- or long-day plants. One exception to this may be sorghum in parts of West Africa (Dingkuhn *et al.*, 2008).

Most whole-plant crop models (*e.g.* DSSAT, APSIM) assume that photoperiod effects are additive to those of temperature, which is the basic underlying driver of development, though others have argued from observations that there is a photoperiod by temperature (P x T) interaction (Wallace and Yan, 1998). The P x T interaction manifests itself as a hyperbolic response to photoperiod (Vaksmann *et al.*, 1998; Wallace and Yan, 1998), variation in the critical photoperiod with temperature (Roberts and Summerfield, 1987), or variation in the optimum temperature with photoperiod (Wallace and Yan, 1998). Sensitivity analyses, however, generally show that fixing critical photoperiods and optimum temperatures does not significantly reduce the accuracy of predictions (Carberry *et al.*, 2001). More recently, a threshold- or appetence-type response has been proposed wherein the threshold or target for floral initiation to occur in reduced time (Dingkuhn *et al.*, 2008). An effect of rate of change of photoperiod has also been proposed (Clerget *et al.*, 2004), and was included in the AFRC wheat model (Weir *et al.*, 1984). As a general rule, all models predict flowering time fairly accurately in crops growing in their normal growing season where days are becoming more inductive and temperatures are favorable. March to July in temperate latitudes north of the equator; June to October in tropical/subtropical latitudes north of the equator; December to April in tropical and subtropical latitudes south of the equator; they work less well outside these norms when, for example, days are becoming less inductive (*e.g.*, post-rainy season crops) or crops are planted before the longest day [e.g. sorghum in West Africa (Clerget *et al.*, 2004; Dingkuhn *et al.*, 2008)].

Most crop models assume that photoperiod only affects rate of development at and below T_0; above T_0, only temperature affects the rate. However, the effect of photoperiod at temperatures $>T_0$ has not been studied extensively and this is clearly of importance for accurately predicting phenology in future climates. Crop models such as DSSAT and APSIM effectively assume that at temperatures $>T_0$ the rate of development is only affected by temperature. In most current seasons and environments there will not be many days where temperature exceeds T_0 and therefore this approach broadly works. However, in future climates where mean temperatures will be higher, the optimum temperature is likely to be exceeded more frequently, so interactions $>T_0$ may be more significant. This is likely to be more so in crop species such as wheat (*Triticum aestivum*) and common bean (*Phaseolus vulgaris*) that originated in temperate climates but are now widely grown in more tropical environments and which have comparatively low values for T_0 of <25 °C (Wallace and Yan, 1998). Crop species that originated in the tropics (e.g. rice, sorghum, millet, and peanut) generally have higher values for T_0, often between 25 °C and 30 °C, and sometimes as high as 35°C (Yin *et al.*, 1997*b*; Clerget *et al.*, 2004).

Only in controlled environments, where temperature and photoperiod can be controlled independently and supra-optimal temperatures can be applied, that these interactions can be investigated. In a series of experiments with maize (*Zea mays*) over several years (Ellis *et al.*, 1992), genotype Tuxpeòo was grown at constant temperatures and photoperiods ranging from 12°C to 36°C and 9 h to 16 h, respectively. The duration from sowing to tassel initiation (the first easily observed sign of reproductive apical development, cf double ridges in wheat) was recorded in 67 P x T treatments and modeled using a linear, additive regression technique assuming plants were sensitive to photoperiod from emergence (Roberts and Summerfield, 1987).

Tassel initiation occurred between 18 d and 111 d after sowing, and the cardinal temperatures were typical for maize (T_b= 6.7, T_o=28.3, and T_m= 44.6°C). The genotype was sensitive to photoperiod; at 25°C, days to tassel initiation increased from 22 d to 36 d as the photoperiod increased from 12 h to 16 h, *i.e.* an increase in duration of ~60%. This large and comprehensive data set suggests that photoperiod has an effect on rate of development up to and beyond T_o in addition to any negative effect of supra-optimal temperatures. So, for example at 30° C and 11 h photoperiod, tassel initiation will occur ~20 d after sowing, whereas at 15 h and the same temperature tassel initiation will occur after 26 d. While these effects may seem small, this delay is equivalent to a 30% increase in duration to the first sign or reproductive development; were days to flowering to be modeled these delays would be much more marked (Craufurd and Wheeler, 2009).

4.7.4. Effect of Salinity

Wetland plants are heavily impacted by saltwater intrusion, which can destroy or damage sensitive populations (Krauss *et al.*, 2000). However, some wetland species exhibit intra specific variation in salinity tolerance, and it is particularly important to understand the effects of salinity on plants that can survive saline conditions (Allen *et al.*, 1997; Hester *et al.*, 1998; Seliskar and Gallagher, 2000). Salinity is a growing environmental concern in wetland communities (Twilley *et al.*, 2001), but very little is known about how salt exposure affects flowering in natural plant populations. Drought stress has been observed to delay flowering (Fox, 1990a, 2001; Bram and Quinn, 2000; Stanton *et al.*, 2000); since salinity and drought cause similar osmotic responses in plants (Levitt, 1980). It was hypothesized that salinity should produce flowering delays. In Louisiana, salinity stress is often episodic because it results from tropical storms. Because clonal plants can retain the effects of previous environmental conditions for subsequent generations of ramets (Schwaegerle *et al.*, 2000).

4.8. Molecular/Genetic Aspects of Flowering Response to Climate Change

Little is known about the molecular mechanisms that control flowering times in response to ambient temperature, other than for the vernalization response, even in *Arabidopsis* (Baurle and Dean, 2006). Similarly little is known about the response of flowering pathway genes to elevated CO_2. Flowering time (usually measured by leaf number) in wild-type *Arabidopsis* responds to temperature in a similar manner to other plants, occurring sooner at warm (up to ~27°C) than at cool (16^{0}C) temperatures (Blazquez *et al.*, 2003; Thingnaes *et al.*, 2003; Balasubramanian *et al.*, 2006). Comparisons of mutants of the photoperiod, gibberellin, and autonomous pathways have shown that ambient temperature (growth temperature) is sensed through genetic pathways involving *FCA* and *FVA,* and integrated through *FT* (Flowering Locus T) (Blazquez *et al.*, 2003; Lee *et al.*, 2007), and that temperature leads to a photoperiod-independent activation of *FT* (Balasubramanian *et al.*, 2006). Temperature also affects responses of phytochrome mutants (Blazquez *et al.*, 2003; Halliday *et al.*, 2003); for example, *PHYA* (Phytochrome A) is not able to promote flowering at cooler temperatures, a response that might contribute to P×T interaction. However, while warm/short days (27°C / 8 h) apparently induce flowering at the same time as cool/long days (16^{0}C / 16h) in *Arabidopsis* (Balasubramanian *et al.*, 2006), temperature cannot usually compensate for non-inductive photoperiod in most crop plants.

In *Arabidopsis*, Springer *et al.* (2008) examined the response to elevated CO_2 in two genotypes, one selected for high seed number at elevated CO_2 and one a random control. They found that flowering was not affected by CO_2 in the control genotype and unsurprisingly the down-regulation of *FLC* (Flowering Locus C) and up-regulation of *SOC1* and *LFY* (Leafy) was similar at ambient and elevated CO_2. However, in the adapted genotype, flowering was delayed by 7–9 d at elevated CO_2 and this was associated with no down-regulation of *FLC* over the course of the experiment and, as a result, later up-regulation of *SOC1 and LFY expression.* Elevated CO_2, as expected, increased plant size, and relative increases in the adapted and control genotypes at ambient and elevated CO_2 were proportional to the delay in flowering. It was suggest that higher sucrose levels may act as a signal to influence flowering. However, it may be that selection for high seed number at elevated CO_2 was associated with later flowering and hence larger plant size and yield.

Models predicting flowering time based on flowering pathway genes have been proposed for *Arabidopsis* (Welch *et al.*, 2003, 2004; Van Oosterom *et al.*, 2004) and these models do predict flowering time in constant photoperiod and temperature environments. Furthermore, these simple gene network models can generate important physiological parameters, such as critical photoperiods. However, none of these models currently quantifies basic (suboptimal) temperature responses and certainly not supra-optimal temperatures; nor do they model temperature x photoperiod interactions.

The amount of water consumed by ornamental plants depends on the particular species and cultivar, the cultivation system, and the plant growing season. It has been estimated that, on average, 100-350 kg of water is needed to produce 1 kg of plant dry matter (Jiménez and Caballero, 1990). Water deficit affects negatively the process of flowering in many plant species by reducing the fertility of newly formed flowers (Slawinska and Obendorf, 2001). The growth of salt-treated plants is often limited by the ability of roots to extract water from the soil and transport it to the shoot due to the osmotic component of salinity (Rodríguez *et al.*, 1997). Sepaskhah and Yarami (2009) indicated that saffron (*Crocus sativus* L.) flower and corn were the most and the least sensitive organ to soil water depletion, respectively. Furthermore, Shillo *et al.* (2002) reported two bulb species, namely, *Hippeastrum×hybridum* Hort. and *Ornitathogalum arabicum* L. that were very sensitive to salinity; and the degree of damage was correlated to the salinity level and this response was expressed as weight reduction in all the plant organs. Moftah and Al-Humaid (2006) indicated that plant biomass, number of leaves, length, and weight of marketable inflorescences and bulb yield of tuberose (*Polianthes tuberosa* L.) were significantly reduced by water deficit (Bahadoran and Salehi, 2015).

4.9. Simulation of Phenology for Future Climate

One of the most commonly documented impacts of climate change on crops is the shortening of development stages in a warmer climate and the change in areas of crop suitability that result. In many studies this response dominates the impact of climate change on yield. Many simulation studies of changes in areas of crop suitability use simple thermal time relationships to represent the rate of development (Fischer *et al.*, 2005), and only a few model the effects of supra-optimal temperatures on the rate of development (Challinor and Wheeler, 2008). Given that we know that photoperiod sensitivity of duration to flowering is a key determinant of crop adaptation to climate, and that many crop varieties are photoperiod sensitive to some extent, it seems important to explore how photoperiod sensitivity may affect the response of crop phenology to temperature warming. For this, the photothermal model of flowering time was used to simulate duration from sowing to tassel initiation in maize cv. Tuxpeño in the current climate at one maize-growing location and at a range of prescribed temperature increases. Specifically, the aim was to determine whether photoperiod sensitivity affected the response of duration from sowing to tassel initiation to increases in mean temperature and to a more variable temperature regime.

One hundred years of current climate at Zaria, Nigeria (11.1°N, 7.7°E), were generated using the MarkSim weather generator (Jones and Thornton,

2000). Another set of 100 years was generated with an increase in temperature variability that was simulated by increasing the diurnal temperature range by 50% compared with the current climate simulations. Temperature increases of +1, +2, to +6°C were added separately to the daily mean temperature of these two sets of current climates to provide a total of 14 sets of 100 years of climate. The photoperiod on each day was calculated from standard astronomical day length equations (Keisling, 1982). A sowing date of 15 May, the average date of the start of the rains at this location (Craufurd and Qi, 2001), was used in all simulations. Time to tassel initiation was simulated using the photothermal time model with the appropriate temperature and photoperiod sensitivity for maize cv. Tuxpeño (PT model), and with the same temperature sensitivity alone (T model, *i.e.* no response to photoperiod). The mean and coefficient of variation of duration from sowing to tassel initiation was calculated for each set of 100 years of simulations.

Mean duration from sowing to tassel initiation became progressively shorter with an increase in mean temperature until an apparent optimum, beyond which duration progressively increased, indicating that current mean temperatures at Zaria are below the optimum. Photoperiod sensitivity (PT model) delayed tassel initiation by 6 d in the current climate (zero change in temperature) and changed the response to temperature warming. Most notably, the apparent optimum for this development stage increased from +2°C without photoperiod sensitivity (T model) to +4 °C with photoperiod sensitivity. The inter annual variability in the duration from sowing to tassel initiation was similar between the simulations with and without photoperiod sensitivity until +2°C, beyond which inter annual variability in development was greater at a given temperature warming in the absence of photoperiod sensitivity. Also, inter annual variability in development was greater when the diurnal temperature variability was increased. To explore this further, the change with temperature in the coefficient of variation of duration to tassel initiation between the climate simulations with and without extra temperature variability was examined. Again, photoperiod sensitivity affected the simulated response to temperature warming. The increase in variability of the timing of tassel initiation due to increased temperature variability at +3°C and warmer was reduced with photoperiod sensitivity.

References

Abdulmalik, M. M., Olarewaju, J.D., Usman, I.S. and Ibrahim, A 2012. Effects of moisture stress on flowering and fruit set In sweet pepper (*Capsicum annum L.*) Cultivars. *Production Agriculture and Technology,* 8(1): 191 -198.

Ahmad, A., Diwan, H. and Abrol, Y.P. 2010. Global climate change, stress and plant productivity, In, "*Abiotic Stress Adaptation in Plants: Physiological, Molecular and Genome Foundation,*" Eds A. Pareek, S.K. Sopory, H.J. Bohnert, Govindjee, Springer Science+Business Media B.V. pp. 503–521.

Ahn, Y.-J. and Zimmerman, J.L., 2006. Introduction of the carrot HSP17.7 into potato (*Solanum tuberosum* L.) enhances cellular membrane stability and tuberization *in vitro. Plant Cell Environ.*, 29: 95–104.

Ainsworth, E. and Long, S.P. 2005. What have we learned from 15 years of free-air CO_2 enrichment (FACE)? A meta-analytic review of the responses of photosynthesis, canopy properties and plant production to rising CO_2. *New Phytologist,* 165*:* 351-372.

Ainsworth, E.A. Beier, C., Calfapietra, C., Ceulemans, R., Durand-Tardif, M., Farquhar, G.D., Godbold, D.L., Hendrey, G.R., Hickler, T., Kaduk. J., Karnosky, D.F., Kimball, B.A., Körner, C., Koornneef, M., Lafarge, T., Leakey, A.D.B., Lewin, K.F., Long, S.P., Manderscheid, R., Mcneil, D.L. Mies, T.A., Miglietta, F., Morgan, J.A., Nagy, J., Norby, R.J., Norton, R.M., Percy, K.E., Rogers, A., Soussana, J.F., Stitt, M., Weigel, H.J., and White, J.W. 2008a. Next generation of elevated CO_2 experiments with crops: a critical investment for feeding the future world. *Plant Cell Environ.,* 31: 1317–1324.

Ainsworth, E., Leakey, A., Ort, D. and Long, S.P. 2008b. FACE-ing the facts: inconsistencies and inter-dependence among field, chamber and modeling studies of elevated [CO_2] impacts on crop yield and food supply. *New Phytologist,* 179*:* 5-9.

Akman, Z. 2009. Comparison of high temperature tolerance in maize, rice and sorghum seeds by plant growth regulators. *J. Anim. Vet. Adv.,* 8: 358–361.

Allakhverdiev, S.I., Kreslavski, V.D., Klimov, V.V., Los, D.A., Carpentier, R. and Mohanty, P. 2008. Heat stress: an overview of molecular responses in photosynthesis. *Photosynth. Res.,* 98: 541–550.

Allen, J.A., Chambers, J.L. and Pezeshki, S.R. 1997. *Effects of salinity on baldcypress seedlings: physiological responses and their relation to salinity tolerance. Wetland,* 17: 310-320.

Allen Jr., L.H. and Boote, K.J. 2000. Crop ecosystem responses to climate change: soybean. In, "*Climate Change and Global Crop Productivity,"* Eds. K.R. Reddy, and H.F. Hodges, CABI Publishing,Oxon, 133-160.

Andaya, V.C. and Mackill, D,J. 2003. Mapping of QTLs associated with cold tolerance during the vegetative stage in rice. *Journal of Experimental Botany*, 54: 2579–2585.

Andersen, M.N., Asch, F., Wu, Y., Jensen, C.R., Naested, H., Mogensen, V.O. and Koch, K.E. 2002. Soluble invertase expression is an early target of drought stress during the critical, abortion-sensitive phase of young ovary development in maize. *Plant Physiol.*, 130*:* 591–604.

Amthor, J.S. 2001. Effects of atmospheric CO_2 concentration on wheat yield: review of results from experiments using various approaches to control CO_2 concentration. *Field Crops Research*, 73: 1-34.

An, H, and Shangguan, Z.P. 2009. Effects of light intensity and nitrogen application on the growth and photosynthetic characteristics of *Trifolium repens* L. *Shengtai Xuebao/Acta Ecologica Sinica.*29: 6017-6024.

Ashraf, M., and Hafeez, M., 2004. Thermotolerance of pearl millet and maize at early growth stages: growth and nutrient relations. *Biol. Plant,* 48: 81–86.

Ashraf, M., Athar, H.R., Harris, P.J.C. and Kwon, T.R. 2008. Some prospective strategies for improving crop salt tolerance. *Adv Agron*., 97: 45–110.

Athar, H.R. and Ashraf, M. 2009. Strategies for crop improvement against salinity and drought stress: an overview. In "*Salinity and water stress: improving crop efficiency*," Eds. M. Ashraf, M. Ozturk, H.R. Athar, vol 44, 1st edn. Springer, New York, pp 1–16.

Atteya, A.M. 2003. Alteration of water relations and yield of corn genotypes in response to drought stress. *Bulg J Plant Physiol*., 29: 63–76.

Ayala, F. and O'leary, 1995. Growth and physiology of *Salicornia bigelovii* Torr. at suboptimal salinity. *International Journal of Plant Science,* 156: 197–205.

Bahadoran, M. and Salehi, H. 2015. Growth and flowering of two tuberose (*Polianthes tuberosa* L.) cultivars under deficit irrigation by saline water. *J. Agr. Sci. Tech.*, 17: 415-426

Baker, J.T., L.H. Allen Jr., K.J. Boote, P. Jones and J.W. Jones, 1990. Developmental responses of rice to photoperiod and carbon dioxide concentration. *Agric. For. Meteorol.*, 50: 201-210.

Bakht, J., Bano, A. and Dominy, P. 2006. The role of abscisic acid and low temperature in chickpea (*Cicer arietinum*) cold tolerance. II. Effects on plasma membrane structure and function. *Journal of Experimental Botany*, 57:707–715.

Balasubramanian, S., Sureshkumar, S., Lempe, J. and Weigel, D. 2006. Potent induction of *Arabidopsis thaliana* flowering by elevated growth temperature. *PLoS Genetics*, 2: 980-989.

Barnabas, B., Jagner, K. and Feher, A. 2008 The effect of drought and heat stress on reproductive processes in cereals. *Plant Cell Environ.*31: 11–38.

Banowetz, G., Ammar, K. and Chen, D. 1999. Temperature effects on cytokinin accumulation and kernel mass in a dwarf wheat. *Ann. Bot.*, 83: 303–307.

Baurle, I. and Dean, C. 2006. The timing of developmental transitions in plants. *Cell*, 125: 655-664.

Beltrão, J., Neves, A., de Brito, J.C. and Seita, J. 2009. Salt removal potential of turfgrasses in golf courses in the Mediterranean Basin. *WSEAS Transactions on Environment and Development*, 5(5): 394–403.

Ben-Zioni, A., Itai, C. and Vaadia, Y. 1967. Water and salt stresses, kinetin and protein synthesis in tobacco leaves. *Plant Physiol.*,42: 361-365.

Berger, J.D. 2007. Eco-geographic and evolutionary approaches to improving adaptation of autumn-sown chickpea (*Cicer arietinum* L.) to terminal drought: The search for reproductive chilling tolerance. *Field Crops Research*, 104: 112–122.

Bemis, S.M. and Torii, K.U. 2007. Autonomy of cell proliferation and developmental programs during *Arabidopsis* aboveground organ morphogenesis. *Dev Biol., 304(1): 367-381.*

Bindi, M., Ferrini, F. and Miglietta, F. 1992. Climatic change and the shift in the cultivated area of olive trees. *Journal of Agricultural Mediterranea*, 22: 41–44.

Bindi, M., Miglietta, F., Gozzini, B., Orlandini, S., and Seghi, L., 1997. A simple model for simulation of growth in grapevine (*Vitis vinifera* L.). I Model development. *Vitis*, 36: 67-71.

Bindi, M. and Moriondo, M. 2005. Impact of a 2^0 C global temperature rise on the Mediterranean region: Agriculture analysis assessment. In: Climate change impacts in the Mediterranean resulting from a 2°C global temperature rise. A WWF Report.

Blazquez, M., Hoon Ahn, J. and Weigel, D. 2003. A thermosensory pathway controlling flowering time in *Arabidopsis thaliana. Nature Genetics*, 33: 168-171.

Bond, W.J 1998. Effluent irrigation – An environmental challenge for soil science. *Aust. J. Soil Res.*,36: 543-555.

Boursiac, Y., Boudet, J., Postaire, O., Luu, D.T., Tournaire-Roux, C. and Maurel, C. 2008. Stimulus-induced downregulation of root water transport involves reactive oxygen species-activated cell signalling and plasma membrane intrinsic protein internalization. *Plant J.*, 56(2): 207-218.

Boyer, J. S. 1982. Plant productivity and environment. *Science,* 218: 443–448.

Boyer, J.S. 2009. Evans Review: Cell wall biosynthesis and the molecular mechanism of plant enlargement. *Funct Plant Biol,* 36(5): 383-394.

Bram, M.R. and Quinn, J.A. 2000. Sex expression, sex-specific traits, and the effects of salinity on growth and reproduction of *Amaranthus cannabinus (Amaranthaceae), a* dioecious annual. *American Journal of Botany,* 87: 1609-1618.

Bray, E.A., Bailey-Serres, J. and Weretilnyk, E. 2000a. Responses to abiotic stress. In "*Biochemistry and molecular biology of plants*," Eds. B.B. Buchanan, W. Gruissenand R.L. Jones, 1st edn. Wiley, Rockville, pp 1158–1249.

Bray, E.A. Bailey-Serres, J. and Weretilnyk, E. 2000b. *In "Responses to abiotic stresses, in Bioche-mistry and Molecular Biology of Plants,"* Eds. B. B. Buchanan, W. Gruissem, and R. L. Jones, American Society of Plant Biologists, Rockville, Md, USA, pp. 1158–1203.

Breshears, D.D., Cobb, N.S., Rich, P.M., Price, K.P., Allen, C.D., Balice, R.G., Romme, W.H., Kestens, J.H., Floyed, M.L., Belnap, J., Anderson, J.J., Myers, O.B. and Meyer, C.W. 2005. Regional vegetation die-off in response to global-change-type drought. *Proc. Natl. Acad. Sci. USA* , 102*:* 15144–15148.

Bui, E.N., K.R.J. Smettem, K.R.J., Moran, C.J. and Williams, J. 1996. Use of soil survey information to assess regional salinization risk using geographical information systems. *Environ. Qual.*, 25: 433-439.

Bunce, J.A. 2005. Seed yield of soybeans with daytime or continuous elevation of carbon dioxide under field conditions. *Photosynthetica*, 43: 435-438.

Busch, W., Wunderlich, M. and Schöffl, F. 2005. Identification of novel heat shock factor-dependent genes and biochemical pathways in *Arabidopsis thaliana. Plant J.*, 41*:* 1–14.

Camejo, D., Rodr´ýguez, P., Morales, M.A., Dell'amico, J.M., Torrecillas, A. and Alarc´on, J.J., 2005. High temperature effects on photosynthetic activity of two tomato cultivars with different heat susceptibility. *J. Plant Physiol.*, 162: 281–289.

Camejo, D., Jiménez, A., Alarcón, J.J., Torres, W., Gómez, J.M. and Sevilla, F. 2006. Changes in photosynthetic parameters and antioxidant activities following heat-shock treatment in tomato plants. *Funct. Plant Biol.,* 33: 177–187.

Carberry, P.S., Ranganathan, R., Reddy, L., Chauhan, Y.S. and Robertson, M.J. 2001. Predicting growth and development of pigeonpea: flowering response to photoperiod. *Field Crops Research*, 69*:* 151-162.

Challinor, A.J., Wheeler, T.R. and Slingo, J.M., 2005. Simulation of the impact of high temperature stress on the yield of an annual crop. *Agric Forest Meteorol.*, 135: 180-189.

Challinor, A. and Wheeler, T.R. 2008. Crop yield reduction in the tropics under climate change: processes and uncertainties. *Agricultural and Forest Meteorology*, 148*:* 343-356.

Chaves, M.M., Flexas, J. and Pinheiro, C. 2009. Photosynthesis under drought and salt stress: Regulation mechanisms from whole plant to cell. *Ann. Bot. (Lond.)*, 103*:* 551–560.

Chen, J., Wang, P., Mi, H.L., Chen, G.Y. and Xu, D.Q. 2010. Reversible association of ribulose-1, 5-bisphosphate carboxylase/oxygenase activase with the thylakoid membrane depends upon the ATP level and pH in rice without heat stress. *J. Exp. Bot.,* 61: 2939–2950.

Chmielewski, F.M., Mu¨ ller, A. and Bruns, E., 2004. Climate changes and trends in phenology of fruit trees and field crops in Germany, 1961–2000. *Agric. For. Meteorol.*, 121: 69–78.

Christensen, J.H. and Christensen, O.B. 2007. A summary of the PRUDENCE model projections of changes in European climate by the end of this century. *Clim. Change,* 81: 7–30.

Clark, H. 2001. Improving tolerance to low temperature in chickpea. Presented at the 4th European Conference on Grain Legumes, Cracow, Poland.

Clerget, B., Dingkuhn, M., Chantereau, J., Hemberger, J., Louarn, G. and Vaksmann, M. 2004.Does panicle initiation in tropical sorghum depend on day-to-day change in photoperiod? *Field Crops Research*, 88*:* 21-37.

Collinson, S.T., Ellis, R.H., Summerfield, R.J. and Roberts, E.H. 1992. Durations of the photoperiod-sensitive and photoperiod-insensitive phases of development to flowering in four cultivars of rice (*Oryza sativa* L.). *Annals of Botany*, 70*:* 339-346.

Collinson, S.T., Ellis, R.H., Summerfield, R.J. and Roberts, E.H. 1995. Relative importance of air and floodwater temperatures on the development of rice (*Oryza sativa*). *Experimental Agriculture*, 31*:*151-160.

Conroy, J.P. 1992. Influence of elevated atmosphere CO_2 concentration on plant nutrition. *Australian Journal of Botany*, 40: 445-456.

Cook, B.I., Ault, T.R., and Smerdon, J.E. 2015. Unprecedented 21st century drought risk in the American Southwest and Central Plains. *Sci. Adv.,* 1: e1400082.

Covington, M.F., Maloof, J.N., Straume, M., Kay, S.A., and Harmer, S. L. 2008. Global transcriptome analysis reveals circadian regulation of key pathways in plant growth and development. *Genome Biol.,* 9: R130.

Cox, P.M., Betts, R.A., Jones, C.D., Spall, S.A. and Totterdell, I.J. 2000. Acceleration of global warming due to carbon-cycle feedbacks in a coupled climate model. *Nature*, 408: 184-187.

Cramer, G.R,, Alberico, G.J. and Schmidt, C. 1994. Leaf expansion limits dry matter accumulation of salt-stressed maize. *Aust J Plant Physiol.,* 21: 663-674.

Cramer, W., Bondeau, A., Woodward, F.I., Prentice, I.C., Betts, R.A., Brovkin, V., Cox, P.M., Fisher, V., Foley, J., Friend, A.D., Kucharik, C., Lomas, M.R., Ramankutty, N., Sitch, S., Smith, B., White, A. and Young-Molling, C. 2001. Global response of terrestrial ecosystem structure and function to CO_2 and climate change: results from six dynamic global vegetation models. *Global Change Biol.*, 7(4): 357–373.

Cramer, G.R., Ergul, A., Grimplet, J., Tillett, R.L., Tattersall, E.A., Bohlman, M.C., Vincent, D., Sonderegger, J., Evans, J., Osborne, C., Quilici, D., Schlauch, K.A., Schooley, D.A. and Cushman, J.C. 2007. Water and salinity stress in grapevines: early and late changes in transcript and metabolite profiles. *Funct Integr Genomics,* 7(2): 111-134.

Cramer, G.R. 2010. Abiotic stress & plant responses from the whole vine to the genes. *Aust J Grape Wine Res.,* 16: 86-93.

Craufurd, P.Q. and Qi, A. 2001 Photothermal adaptation of sorghum (*Sorghum bicolor*) in Nigeria. *Agricultural and Forest Meteorology*, 108*:* 199-211.

Craufurd, P.Q. and Wheeler, T.R. 2009. Climate change and the flowering time of annual crops. *Journal of Experimental Botany*, 60(9): 2529-2539.

Cure, J.D. and Acock, B. 1986. Crop responses to carbon dioxide doubling: A literature survey. *Agricultural and Forest Meteorology*, 38: 127-145.

Curtis, D.L. 1968. The relation between the date of heading of Nigerian sorghums and the duration of the growing season. *Journal of Applied Ecology,* 5*:* 215-226.

Cutler, S.R., Rodriguez, P.L., Finkelstein, R.R. and Abrams, S.R. 2010. Abscisic acid: Emergence of a core signaling network. *Annu. Rev. Plant Biol.*, 61*:* 651–679.

Dasberg, S., Bielorai, H., Haimowitz, A. and Erner, Y. 1991. The effect of saline irrigation water on Shamouti' orange trees. *Irrig. Sci.*, 12: 205-211.

De Carvalho Gonçalves, J.F., De Sousa Barreto, D.C., Dos Santos Jr. U.M., Fernandes, A.V., Barbosa Sampaio, P.D.T. and Buckeridge, M.S. 2005. Growth, photosynthesis and stress indicators inyoung rosewood plants (*Aniba rosaeodora* Ducke) under different light intensities. *Braz. J. Plant Physiol.* 17:325-334.

Dagdelen, N. Yilmaz, E. Sezg Yn, F. and Gurbuz, T. 2004. Effects of water stress at different growth stages on processing pepper (*Capsisum annuum* cv. Kapija) yield, water use and quality characteristic. *Pakistan Journal of Biological Science,* 7(12): 2167-2172.

Dhindsa, R.S. and Cleland, R.E. 1975. Water stress and protein synthesis: I. Differential inhibition of protein synthesis. *Plant Physiol.,* 55(4): 778-781.

Dingkuhn, M., Kouressy, M., Vaksmann, M., Clerget, B. and Chantereau, J. 2008. A model of sorghum photoperiodism using the concept of threshold-lowering during prolonged appetence. *European Journal Agronomy*, 28*:* 74-89.

Dinneny, J.R., Long, T.A., Wang, J.Y., Jung, J.W., Mace, D., Pointer, S., Barron, C., Brady. S.M., Schiefelbein, J. and Benfey, P.N. 2008. Cell identity mediates the response of *Arabidopsis* roots to abiotic stress. *Science, 320(5878):* 942-945.

Dodd, A.N., Salathia, N., Hall, A., Kevei, E., Toth, R., Nagy, F., Julian, M.H., Miller, A.J., Webb, A.A.R. 2005. Plant circadian clocks increase photosynthesis, growth, survival, and competitive acvantage. *Science,* 309: 630–633.

Dolatabadian, A., Sanavy, S.A.M.M. and Ghanati, F. 2011. Effect of salinity on growth, xylem structure and anatomical characteristics of soybean, *Notulae Scientia Biologicae,* 3(1): 41–45.

Drake, B.G. 1997. More Efficient Plants: A Consequence of Rising Atmospheric CO_2? *Annals Review of Plant Physiology,*48: 609-639.

Duchêne, E. abd Schneider, C., 2005. Grapevine and climatic changes: a glance at the situation in Alsace. *Agron Sustain Dev,* 25: 93–99.

Edward, P.G. and Brown, J.J. 1998. Effects of soil salt levels on the growth and water use efficiency of *Atriplex canescens* (chenopodiaceae) varieties in drying soil. Am. J. Bot., 85(1): 10–16.

Edwards, K.D., Anderson, P.E., Hall, A., Salathia, N.S., Locke, J.C., Lynn, J.R., Straume, M., Smith, J.Q. and Millar, A.J. 2006. FLOWERING LOCUS C mediates natural variation in the high-temperature response of the *Arabidopsis* circadian clock. *Plant Cell,* 18: 639–650.

Ehlers, J.D., Hall, A.E., 1998. Heat tolerance of contrasting cowpea lines in short and long days. *Field Crops Res.,* 55: 11–21.

Ellis, R.H., Summerfield, R.J., Edmeades, G.O. and Roberts, E.H. 1992. Photoperiod, temperature, and the interval from sowing to tassel initiation in diverse cultivars of maize. *Crop Science,* 32: 1225-1232.

Ellis, R.H., Craufurd, P.Q., Summerfield, R.J. and Roberts, E.H. 1995. Linear relations between carbon dioxide concentration and rate of development towards flowering in sorghum, cowpea and soybean. *Annals of Botany,* 75: 193-198.

Endo, M., Tsuchiya, T., Hamada, K., Kawamura, S., Yano, K., Ohshima, M., Hingashitani, A., Watanabe, M., and Kawagishi-Kobayashi, M. 2009. High temperatures cause male sterility in rice plants with transcriptional alterations during pollen development. *Plant Cell Physiol.,* 50: 1911–1922.

Estrella, N., Sparks, T. and Menzel, A. 2007. Trends and temperature response in the phenology of crops in Germany. *Global Change Biology,* 13: 1737-1747.

European Commission. 2007. Addressing the challenge of water scarcity and droughts in the European Union. *Communication from the Commission to the European Parliament and the Council,* 414, Brussels.

Evans, L.T. 1993. Crop evolution, adaptation and yield. Cambridge: Cambridge University Press;

Ewert, F., Rodriguez, D., Jamieson, P., Semenov, M.A., Mitchell, R.A.C., Goudriaan, J., Porter, J.R., Kimball, B.A., Pinter Jr., P.J., Manderscheid, R., Weigel, H.J., Fangmeier, A., Fereres, E. and Villalobos, F. 2002. Effects of elevated CO_2 and drought on wheat: testing crop simulation models for different experimental and climatic conditions. *Agriculture, Ecosystems and Environment,* 93: 249-266.

Facella, P., Lopez, L., Carbone, F., Galbraith, D.W., Giuliano, G., and Perrotta, G. 2008. Diurnal and circadian rhythms in the tomato transcriptome and their modulation by cryptochrome photoreceptors. *PLoS ONE,* 3: e2798.

Farage, P.K., Mckee, I.F. and Long, S.P. 1998. Does a low nitrogen supply necessarily lead to acclimation of photosynthesis to elevated CO_2. *Plant Physiology,* 118:, 573-580.

Farooq, M., Wahid, A., Kobayashi, N.D.F. and Basra, S.M.A. 2009. Plant drought stress: Effects, mechanisms and management. *Sustainable Agric.,* 29: 185–212.

Farrell, A.D. and Gilliland, T.J. 2011 Yield and quality of forage maize grown under marginal climatic conditions in Northern Ireland. *Grass Forage Sci.,* 66: 214-223.

Filichkin, S.A., Breton, G., Priest, H.D., Dharmawardhana, P., Jaiswal, P., Fox, S.E., Michael, T.P., Chory, J., Kay, S.A.and Mockler, T.C. 2011. Global profiling of rice and poplar transcriptomes highlights key conserved circadian-controlled pathways and *cis*-regulatory modules. *PLoS ONE,* 6: e16907.

Finger, R. and S. Schmid, 2008. Modeling agricultural production risk and the adaptation to climate change. *Agric. Finance Rev.*, 68: 25-41.

Fischer, G., Shah, M., Tubiello, F.N. and van Velhuizen, H. 2005. Socio-economic and climate change impacts on agriculture: an integrated assessment, 1990–2080. *Philosophical Transactions of the Royal Society B: Biological Sciences,* 360: 2067-2083.

Fitter, A. and Fitter, R. 2002. Rapid changes in flowering time in British plants. *Science*, 296: 1689-1691.

Fitzpatrick, D. 1994. Money trees on your property. *Inkata Press*, Australia.

Flowers, T.J. and Yeo, A.R. 1995. Breeding for salinity resistance in crop plants: Where next? *Aust. J. Plant Physiol.*, 22:, 875-884..

Flowers, T. and Yeo, A. 1986. Ion relations of plants under drought and salinity. *Aust. J. Pl. Physiol.,* 13: 75–91.

Fox, G. A. 1990a. Components of flowering time variation in a desert annual. *Evolution*, 44: 1404-1423.

Fox, G. A. 1990b. Drought and the evolution of flowering time in desert annuals. *American Journal of Botany*, 77: 1508-1518.

Fox, G. A. 2001. Failure-time analysis: studying times-to-events and rates at which events occur. In, *"Design and analysis of ecological experiments," Eds. S.M. Scheiner and J. Gurevitch, Oxford* University Press, Oxford, UK.

Frank G., Pressman E., Ophir R., Althan L., Shaked R., Freedman M., Shen, S. and Firon, N. 2009. Transcriptional profiling of maturing tomato (*Solanum lycopersicum* L.) microspores reveals the involvement of heat shock proteins, ROS scavengers, hormones, and sugars in the heat stress response. *J. Exp. Bot.,* 60(13): 3891–3908.

Fujii, H., Chinnusamy, V., Rodrigues, A., Rubio, S., Antoni, R., Park S.Y., Cutler S.R., Sheen J., Rodriguez, P.L. and Zhu, J.K. 2009. In vitro reconstitution of an abscisic acid signalling pathway. *Nature*, 462: 660–664.

Fujita, M., Fujita, Y., Noutoshi, Y., Takahashi, F., Narusaka, Y., Yamaguchi-Shinozaki, K. and Shinozaki, K. 2006. Crosstalk between abiotic and biotic stress responses: A current view from the points of convergence in the stress signaling networks. *Curr. Opin. Plant Biol.*, 9: 436–442.

Gass, T., Schori, A., Fossati, A., Soldati, A. and Stamp, A.P. 1996. Cold tolerance of soybean (*Glycine max* (L.) Merr.) during the reproductive phase. *European Journal of Agronomy*, 5: 71–88.

Gerland, P., Raftery, A.E., Ševèíková, H., Li, N., Gu, D., Spoorenberg, T., Alkema, L., Fosdick, B.K., Chunn, J., Lalic, N., Bay, G., Buettner, T., Heilig, G.K. and Wilmoth, J. 2014. World population stabilization unlikely this century. *Science*, 346:,234–237.

Gifford, R M ., Barrett, D J. and Lutze, J.L. 2000 The effect of elevated CO_2 on the C: N and C:P mass ratio of plant tissues. *Plant and Soil*, 224: 1-14.

Gill, S.S., and Tuteja, N. 2010. Reactive oxygen species and antioxidant machinery in abiotic stress tolerance in crop plants. *Plant Physiol. Biochem.,* 48: 909–930.

Good, A.G. and Zaplachinski, S.T. 1994. The effects of drought stress on free amino acid accumulation and protein synthesis in *Brassica napus*. *Physiol Plant,* 90(1): 9-14.

Graf, A., Schlereth, A., Stitt, M., and Smith, A.M. 2010. Circadian control of carbohydrate availability for growth in*Arabidopsis* plants at night. *Proc. Natl. Acad. Sci. U.S.A.* 107: 9458–9463.

Grattan, S.R. and Grieve, C.M. 1999. Salinity-mineral nutrient relations in horticultural crops. *Sci. Hort.*, 78: 127-157.

Green, R.M., Tingay, S., Wang, Z.-Y., and Tobin, E.M. 2002. Circadian rhythms confer a higher level of fitness to *Arabidopsis*plants. *Plant Physiol.,* 129: 576–584.

Grouzis, M. and Sicot, M. 1980. A method for the phenological study of browse populations in the Sahel: the influences of some ecological factors. In, "*Browse in Africa, the current state of knowledge*," Ed. H.N. Le Houerou, Adddis Ababa: ILCA.

Gudkova, T.I. 1980. Physiological and cytological studies of the reasons for pollen sterility in spring wheat under low temperature. *Nauch. Tr. Linengar. S. Kh. In. Ta.* 394:103–108.

Guilioni, L., Wery, J. and Tardieu, F. 1997. Heat stressinduced abortion of buds and flowers in pea. Is sensitivity linked to organ age or to relations between reproductive organs? *Annals of Botany*, 80:159–168.

Guo, T.R., Zhang, G.P. and Zhang, Y.H. 2007. Physiological changes in barley plants under combined toxicity of aluminum, copper and cadmium. *Colloids Surf. Biointerfaces, 57:* 182–188.

Guy, C. 1999. Molecular responses of plants to cold shock and cold acclimation. *J. Mol. Microbiol. Biotechnol.,* 1: 231–242.

Habte, E., Muller, L.M., Shtaya, M., Davis, S.J., and Von Korff, M. 2014. Osmotic stress at the barley root affects expression of circadian clock genes in the shoot. *Plant Cell Environ.,* 37: 1321–1327.

Halliday, K., Salter, M., Thingnaes, E. and Whitelam, G. 2003. Phytochrome control of flowering is temperature sensitive and correlates with expression of the floral integrator FT. *The Plant Journal,* 33*:* 875-885.

Hall, A. E. 1992. Breeding for heat tolerance.*Plant Breeding Review.* 10: 129–168.

Hall, A.E., 2001. Crop Responses to Environment. CRC Press LLC, Boca Raton, Florida.

Hall, A. 2001. Crop developmental responses to temperature, photoperiod, and light quality, In, "*Crop Response to Environment*," Ed. E. Anthony and A. Hall, Boca Raton: CRC; pp. 83–87

Hamidov, A., Beltrao, J., Neves, A., Khaydarova, V. and Khamidov, M. 2007. Apocynum lancifolium and Chenopodium album—potential species to remediate saline soils. *WSEAS Transactions on Environment and Development*, 3(7): 123–128.

Hansen, W.J., Sato, M., Ruedy, R., Lacis, A. and Oinas, V. 2000. Global warming in the Twenty-First Century: An alternative scenario. *Proceedings of National Academy of Science*, 97: 9875-9880.

Hanson, C., Palutikof, J. and Perry, A.H. 2006. Modelling the Impact of Climate Extremes. *Clim Res.*, 31: 1-133.

Hansen, J., Ruedy, R., Sato, M., and Lo, K. 2010. Global surface temperature change. *Rev. Geophys.* 48, RG 4004.

Hasegawa, P.M., Bresan, R.A., Zhu, J.K. and Bohnert: H.J. 2000. Plant cellular and molecular responses to high salinity. *Ann. Rev. Plant Physiol. Plant Mol. Biol.*, 51: 4632-499.

Hastenrath, S 1991. Climate dynamics of the tropics. Kluwer, Dordrecht, Boston, London.

Hedhly, A., Hormaza, J.I. and Herrero, M. 2009. Global warming and sexual plant reproduction. *Trends Plant Sci.,* 14: 30–36.

Hedhly, A. 2011. Sensitivity of flowering plant gametophytes to temperature fluctuations. *Environ. Exp. Bot.,* 74: 9–16.

Hester, M. Mendelssohn, W.I.A. and McKee, K.L. 1998. Intraspecific variation in salt tolerance and morphology in Panicum hemitomon and Spartina lterniflora (Poaceae). *International Journal of Plant Sciences*, 159: 127-138.

Hinson K. 1989. Use of a long juvenile trait in cultivar development. *In,* "Proceedings of the World Soybean Research Conference IV," Ed. A.J. Pascale, *pp.* 983-1000.

Hirayama, T. and Shinozaki K. 2007. Perception and transduction of abscisic acid signals: Keys to the function of the versatile plant hormone ABA. *Trends Plant Sci.* 12*:* 343–351.

Hoerling, M., Schubert, S., Mo, K., Aghakouchak, A., Berbery, H., Dong, J., Hoerling, M., Kumar, A., Lakshmi, V., Leung, R., Li, J., Liang, X., Luo, L., Lyon, B., Miskus, D., Mo,

K., Quan, X., Schubert, S., Seager, R., Sorooshian, S., Wang, H., Xia, Y. and Zeng, N. 2013. An interpretation of the origins of the 2012 central great plains drought. *Assessment Report from the NOAA Drought Task Force Narrative Team.*

Horváth I., Glatz A., Nakamoto H., Mishkind M. L., Munnik T., Saidi Y., Goloubinoff, P., Harwood, J.L. and Vigh, L. 2012. Heat shock response in photosynthetic organisms: membrane and lipid connections. *Prog. Lipid Res.,* 51: 208–220.

Houghton, J. T. 2001. The scientific basis: contribution of working group I to the third assessment report of the intergovernmental panel on climate change. Climate Change, Cambridge University Press, pp 525–582.

Howarth, C.J., 2005. Genetic improvements of tolerance to high temperature. In, "*Abiotic Stresses: Plant Resistance Through Breeding and Molecular Approaches*," Eds. M. Ashraf and P.J.C. Harris, Howarth Press Inc., New York.

Hu, Q., Weiss, A., Feng, S. and Baenziger, P. 2005. Earlier winter wheat heading dates and warmer spring in the U.S. Great Plains. *Agricultural and Forest Meteorology*, 135*:* 284-290.

Hua, J. 2009 From freezing to scorching, transcriptional responses to temperature variations in plants. *Curr. Opin. Plant Biol.,* 12: 568–573.

Huang, J. and Redmann, R.E. 1995. Responses of growth, morphology, and anatomy to salinity and calcium supply in cultivated and wild barley, *Can. J. Bot.*, 73(12): 1859–1866,

Huang, Y., W. Zhang, W. Sun and X. Zheng, 2007. Net primary production of Chinese croplands from 1950 to 1999. *Ecol. Applic.*, 17: 692-701.

Hummel, I., Pantin, F., Sulpice, R., Piques, M., Rolland, G., Dauzat, M., Christophe, A., Pervent, M., Bouteille, M., Stitt, M., Gibon, Y. and Muller, B. 2010. *Arabidopsis* plants acclimate to water deficit at low cost through changes of carbon usage: an integrated perspective using growth, metabolite, enzyme, and gene expression analysis. *Plant Physiol.,* 154(1): 357-372.

Iba, K., 2002. Acclimative response to temperature stress in higher plants: approaches of gene engineering for temperature tolerance. *Annu. Rev. Plant Biol.*, 53: 225–245.

Imin, N., Kerim, T., Rolfe, B.G. and Weinman, J.J.. 2004. Effect of early cold stress on maturation of rice anthers. *Proteomics*, 4: 1873–1882.

IPCC 2007 Climate Change 2007: The Physical Science Basis. Contribution of Working Group I to the Fourth Assessment Report of the Intergovernmental Panel on Climate Change. Cambridge University Press, Cambridge.

Ismail, A.M. and Hall, A.E. 1999. Reproductive-stage heat tolerance, leaf membrane thermostability and plant morphology in cowpea. *Crop Sci.,* 39: 1762–1768.

Iwata, F. 1975. Heat unit concept of crop maturity. In, "*Physiological aspects of dryland farming*," Ed. V.S. Gupta, New Delhi and Oxford: I.B.H. Publishing Co.

Jablonski, L.M., Wang, X.Z. and Curtis, P.S. 2002. Plant reproduction under elevated CO_2 conditions: A meta-analysis of reports on 79 crop and wild species. *New Phytologist*, 156: 9-26.

Jaggard, K.W., Qi, A. and Ober, E.S. 2010. Possible changes to arable crop yields by 2050. *Philos. Trans. R. Soc. Lond. B Biol. Sci.*, 365: 2835–2851.

Jain, M., Prasad, P.V.V., Boote, K.J., Hartwell, A.L. and Chourey, P.S. 2007. Effects of season-long high temperature growth conditions on sugar-to-starch metabolism in developing microspores of grain *Sorghum (Sorghum bicolor* L. Moench). *Planta,* 227: 67–79.

Jagadish, S.V.K., Craufurd, P.Q. and Wheeler, T.R. 2008. Phenotyping parents of mapping populations of rice for heat tolerance during anthesis. *Crop Science,* 48*:*1140-1146.

Jampeetong, A. and Brix, H. 2009. Effects of NaCl salinity on growth, morphology, photosynthesis and proline accumulation of *Salvinia natans*. *Aqua. Bot.*, 91(3): 181–186.

Jamieson, P.D., Brooking, I.R., Porter, J.R. and Wilson, D. 1995. Prediction of leaf appearance in wheat: a question of temperature. *Field Crops Research*, 41*:* 35-44.

Jiménez, R. and Caballero, M. 1990. El riego y la Fertilización. In, "*El Cultivo Industrial de Plantas en Maceta", Eds. R.* Jiménez, and M. Caballero, Barcelona, Spain, PP. 101–122.

Jones, P.G. and Thornton, P.K. 2000. MarkSim. Software to generate daily weather data for Latin America and Africa. *Agronomy Journal*, 9*:* 445-453.

Kakumanu, A., Ambavaram, M.M., Klumas, C., Krishnan, A., Batlang, U., Myers, E., Grene, R. and Pereira, A. 2012. Effects of drought on gene expression in maize reproductive and leaf meristem tissue revealed by RNA-Seq. *Plant Physiol.*, 160*:* 846–867.

Kaur, G., Kumar, S., Nayyar, H. and Upadhyaya, H.D. 2008. Cold stress injury during the pod-filling phase in chickpea (*Cicer arietinum* L.). Effects on quantitative and qualitative components of seeds. *Journal of Agronomy and Crop Science*, 194: 457–464.

Kawasaki, S., Borchert, C., Deyholos, M., Wang, H., Brazille, S., Kawai, K., Galbraith, D. and Bohnert, H.J. 2001. Gene expression profiles during the initial phase of salt stress in rice. *Plant Cell* 13*:* 889–905.

Keeling, R.F. and Piper, S.C. 2009 Atmospheric CO_2 Records from Sites in the SIO Air Sampling Network. In, "*Trends: A Compendium of Data on Global Change,"* Eds. T.A. Boden, D.P.

Kaiser, R.J. Sepanki, and F.W. Stoss, Carbon Dioxide Information Analysis Center, Oak Ridge National Laboratory, U.S. Department of Energy, Oak Ridge, 16-26.

Keisling, T.C. 1982. Calculation of the length of day. *Agronomy Journal*, 74*:* 758-759.

Khan, S., Rowe, S., and Harmon, F. 2010. Coordination of the maize transcriptome by a conserved circadian clock. *BMC Plant Biol.,* 10: 126.

Kiehl, J. 2011. Lessons from Earth's Past. *Science,*,331: 158-159. Springer, C.J. and Ward, J.K. 200719

Kilian, J., Whitehead, D., Horak, J., Wanke, D,, Weinl, S., Batistic, O., D'Angelo, C., Bornberg-Bauer, E.,

Kudla, J. and Harter, K. 2007. The At GenExpress global stress expression data set: protocols, evaluation and model data analysis of UV-B light, drought and cold stress responses. *Plant J.,* 50(2): 347-363.

Kimball, B.A. 1983. Carbon dioxide and agricultural Yield: An assemblage and analysis of 430 Prior Observations. *Agronomy Journal*, 75: 779-788. .

Kimball, B.A. 2011. Lessons from FACE: CO_2 effects and interactions with water, nitrogen, and temperature. In, "*Handbook of Climate Change and Agroecosystems: Impacts, Adaptation, and Mitigation*" Eds. D. Hillel, and C. Rosenzweig, Imperial College Press, London, 87-107.

Klingler, J.P., Batelli, G. and Zhu, J.K. 2010. ABA receptors: The START of a new paradigm in phytohormone signalling. *J. Exp. Bot.*, 61*:* 3199–3210.

Kliewer, W.M. and Torres, R.E., 1972. Effect of controlled day and night temperatures on grape coloration. *Am J Enol Viticult.*, 23: 71–77.

Kokubun, M., Shimada, S., and Takahashi, M. 2001. Flower abortion caused by pre-anthesis water deficit is not attributed to impairment of pollen in soybean. *Crop Science*, 41: 1517–1521.

Krasensky, J., and Jonak, C. 2012. Drought, salt, and temperature stress-induced metabolic rearrangements and regulatory networks. *J. Exp. Bot.,* 63: 1593–1608.

Krauss, K., Chambers, W.J.L., Allen, J.A., Soileau Jr., D.M. and DeBosier, A.S. 2000. Growth and nutrition of baldcypress under varying salinity regimes in Louisiana, USA. *Journal of Coastal Research,* 16: 153-*163.*

Kreps, J.A., Wu, Y., Chang, H.S., Zhu, T., Wang, X. and Harper, J.F. 2002. Transcriptome changes for Arabidopsis in response to salt, osmotic, and cold stress. *Plant Physiol.,* 130*:* 2129–2141.

Kumar, S., Nayyar, H., Bains, T.S. and Kaur, G. 2005. Low temperature effects on early maturing chickpea genotype ICCV-96029. *International Chickpea and Pigeonpea Newsletter*, 12:19–22.

LaDeau, S.L. and Clark, J.S. 2006a. Elevated CO_2 and tree fecundity: The role of tree size, inter annual variability, and population heterogeneity. *Global Change Biology*, 12: 822-833.

LaDeau, S.L. and Clark, J.S. 2006b. Pollen production by *Pinus taeda* growing in elevated atmospheric CO_2. *Functional Ecology*, 20: 541-547.

Lafitte, H.R., Yongsheng, G., Yan, S. and Li, Z.K. 2007. Whole plant responses, key processes, and adaptation to drought stress: the case of rice. *J Exp Bot.*, 58: 169–175.

Lalonde, S., Beebe, D.U. and Saini, H.S. 1997. Early signs of disruption of wheat anther development associated with the induction of male sterility by meiotic-stage water deficit. *Sex. Plant Reprod.* 10: 40–48.

Lawrimore, J. H., Menne, M.J., Gleason, B.E., Williams, C.N., Wuertz, D.B., Vose, R.S. and Rennie, J. 2011. An overview of the Global Historical Climatology Network monthly mean temperature data set, version 3. *J. Geophys. Res. Atmos.*, 116: D19121.

Leakey, A., Uribelarrea, M., Ainsworth, E., Naidu, S., Rogers, A., Ort, D. and Long, S.P. 2006. Photosynthesis, productivity, and yield of maize are not affected by open-air elevation of CO_2 concentration in the absence of drought. *Plant Physiology,* 140: 779-790.

Lee, J., Yoo, S., Park, S., Hwang, I., Lee, J., Ahn, J. 2007. Role of SVP in the control of flowering time by ambient temperature in *Arabidopsis. Genes and Development,* 21: 397-402.

Levitt, J. 1980. *Responses of plants to environmental stresses, vol. II. Academic Press, New York, New York, USA.*

Lin, S.S. and Peterson, M.L. 1975. Low temperature induced floret sterility in rice. *Crop Science* 15: 657–660.

Liu, J.X. and Howell, S.H. 2010. Endoplasmic reticulum protein quality control and its relationship to environmental stress responses in plants. *Plant Cell,* 22(9): 2930-2942.

Lobell, D.B. and Asner, G.P., 2003. Climate and management contributions to recent trends in US Agricultural Yields. *Science,* 299: 1032.

Lobell, D.B., and Gourdji, S.M. 2012. The influence of climate change on global crop productivity. *Plant Physiol.,* 160: 1686–1697.

Long, S.P., Ainsworth E., Leakey, A., Nosberger, J. and Ort, D. 2006. Food for thought: lower-than-expected crop yield stimulation with rising CO_2 concentrations. *Science*, 312: 1918-1921.

Long, S.P. and Ort, D.R. 2010 More than taking the heat: Crops and global change. *Curr Opin Plant Biol.,* 13: 240-247.

Lotze-Campen, H. and Schellnhuber, H.J. 2009. Climate impacts and adaptation options in agriculture: What we know and what we don't know. *Journal für Verbraucherschutz und Lebensmittelsicherheit*, 4: 145-150.

Lu, P.L., Q. Yu, E. Wang, J.D. Liu and S.H. Xu, 2008. Effects of climatic variation and warming on rice development across South China. *Climate Res.*, 36: 79-88.

Maestri, E., Klueva, N., Perrotta, C., Gulli, M., Nguyen, H.T. and Marmiroli, N. 2002. Molecular genetics of heat tolerance and heat shock proteins in cereals. *Plant Mol. Biol.,* 48: 667–681.

Marcum, K.B. 2006. Use of saline and non-potable water in the turfgrass industry: constraints and developments. *Agric. Water Manage.*, 80(1–3): 132–146.

Manderscheid, R., Bender, J., Jager, H.J. and Weigel, H.J. 1995. Effects of season long CO_2 enrichment on cereals. II. Nutrient concentrations and grain quality. *Agriculture, Ecosystems & Environment*, 54: 175-185.

Marcolino-Gomes, J., Rodrigues, F.A., Fuganti-Pagliarini, R., Bendix, C., Nakayama, T.J., Celaya, B., Molinari, H.B., de Oliveira, M.C., Harmon, F.G. and Nepomuceno, A. 2014. Diurnal oscillations of soybean circadian clock and drought responsive genes. *PLoS ONE,* 9(1): e86402.

McCree, K. and Richardson, S. 1987. Salt increases the water use efficiency in water stressed plants. *Crop Science,* 27: 543–547.

McMaster, G.S., Edmunds, D.A., Wilhelm, W.W., Nielsen, D.C., Prasad, P.V.V. and Ascough J.C. 2011. PhenologyMMS: A program to simulate crop phenological responses to water stress. *Computers and Electronics in Agriculture,* 77: 118-125.

McNaughton, S. 1991. Dryland herbaceous perennials. In "*Response of plants to multiple stresses,*" Eds. H. Mooney, W. Winner, and E. Pell, Academic Press, New York, NY. pp 307-328.

Melcher, K., Ng, L.-M., Zhou, X.E., Soon, F.F., Xu, Y., Powell, K.M.S., Park, S.Y., Weiner, J.J., Fujii, H., Chinnaswamy, V., Kovach, A., Li, J., Wang, Y., Li, J., Peterson, F.C., Jensen, D.R., Yong, E.L., Volkman, B.F., Cutler, S.R., Zhu, J.K. and Xu, H.E. 2009. A gate-latch-lock mechanism for hormone signalling by abscisic acid receptors. *Nature*, 462: 602–608.

Menzel, A., Estrella, N. and Fabian, P., 2001. Spatial and temporal variability of the phenological seasons in Germany from 1951 to 1996. *Glob. Change Biol.*, 7: 657–666.

Menzel, A., Jakobi, G., Ahas, R., Scheifinger, H. and Estrella, N. 2003. Variations of the climatological growing season (1951–2000). in Germany compared with other countries. *International Journal of Climatology*, 23: 793-812.

Menzel, A., Sparks, T., Estrella, N., Koch, E., Aasa, A., Ahas, R. Alm-kübler, K., Bissolli, P., Braslavská, O., Briede, A., Chmielewski, F.M., Crepinsek, Z., Curnel, Y., Dahl, A., Defila, C., Donnelly, A., Filella, Y., Jatczak, K., Måge, F., Mestre, A., Nordli, O., Peñuelas, J., Pirinen, P., Remišová, V., Scheifinger, H., Striz, M., Susnik, A., Van Vliet, A.J.H., Wielgolaski, F-E., Zach, S. and Zust, A. 2006. European phenological response to climate change matches the warming pattern. *Global Change Biology*, 12: 1969-1976.

Michael, T.P., Mockler, T.C., Breton, G., Mcentee, C., Byer, A., Trout, J.D., Hazen, S.P., Shen, R., Priest, H.D., Sullivan, C.M., Givan, S.A., Yanovsky, M., Hong, F., Kay, S.A. and Chory, J. 2008. Network discovery pipeline elucidates conserved time-of-day-specific *cis*-regulatory modules. *PLoS Genet.,* 4: e14.

Mielke, M.S. 2010. Schaffer B. Photosynthetic and growth responses of *Eugenia uniflora* L. seedlings to soil flooding and light intensity. *Environ. Exper. Bot.* 68: 113-121.

Miglietta, F., Magliulo, V., Bindi, M., Cerio, L., Vaccari, F., Loduca ,V. and Peressotti, A. 1998. Free air CO_2 enrichment of potato (*Solanum tuberosum* L.): development, growth and yield. *Global Change Biology,* 4: 163-172.

Minges, P.A. and Warholic, D. 1973. Pepper Problems in New York. 1:14, In, Proceedings of National Pepper Conference, Packers Intl., St. Charles, U.S.A.

Mittler R., Vanderauwera S., Suzuki N., Miller G., Tognetti V. B., Vandepoele K., Gollery, M., Shulaev, V., Van Breusegem, F. 2011. ROS signaling: the new wave? *Trends Plant Sci.,* 16: 300–309.

Mittler, R., and Blumwald, E. 2015. The roles of ROS and ABA in systemic acquired acclimation. *Plant Cell,* 27: 64–70.

Moftah, A. E. and Al-Humaid, A. I. 2006. Response of Vegetative and Reproductive Parameters of Water Stressed Tuberose Plant to Vapor Grad and Kaolin Antitranspirants. *JKSU*, 18: 127–139.

Mahmood, S., Wahid, A., Javed, F. and Basra S.M.A. 2010. Heat stress effects on forage quality characteristics of maize (*Zea mays*) cultivars. *Int. J. Agric. Biol.,* 12: 701–706.

Molassiotis, A. and Fotopoulos, V. 2011. Oxidative and nitrosative signaling in plants: two branches in the same tree? *Plant Signal Behav.,* 6(2): 210-214.

Momcilovic, I. and Ristic, Z., 2007. Expression of chloroplast protein synthesis elongation factor, EF-Tu, in two lines of maize with contrasting tolerance to heat stress during early stages of plant development. *J. Plant Physiol.*, 164: 90–99.

Monneveux, P., Sa´nchez, C., Beck, D. and Edmeades, G.O. 2006. Drought tolerance improvement in tropical maize source populations: evidence of progress. *Crop Sci.*, 46:180–191.

Morales, D., Rodríguez, P., Dell'Amico, J., Nicolas, E., Torrecillas, A., Sánchez-Blanco, M.J. 2003. High- temperature preconditioning and thermal shock imposition affects water relations, gas exchangeand root hydraulic conductivity in tomato. *Biol. Plant.,* 47: 203–208

Morrison, M.J., and Stewart, D.W., 2002. Heat stress during flowering in summer brassica. *Crop Sci.*, 42:, 797–803.

Morita, S., Siratsuchi, H., Takanashi, J., and Fujita, K., 2004. Effect of high temperature on ripening in rice plant. Analysis of the effect of high night and high day temperature applied to the panicle in other parts of the plant. *Jpn. J. Crop. Sci.*, 73: 77–83.

Moriondo, M., Orlandini, S., De Nuntiis, P. and Mandrioli, P., 2001. Effect of agrometeorological parameters on the phenology of pollen emission and production of olive trees (*Olea europaea* L.). *Aerobiologia*, 7: 225-232

Moriondo, M., 2003. Downscaling delle temperature minime e massime a partire da un GCM: il caso di Firenze. Università degli Studi di Firenze. Master in Meteorologia applicata.

Moriondo, M. and Bindi, M., 2005. Cambiamenti climatici e ambiente vitivinicolo: impatto e strategie di adattamento. Rapporto finale per ARSIA

Moriondo, M. and Bindi, M. 2006. Comparison of temperatures simulated by GCMs, RCMs and statistical downscaling: potential application in studies of future crop development. *Clim Res.,* 30: 149–160.

Moriondo, M. and Bindi, M. 2007. Impact of climate change on the phenology of typical mediterranean crops. *Italian Journal of Agrometeorology*, 30(3): 5-12.

Moriondo, M., Giannakopoulos, C. and Bindi, M. 2011. The role of climate extremes in crop yield climate change impact assessments. *Climatic Change*, 104: 679-701.

Mugford, S.T., Fernandez, O., Brinton, J., Flis, A., Krohn, N., Encke, B., Feil, R., Sulpice, R., Lunn, J.E., Stiit, M. and Smith, A.M. 2014. Regulatory properties of ADP glucose pyrophosphorylase are required for adjustment of leaf starch synthesis in different photoperiods. *Plant Physiol.,* 166: 1733–1747.

Mullins, M.G., Bouquet, A. and Williams, L.E. 1992. Biology of the Grapevine. (Cambridge Univ. Press, London).

Munns, R. and Tester, M. 2008. Mechanisms of salinity tolerance. *An. Rev. Pl. Biol.*, 59: 651–681.

Muraoka, H., Tang, Y., Koizumi, H., and Washitani, I. 2002. Effects of light and soil water availability on leaf photosynthesis and growth of *Arisaema heterophyllum*, a riparian forest under storey plant. *J. Plant Res.* 115::419-427.

Muscolo, A., Panuccio, M.R. and Sidari, M. 2003. Effects of salinity on growth, carbohydrate metabolism and nutritive properties of kikuyu grass (*Pennisetum clandestinum* Hochst), *Plant Sci.,* 64(6): 1103–1110.

Narciso G, Ragni P, and Venturi A. 1992. Agrometeorological aspects of crops in Italy, Spain and Greece. Joint Research Centre, Commission of the European Communities, Brussels, Luxembourg.

Nardini, A,, Lo, G.M.A. and Salleo, S. 2011. Refilling embolized xylem conduits: is it a matter of phloem unloading? *Plant Sci.,* 180(4): 604-611.

Nava, G.A., Dalmago, G.A., Bergamaschi, H., Paniz, R., dos Santos, R.P, Marodin, G.A.B. 2009. Effect of high temperatures in the pre-blooming and blooming periods on ovule formation, pollen grains and yield of 'Granada' peach. *Sci. Hortic.,* 122: 37–44.

Nemani, R.R., Keeling, C.D., Hashimoto, H., Jolly, W.M., Piper, S.C., Tucker, C.J., Myneni, R.B. and Running, S.W., 2003. Climate-driven increases in Global Terrestrial Net Primary Production from 1982 to 1999. *Science*, 300: 1560–1563.

Nayyar, H., Bains, T. and Kumar, S. 2005. Low temperature induced floral abortion in chickpea: relationship to abscisic acid and cryoprotectant in reproductive organs. *Environmental and Experimental Botany,* 53: 39–47.

Nishiyama, I., 1976. Effects of Temperature on the Vegetative Growth of Rice Plants. In: Proceedings of the Symposium on Climate and Rice, International Rice Research Institute (Ed.). International Rice Research Institut, Los Baños, The Philippines, pp: 159-185.

Nishiyama, I. 1995. Damage due to extreme temperatures. In, "*Science of the rice plant,*" Eds. T. Matsuo, K. Kumazawa, R. Ishii, H. Ishihara, and H. Hirata, Tokyo, Japan: Food and Agriculture Policy Research Center, pp. 769–812

Nonami, H., Wu, Y.J. and Boyer, J.S. 1997. Decreased growth-induced water potential: primary cause of growth inhibition at low water potentials. *Plant Physiol.,* 114(2): 501-509.

Ooka, H., Satoh, K., Doi, K., Nagata, T., Murakami, K., Matsubara, K., Osato, N., Kawai, J., Caminci, P., Hayashizaki, Y., Suzuki, K., Kojima, K., Takahara, Y., Yamamoto, K. and Kikuchi, S. 2003. Comprehensive analysis of NAC family genes in *Oryza sativa and Arabidopsis thaliana. DNA Res.*, 10*:* 239–247.

Osborne, C.P., Chuine, I., Viner, D. and Woodword, F.I. 2000. Olive phenology as a sensitive indicator of future climatic warming in the Mediterranean. *Plant Cell Environ.*, 23: 701–710.

Oshino, T., Abiko, M., Saito, R., Ichiishi, E., Endo, M., Kawagishi-Kobayashi M., and Higashitani, A. 2007. Premature progression of anther early developmental programs accompanied by comprehensive alterations in transcription during high-temperature injury in barley plants. *Mol. Genet. Genomics,* 278(1): 31–42.

Overdieck, D. 1993 Elevated CO_2 and mineral content of herbaceous and woody plants. *Vegetation,* 104-105: 403-411.

Pachauri, R.K. 2007. Sustainable well-being. *Science,* 315(5814): 9.

Parent, B., Hachez, C., Redondo, E., Simonneau, T., Chaumont, F. and Tardieu, F. 2009. Drought and abscisic acid effects on aquaporin content translate into changes in hydraulic conductivity and leaf growth rate: a trans-scale approach. *Plant Physiol,* 149(4): 2000-2012.

Parida, A.K. and Das, A.B. 2005. Salt tolerance and salinity effects on plants: a review. *Ecotoxicology and Environmental Safety*, 60(3): 324–349.

Parish, R.W., Phan, H.A., Iacuone, S. and Li, S.F. 2012. Tapetal development and abiotic stress: a centre of vulnerability. *Funct. Plant Biol.,* 39: 553–559.

Park, S.O., Hwang, S. and Hauser, B.A. 2004. The phenotype of *Arabidopsis* ovule mutants mimics the morphology of primitive seed plants. *Proc. Biol. Sci.*, 271*:* 311–316.

Park, S.Y., Fung, P., Nishimura, N., Jensen, D.R., Fujii, H., Zhao, Y., Lumba, S., Santiago, J., Rodrigues, A., Chow, T.F., Alfred, S.E., Sonetta, D., Finkelstein, T., Provart, N.J., Desveaux, D., Rodriquez, P.L., McCourt, P., Zhu, J.K., Schroeder, J.I., Volkman, B.F. and Cutler, S.R. 2009. Abscisic acid inhibits type 2C protein phosphatases via the PYR/PYL family of START proteins. *Science*, 324: 1068–1071.

Parry, M.L., Rosenzweig, C., Iglesias, A., Livermore, M. and Fischer, G. 2004. Effects of climate change on global food production under SRES emissions and socio-economic scenarios. *Glob. Environ. Change*, 14: 53–67.

Patel, P.N., and Hall, A.E., 1990. Genotypic variation and classification of cowpea for reproductive responses to high temperatures under long photoperiods. *Crop Sci.*, 30: 614–621.

Pearce, R.S. 2001. Plant freezing and damage. *Ann. Bot.,* 87: 417–424.

Peet, M., Sato, S. and Gardner, R. 2002. Comparing heat stress effects on male-fertile and male-sterile tomatoes. *Plant Cell Environ.,* 21: 225–231.

Peng, S., Huang, J., Sheehy, J.E., Laza, R.C., Visperas, R.M., Zhong, X., Centeno, G.S., Khush, G.S. and Cassman, K.G., 2004. Rice yields decline with higher night temperature from global warming. *Proc. Natl. Acad. Sci. U.S.A.*, 101(27): 9971–9975.

Pickering, N.B., Allen Jr., F.L.H., Albrecht, S.L., Jones, P., Jones, J.W. and Baker, J.T. 1994. Environmental plant chambers: controls and measurements using CR-10T data loggers. In, "*Computers in Agriculture: Proceedings of the 5th International Conference,*" Eds. D.G. Watson, F.S. Zuzueta, and T.V. Harrison, American Society of Agricultural Engineers, Orlando, 5- 9 February 1994, 29-35.

Pimmongkol, A., Terapongtanakhon, S. and Udonsirichakhon, K. 2002. Anatomy of salt-and non-salt-tolerant rice treated with NaCl, In "*Proceedings of the 28th Congress on Science and Technology of Thailand,*" Bangkok, Thailand.

Pinheiro, C. and Chaves, M.M. 2011. Photosynthesis and drought: can we make metabolicconnections from available data? *J Exp Bot,* 62(3): 869-882.

Pitman, M.G. and Lauchli, A. 2002. Global impact of salinity and agricultural ecosystems. In, "*Salinity: environmental, plants, molecules,*" Eds. A. Lauchli, U. Luttge, Springer, Dotrecht, pp 3–20.

Ponnamperuma, F.N. 1984. Role of cultivar tolerance in increasing rice production in saline lands. In, "*Salinity tolerance in plants: Strategies for crop improvement,*" Eds. R.C. Staples and G.H. Toenniessen, John Wiley and Sons, New York. pp. 255-271.

Poorter, H., Berkel, Y.V., Baxter, R., Den Hertog, J., Dijkstra, P., Gifford, R.M., Griffin, K.L., Roumet, C., Roy, J. and W ong, S.C. 1997. The effect of elevated CO_2 on the chemical composition and construction costs of leaves of 27 C3 species. *Plant, Cell and Environment,* 20: 472-482.

Porter, J.R. and Gawith, M. 1999. Temperatures and the growth and development of wheat: a review. *Eur J Agron.,* 10: 23–36.

Potters, G., Pasternak, T.P., Guisez, Y and Jansen, M.A.K. 2008. Different stresses, similar morphogenic responses: integrating a plethora of pathways. *Plant Cell Environ.,32:* 58–169.

Prasad, N.G., Dey, S., Shakarad, M. and Joshi, A. 2003. The evolution of population stability as a by-product of life-history evolution. *Proc.Biol. Sci.,* 270(Suppl. 1): S84–S86.

Prasad, P., Boote, K., Allen, L., Sheehy, J., Thomas, J. 2006. Species, ecotype and cultivar differences in spikelet fertility and harvest index of rice in response to high temperature stress. *Field Crops Res.,* 95: 398–411.

Rasheed, R. 2009. *Salinity and Extreme Temperature Effects on Sprouting Buds of Sugarcane (Saccharum officinarum L.): Some Histological and Biochemical Studies.* Ph. D. thesis, Department of Botany, University of Agriculture; Faisalabad.

Reddy, K.R., Kakani, V.G., Zhao, D., Koti, S. and Gao, W. 2004. Interactive effects of ultraviolet-B radiation and temperature on cotton physiology, growth, development and hyperspectral reflectance. *Photochemistry and Photobiology,* 79: 416-427.

Reekie, J., Hicklenton, P. and Reekie, E. 1997. The interactive effects of carbon dioxide enrichment and day length on growth and development in *Petunia hybrida. Annals of Botany,* 80: 57-64.

Rehman, H., Malik, S.A., and Saleem, M., 2004. Heat tolerance of upland cotton during the fruiting stage evaluated during cellular membrane thermostability. *Field Crops Res.,* 85: 149–158.

Rehman, S., Harris, P.J.C. and Ashraf, M. 2005. Stress environments and their impact on crop production. In "*Abiotic stresses: plant resistance through breeding and molecular approaches,*" Eds. M. Ashraf and P.J.C. Harris, 1st edn. Haworth Press, New York, pp 3–18.

Ren, C., Bilyeu, K.D. and Beuselinck, P. 2009. Composition, vigor, and proteome of mature soybean seeds developed under high temperature. *Crop Sci.,* 49: 1010–1022.

Richards, R. 1992. Increasing salinity tolerance of grain crops: is it worthwhile? *Plant and Soil,* 146: 89–98.

Richards, R.A. 2006. Physiological traits used in the breeding of new cultivars for water-scarce environments. *Agricultural Water Management,* 80: 197-211.

Roberts, E.H., Summerfield, R.J., Muehlbauer, F.J. and Short, R.W. 1986. Flowering in lentil (*Lens culinaris* Medic.): the duration of the photoperiodic inductive phase as a function of accumulatedday length above the critical photoperiod. *Annals of Botany,* 58: 235-248.

Roberts, E.H. and Summerfield, R.J. 1987. Measurement and prediction of flowering in annual crops. *In:* Atherton JG, editor. *Manipulation offlowering.* London: Butterworths,*pp.* 17-50.

Roberts, E.H., Qi ,A., Ellis, R., Summerfield, R.J., Lawn, R.J. and Shanmugasundaram, S. 1996. Use of field observations to characterise genotypic flowering responses to photoperiod and temperature: a soybean exemplar. *Theoretical and Applied Genetics,* 93: 519-533.

Rodríguez, P., Dell Amico, J., Morales, D., Sánchez-Blanco, M. J. and Alarcón, J. J. 1997. Effects of Salinity on Growth, Shoot Water Relations and Root Hydraulic Conductivity in Tomato Plants. *J. Agron.Crop. Sci.*, 128: 439–444.

Rodziewicz, P., Swarcewicz, B., Chmielewska, K., Wojakowska, A. and Stobiecki, M. 2014. Influence of abiotic stresses on plant proteome and metabolome changes. *Acta Physiol Plant,* 36: 1–19.

Rogers, G.S., Milham, P.J., Thibaud, M.C. and Conroy, J.P. 1996. Interaction between rising CO_2 concentration and nitrogen supply in cotton. I. Growth and leaf nitrogen concentration. *Australian Journal of Plant Physiology*, 23: 119-125.

Root, T.L., MacMynowski, D.P., Mastrandrea, M.D. and Schneider, S.H. 2005. Human-modified temperatures include species changes: joint attribution. *Proceedings of the National Academy of Sciences, USA,* 102: 7465-7469.

Rosenzweig, C., A. Iglesias, X.B. Yang, P.R. Epstein and E. Chivian, 2001. Climate change and extreme weather events; implications for food production, plant diseases and pests. *Global Change Hum.Health*, 2: 90-104.

Rylski, I. 1979. Fruit set and development of seeded and seedless tomato fruits under diverse regimes of temperature and pollination. *Journal of American Society of Horticultural Science*, 104: 835–839.

Sadras, V.O. and Connor, D.J., 1991. Physiological basis of the response of harvest index to the fraction of water transpired after anthesis A simple model to estimate harvest index for determinate species. *Field Crops Res.*, 26: 227–239.

Sadras, V.O. and Monzon, J.P., 2006. Modelled wheat phenology captures rising temperature trends: Shortened time to flowering and maturity in Australia and Argentina. *Field Crops Res.,* 99: 136–146.

Saidi, Y., Finka, A., Muriset, M., Bromberg, Z., Weiss, Y.G., Maathuis, F.J. and Goloubinoff, P. 2009. The heat shock response in moss plants is regulated by specific calcium-permeable channels in the plasma membrane. *Plant Cell,* 21: 2829–2843.

Saini, H. S. and Aspinall, D. 1982. Abnormal sporogenesis in wheat (Triticumaestivum L.) induced by short periods of high temperature. *Ann Bot.*, 49(6): 835-846.

Saini, H.S. 1997. Effects of water stress on male gametophyte development in plants. *Sex. Plant Reprod.* 10: 67–73.

Sakata, T. and Higashitani, A. 2008. Male sterility accompanied with abnormal anther development in plants–genes and environmental stresses with special reference to high temperature injury. *Int. J. Plant Dev. Biol.,* 2: 42–51.

Sakata, T., Oshino, T., Miiura, S., Tomabechi, M., Tsunaga, Y., Higashitani, N., Miyazawa, Y., Takahashi, H., Watanabe, M. and Higashitani, A. 2010.Auxins reverse plant male sterility caused by high temperatures. *Proc. Natl. Acad. Sci. U.S.A.* 107: 8569–8574. .

Salvucci, M.E., and Crafts-Brandner, S. J. 2004. Inhibition of photosynthesis by heat stress: the activation state of Rubisco as a limiting factor in photosynthesis. *Physiol. Plant.* 120: 179–186.

Sandhu, J.S,, Gupta. S.K., Singh, S. and Dua, R.P. 2007. Genetic variability for cold tolerance in pigeonpea. *Journal of SAT Agricultural Research*, 5: 1–3.

Saxena, N.P. and Johansen, C. 1990. Realized yield potential in chickpea and physiological considera-tions for further genetic improvement. In, "*Proceedings of International Congress on Plant Physiology*," Eds. S.K. Sinha, P.V. Sane, S.C. Bhargava, P.K. Agarwal , New Delhi, India: Indian Society of Plant Physiology and Biochemistry , pp. 279–288

Sch¨offl, F., Prandl, R., Reindl, A., 1999. Molecular responses to heat stress. In, "*Molecular Responses to Cold, Drought, Heat and Salt Stress in Higher Plants*," Eds. K. Shinozaki and K. Yamaguchi-Shinozaki, R.G. Landes Co.,Austin, Texas, pp. 81–98.

Schwaegerle, K.E.H., McIntyre C. and Swingley 2000. *Quantitative genetics and the persistence of environmental effects in clonally propagated organisms. Evolution* 54*:* 452*-461.*

Schwabele, K.A., Iddo, K. and Knap, K.C. 2006. Drain water management for salinity mitigation in irrigated agriculture. *Am. J. Agric. Eco.,* 88: 133–140.

Scialdone, A., Mugford, S.T., Feike, D., Skeffington, A., Borrill, P., Graf, A., Smith, A.M. and Howard, M. 2013. *Arabidopsis* plants perform arithmetic division to prevent starvation at night. *Elife* 2: e00669.

Schröter, D., Carmer, W., Leemans, R., Prentice, I.C., Araújo, M.B., Arnell, N.W., Bondeau, A., Bugmann, H., Carter, T.R., Gracia, C.A., de la Vega-Leinert, A.C., Erhard, M., Evert, F., Glendining, M., House, J.J., Kankaanpää, S., Klein, R.J., Lavorel, S., Lindner, M., Metzger, M.J., Meyer, J., Mitchel, T.D., Reginster, J., Rounsevell, M., Sabatì, S., Sitch, B., Smith, J., Sykes, M.T., Thonicke, K., Thuiller, W., Tuck, G., Zaehle, S. and Zierl, B. 2005. Ecosystem service supply and vulnerability to global change in Europe. *Science*, 310*:* 1333–1337.

Siddique, K.H.M., Loss, S.P., Regan, K.L., and Jettner, R.L., 1999. Adaptation and seed yield of cool season grain legumes in Mediterranean environments of south-western Australia. *Aust. J. Agric. Res.*, 50: 375–387.

Seliskar, D. M. and J. L. Gallagher, J.L. 2000. Exploiting wild population diversity and somaclonal variation in the salt marsh grass Distichlis spicata (Poaceae) for marsh creation and restoration. *American Journal of Botany,* 87: 141-146.

Sepaskhah, A. R. and Yarami, N. 2009. Interaction Effects of Irrigation Regime and Salinity on Flower Yield and Growth of Saffron. *J. Hort. Sci. Biotechnol.*, 84: 216- 222.

Shah, N. and Paulsen, G. 2003. Interaction of drought and high temperature on photosynthesis and grain-filling of wheat. *Plant Soil,* 257: 219–226.

Saha, S., Hossain, M., Rahman, M., Kuo, C. and Abdullah, S. 2010. Effect of high temperature stress on the performance of twelve sweet pepper genotypes. *Bangladesh J. Agric. Res.,* 35: 525–534.

Shalhevet, J. 1993. Plants under water and salt stress. In "*Plant adaptation to environmental stress*", Eds. L. Fowden, T. Mansfield, and J. Stoddart, Chapman and Hall, New York, NY. 133-154.

Sharkey, T.D. 2005. Effects of moderate heat stress on photosynthesis: importance of thylakoid reactions, rubisco deactivation, reactive oxygen species, and thermotolerance provided by isoprene. *Plant Cell Environ.,* 28: 269–277.

Sharkey, T.D. and Zhang, R. 2010. High temperature effects on electron and proton circuits of photosynthesis. *J. Integr. Plant Biol.,* 52: 712–722.

Sheoran, I.S. and Saini, H.S. 1996. Drought-induced male sterility in rice: Changes in carbohydrate levels and enzyme activityies associated with the inhibition of starch accumulation in pollen. *Sex. Plant Reprod.* 9*:* 161–169.

Shillo, R., Ding, M., Pasternak, D. and Zaccai, M. 2002. Cultivation of Cut Flower and Bulb Species with Saline Water. *Sci. Hortic*., 92: 41–54.

Shimono, H., Okada, M., Yamakawa, Y., Nakamura, H., Kobayashi, K. and Hasegawa, T. 2009. Genotypic variation in rice yield enhancement by elevated CO_2 relates to growth before heading, and not to maturity group. *Journal of Experimental Botany,* 60*:* 523-532.

Shimono, H., 2011. Earlier rice phenology as a result of climate change can increase the risk of cold damage during reproductive growth in northern Japan. *Agric. Ecosyst. Environ.*, 144: 201-207.

Singh, K.B., Malhotra, R.S. and Saxena, M.C. 1990. Sources for tolerance to cold in *Cicer* species. *Crop Science*, 30:1136–1138.

Skirycz, A,and Inze, D. 2010. More from less: plant growth under limited water. *Curr Opin Biotechnol,* 21(2): 197-203.

Slawinska, J. and Obendorf, R. L. 2001. Buckwheat Seed Set in Plant and during *iIn vitro* Inflorescence Culture: Evaluation of Temperature and Water Deficient Stress.Seed. *Sci. Res.*, 11: 223–233.

Smertenko, A., Draber, P., Viklicky, V. and Opatrny, Z., 1997. Heat stress affects the organization of microtubules and cell division in *Nicotiana tabacum* cells. *Plant Cell Environ.*, 20: 1534–1542.

Sousa Paiva, É.A., Dos Santos Isaias, R.M., Aguiar Vale, F.H., De Senna Queiroz, C.G. 2003. The influence of light intensity on anatomical structure and pigment contents of *Tradescantia pallid* (Rose) Hunt. cv. *purpurea* boom (Commelinaceae) leaves. *Braz. Arch. Biol. Techn.* 46: 617-624.

Springer, C.J. and Ward, J.K. 2007. Flowering time and elevated atmospheric CO_2. *New Phytologist,* 176: 243-255.

Springer, C., Orozco, R., Kelly, J. and Ward, J. *2008.* Elevated CO_2 influences the expression of floral-initiation genes in *Arabidopsis thaliana. New Phytologist,* 178*:* 63-67.

Srivastava, A.C., Sengupta, U.K. and Pal, M. 2001 Growth, CO_2 exchange rate and dry matter partitioning in mungbean grown under elevated CO_2. *Indian Journal Experimental Biology,* 39: 572-577.

Srivastava, A.C., Khanna, Y.P., Meena, R.C., Pal, M. and Sengupta, U.K. 2002. Diurnal changes in photosynthesis,sugars,and nitrogen of wheat and mungbean grown under elevated CO_2 concentration. *Photosynthetica*, 40: 221-225.

Stanton, M.L., Roy, B.A. and Thiede, D.A. 2000. Evolution in stressful environments. I. Phenotypic variability, phenotypic selection, and response to selection in five distinct environmental stresses. *Evolution* 54*:* 93-111.

Stockle, C.O., Donatelli, M. and Nelson, R. 2003. CropSyst, a cropping systems simulation model. *Eur J Agron*., 18: 289-307.

Stone, P.J., and Nicol´as, M.E., 1994. Wheat cultivars vary widely in their responses of grain yield and quality to short periods of post-anthesis heat stress. *Aust. J. Plant Physiol*., 21: 887–900.

Sultana, N., Ikeda, T. and Kashem, M.A. 2001. Effect of foliar spray of nutrient solutions on photosynthe-sis, dry matter accumulation and yield in seawater-stressed rice. *Environ. Exp. Bot*., 46: 129-140.

Sumesh, K., Sharma-Natu, P. and Ghildiyal, M. 2008. Starch synthase activity and heat shock protein in relation to thermal tolerance of developing wheat grains. *Biol. Plant.,* 52: 749–753.

Summerfield, R.J., Lawn, R.J. Qi, A., Ellis, R.H., Roberts, E.H., Chay, P.M., Brouwer, J.B., Rose, J.L., Shanmugasundaramu, S., Yeates, S.J. and Sandover, S. 1993. Towards the reliable prediction of time to flowering in six annual crops. II. Soybean (*Glycine max*). *Experimental Agriculture,* 29*:* 253-289.

Summerfield, R.J., Asumadu, H., Ellis, R.H. and Qi, A. 1998. Characterization of the photoperiod response of post-flowering development in maturity isolines of soybean [*Glycine max* (L.) Merrill] 'Clark'. *Annals of Botany,* 82*:*775-771.

Sun, K., Hunt, K. and Hauser, B.A. 2004. Ovule abortion in *Arabidopsis* triggered by stress. *Plant Physiol.*, 135*:* 2358–2367.

Sun, K., Cui, Y. and Hauser, B.A. 2005. Environmental stress alters genes expression and induces ovule abortion: Reactive oxygen species appear as ovules commit to abort. *Planta,* 222*:* 632–642.

Sun, W. and Y. Huang, 2011. Global warming over the period 1961-2008 did not increase high-temperature stress but did reduce low-temperature stress in irrigated rice across China. *Agric. For. Meteorol.*, 151: 1193-1201.

Suzuki, K., Takeda, H., Tsukaguchi, T. and Egawa, Y. 2001. Ultrastructural study on degeneration of tapetum in anther of snap bean (*Phaseolus* vulgaris L.) under heat stress. *Sex. Plant Reprod.,* 13: 293–299.

Suzuki, N., Koussevitzky, S., Mittler, R.O.N., and Miller, G.A.D. 2012. ROS and redox signalling in the response of plants to abiotic stress. *Plant Cell Environ.,* 35: 259–270.

Szaboles, I. 1992.Salinization of soils and water and its relation to desertification. *Desertification Control Bulletin*,, 21: 32-37.

Szaboles, I. 1994. Soils and salinization. In "Handbook of plant and crop stress," Ed. M. Pessarakli. Marcel Dekker, New York. pp. 3-11.

Taiz, L. and Zeiger, E. 2006. Stress physiology. In, "*Plant Physiology*," Eds. L. Taiz, and E. Zeiger, Sunderland: Sinauer Associates, Inc. pp. 671–681

Takahashi, S., Seki, M., Ishida, J., Satou, M., Sakurai, T., Narusaka, M., Kamiya, A., Nakajima, M., Enju, A., Akiyama, K., Yamaguchi-Shinozaki, K., Shinozaki, K. 2004. Monitoring the expression profiles of genes induced by hyperosmotic, high salinity, and oxidative stress and abscisic acid treatment in *Arabidopsis* cell culture using a full-length cDNA microarray. *Plant Mol Biol.,* 56(1): 29-55.

Takemura, T., Hanagata, N., Dubinsky, Z. and Karube, I. 2002. Molecular characterization and response to salt stress of mRNAs encoding cytosolic Cu/Zn superoxide dismutase and catalase from *Bruguiera gymnorrhiza*, *Trees*, 16(2-3): 94–99.

Takeoka, Y, Mamun, A.A., Wada, T. and Kaufman, P.B. 1992. Reproductive adaptation of rice to environmental stress. Tokyo, Japan: Japan Scientific Societies Press/ Elsevier. Pp. 237.

Tang, A.C. and Boyer, J.S. 2002. Growth-induced water potentials and the growth of maize leaves. *J Exp Bot.,* 53(368): 489-503.

Tao, F., Xiong, W., Xu, Y. and Lin, E., 2000. Simulation of peanut potential yield under future climate scenario in China. *Chin. Environ. Sci.*, 20: 392–395.

Tao, F., Yokozawa, M., Hayashi, Y. and Lin, E., 2003. Changes in agricultural water demands and soil moisture in China over the last half-century and their effects on agricultural production. *Agric. For. Meteorol.*, 118: 251–261.

Tao, F., Yokozawa, M., Zhang, Z., Hayashi, Y., Grassl, H. and Fu, C. 2004. Climatological and agricultural production variability in China in association with East Asia Monsoon and El Nin˜o Southern Oscillation. *Clim. Res.* ,28: 23–30.

Tao, F., Yokozawa, M., Xu, Y., Hayashi, Y. and Zhang, Z. 2006. Climate changes and trends in phenology and yields of field crops in China, 1981–2000. *Agricultural and Forest Meteorology,* 138: 82–92.

Tauber, E., Zordan, M., Sandrelli, F., Pegoraro, M., Osterwalder, N., Breda, C., Daga, A., Selmin, A., Monger, K., Benna, C., Rosato, E., Kyriacou, C.P. and Costa, R. 2007.Natural selection favors a newly derived timeless allele in *Drosophila melanogaster*. *Science,* 316: 1895–1898.

Tsuji, H., Tamaki, S. and Komiya, R. 2008. Florigen and the photoperiodic control of flowering in rice. *Rice,* 1:25-35.

Tattersall, E.A., Grimplet, J., Deluc, L., Wheatley, M.D., Vincent, D., Osborne, C., Ergul, A., Lomen, E., Blank, R.R., Schlauch, K.A., Cushman, J.C. and Cramer, G.R. 2007.Transcript abundance profiles reveal larger and more complex responses of grapevine to chilling compared to osmotic and salinity stress. *Funct Integr Genomics,* 7(4): 317-333.

Terres, A.L.1991. Melhoramento de arroz irrigado para tolerância ao frio no Rio Grande do Sul Brasil. Pages 91–103 *in* Reunión Sobre Mejoramiento De Arroz En El Cono Sur, Goiânia. Trabajos: Montevideo, IICA Procisur

Thakur, P., Kumar, S., Malik, J.A., Berger, J.D. and Nayyar, H. 2010. Cold stress effects on reproductive development in grain crops, an overview. *Environ. Exp. Bot.,* 67: 429–443.

Thingnaes, E., Torre, S., Ernsten, A. and Moe, R. 2003. Day and night temperature responses in *Arabidopsis*: effects of gibberellin and auxin content, cell size, morphology and flowering time. *Annals of Botany,* 92*:* 601-612.

Tian, J., Belanger, F.C. and Huang, B. 2009. Identification of heat stress-responsive genes in heat-adapted thermal *Agrostis scabra* by suppression subtractive hybridization. *J. Plant Physiol.* 166: 588–601.

Thompson, L.M., 1986. Climatic change, weather variability and corn production. *Agron. J.*, 78: 649–653.

Tilman, D., Cassman, K.G., Matson, P.A., Naylor, R. and S. Polasky, S. 2002. Agricultural sustainability and intensive production practices. *Nature*, 418: 671-677.

Todorov, D., Karanov, E., Smith, A.R. and Hall, M.A. 2003. Chlorophyllase activity and chlorophyll content in wild and mutant plants of *Arabidopsis thaliana. Biol. Plant., 46:* 125–127.

Toriyama, K. and Hinata, K. 1984. Anther respiratory activity and chilling resistance in rice. *Plant Cell Physiology* , 25: 1215–1217.

Tubiello, F.N., J.F. Soussana and S.M. Howden, 2007. Crop and pasture response to climate change. *Proc. Natl. Acad. Sci. USA.*, 104: 19686-19690.

Twilley, R. Barron, R.E.J., Gholz, H.L. Harwell, M.A., Miller, R.L., Reed, D.J., Rose, J.B., Siemann,, E.H., *Wetzel, R.G. and Zimmerman, R.J.* 2001. *Confronting climate change in the Gulf Coast region.* Prospects for sustaining ecological heritage. Union of Concerned Scientists, Cambridge, *Massachusetts, and Ecological Society of America, Washington, D.C., USA.*

Ulman, P., Catsky, J. and Pospisilova, J. 2000 Photosynthetic traits in wheat grown under decreased and increased CO_2 concentration, and after transfer to natural CO_2 concentration. *Biologia Plantarum*, 43: 227-237.

Umezawa, T., Sugiyama, N., Mizoguchi, M., Hayashi, S., Myouga, F., Yamaguchi-Shinozaki, K. Shinozaki, K., Ishihama, Y., Hirayama, T. and Shinozaki, K. 2009. Type 2C protein phosphatases directly regulate abscisic acid-activated protein kinases in *Arabidopsis. Proc. Natl. Acad. Sci. USA* 106*:* 7588–17593.

Vaksmann, M., Traore, S.B., Kouressy, M., Coulibaly. H. and Reyniers, F.-N. 1998. Etude du developpement d'un sorgho photoperiodique du Mali. *In:* Bacci L, Reyniers F-N, *editors.* Le futur des cereales photoperiodiques pour une production durable en Afrique tropicale semi-aride. Florence*:* Ce.S.I.A. and CIRAD*; pp.* 109-122.

Vandeleur, R.K., Mayo, G., Shelden, M.C., Gilliham, M., Kaiser, B.N. and Tyerman, S.D. 2008. The role of plasma membrane intrinsic aquaporins in water transport through roots: diurnal and drought stress responses reveal different strategies between isohydric and anisohydric cultivars of 3 grapevine. *Plant Physiol.,149(1): 445-460.*

Van Oosterom, E.J., Hammer, G. and Chapman, S.C. 2004. *Can transition to flowering be modeled* dynamically from the gene level. 4th International Crop Science Congress. Brisbane, Australia, *pp. 7.*

Vara Prasad, P.V., Craufurd, P.Q., Summerfield, R.J., and Wheeler, T.R., 2000. Effect of short epidosed of heat stress on flower production and fruit set of groundnut (*Arachis hypogea* L.). *J. Exp. Bot.*, 51: 777–784.

Vergara, B.S. and Chang, T.T. 1976. The flowering response of the rice plant to photoperiod: a review of the literature. Los Banos*:* IRRI.

Vincent, D., Ergul, A., Bohlman, M.C., Tattersall, E.A., Tillett, R.L., Wheatley, M.D., Woolsey, R., Quilici, D.R., Joets, J., Schlauch, K., Schooley, D.A., Cushman, J.C. and Cramer, G.R. 2007. Proteomic analysis reveals differences between *Vitis vinifera* L. cv. Chardonnay and cv. Cabernet Sauvignon and their responses to water deficit and salinity *J. Expt. Bot.,* 58(7): 1873-1892.

Visser, M.E. and Both, C., Shifts in phenology due to global climate change: the need for a yardstick. *Proc R Soc. B*, 272: 2561–2569.

Vollenweider, P. and Günthardt-Goerg, M.S. 2005 Diagnosis of abiotic and biotic stress factors using the visible symptoms in foliage. *Environ. Pollut.,* 137: 455–465.

Wahid, A. and Shabbir, A. 2005. Induction of heat stress tolerance in barley seedlings by pre-sowing seed treatment with glycinebetaine. *Plant Growth Regul.,* 46: 133–141.

Wahid, A., Gelani, S.., Ashraf, M. and Foolad. M.R. 2007. Heat tolerance in plants: An overview. *Environmental and Experimental Botany*, 61: 199–223.

Wallace, D.H. and Yan, W. 1998. Plant breeding and whole-system physiology. Improving adaptation, maturity and yield. Wallingford, UK: CABI.

Wang, Y., Guo, Q. and Jin, M. 2009. Effects of light intensity on growth and photosynthetic characteristics of *Chrysanthemum morifolium. Zhongguo Zhongyao Zazhi.* 34:1633-1635.

Wanjogu, S.N., Muya, E.M., Gicheru, P.T. and Waruru, B.K. 2001. Soil degradation: Management and rehabilitation in Kenya. In "Proceedings of the FAO/ISCW expert consultation on Management of Degraded Soil in Southern and Eastern Africa (MADS-SEA) 2nd Networking meeting, Pretoria, South Africa, PR102-113.

Wardlaw, I. F. and Moncur, L. 1995. The response of wheat to high temperature following anthesis. I The rate and duration of kernel filling. *Aust. J. Plant Physiol.*, 22: 391–397.,

Wassmann, R., Jagadish, S., Sumfleth, K., Pathak, H., Howell, G., Ismail, A., Serraj, R., Rredona, E., Singh, R.K. and Heuer, S. 2009. Regional vulnerability of climate change impacts on Asian rice production and scope for adaptation. *Adv. Agron.*, 102: 91–133.

Watts Padwick, G. 1979. Growth phases in plants and their bearing on agronomy. *Exp. Agric*, 15:15-26. Webb, A.A.R., and Satake, A. 2015. Understanding circadian regulation of carbohydrate metabolism in *Arabidopsis* using mathematical models. *Plant Cell Physiol.,* 56: 586–593.

Weir, A.H., Bragg, P.L., Porter, J.R., Rayner, J.H. 1984. A winter wheat crop simulation model without water or nutrient limitations. *Journal of Agricultural Science,* 102: 371-382.

Weis, E. and Berry, J.A. 1988. Plants and high temperature stress. *Symp. Soc. Exp. Biol., 42:* 329–346.

Welch, S., Roe, J., Dong, Z. 2003. A genetic neural network of flowering time control in *Arabidopsis thaliana. Agronomy Journal,* 95: 71-81.

Welch, S., Dong, Z., Roe, J. 2004. *Modeling gene networks controlling transition to flowering in Arabidopsis. 4th International Crop Science Congress.* Brisbane: Australia; pp. 20.

Welch, J.R., J.R. Vincent, M. Auffhammer, P.F. Moya, A. Dobermann and D. Dawe, 2010. Rice yields in tropical/subtropical Asia exhibit large but opposing sensitivities to minimum and maximum temperatures. *Proc. Natl. Acad. Sci. USA.*, 107: 14562-14567.

Wheeler, T.R., Craufurd, P.Q., Ellis, R.H., Porter, J.R., Vara Prasad, P.V. 2000. Temperature variability and the yield of annual crops. *Agric. Ecosys. Environ.,* 82: 159-167.

White, M.A., Running, S.W., Thornton, P.E. 1999. The impact of growing-season length variability on carbon assimilation and evapotranspiration over 88 years in the eastern US deciduous forest. *International Journal of Biometeorology,* 42: 139-145.

Wien, H.C. 1990. Screening pepper cultivars for resistance to flower abscission: A comparison of techniques. *Hort. Science,* 25: 1634-1636.

Williams, T. and Abberton, M. 2004. Earlier flowering between 1962 and 2002 in agricultural varieties of white clover. *Oecologia,* 138: 122-126.

Wolfe, D.W., Schwartz, M.D., Lakso, A.N., Otsuki, Y., Pool, R.M. and Shaulis, N.J., 2005. Climate change and shifts in spring phenology of three horticultural woody perennials in northeastern USA. *Int J Biometeorol.*, 49: 303–309.

Yamaguchi-Shinozaki, K. and Shinozaki, K. 2006. Transcriptional regulatory networks in cellular responses and tolerance to dehydration and cold stresses. *Annu. Rev. Plant Biol.*, 57: 781–803.

Yamazaki, T., Kawamura, Y., and Uemura, M. 2008. Cryobehavior of the plasma membrane in protoplasts isolated from cold-acclimated *Arabidopsis* leaves is related to surface area regulation. *Plant Cell Physiol.,* 49: 944–957.

Yamazaki, T., Kawamura, Y., and Uemura, M. 2009. Extracellular freezing-induced mechanical stress and surface area regulation on the plasma membrane in cold-acclimated plant cells. *Plant Signal. Behav.,*4: 231–233.

Yan, W.K., Wallace, D.H. 1998. Simulation and prediction of plant phenology for five crops based on photoperiod x temperature interaction. *Annals of Botany,* 81: 705-716.

Yancey, P.H. 2005. Organic osmolytes as compatible, metabolic and counteracting cytoprotectants in high osmolarity and other stresses. *J. Exp. Biol.*, 208: 2819–2830.

Yang, X.Y., Ye, X.F., Liu, G.S., Wei, H.Q. and Wang, Y. 2007. Effects of light intensity on morphological and physiological characteristics of tobacco seedlings. *Chin. J. Appl. Ecol.* 18: 2642-2645.

Yeo, A.R. 1998. Molecular biology of salt tolerance in the context of whole plant physiology. *J. Exp. Bot.,* 49: 915-929.

Yin, X., Kropff, M.J., Nakagawa, H., Horie, T. and Goudriaan, J. 1997a. A model for photothermal responses of flowering in rice, II. Model evaluation. *Field Crops Research,* 51: 201-211.

Yin, X., Kropff, M. and Ynalvez, M.A. 1997b. Photoperiodically sensitive and insensitive phases of preflowering development in rice. *Crop Science*, 37: 182-190.

Young, M.W., and Kay, S.A. 2001. Time Zones: a comparative genetics of circadian clocks. *Nat. Rev. Gen.* 2: 702–715.

Young, L.W., Wilen, R.W. and Bonham-Smith, P.C. 2004. High temperature stress of *Brassica napus*during flowering reduces micro-and megagametophyte fertility, induces fruit abortion, and disrupts seed production. *J. Exp. Bot.,* 55: 485–495.

Yoshida, S., 1981. Fundamentals of Rice Crop Science. International Rice Research Institute, Philippines, ISBN-13: 9789711040529, Pages: 269.

Yuan, J.S., Galbraith, D.W., Dai, S.Y., Griffin, P. and Stewart, C.N. Jr. 2008. Plant systems biology comes of age. *Trends Plant Sci,* 13: 165–171.

Zahedi, M. and Jenner, C. F. 2003. Analysis of effects in wheat of high temperature on grain filling attributes estimated from mathematical models of grain 1310 filling. *J. Agric. Sci.*, 141: 203–212.

Ziska, L.H., Manalo, P.A. and Ordonez, R.A. 1996. Intraspecific variation in the response of rice (*Oryza sativa* L.) to increased CO_2 and temperature: growth and yield response of 17 cultivars. *Journal of Experimental Botany*, 47:1353-1359.

Zhai, P., Sun, A., Ren, F., Liu, X., Gao, B. and Zhang, Q., 1999. Changes of climate extremes in China. *Clim. Change*, 42: 203–218.

Zhang, S., Ma, K., and Chen, L. 2003. Response of photosynthetic plasticity of *Paeonia suffruticosa* to changed light environments. *Environ. Exper. Bot.* 49:121-133.

Zhang, X., Friedl, M.A., Schaaf, C.B. and Strahler, A.H., 2004. Climate controls on vegetation phenological patterns in northern mid- and high latitudes inferred from MODIS data. *Glob. Change Bio*, 10(7): 1133–1145.

Zheng, J., Ge, Q. and Hao, Z., 2002. Impacts of climate warming on plants phenophases in China for the last 40 years. *Chin. Sci. Bull.*, 47(21): 1826–1831.

Zhou, L., Kaufmann, R.K., Tian, Y., Myneni, R.B. and Tucker, C.J., 2003. Relation between inter-annual variations in satellite measures of northern forest greenness and climate between 1982 and 1999. *J. Geophys. Res.*, 108(D1): 4004.

Zhu, J.K. 2001. Plant salt tolerance. *Trends Plant Sci.*, 6: 66–71.

Zhu, J.K. 2002. Salt and drought stress signal transduction in plants. *Annu. Rev. Plant Biol.*, 53: 247–273.

Zinn, K.E., Tunc-Ozdemir, M. and Harper, J.F. 2010. Temperature stress and plant sexual reproduction: uncovering the weakest links. *J. Exp. Bot.*, 61: 1959–1968.

5

Photosynthesis Under Abiotic Stress Conditions

A. Bhattacharya

Photosynthesis is the process converting light energy to chemical energy and storing the energy in the bonds of sugars and other organic compounds. Carbohydrates are synthesized from water and carbon dioxide and oxygen is released as a by-product. In plants and algae, the process of photosynthesis occurs in specialized organelles known as chloroplasts which contain the photosynthetic apparatus. Photosynthesis mainly occurs in mesophyll cells in plant leaves.

Photosynthesis is one of the major metabolic processes which is sensitive to environmental stresses. It is important to regulate the energy flow to optimise carbon fixation and prevent light induced damage. Photosynthesis consists of two major processes: the light-dependent electron transport chain and the light-independent carbon fixation cycle or the Calvin-Benson-Bassham cycle.

Plants are subjected to several harsh environmental stresses that adversely affect growth, metabolism, and yield. Drought, salinity, low and high temperatures, flood, pollutants, and radiation are the important stress factors limiting the productivity of crops (Lawlor, 2002). Several biotic (insects, bacteria, fungi, and viruses) and abiotic (light, temperature, water availability, nutrients, and soil structure) factors affect the growth in higher plants (as reviewed by Lichtenthaler, 1996, 1998). Although the plant growth is controlled by a multitude of physiological, biochemical, and molecular processes, photosynthesis is a key phenomenon, which contributes substantially to the plant growth and development. The chemical energy expended in a number of metabolic processes is, in fact, derived from the process of photosynthesis, which is capable of converting light energy into a usable chemical form of energy. This key process occurs in all green plants, whether lower or higher, occurring in oceans or on land as well as in photosynthetic bacteria (Taiz and Zeiger 2010, Pan *et al.*,

2012). However, stressful environments, including drought, salinity, and unfavourable temperatures, considerably hamper the process of photosynthesis in most plants by altering the ultrastructure of the organelles and concentration of various pigments and metabolites including enzymes involved in this process as well as stomatal regulation.

Environmental stresses have a direct impact on the photosynthetic apparatus, essentially by disrupting all major components of photosynthesis including the thylakoid electron transport, the carbon reduction cycle and the stomatal control of the CO_2 supply, together with an increased accumulation of carbohydrates, peroxidative destruction of lipids and disturbance of water balance (Allen and Ort, 2001).

The ability of crop plants to acclimate to different environments is directly or indirectly associated with their ability to acclimate at the level of photosynthesis, which in turn affects biochemical and physiological processes and, consequently, the growth and yield of the whole plant (Chandra, 2003). Drought stress severely hampered the gas exchange parameters of crop plants and this could be due to decrease in leaf expansion, impaired photosynthetic machinery, premature leaf senescence, oxidation of chloroplast lipids and changes in structure of pigments and proteins (Menconi *et al.*, 1995). Anjum *et al.* (2011a) indicated that drought stress in maize led to considerable decline in net photosynthesis (33.22%), transpiration rate (37.84%), stomatal conductance (25.54%), water use efficiency (50.87%), intrinsic water use efficiency (11.58%) and intercellular CO_2 (5.86%) as compared to well water control. Many studies have shown the decreased photosynthetic activity under drought stress due to stomatal or non-stomatal mechanisms (Del Blanco *et al.*, 2000; Samarah *et al.*, 2009). Stomata are the entrance of water loss and CO_2 absorbability and stomatal closure is one of the first responses to drought stress which result in declined rate of photosynthesis. Stomatal closure deprives the leaves of CO_2 and photosynthetic carbon assimilation is decreased in favor of photorespiration. Considering the past literature as well as the current information on drought-induced photosynthetic responses, it is evident that stomata close progressively with increased drought stress (Anjum *et al.,* 2011b). It is well known that leaf water status always interacts with stomatal conductance and a good correlation between leaf water potential and stomatal conductance always exists, even under drought stress. It is now clear that there is a drought-induced root-to-leaf signaling, which is promoted by soil drying through the transpiration stream, resulting in stomatal closure. The "non-stomatal" mechanisms include changes in chlorophyll synthesis, functional and structural changes in chloroplasts, and disturbances in processes of accumulation, transport, and distribution of assimilates.

5.1. Chlorophyll

Chlorophyll is one of the major chloroplast components for photosynthesis, and relative chlorophyll content has a positive relationship with photosynthetic rate. The decrease in chlorophyll content under drought stress has been considered a typical symptom of oxidative stress and may be the result of pigment photo-oxidation and chlorophyll degradation. Photosynthetic pigments are important to plants mainly for harvesting light and production of reducing powers. Both the chlorophyll a and b are prone to soil dehydration (Farooq *et al.*, 2009). Decreased or unchanged chlorophyll level during drought stress has been reported in many species, depending on the duration and severity of drought (Kpyoarissis *et al.*, 1995; Zhang and Kirkham, 1996). Drought stress caused a large decline in the chlorophyll a content, the chlorophyll b content, and the total chlorophyll content in different sunflower varieties (Manivannan *et al.*, 2007). Exposure of two olive cultivars to reduced irrigation led to lower chlorophyll (a + b) contents. These reductions were 29 and 42% for Chemlali and Chetoui, respectively (Guerfel *et al.*, 2009). Loss of chlorophyll contents under water stress is considered a main cause of inactivation of photosynthesis. Furthermore, water deficit induced reduction in chlorophyll content has been ascribed to loss of chloroplast membranes, excessive swelling, distortion of the lamellae vesiculation, and the appearance of lipid droplets (Kaiser *et al.*, 1981). Low concentrations of photosynthetic pigments can directly limit photosynthetic potential and hence primary production. From a physiological perspective, leaf chlorophyll content is a parameter of significant interest in its own right. Studies by majority of chlorophyll loss in plants in response to water deficit occurs in the mesophyll cells with a lesser amount being lost from the bundle sheath cells.

In the process of photosynthesis (Fig. 1), two key events occur mandatorily; light reactions, in which light energy is converted into ATP and NADPH and oxygen is released, and dark reactions, in which CO_2 is fixed into carbohydrates by utilizing the products of light reactions, ATP and NADPH (Lawlor, 2001, Taiz and Zeiger, 2010, Dulai *et al.,* 2011). There are two main pathways of CO_2 fixation, C3 and C4. Plants have been categorized into C3, C4, or C3-C4 intermediate plants depending on the spatial distribution of these two pathways within leaf tissues or as crassulacean acid metabolism plants with a temporal distribution (Doubnerová and Ryšlavá, 2011, Freschi and Mercier, 2012). Plants possessing the different types of photosynthetic mechanisms are adapted to specific climatic zones. For example, C3 plants, representing over 95% of the Earth plant species, thrive well in cool and wet climates, with usually low light intensity. In contrast, C4 plants occur in hot and dry climatic conditions with

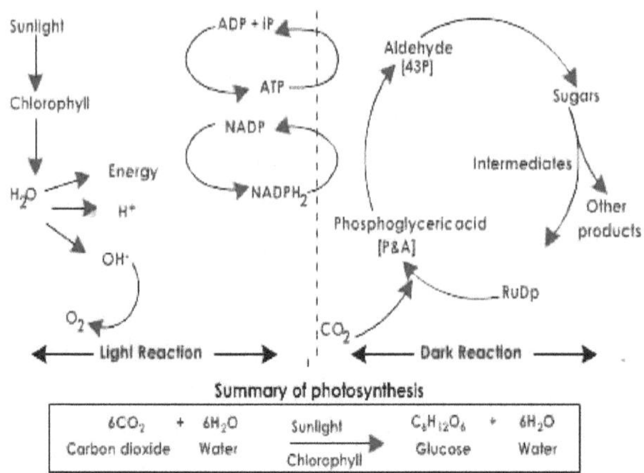

Fig. 1. Light and dark reaction of photosynthesis

usually high light intensity. Generally, C4 and crassulacean acid metabolism plants are the best adapted to arid environments, because they have higher water-use efficiency than that of C3 plants. C4 plants have higher photosynthetic efficiency than C3 plants, namely in arid, hot, and under high-light conditions, because they possess an additional carbon fixation pathway and characteristic anatomy to limit photorespiration. Furthermore, crassulacean acid metabolism plants can effectively save metabolic energy and water during harsh environmental conditions by closing their stomata during the day (Taiz and Zeiger 2010). Different growth and development related processes depend on the interplay of intracellular organelles. The chloroplast is the key site for photosynthesis, in which both light and dark reactions of photosynthesis take place. However, this organelle is highly sensitive to different stressful environments such as salinity, drought, extremes of temperature, flooding, varying light intensity, and UV radiation, and it plays a premier role in the modulation of stress reesponses (Biswal *et al.*, 2008, Saravanavel *et al.*, 2011). All these stresses reduce the photosynthetic rate by stress-induced stomatal or nonstomatal limitations (Saibo *et al.,* 2009, Rahnama *et al.,* 2010). For example, drought stress, particularly at its mild intensity, can inhibit leaf photosynthesis and stomatal conductance in most green plants (Medrano *et al.,* 2002). A number of reports showed that stomata usually close during the initial stages of drought stress resulting in increased water use efficiency (net CO_2 assimilation rate/ transpiration). Stomata closure is known to have a more inhibitory effect on transpiration of water than that on CO_2 diffusion into the leaf tissues (Chaves *et al.,* 2009, Sikuku *et al.*, 2010). However, in contrast, under severe drought stress, dehydration of mesophyll cells takes place causing a marked inhibition of basic metabolic processes of photosynthesis as well as a reduction of plant water use efficiency (Damayanthi *et al.,* 2010, Anjum *et al.*, 2011). Drought stress also reduces the efficiency of mesophyll cells to utilize the available CO_2 (Karaba *et al.*, 2007, Dias and Bruggemann, 2010a, b).

The regulation of leaf stomatal conductance (*gs*) is a key phenomenon in plants as it is vital for both a prevention of desiccation and CO_2 acquisition (Dodd 2003, Medici *et al.*, 2007). Stomata closure in response to drought and salinity stress generally occurs due to decreased leaf turgor and atmospheric vapour pressure along with root-generated chemical signals (Chaves *et al.*, 2009). Thus, the decrease in photosynthetic rate under stressful conditions (salinity, drought, and temperature) is normally attributed to a suppression in the mesophyll conductance and the stomata closure at moderate and severe stress (Flexas *et al.*, 2004, Chaves *et al.*, 2009). The effects of salinity and drought on photosynthesis are attributed directly to the stomatal limitations for diffusion of gases, which ultimately alters photosynthesis and the mesophyll metabolism (Parida *et al.*, 2005, Chaves *et al.*, 2009). However, of various physiological processes, the accumulation of a plant hormone, abscisic acid (ABA), shows a vitally important role in the plant growth and metabolism under stress conditions. ABA is usually known as a stress hormone due to its high accumulation under stress environments (Kempa *et al.*, 2008, Melcher *et al.,* 2009). One of the immediate responses to water stress is stomata closure, which is caused mainly due to the action of ABA. High ABA level has been reported to cause an increase in cytosolic Ca^{2+} and activation of plasma membrane-localized anion channels (Hamilton *et al.,* 2000, Kohler and Blatt 2002). This, in turn, causes potassium efflux, guard cell depolarization, loss of guard cell volume and turgor, high H_2O_2 production, and finally the stomata closure (Zhang *et al.*, 2006, Wang *et al.*, 2012).

The foliar photosynthetic rate, A, of higher plants is known to decrease as the relative water content (RWC) and leaf water potential decrease (Lawlor and Cornic, 2002). However, the debate continues as to whether drought mainly limits photosynthesis through stomatal closure or through metabolic impairment (Lawson *et al.*, 2003). Stomatal limitation was generally accepted to be the main determinant of reduced photosynthesis under drought stress (Cornic, 2005). This has been attributed to decreases in both A and internal CO_2 concentration, which finally inhibits total photosynthetic metabolism. Several nonstomatal effects are also attributed for stomatal closure during drought. These include photophosphorylation (Meyer and Genty, 1999), ribulose-1,5-bisphosphate (RuBP) regeneration (Lawlor, 2002), rubisco activity (Medrano *et al.*, 1997) and ATP synthesis (Tezara *et al.*, 1999). Considering the past literature as well as the current information on drought-induced photosynthetic responses, it is evident that stomata close progressively with increased drought stress, followed by reduced net photosynthetic rates. It is well known that leaf water status always interacts with stomatal conductance and a good correlation between leaf water potential and stomatal conductance always exists, even under drought stress. It is now clear that there is a drought-induced root-to-leaf signaling, which is promoted by soil drying through the transpiration stream, resulting in stomatal closure. This chemical signal is now known to be abscisic acid (ABA) and a direct correlation between the xylem ABA content and stomatal conductance has been demonstrated (Socias *et al.*, 1997). The closure of stomata under drought has also been implicated due to changes in plant nutritional status, xylem sap pH, farnesyl tranferase activity, xylem hydrolic conductivity, and leaf-to-air vapor pressure deficit (Oren *et al.*, 1999). However, differences in the complex regulation of stomatal conductance among several species and genotypes in response to leaf water potential and ABA makes it difficult to see a clear pattern in photosynthetic responses to drought. Decrease in RWC has been known to induce stomatal closure and thus a parallel decrease. Nevertheless, it has been emphasized that a high degree of co-regulation exists between stomatal opening and photosynthesis (Farquhar *et al.*, 2001; Hubbard *et al.*, 2001). Thus, stomatal movements are very dynamic, involving complex regulation by several environmental factors, and stomatal conductance should be taken as an integrative parameter to assess photosynthetic responses under drought.

5.2. Response to Drought Stress

The responses of plants to drought stress are highly complex, involving deleterious and/or adaptive changes. In particular, if the drought stress is under field conditions, the plant responses can be modified synergistically or antagonistically. Early responses of plants to drought stress usually help the plant to survive for some time, while the acclimation of the plant subjected to drought is indicated by the accumulation of certain new metabolites associated with the structural capabilities to improve plant functioning under drought stress (Pinhero *et al.*, 2001). Acclimation-related changes in the root/shoot ratio or the temporary changes in accumulation of reserves in the stem under drought stress are accompanied by the changes in carbon and nitrogen metabolism (Noctor *et al.*, 2002). A continuous oxidative assault on plants during drought stress has led to the presence of an arsenal enzymatic (discussed above) and nonenzymatic antioxidant defenses to counter the phenomenon of oxidative stress in plants. The non-enzymatic plant antioxidants can be classified into two major types:

1. AA-like scavengers, and
2. Pigments such as carotenoids (Conklin, 2001).

AA is an important antioxidant, which reacts not only with H_2O_2 but also with O_2^-, OH and lipid hydroperoxidases. Protective functions of AA are also assayed in a wide variety of common human ailments and diseases. In recent years, clear evidence has emerged that elevated dietary intake of AA lowers the incidence of cancer, cardiovascular disease, and other oxidative stress disorders (Grassmann *et al.*, 2002; Munne-Bosch and Algere, 2003). On the other hand, AA has been implicated in several types of biological activities in plants:

- As an enzyme co-factor,
- As an antioxidant, and
- As a donor/ acceptor in electron transport at the plasma membrane or in the chloroplasts, all of which are related to oxidative stress resistance (Conklin, 2001).

In chloroplasts, the so-called "Halliwell-Asada" pathway indicates that APX uses AA and oxidizes it to monodehydroascorbate (MDA). MDA may give rise to dehydroascorbate (DHA). Both MDA and DHA will then be reduced to regenerate the ascorbate pool. This type of scavenging is thought to occur near PSI, thereby minimizing the risk of escape and reaction of ROS with each other (Foyer and Noctor, 2000). AA is water-soluble and also has an additional role in protecting or regenerating oxidized carotenoids or tocopherols (Imai *et al.*, 1999). AA is a major metabolite in chloroplasts of higher plants and represents about 10% of the soluble carbohydrate pool in leaves (Noctor and Foyer, 1998). AA is also known to function as the "terminal antioxidant" because the redox potential of AA/MDA pair (þ280 mv) is lower than that of most of the bioradicals (Scandalios *et al.*, 1997). The levels of glutahione reductase and DHA reductase increased upon water stress in Sporobolus stapfianus (Conklin, 2001). An increase in MVdependent photoproduction of MDHA was also observed in lettuce leaves due to water stress (Conklin, 2001). However, very little is known about the regulation of AA biosynthesis in higher plants. Nevertheless, AA concentrations are reported to increase in response to wounding (Robinson and Bunce, 2000). It is conceivable that adaptations of intracellular AA to drought stress might clearly depend on the balance between the rates and capacity of AA biosynthesis and turnover related to antioxidant demand (Chaves *et al.*, 2002). The biosynthesis of AA from hexose phosphate and its involvement in protection against photooxidative stress suggest that there may be links between photosynthesis and the AA pool size.

Although the antioxidant defense system is impaired under stressful conditions, plants are able to get rid of excessive energy by thermal dissipation associated with an increase in the concentration of xanthophyll pigments, zeaxanthin, and antheraxanthin, at the expense of violoxanthin in water stressed plants (Alonso *et al.*, 2001). The high proportion of xanthophylls under stress conditions may serve as a protection mechanism in leaves. The activation of the xanthophyll cycle has also been detected in several other plants subjected to drought stress (Foyer, 2001). Mechanisms are not known of a precise regulation to prevent potentially destructive side reactions involving AOS that damage photosynthetic pigments and proteins. In addition, the xanthophyll cycle may be involved in lowering the yield of triplet chlorophyll formation by pre-emptive quenching of the excited singlet state of chlorophyll (Rontein *et al.*, 2002). It is thus presumed that the role of antioxidants and carotenoid pigments in regulating photosynthetic electron transport is crucial. Thus, a complex relationship between xanthophyll cycle-dependent energy quenching and formation of AOS exists in photosynthetic systems of plants under drought stress. a-tocopherol, found in green parts of plants also scavenges lipid peroxy radicals through the concerted action of other antioxidants (Munne-Bosch and Alegre, 2002). Further, tocopherols were also known to protect lipids and other membrane components by physically quenching and chemically reacting with $^1O_2$ in chloroplasts, thus protecting the structure and function of PSII (Trebst *et al.*, 2002).

Many plants and other organisms cope with osmotic stress by synthesizing and accumulating some compatible solutes, which are termed as osmoprotectants or osmolytes. These compounds are small, electrically neutral molecules, which are non-toxic even at molar concentrations (Alonso *et al.*, 2001). During osmotic stress, plant cells accumulate solutes to prevent water loss and to reestablish cell turgor. The solutes that accumulate during the osmotic adjustment include ions such as K^+, Na^+, and Cl^- or organic solutes that include nitrogen-containing compounds, such as proline and other amino acids, polyamines and quaternary ammonium compounds like glycine betaine (GlyBet) (Tamura *et al.*, 2003). Other osmolytes, that are produced in response to stress, include sucrose, polyols, sugar alchohols (pinitol), and oligosacharides. The organic solutes are compatible with cellular processes and accumulate to high levels in the cytosol with increasing drought. Production of osmolytes is a general way to stabilize membranes and maintain protein conformation at low leaf water potentials. The synthesis and accumulation of osmolytes varies among plant species as well as among different cultivars of the same species. Osmolytes play a major role in osmotic adjustment and also protect the cells by scavenging ROS (Pinhero *et al.*, 2001). Proline is also known to be involved in reducing the photo damage in the thylakoid membranes by scavenging and/or reducing the production of $^1O_2$. Proline accumulation in plants is caused, not only by the activation of proline biosynthesis, but also by the inactivation of proline degradation, thereby resulting in a decrease in the level of accumulated proline in rehydrated plants. Proline degradation to glutamic acid via D1 pyrroline-5-carboxylate in higher plants is catalyzed by proline dehydrogenase and D1 pyrroline-5-carboxylate dehydrogenase in mitochondria (Bohnert and Jenson, 1996). It can also be inferred that proline acts as a free radical scavenger and may be more important in overcoming stress than in acting as a simple osmolyte. Such studies open a new avenue of research for metabolic engineering in several agriculturally important crop plants for drought resistance. Accumulation of other amino acids like glycine, serine, and glutamate are known to regulate and integrate the metabolism in stressed photosynthetic tissues (Lawlor and Cornic, 2002). Proline concentrations increase many fold with reduced leaf water potentials and at this stage photosynthesis is known to be quite reduced (Morot-Guadry *et al.*, 2001). A more common explanation for the accumulation of proline is that it confers advantages by protecting membranes and proteins when RWC decreases. In plant cells, osmolytes are typically confined to the chloroplasts and cytoplasmic compartments that together occupy 20% or less of the volume of mature cells (Ain-Lhout *et al.*, 2001). Natural osmolyte concentrations in plant cells can reach 200mM or more and such concentrations are osmotically significant and have pivotal roles in maintaining cell turgor and driving the gradient for water uptake under stress (Rhodes and Samaras, 1994).

Accumulation of GlyBet occurs in some, but not in all, higher plants. Most of the GlyBet is synthesized in chloroplasts as two enzymes, namely choline monoxygenase (CMO) and betaine aldehyde dehydrogenase, are responsible for GlyBet synthesis chloroplastically. GlyBet synthesis can be induced by both drought and salt stress (NaCl, KCl, $MgCl_2$, Na_2SO_4) by over-expression of CMO, and betaine aldehyde dehydrogenase (Nakamura *et al.*, 2001). Progressive drought and salinity were also known to induce late embryogenesis abundant (LEA) proteins in vegetative organs, which can stabilize enzyme complexes and the structure of cell membranes (Chourey *et al.*, 2003; Liang *et al.*, 2003). The increased concentrations of GlyBet under drought stress situations clearly suggest that this osmolyte has an important role in protecting plant cell mechanisms under conditions of drought. A physiological role of GlyBet in alleviating osmotic stress was proposed based on accumulation of GlyBet in plants subjected to drought (Jun *et al.*, 2000). GlyBet has been shown to protect enzymes and membranes and also to stabilize PSII protein pigment complexes under stressful conditions (Papageorgiou and Morata, 1995). A moderate stress tolerance, as shown by dry weight production in transgenic plants, was noticed based on relative shoot growth studies under stress conditions, like drought (Jun *et al.*, 2000).

The limitation of photosynthesis under drought through metabolic impairment is a more complex phenomenon than stomatal limitation. Changes in the cellular carbon metabolism are likely to occur early in the dehydration processes. Drought generally reduces the biochemical capacity for carbon assimilation and utilization. The rate of photosynthesis in higher plants depends on the activity of ribulose-1, 5-bisphosphate carboxylase/oxygenase (rubisco) as well as synthesis of RuBP (Chaitanya *et al.*, 2002a; Parry *et al.*, 2002). Despite several studies on photosynthetic carbon assimilation under drought, a definitive conclusion regarding the most sensitive changes in rubisco metabolism remains elusive. Rubisco quantity limits photosynthesis in most cases, although in some cases the regeneration of RuBP has been shown to limit the photosynthetic capacity (Vu *et al.*, 1999). Studies on transgenic plants suggest that rubisco may not be the main limitation in chloroplast metabolism (Tezara *et al.*, 1999). However, most of the evidence on drought-induced changes indicates that the amount and activity of rubisco really control photosynthetic carbon assimilation. The quantity of rubisco in leaves is controlled by the rate of synthesis and degradation of the enzyme, even in stressful environments. Decreased synthesis of rubisco under drought was evidenced by a rapid decrease in the abundance of rubisco small subunit (rbcS) in tomato (Vu *et al.*, 1999). Parry *et al.* (2002) suggested that rubisco activity is regulated to match the capacity of the leaf to regenerate RuBP. Loss of rubisco activity has been reported in several plants under drought (Parry *et al.*, 2002; Chaitanya *et al.*, 2003). In tobacco, initial and total activities of rubisco were decreased under drought and this decrease was due to the reduction in the apparent Kcat; rather than the changes in the activation state of the enzyme (Parry *et al.*, 2002). These studies indicate that irreversible damage to rubisco should be taking place in the leaves, or there is a reduction in the total soluble protein. Inhibition of rubisco activity was due to binding of inhibitors like CA1P with rubisco. Total activity of rubisco has been used as an indicator to estimate the percentage of catalytic sites that are blocked by inhibitors. The decrease in the initial and total activities of rubisco with an increase in drought stress was related to an increase in CA1P (Parry *et al.*, 2002). However, the interactions of rubisco with tight binding inhibitors have been known to be advantageous in vivo, as they could prevent rubisco from degradation by proteases. It is thus believed that although drought stress reduces rubisco activity, the carboxylating enzyme is otherwise protected from protease degradation. It is quite interesting to know that the release of tight binding receptors requires the participation of rubisco activase and ATP hydrolysis. Rubisco activase is an abundant protein, which regulates the active site conformation of rubisco, removes the tight-bound inhibitors, and then allows rubisco to undergo rapid carboxylation. This suggests that there is now evidence that rubisco activase activity also decreases with increasing drought stress

(Chaves *et al.*, 2002). Under these conditions, the removal of inhibitors from rubisco binding sites by rubisco activase is now known to be impaired due to reduced ATP concentrations under drought stress conditions (Tezara *et al.*, 1999). The rate of photosynthesis also depends on the synthesis of RuBP. The response of photosynthesis to RuBP content was linear, indicating that the supply of RuBP determines the net photosynthetic activity. However, limited data are available on the actual role of RuBP regeneration in relation to the rubisco activity. Tezara *et al.* (1999) reported that limitations in RuBP regeneration could result from an inadequate supply of ATP and/or NADPH to the PCR cycle or from a decreased rate of turn over caused by low enzyme activity either in stressed or non-stressed plants. The reactions of RuBP to 3-PGA always decreased with reducing RWC, which suggest that regeneration of RuBP was inhibited under drought. RuBP contents and synthesis are presumed to be under the control of the PCR cycle or the supply of ATP and NADPH to the PCR cycle. ATP synthesis was known to limit A at low RWC because of the inhibition of photophosphorylation in water stressed sunflower leaves (Lawlor, 2002). Under conditions of drought, reduction in chloroplast volume might also lead to desiccation within the chloroplast, which in turn leads to conformational changes in rubisco. Drought stress conditions are also known to acidify the chloroplast stroma, resulting in inhibited rubisco activity (Meyer and Genty, 1999). However, C4 plants are known to use water more efficiently than C3 plants, but they also need rubisco to achieve a given rate of photosynthesis. The primary carboxylating enzyme in C4 plants, phosphoenol pyruvate carboxylase (PEPcase), is also inhibited under drought (Boyer *et al.*, 1997).

It is also known that the limitation of PCR cycles *per se* under drought stress conditions is due to partial reduction in the activities of the other enzymes involved in the cycle. The activities of phosphoribulokinase and fructose-1,6-bisphosphatase (FBPase) were also reduced with decreases in RWC (Haupt-Herting and Fock, 2002). Drought also results in changes in the ratio of the end products of photosynthesis, starch, and sucrose. With reduced water potentials, the activities of FBPase and sucrose phosphate synthase (SPS) declined, indicating that the rate of sucrose synthesis is strongly influenced by drought stress (Haupt-Herting and Fock, 2002). The inhibited activities of FBPase and SPS might regulate the synthesis of sucrose and starch and as well as their partitioning under drought stress. The altered starch/sucrose ratios under drought stress should certainly cause changes in the Pi flux across the chloroplast membrane. Reduced levels of Pi in the chloroplasts inhibit ATP synthesis, which in turn has a greater impact on photophosphorylation and PCR cycle (Tezara *et al.*, 1999). Sometimes the combined effects of heat and light stress on photosynthesis superimposed with drought will be more complex (Wingler *et al.*, 1999). Under these unfavorable conditions, plants are known to lose chlorophyll (Havaux and Tardy, 1999) or divert the absorbed light to other processes, like thermal dissipation to protect the photosynthetic apparatus (Deming-Adams and Adams, 1996). Photochemical efficiency under drought can also be monitored by measuring chlorophyll fluorescence to assess the photosynthetic functioning (Osmond *et al.*, 1999), while multispectral fluorescence images are also currently being used to study plant responses to drought stress (Lichtenthaler, 1997).

5.3. CO_2 Fixation Under Drought

Both higher temperatures and periods with limited water supply probably alter the ability of terrestrial ecosystems to act as carbon sinks (Schimel, 1995; IPCC, 2007). During drought, the net primary productivity of crops decreased because of a large reduction in photosynthesis and a relatively smaller decrease in plant and soil microbial CO_2 efflux (Ciais *et al.*, 2005). During a drought period, plants face the dilemma 'to lose water to gain carbon' (Chaves *et al.*, 2003). Under water stress conditions, increased stomatal closure and reduced mesophyll conductance are the main causes of decreased photosynthesis rates, by reducing CO_2 diffusion from the atmosphere to the carboxylation sites (Chaves *et al.*, 2003; Flexas *et al.*, 2006). Whereas drought-induced regulation of assimilation is relatively well known, there is not much information available on carbon export from the leaf and on its transport within the plant under drought conditions (Ruehr *et al.*, 2009).

Some studies, focusing on the transport of recent assimilates under water stress within plants from leaves to sink organs (*e.g.* Deng *et al.*, 1990; Li *et al.*, 2003), found the translocation of newly assimilated carbon from source leaves to be delayed under severe water stress. While Wardlaw (1969) pointed out that the effects of water stress on carbon translocation may result indirectly from effects on growth, other studies (Deng *et al.*, 1990) have proposed that drought may directly affect the translocation pathway, for example by impaired phloem loading or by reduced transport velocity in the sieve tubes. However, the interest in ecosystem, carbon dynamics under future climate conditions, and particularly under drought conditions, is rising. In addition, about half of the soil CO_2 efflux is thought to originate from recent (*i.e.* several days old) photosynthates (Hogberg & Read, 2006), thus indicating a close coupling of above-ground and below-ground processes. Unfortunately, there is only little, sometimes contradictory, information about such coupling under drought conditions. In a study on wheat plants, Palta & Gregory (1997) found the allocation of recent carbon to roots and to soil CO_2 efflux to be higher under drought, 2 day after ^{13}C pulse-labelling. By contrast, in European shrub lands

Gorissen *et al.* (2004) measured a significant decrease in the allocation of recent carbon to the soil and to soil microorganisms under drought, 3 day after ^{14}C pulse-labelling. However, neither of these studies accounted for the time course of the allocation of recent carbon from the plants to the soil. The ^{14}C and ^{13}C pulse-labelling studies that have traced the fate of newly assimilated carbon over time in trees and ecosystems (*e.g.* Carbone *et al.*, 2007; HoGberg *et al.*, 2008) have not yet investigated the effects of changes in water availability.

In general it has been documented that drought stress decreases the rate of photosynthesis (*e.g.*, Kawamitsu *et al.*, 2000, Serraj *et al.*, 2004). Plants grown under drought condition have a lower stomatal conductance in order to conserve water. Consequently, CO_2 fixation is reduced and photosynthetic rate decreases, resulting in less assimilate production for growth and yield of plants. Diffusive resistance of the stomata to CO_2 entry probably is the main factor limiting photosynthesis under drought (Boyer, 1970). Certainly under mild or moderate drought stress stomatal closure (causing reducted leaf internal CO_2 concentration (Ci) is the major reason for reduced rates of leaf photosynthetic (Cornic, 2000; Flexas *et al.*, 2004). Severe drought stress also inhibits the photosynthesis of plants by causing changes in chlorophyll content, by affecting cholorophyll components and by damaging the photosynthetic apparatus (Iturbe Ormaetxe *et al.*, 1998). Ommen *et al.* (1999) reported that leaf chlorophyll content decreases as a result of drought stress. Drought stress caused a large decline in the chlorophyll 'a' content, the chlorophyll 'b' content, and the total chlorophyll content in all sunflower varieties investigated (Manivannan *et al.*, 2007). The decrease in chlorophyll under drought stress is mainly the result of damage to chloroplasts caused by active oxygen species (Smirnoff 1995). Plants can partly protect themselves against mild drought stress by accumulating osmolytes. Proline is one of the most common compatible osmolytes in drought stressed plants. For example, the proline content increased under drought stress in pea (Sanchez *et al.*, 1998; Alexieva *et al.*, 2001). Proline accumulation can also be observed with other stresses such as high temperature and under starvation (Sairam *et al.*, 2002). Proline metabolism in plants, however, has mainly been studied in response to osmotic stress (Verbruggen and Hermans 2008). Proline does not interfere with normal biochemical reactions but allows the plants to survive under stress (Stewart, 1981). The accumulation of proline in plant tissues is also a clear marker for environmental stress, particularly in plants under drought stress (Routley, 1966). Proline accumulation may also be part of the stress signal influencing adaptive responses (Maggio *et al.* 2002). The purpose of the present study was to contribute to a better understanding of the physiology responses of chickpea plants to drought stress. Mafakheri *et al.*(2010) have investigated the influence of four types of drought stress on the chlorophyll (a, b, a/b) content, proline content, photosynthesis, transpiration and stomatal conductance in chickpea varieties differing in drought tolerance and showed that mesophyll resistance is the basic determinate of rate of phototosynthesis under drought stress conditions.

5.4. Response to High Temperature Stress

Alterations in various photosynthetic attributes under heat stress are good indicators of thermo tolerance of the plant as they show correlations with growth. Any constraint in photosynthesis can limit plant growth at high temperatures. Photochemical reactions in thylakoid lamellae and carbon metabolism in the stroma of chloroplast have been suggested as the primary sites of injury at high temperatures (Wise *et al.*, 2004). Chlorophyll fluorescence, the ratio of variable fluorescence to maximum fluorescence (*F*v/*F*m), and the base fluorescence (*F0*) are physiological parameters that have been shown to correlate with heat tolerance (Yamada *et al.*, 1996). Increasing leaf temperatures and photosynthetic photon flux density influence thermo tolerance adjustments of PSII, indicating their potential to optimize photosynthesis under varying environmental conditions as long as the upper thermal limits do not exceed (Salvucci and Crafts-Brandner, 2004b; Marchand *et al.*, 2005). In tomato genotypes differing in their capacity for thermotolerance as well as in sugarcane, an increased chlorophyll *a*:*b* ratio and a decreased chlorophyll : carotenoids ratio were observed in the tolerant genotypes under high temperatures, indicating that these changes were related to thermo tolerance of tomato (Camejo *et al.*, 2005; Wahid and Ghazanfar, 2006). Furthermore, under high temperatures, degradation of chlorophyll *a* and *b* was more pronounced in developed compared to developing leaves (Karim *et al.*, 1997, 1999). Such effects on chlorophyll or photosynthetic apparatus were suggested to be associated with the production of active oxygen species (Camejo *et al.*, 2006; Guo *et al.*, 2006). PSII is highly thermolabile, and its activity is greatly reduced or even partially stopped under high temperatures (Camejo *et al.*, 2005), which may be due to the properties of thylakoid membranes where PSII is located (Mcdonald and Paulsen, 1997). Heat stress may lead to the dissociation of oxygen evolving complex (OEC), resulting in an imbalance between the electron flow from OEC toward the acceptor side of PSII in the direction of PSI reaction center (Fig. 2) (De Ronde *et al.*, 2004). Heat stress causes dissociation of manganese (Mn)-stabilizing 33-kDa protein at PSII reaction center complex followed by the release of Mn atoms (Yamane *et al.*, 1998). Heat stress may also impair other parts of the reaction center, *e.g.*, the D1 and/or the D2 proteins (De Las Rivas and Barber, 1997). In wheat, high temperatures and excessive light dam aged different sites of PSII, which implied different pathways for the recovery of its functional activity (Sharkova, 2001). In barley, heat pulses abruptly damaged the PSII units and caused loss of their capacity of oxygen evolution leading to a restricted electron transport, which was totally abolished after four hours (Toth *et al.*, 2005). This implied that the degradation of the impaired PSII units occurred in the light during this period of time. Following this, *de novo* synthesis of PSII units in the light gave a gradual rise to the observed PSII activities. These effects can result from different events, including inhibition of electron transport activity and limited generation of reducing powers for metabolic functions (Allakhverdieva *et al.*, 2001). In field-grown Pima cotton under high temperatures, leaf photosynthesis was functionally limited by photosynthetic electron transport and ribulose-1,5-bisphosphate (RuBP) regeneration capacity, but not rubisco activity (Wise *et al.*, 2004). On the other hand, under high temperatures, PSI stromal enzymes and chloroplast envelops are thermostable and in fact PSI driven cyclic electron pathway, capable of contributing to thylakoid proton gradient, is activated (Bukhov *et al.*, 1999).

High temperature influences the photosynthetic capacity of C3 plants more strongly than in C4 plants. It alters the energy distribution and changes the activities of carbon metabolism enzymes, particularly the rubisco, thereby altering the rate of RuBP regeneration by the disruption of electron transport and inactivation of the oxygen evolving enzymes of PSII (Salvucci and Crafts-Brandner, 2004b). Heat shock reduces the amount of photosynthetic pigments (Todorov *et al.*, 2003), soluble proteins, rubisco binding proteins (RBP) and large- (LS) and smallsubunits (SS) of rubisco in darkness but increases them in light, indicating their roles as chaperones and HSPs (Kepova *et al.*, 2005). Moreover, under heat stress, starch or sucrose synthesis is greatly influenced as observed from reduced activities of sucrose phosphate synthase (Chaitanya

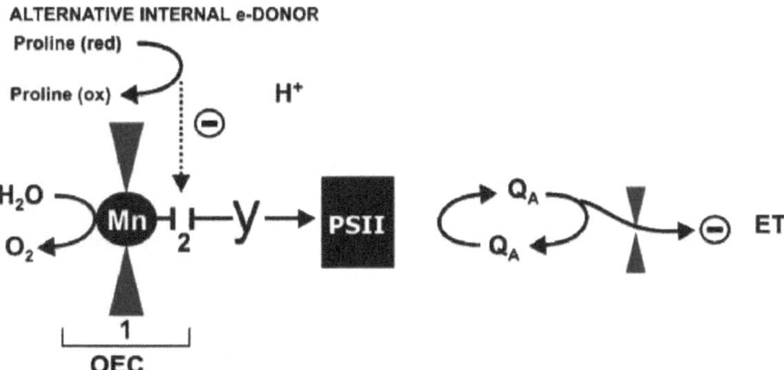

Fig. 2: Heat induced inhibition of oxygen evolution and PSII activity. Heat stress leads to either (1) dissociation or (2) inhibition of the oxygen evolving complexes (OEC). This enables an alternative internal e”-donor such as proline instead of H_2O to donate electrons to PSII. Reproduced with permission from De Ronde *et al.* (2004).

et al., 2001), ADPglucose pyrophosphorylase and invertase (Vu *et al*., 2001). In any plant species, the ability to sustain leaf gas exchange under heat stress has a direct relationship with heat tolerance. During the vegetative stage, high day temperature can cause damage to compensated leaf photosynthesis, reducing CO_2 assimilation rates (Hall, 1992). Increased temperatures curtail photosynthesis and increase CO_2 transfer conductance between intercellular spaces and carboxylation sites. Stomatal conductance (gs) and net photosynthesis (Pn) are inhibited by moderate heat stress in many plant species due to decreases in the activation state of rubisco (Crafts-Brander and Salvucci, 2002; Morales *et al*., 2003). Although with an increase in temperature rubisco catalytic activity increases, a low affinity of the enzyme for CO_2 and its dual nature as an oxygenase limits the possible increases in Pn. For example, in maize the Pn was inhibited at leaf temperatures above 38^{0}C and inhibition was much more severe when temperature was increased abruptly rather than gradually. However, this inhibition was independent of stomatal response to high temperature (Crafts-Brander and Salvucci, 2002). Despite observed negative effects of high temperature, the optimum temperature for leaf photosynthesis is likely to increase with elevated levels of atmospheric CO_2. Several studies have concluded that CO_2-induced increases in crop yields are much more plausible in warm- than in cool-season crops. Thus, despite its other potential negative implications, global warming may not greatly affect the overall Pn. A well-known consequence of elevated temperature in plants is the damage caused by heat-induced imbalance in photosynthesis and respiration; in general the rate of photosynthesis decreases while dark- and photo-respiration rates increase considerably under high temperatures. Also, rate of biochemical reactions decreases and enzyme inactivation and denaturation take place as the temperature increases leading to severely reduced photosynthesis (Nakamoto and Hiyama, 1999). However, the magnitude of such alterations in response to heat stress differs with species and genotypes (Mcdonald and Paulsen, 1997). Furthermore, it has been determined that the photosynthetic CO_2 assimilation rate is less affected by heat stress in developing leaves than in completely developed leaves. Heat stress normally decreases the duration of developmental phases leading to smaller organs, reduced light perception and carbon assimilation processes including transpiration, photosynthesis and respiration (001). Nonetheless, photosynthesis is considered as the physiological process most sensitive to high temperatures, and that rising atmospheric CO_2 content will drive temperature increases in many already stressful environments. This CO_2-induced increase in plant high-temperature tolerance may have a substantial impact on both the productivity and distribution of many crop species in future.

5.5. Response to Low Temperature Stress

Among various abiotic stresses, low temperature (LT) is one of the most important factors limiting the crop productivity and distribution of plants. Low temperatures are defined as low but not freezing temperatures from 0°C to -15°C; they are common in nature and can damage many plant species. Chilling temperatures range from 0°C to 15°C, whereas freezing temperature starts from 0°C and further. Chilling and freezing temperatures are different from each other and have different tolerance mechanisms. Low temperature stress is a major factor that limits the agricultural productivity in hilly areas and at high altitude areas especially (Xin and Browse, 2001). Plants of tropical and subtropical origins are sensitive to chilling temperature (0-10°C) and are incapable of cold acclimation. Whereas plants belonging to temperate climatic regions are considered to be chilling tolerant with variable degrees of temperatures and can increase their freezing tolerance by being exposed to chilling, nonfreezing temperatures, this is known as cold acclimation (Zhu *et al.*, 2007). Breif exposure of plants to low temperature stress may cause necrosis, wilting and chlorosis and finally leads to death of plants.

Various efforts to improve crop performance and productivity under low temperature stress were not yet so successful, because of two main reasons: (1) incomplete understanding of underlaying fundamental mechanism of stress tolerance and (2) lack of knowledge about the interactions of different stressers. As a result of exposure to low temperature, many physiological and morphological changes have been correlated with visible symptoms like wilting, chlorosis and necrosis that lead to death of plants (Adam and Murthy, 2014)..

Plants ultimately show marked changes in biochemical and cell functions such as the following:

Biomembrane lipid composition,

- Small molecular accumulation (Gilmour, 2000; Yamaguchi-Shinozaki, 2006).
- Amino acid (Wanner and Junttila, 1999; Shao *et al.*, 2006).
- Redistribution of intercellular calcium ions (Knight *et al.*, 1998).
- Cellular leakage of electrolytes.
- Diversion of electron flow to alternate pathways (Seo *et al.*, 2010).
- Enzyme activities (Ruelland and Zachowski, 2010).

Each plant has its unique set of temperature requirements, which are optimum for its proper growth and development. A set of temperature conditions, which are optimum for one plant may be stressful for another plant. Many plants, especially those, which are native to warm habitat, exhibit symptoms of injury when exposed to low non-freezing temperatures (Lynch, 1990). These plants including maize (*Zea mays*), soybean (*Glycine max*), cotton (*Gossypium hirsutum*), tomato (*Lycopersicon esculentum*) and banana (*Musa* sp.) are in particular sensitive to temperatures below 10–15°C and exhibit signs of injury (Lynch, 1990; Guy, 1990; Hopkins, 1999) . The symptoms of stress induced injury in these plants appear from 48 to 72 hours; however, this duration varies from plant to plant and also depend upon the sensitivity of a plant to cold stress. Various phenotypic symptoms in response to chilling stress include reduced leaf expansion, wilting, chlorosis (yellowing of leaves) and may lead to necrosis (death of tissue). Chilling also severely hampers the reproductive development of plants for example exposure of rice plants to chilling temperature at the time of anthesis (floral opening) leads to sterility in flowers (Jiang *et al.*, 2002).

The major malicious effect of freezing is that it induces severe membrane damage (Steponkus *et al.*, 1993). This damage is largely due to the acute dehydration associated with freezing. Membrane lipids are primarily composed of two kinds of fatty acids unsaturated as well as saturated fatty acids. Unsaturated fatty acids have one or more double bonds between two carbon atoms (-CH=CH-) whereas saturated fatty acids are fully saturated with hydrogen atoms ($-CH_2-CH_2-$). It is a well-known fact that lipids containing saturated fatty acids solidify at temperatures higher than those containing unsaturated fatty acids. Therefore, the relative proportion of unsaturated fatty acids in the membrane strongly influences the fluidity of the membrane. The temperature at which a membrane changes from semi fluid state to a semi crystalline state is known as the transition temperature. Chilling sensitive plants usually have a higher proportion of saturated fatty acids and, therefore, a higher transition temperature. Chilling resistant species on the other hand are marked by higher proportion of unsaturated fatty acids and correspondingly a lower transition temperature (Mahajan and Tuteja, 2005).

The success of many crops rests on their ability to withstand the freezing temperature of late spring or early autumn frost. Therefore tolerance to freezing temperatures is in particular important for the sustainability of agricultural crops. As understanding the basics of a disease is essential for its cure, in the same way understanding of low freezing induces its injurious effects on plants is essential for the development of frost tolerant crops. The real cause of freeze-induced injury to plants is the ice formation rather than low temperatures. It is noteworthy to mention here that dehydrated tissues such as seeds and fungal spores can survive at very low temperatures without any symptoms of injury. Even cryopreservation is a common method for storage of seeds and other biological materials, which is based on the fact that water essentially solidiWes without the formation of ice crystals.

Ice formation in plants, begins in the apoplastic space as it has relatively lower solute concentration. As the vapor pressure of ice is much lower than water at any given temperature, ice formation in the apoplast establishes a vapor pressure gradient between the apoplast and surrounding cells. The unfrozen cytoplasmic water migrates down the gradient from the cell cytosol to the apoplast, which contributes to the enlargement of existing ice crystals and causes a mechanical strain on the cell wall and plasma membrane leading to cell rupture (McKersie and Bowley, 1997; Olien, 1997). Freeze induced cellular dehydration results in multiple forms of membrane damage including expansion-induced-cell lyses and fracture lesions (McKersie and Bowley, 1997; Uemura and Steponkus, 1997) and lamellar-to-hexagonal-II phase transition. Although freeze exerts its effect largely by membrane damage due to severe cellular dehydration, certain additional factors may also contribute to damage induced by freeze.

ROS produced in response to freeze stress contributes to membrane damage. Chilling sensitive plants characteristically exhibits structural injuries and may suff er from metabolic dysfunction when chilled (Kacperska, 1989). Overall, chilling ultimately results in loss of membrane integrity, which leads to solute leakage. The integrity of intracellular organelles is also disrupted leading to the loss of compartmentalization, reduction and impairing of photosynthesis, protein assembly and general metabolic processes. The primary environmental factors responsible for triggering increased tolerance against freezing, is the phenomenon known as 'cold acclimation.' It is the process where certain plants increase their freezing tolerance upon prior exposure to low non-freezing temperatures.

5.6. Effect of Cold Stress on CO_2 Fixation

Low temperature affects different aspects of photosynthesis such as sucrose synthesis in the cytosol leading to accumulation of phosphorus related inter-mediates. This results in the depletion of the available inorganic phosphate and in the decreased cycling of inorganic phospjate between the cytosol and chloroplast (Hurry *et al.*, 2000). This in turn impedes synthesis of the ATP necessary for the regeneration of ribulose 1-5 bis-phosphate to support CO_2 fixation. Carbohydrate generation has been reported to have greater instaneous low temperature sensitivity than other components of photosynthesis (Leegood and Edwards, 1996). Pollock and Lloyed (1987) have shown that starch synthesis is more sensitive to low temperature than sucrose and fructose synthesis in cood tolerant species. Growth of winter rye at 5^0C results in an increased capacity for CO_2-saturated O_2 evolution, dampening of the fluorescent induction and reductions in the lag time to reach steady-state photosynthesis. Low temperature stress inhibits sucrose synthesis in the cytosol coursing decreased inorganic phosphate cycling between cytosol and chloroplast (Hurry *et al.*, 2000).

Concomitant with the upgradation of carbon metabolism, cold tolerant winter wheat also exhibits stimulation of carbon export from the leaf compared to cold stressed plants. Cold induced tolerance includes increase in enzyme activities (*e.g., from the Calvin cycle and sugar metabolism*), reinforcement of energy dissipation mechanism and antioxidative molecules and qualitative and quantitative changes in lipds and protein membrane composition and protects the cell structure by keeping metabolic pathway regulation mechanism intact (Ramalho *et al.*, 2003; Quartin *et al.*, 2004).

In contrast, growth and development at low temperature increased the expression and subsequent activity of Calvin cycle enzymes. (Hurry *et al.* 2000). Decline in photosynthesis after chilling during dark and light photoperiods attributes loss in Rubisco activity. It has been suggested that chilling damages the Rubisco protein itself. In addition cold-tolerant, herbaceous plants grown at cold temperatures exhibit an increase in inorganic phosphorous availability in the chloroplast (Stitt and Hurry 2002) as well as in adenylates and phosphorylated intermediates and in the capacity for the regeneration of RUBP (Hurry *et al.* 1994)

5.7. Cold Acclimation and its Role in Providing Freeze-tolerance

The primary function of cold acclimation is to stabilize the membranes against freeze injury. Acclimation results in increase in proportion of unsaturated fatty acids and thereby a drop in transition temperature (Raison and Orr, 1986; Williams *et al.*, 1988). It functions to prevent the expansion-induced lyses and formation of hexagonal II phase lipids in rye and other plants (Steponkus *et al.*, 1993; Uemura and Steponkus, 1997). Cold acclimation results in physical and biochemical restructuring of cell membranes through changes in the lipid composition and induction of other non-enzymatic proteins that alter the freezing point of water. Addition of solutes decreases the freezing point of water to a more negative value, thus preventing ice formation.

Low temperatures induce a number of alterations in cellular components, including the extent of unsaturated fatty acids (Cossins, 1994), the composition of glycerolipids (Lynch and Thompson Jr, 1982), changes in protein and carbohydrate composition and the activation of ion channels (Knight and Trewavas, 1996). Accumulation of sucrose and other simple sugars that occurs with cold acclimation also contributes to the stabilization of membrane as these molecules can protect membranes against freeze-damage. Freezing tolerance is a multigenic trait. Low temperatures activate a number of cold-inducible genes (Jones and Inouye, 1994), such as those that encode dehydrins, lipid transfer proteins, translation elongation factors and the late-embryogenesis-abundant proteins (Nishida and Murata, 1996). Moreover, intercellular ice formation can cause a mechanical strain on cell wall and membrane leading to cell rupture (McKersie and Bowley, 1997; Olien, 1997). There is also substantiation that protein denaturation occurs in plants at low temperature which could also result in cellular damage (Guy *et al.*, 1998).

Overall, cold acclimation results in protection and stabilization of the integrity of cellular membranes, enhancement of the antioxidative mechanisms, increased intercellular sugar levels as well as accumulation of other cryoprotectants including polyamines that protect the intracellular proteins by inducing the genes encoding molecular chaperones (Guy and Li, 1998). All these modifications help the plant to withstand and surpass the severe dehydration associated with freezing stress.

5.8. Response to Salinity Stress

Salinisation is the accumulation of soluble salts including sodium, magnesium and calcium in the soil, which decrease the fertility of that soil. Salt affected lands can be found in all climatic regions, from the humid tropics to the polar regions. Saline soil is also found in places below sea level, such as the area around the Dead Sea and in high mountainous regions, such as the Tibetan Plateau and the Rocky Mountains (Pitman and Lauchli, 2002; Manchanda and Garg, 2008). Salinity can be classified as primary or secondary. Primary salinity is the accumulation of soluble salts over a long period of time through weathering of rocks containing salts and deposition of salts from oceans carried by wind and rain (Manchanda and Garg, 2008). Secondary salinity is the accumulation of soluble salt in the surface of the soil, due to rises in the water table by the exploitation of the land and water through agricultural practices (Manchanda and Garg, 2008). Soil salinity can also be categorized as either irrigated land salinity or dry land salinity. Irrigated land salinity is the rise of salt-contaminated ground water and accumulation of soluble salt in crop fields. The major causes of irrigated land salinity are over-irrigation of farm lands, inefficient water use, poor drainage and irrigation on unsuitable soils which make salt 'leak' and become deposited in water channels, drains and water stores (Irrigation salinity, 2011). Dry land salinity mainly occurs in the arid regions of the world, where salt moves to the soil surface and concentrates due to evapotranspiration, making the land unsuitable for agricultural purposes (Salinity Factsheet, 2011).

Salinity is one of the major abiotic stresses which cause crop losses every year and it is predicted to get considerably worse over the next 30-50 years (Tuteja, 2007). Salt in soil affects photosynthetic organisms including plants, algae and cyanobacteria in two ways. First, salt triggers osmotic stress in plants or other photosynthetic organisms, due to low water availability resulting from the negative water potential of the soil. Second, ionic stress occurs through solute imbalance, due to the high levels of Na^+ and Cl^- in the cytosol and changes to the intracellular K^+/Na^+ ratio (Blumwald *et al*., 2000; Conde *et al*., 2011). Research has been carried in to identify genes responsible for stress tolerance in photosynthetic organisms. Stress tolerance is a multigenic trait, therefore it is difficult to understand the involvement of genes in stress tolerance through a gene-by-gene approach (Brinker *et al*., 2010). Projects, including complete genome sequencing and large-scale expressed sequence tag (EST) sequencing have been performed to identify gene functions in stress tolerance (Ahmad *et al*., 2013). Work from Hatzimanikatis and Lee (1999) showed how the interaction between protein abundance and mRNA levels is important to understand gene expression and protein function in stress tolerance. Many genes and proteins including ferritin, HSPs (heat stress proteins), FtSH (the ATP-dependent integral membrane protease), GST (glutathione S-transferase) and proteasome proteins responsible for stress responses and tolerance have been identified through proteomic and transcriptomic analysis (Ahmad *et al*., 2013).

Photosynthesis is one of the major physiological processes affected by salt stress. Salinity in soil either causes short-term or long-term effects on photosynthesis (Parida and Das, 2005). Salinity in soil prevents water uptake by plants, which results in drought stress. This causes stomatal closure in leaves to reduce water release from transpiration (Fricke *et al*., 2004; Munns and Tester, 2008; Negi *et al*., 2014). Stomatal closure under salinity occurs due to loss of leaf turgor which is generated through either high vapour pressure deficit or ABA mediated signal (Munns and Tester, 2008; Chaves *et al*., 2009). As a result, photosynthesis is inhibited, due to low CO_2 assimilation. Although low photosynthetic rates initially occur due to the stomatal closure, as salt stress becomes more severe, photosynthesis is inhibited due to metabolic impairments (Cornic and Briantais, 1991; Pankoviæ *et al*., 1999). In addition to reduced CO_2 diffusion through stomata, salt restricts CO_2 diffusion through the leaf mesophyll (gm) (Flexas *et al*., 2004, 2007). This might be due to several reasons, like physical alterations in the structure of intercellular spaces caused by leaf shrinkage (Lawlor and Cornic, 2002) or changes in the biochemistry or membrane permeability (Gillon and Yakir, 2000; Flexas *et al*., 2008). A study by Kao *et al*., (2001) pointed out that photosynthetic rate and stomatal conductance were decreased in seedlings of a mangrove species, *Kandelia candel* when salt concentration increased up to 430 mM. Work from Stepien and Johnson (2009) showed that salt affects PSII and PSI photochemistry and the total leaf chlorophyll content in *A. thaliana* when exposed to high salt concentrations while the halophyte, *Thellungiella salsuginea* (previously *Thellungiella halophila*) was unaffected.

5.9. CO_2 Fixation Under Salinity Stress

Salinity tolerance is a complex phenomenon, brought about by adaptations in a range of physiological processes. Plants have developed a complex defense system, including ion homeostasis, osmolyte biosynthesis, compartmentation of toxic ions, and reactive oxygen species (ROS) scavenging systems (Hasegawa *et al.*, 2000; Mittova *et al.*, 2004; Stepien and Klobus, 2005; Flowers and Colmer, 2008). Of paramount importance is the process of photosynthesis that is well established as a primary target of many forms of environmental stress, including salinity (Garcia-Sanchez *et al.*, 2002; Liska *et al.*, 2004; Stepien and Klobus, 2006). Soil salt prevents plants from taking up water, exposing them to drought stress. To conserve water, they close their stomata. This simultaneously restricts the entry of CO_2 into the leaf, reducing photosynthesis. At higher concentrations, NaCl may also directly inhibit photosynthesis. When such inhibition occurs, the plant is liable to suffer from oxidative stress. Absorption of sunlight leads to formation of reactive oxygen species, mainly in the chloroplast, either via photoreduction of O_2 to form superoxide (the Mehler reaction) or through the interaction of triplet-excited chlorophyll to form singlet excited oxygen (Asada, 2000; Foyer *et al.*, 2002). ROS are highly reactive and can cause widespread damage to membranes, proteins, and DNA. To prevent such damage, there are a number of enzymatic processes in chloroplast to scavenge reactive oxygen species (Asada, 2000). These are energetically demanding, requiring the synthesis of high concentrations of antioxidants and enzymes. An alternative strategy, placing less of a metabolic burden on plants, would be to avoid the production of reactive oxygen species (Stepien and Johnson, 2009). This can be achieved by regulation of photosynthetic electron transport (Johnson, 2005). Although the role of salt stress in inducing oxidative damage has been widely studied (Herna'ndez *et al.*, 2001; Bor *et al.*, 2003; Stepien and Klobus, 2005), the extent to which regulatory processes are induced under such conditions, and the extent to which variation in their capacity determines the degree of damage incurred by plants exposed to salt have not been widely investigated. Studies so far reported using Thellungiella as a model for salt tolerance have focused on short-term responses to salinity, in particular examining changes in gene expression (Inan *et al.*, 2004; Kant *et al.*, 2006; Wong *et al.*, 2006). Fewer studies have examined the physiology of salt tolerance in this plant and none the effects of salt on leaf physiology.

Salinity is a severe and increasing threat limiting plant growth and crop yields worldwide (Shabala and Cuin, 2007; Zhu, 2001). There are approximately 6% of the world's total land area and 50% irrigated lands being severely affected by salinity (Munns, 2005). It had been shown that salt stress affects CO_2 assimilation in many plants including cotton (*Gossypium hirsutum* L.) (Meloni *et al.*, 2003), bean (*Phaseolus vulgaris* L.) (Bernstein *et al.*, 2010), bell pepper (*Capsicum annuum* L.) (Bethke and Drew, 1992), celery (*Apium graveolens*

L.) (Everard *et al.*, 1994), spinach (*Spinacia oleracea* L.) (Delfine *et al.*, 1999), rice (*Oryza sativa* L.) (Dionisio-Sese and Tobita, 2000), sea barleygrass (*Hordeum marinum* Huds.), cultivated barley (*Hordeum vulgare* L.) (Seckin *et al.*, 2010), and tomato (*Solanum lycopersicum*) (Sade *et al.*, 2010). Meloni *et al.* (2003) observed a lower level of CO_2 assimilation and stomatal conductance with increasing NaCl concentration in cotton (*Gossypium hirsutum* L.) and concluded that stomatal aperture limited leaf photosynthetic capacity in the NaCl-treated plants. Similar results have been obtained for cotton and bean (*Phaseolus vulgaris* L.) (Brugnoli and Lauteri, 1991), spinach (Delfine *et al.*, 1999), rice (Dionisio-Sese and Tobita, 2000) and sea barleygrass and cultivated barley (Seckin *et al.*, 2010). Other studies, however, proved that non-stomatal factors play a key role in the response of leaf CO_2 assimilation to salinity environment, such as ion toxicity (Bethke and Drew, 1992), PSII activity and photophosphorylation activity (Everard *et al.*, 1994), enzyme activity (Seemann AND Sharkey, 1986) and salt tolerant gene expression (Sade *et al.*, 2010). Photochemistry efficiency (F_v/F_m) as an important parameter has been widely used as an indicator of photoinhibition of photosynthesis in higher plants to environmental stresses (Baker, 1991; Havaux, 1992). Therefore, the mechanisms by which salt stress leads to a decrease in CO_2 assimilation are still not yet clearly understood.

Sugars are primary products of photosynthesis in higher plants (Moing *et al.*, 1992). Soluble sugars (*i.e.* sucrose, glucose and fructose) are highly sensitive to environmental stresses and the major forms of carbohydrates. Sugars not only provide energy and solutes for osmotic adjustment, but also modulate expression of multiple genes as regulatory messenger through the sugar sensing and signaling network in many metabolic processes (Fu *et al.*, 2010; Koch, *et al.*, 1992; Koch, 2004). Soluble sugar is a very dynamic cycling process of degradation and synthesis in carbohydrates metabolism (Bläsing *et al.*, 2005; Rolland *et al.*, 2006). Biotic and abiotic stresses alter sugar concentration and metabolic flux (Prado *et al.*, 2000; Thomas, 1999). It is important to determine soluble sugar and sugar flux for understanding feed-forward and feed-back control of photosynthesis in plants response to salinity stress.

Plants can increase a regulatory level of sugar signal molecules by reprogramming the expression of endogenous genes (Smeekens, 2000). Lower levels of transcripts of the genes sucrose phosphate synthase (*SPS*), sucrose synthase (*SS*) and cell wall sucrose invertase (*SI*) were maintained in the WT sugar beet (*Beta vulgaris* L.) compared with the transgenic lines at 300 mM NaCl (Liu *et al.*, 2008). It was reported that sucrose: sucrose 6-fructosyltransferase (*6-SFT*) involved a signal regulating in grass fructan biosynthesis (Amiard *et al.*, 2003). It is important to elucidate the expression of candidate genes involved in the regulation pathways in carbohydrate metabolism for understanding molecular adaptations of plants to salinity stress.

Perennial ryegrass (*Lolium perenne* L.) is an important forage grass and cool-season turfgrass species and cultivated in many countries because of its rapid establishment rate and good wear tolerance (Xiong *et al.*, 2007). Although the effects of salinity stress on germination and growth (Nizam, 2011), chlorophyll content (Liu *et al.*, 2008), transgenic effect (Wu *et al.*, 2005) and the antioxidant system (Hu *et al.*, 2011) of perennial ryegrass have been investigated, there is limited information on photosynthesis and carbohydrate allocation of perennial ryegrass in response to salt stress.

5.10. Response to High Light Intensity Stress

Light can be an unpredictable resource for plants. Sustained changes in irradiance induce alterations in biochemical composition and the morphology of whole plants and leaves. These have the effect of optimizing photosynthetic efficiency and the acclimation of photosynthesis to irradiance in leaves is well documented (Bailey *et al.*, 2001; Walters *et al.*, 2003). In general, low-light responses occur to enhance the efficiency of photon capture, whilst high light responses occur to maximize light-saturated rates of photosynthesis. Under controlled growth conditions it is common to see acclimation maintaining ambient photosynthetic rates at a point below that of light-saturation (Murchie *et al.*, 2005).

Leaf thickness and architecture affect the content of photosynthetic components per unit leaf area. Thus, leaves grown in high irradiance often have higher rates of photosynthesis due to a higher content per unit leaf area of most photosynthetic components, including Rubisco and components of electron transport and ATP synthesis. However, changes also occur on the single chloroplast level: for example, the ratio of PSII to PSI has been shown to vary according to irradiance level (Murchie and Horton, 1998; Yamazaki *et al.*, 1999). Plants grown under high-light conditions have fewer peripheral light-harvesting complexes per PSII reaction centre and a lower amount of Rubisco per unit chlorophyll, and cytochrome b/f complex per unit chlorophyll (Anderson *et al.*, 1995; Murchie and Horton, 1998).

Low-light-grown leaves are generally thin compared with those grown under high light, with a wider overall area and require less investment in terms of nitrogen and carbon. High-light-grown leaves are thicker, have a higher capacity for nitrogen accumulation, and their construction demands a higher investment (Sims and Pearcy, 1994). In the context of maximizing biomass production, such leaf-level changes are often part of a set of integrated mechanisms which include, for example, whole plant biomass partitioning and night-time respiration (Sims and Pearcy, 1994). Morphological adjustments can actually maintain the rate of photosynthesis per unit dry mass at a constant value, despite a change in the rate of photosynthesis per unit leaf area (Sims and Pearcy, 1994; Evans and Poorter, 2001). However, it is unclear whether a low-light-type leaf would be as productive if subsequently transferred to a high light environment. If the reduced leaf thickness is off set by an increase in area, then such leaves may have the same rate of photosynthesis per leaf when exposed to saturating light (Sims and Pearcy, 1992, 1994).

In terms of photosynthetic productivity it is therefore of importance to distinguish two fundamental types of acclimation: that which occurs before or whilst cellular morphology is established and that which occurs afterward. Features of low or high light leaf anatomy are established at an early point during leaf expansion (Sims and Pearcy, 1992; Oguchi *et al*., 2003). Leaf cell size may also be a determinant of chloroplast number (Pyke and Leech, 1991). Although some expansion of mesophyll cells can be induced by light after full leaf extension (Yano and Terashima, 2004), leaves need to be transferred from low to high light prior to, or in the process of, expansion and division in order to undergo large changes in acclimation of leaf anatomy. Parameters may vary in magnitude, rather than possessing a threshold where either a high light leaf or a low-light leaf is formed. (Sims and Pearcy, 1992). There is also evidence for species-specific differences in the relative durations of cell division and cellular expansion during leaf formation (Van Volkenburgh, 1999; Stiles and Van Volkenburgh, 2002).

Oguchi *et al*. (2003) examined the role of leaf anatomy in acclimation to high light in fully expanded leaves of *Chenopodium album* and concluded that while acclimation of Rubisco and photosynthetic rate was seen, it was indeed limited. The most significant post-expansion acclimation feature was an enlarged chloroplast volume and a larger chloroplast surface area exposed to intercellular spaces. Other light acclimation studies have noted, on transfer of mature leaves from low to high irradiance, an increase in light-saturated photosynthesis (Sims and Pearcy, 1992) and measurable Rubisco activity in leaves in pea (Chow and Anderson, 1987).

The extent to which these features of acclimation limit photosynthesis depend on other specific mechanisms of adaptation. For example, in fast-growing species the partitioning of resources into newly synthesized leaves may be of more benefit than investment in already existing leaves. By contrast, the stress-tolerant epiphyte *Guzmania mono-stachia* has long-lived leaves which have a high capacity for reversible acclimation characterized by the loss and gain of entire photosynthetic units according to seasonally dependent variations in light level (Maxwell *et al.*, 1999).

There is little known concerning the interaction between acclimation of photosynthesis to irradiance and processes of leaf ageing following leaf extension. It is likely to be greatly dependent on the species under study. For example, in rice leaves, a decline in Rubisco content and photosynthesis can be observed shortly following full leaf extension (Makino *et al.*,1985). On the other hand it is clear that certain features of chloroplast acclimation can be retained even during advanced leaf senescence (Humbeck and Krupinska, 2003). The above has mostly considered dicotyledons. Rice has been shown to show acclimation according to irradiance level (Murchie *et al.*, 2002). Grasses such as rice have certain developmental features that differ from *Arabidopsis* such as development of leaves within a leaf sheath. How these features affect the dependency of acclimation on leaf development is investigated. The nature of acclimation of photosynthesis post-leaf-extension is also examined by measuring changes in gene expression following a transfer of rice plants from low light to high light.

5.11. Stress-induced Changes

5.11.1. Effects on Photosynthetic Pigments

Salt stress: Different stressful environments have been reported to reduce the contents of photosynthetic pigments. For example, salt stress can break down chlorophyll (Chl), the effect ascribed to increased level of the toxic cation, Na^+ (Pinheiro *et al.,* 2008, Li *et al.*, 2010, Yang *et al.*, 2011). Reduction in photosynthetic pigments, such as chlorophyll *a* and *b* has been reported in some earlier studies on different crops, *e.g.*, sunflower, *Heliantus annuus* (Ashraf and Sultana 2000, Akram and Ashraf, 2011), wheat, *Triticum aestivum* (Arfan *et al.*, 2007, Perveen *et al.*, 2010), and castor bean, *Ricinus communis* (Pinheiro *et al.*, 2008). The salt-induced alterations in leaf chlorophyll content could be due to impaired biosynthesis or accelerated pigment degradation. However, during the process of chlorophyll degradation, chlorophyll *b* may be converted into chlorophyll *a,* thus resulting in the increased content of chlorophyll *a* (Eckardt 2009). A series of experiments with sunflower callus and plants (Santos *et al.*, 2001, Santos, 2004, Akram and Ashraf, 2011) have shown that the important precursors of chlorophyll, *i.e.*, glutamate and 5-aminolaevulinic acid (ALA), decreased in salt-stressed calli and leaves, which indicates that salt stress affects more markedly chlorophyll biosynthesis than chlorophyll breakdown.

Although salt stress reduces the chlorophyll content, the extent of the reduction depends on salt tolerance of plant species. For example, it is generally known that in salt tolerant species, chlorophyll content increases, whereas it decreases in salt-sensitive species under saline regimes (Khan *et al.* 2009, Akram and Ashraf 2011). In view of this, an accumulation of chlorophyll has been proposed as one of the potential biochemical indicators of salt tolerance in different crops, *e.g.*, in wheat (Sairam *et al.* 2002, Raza *et al.* 2006, Arfan *et al.* 2007), pea (Noreen *et al.* 2010), sunflower (Ashraf and Sultana 2000, Akram and Ashraf 2011), and proso millet (*Panicum miliaceum*) (Sabir *et al.* 2009). Since the crops listed here belong to either dicots or monocots, it means that chlorophyll accumulation is not an indicator of salt tolerance of a specific group of plants. Although the above-cited studies suggest that chlorophyll accumulation could be used as biochemical marker for salt tolerance in different crops, in some other studies, chlorophyll accumulation under saline stress is not always associated with salt tolerance. For example, Juan *et al.* (2005) found a weak relationship between leaf Na^+ and photosynthetic pigments in tomato cultivars differing in salinity tolerance. They concluded that chlorophyll *a* and *b* are not good indicators for salt tolerance in tomato. Therefore using chlorophyll accumulation as an indicator of salt tolerance depends on the nature of the plant species or cultivar.

Carotenoids are necessary for photoprotection of photosynthesis and they play an important role as a precursor in signaling during the plant development under abiotic/biotic stress. They have a significant potential to enhance nutritional quality and plant yield. Lately, enhanced carotenoids contents in plants are of considerable attention for breeding as well as genetic engineering in different plants (Li *et al.* 2008). Working with sugar cane, Gomathi and Rakkiyapan (2011) found that imposition of salt stress (7–8 dS m-1) at various plant growth stages caused a marked reduction in chlorophyll and carotenoids contents, but salt-tolerant varieties exhibited higher membrane stability and pigment contents. In another study, Ziaf *et al.* (2009) found significantly higher chlorophyll and carotenoids contents at 60 mM NaCl and they suggested that relative water content (RWC) and carotenoids contents could be used as reliable selection criteria for salt tolerance in hot pepper. In wheat, carotenoids accumulation was less sensitive under high temperature stress (37°C for 24 h and 50°C for 1 h) as compared to that of chlorophyll. A considerable reduction (about 52%) in chlorophyll/Carotinoids ratio was reported in wheat plants under high-temperature stress (Yildiz and Terzi 2008). carotenoids are also present in the plant cellular membranes. They protect the membranes from light-dependent oxidative damage. The role of carotenoids in scavenging reactive oxygen species (ROS) has been well studied (Davison *et al.* 2002, Verma and Mishra 2005). Plants more tolerant to high light and high temperature could be attributed to having reduced lipid peroxidation, necrosis, as well as lower production of another stress indicator, anthocyanins (Davison *et al.* 2002). Growth improvement in plants under stressful environment has been widely reported to be due to the significant role of zeaxanthin in alleviating oxidative damage of membranes (Davison *et al.* 2002, Verma and Mishra 2005, Isaksson and Andersson 2008).

Drought stress: As salinity stress, drought stress causes not only a substantial damage to photosynthetic pigments, but it also leads to deterioration of thylakoid membranes (Huseynova *et al.* 2009, Anjum *et al.* 2011, Kannan and Kulandaivelu 2011). Thus, a reduction in photosynthetic capacity in plants exposed to drought stress is expected. The decrease in chlorophyll content is a commonly observed phenomenon under drought stress (Bijanzadeh and Emam 2010, Mafakheri *et al.* 2010, Din *et al.* 2011). In contrast, Kulshrehtha *et al.* (1987) found no significant effect of drought stress on chlorophyll content in wheat. There are also some reports, which show an enhanced accumulation of chlorophyll under drought stress (Hamada and Al-Hakimi 2001, Pirzad *et al.* 2011). Ashraf and Karim (1991) reported an increase in some cultivars of blackgram (*Vigna mungo*) and a decrease in others under water-deficit conditions (3 and 6 cycles of drought as wilting and rewatering) and suggested that it may be due to variation in chlorophyll synthesis among the cultivars mediated by the variation in the activities of specific enzymes involved in the biosynthesis of chlorophyll. However, studies on chlorophyllase and peroxidase revealed that the decrease may be attributed to accelerated breakdown of chlorophyll rather than its slow synthesis (Harpaz-Saad *et al.* 2007, Kaewsuksaeng 2011).

It is generally known that under drought stress the reduction of chlorophyll *b* is greater than that of chlorophyll *a*, thus, transforming the ratio in favor of chlorophyll *a* (Jaleel *et al.* 2009, Jain *et al.* 2010). For example, in wheat, there were reported a slight rise in chlorophyll *a*/*b* ratio in drought tolerant cultivars and a significant decrease in the susceptible ones under water deficit conditions (PEG-6000 at –0.6 MPa) (Ashraf *et al.* 1994). These differences could be due to a shift in an occurence of photosynthetic systems towards a lower ratio of photosystem (PS) II to PSI. On the other hand, Ashraf and Mehmood (1990) found a decrease in chlorophyll *a*/*b* ratio in three out of four *Brassica* species under water-deficit conditions.

Temperature stress: A number of reports indicate that plants exposed to high-temperature stress show reduced chlorophyll biosynthesis (Efeoglu and Terzioglu 2009, Balouchi 2010, Reda and Mandoura 2011). The impaired chlorophyll biosynthesis is the first of the processes occurring in plastids affected by the high temperature (Dutta *et al.* 2009, Li *et al.* 2010). Lesser accumulation of chlorophyll in high temperature- stressed plants may be attributed to impaired chlorophyll synthesis or its accelerated degradation or a combination of both. The inhibition of chlorophyll biosynthesis under high-temperature regimes results from a destruction of numerous enzymes involved in the mechanism of chlorophyll biosynthesis (Dutta *et al.* 2009, Reda and Mandoura 2011). For example, the activity of 5-aminolevulinate dehydratase (ALAD), the first enzyme of pyrrole biosynthetic pathway, decreased in cucumber and wheat under high-temperature regimes (Tewari and Tripathy 1998, 1999; Mohanty *et al.* 2006).

Tewari and Tripathy (1998) found that chlorophyll synthesis under the low temperature (7°C)- and high temperature (42°C)-stressed cucumber (cv. Poinsette) seedlings was affected by 90 and 60%, respectively.The suppression in chlorophyll biosynthesis was found to be partially due to the inhibition in 5-aminolevulinic acid (ALA) biosynthesis under both low- (78%) and high-temperature (70%) regimes. Furthermore, biosynthesis of protochlorophyllide (Pchlide) in low- and high-temperature-stressed seedlings was impaired by 90 and 70%, respectively. In hexaploid wheat (cv. HD2329) seedlings, protochlorophyllide synthesis, porphobilinogen deaminase, and protochlorophyllide oxidoreductase were affected similarly to that of cucumber, which suggests that temperature stress has generally a similar effect on enzymes involved in chlorophyll biosynthesis in both wheat and cucumber. PSII has been long believed to be a prominent heat sensitive component of photosynthesis (Schrader *et al.* 2004), but it can perform normal functioning up to 45°C (Gombos *et al.* 1994). However, there are some reports showing that moderately high temperature (35–45°C) can induce the cyclic transport of electrons and thylakoid membranes become leaky (Sharkey 2005). Similarly, permeability of the thylakoid membranes is one of the most heat-sensitive components of the photosynthetic apparatus (Havaux *et al.* 1996), which could be counteracted by zeaxanthin. An increased stability of thylakoid membranes was observed at mild heat treatment of potato (35°C for 2 h), which indicated that de-epoxidized xanthophylls maintained thylakoids and thylakoid membranes against heat-induced disorganization (Havaux *et al.* 1996, Brugnoli *et al.* 1998). The deactivation of ribulose-1,5-bisphosphate carboxylase/oxygenase at mild heat stress could be attributed to the deleterious effects of heat on chloroplast reactions (Sharkey 2005, Velikova *et al.* 2012).

Photosynthesis is sensitive to changes in temperature, but a reversible decline has been observed under mild temperature stress, while a permanent impairment was found after an exposure to severe heat stress. At high temperatures, PSII and stroma become oxidized and a significant reduction takes place in PSI. In addition, evidence supports the existence of considerable cyclic electron flow at high temperature, which suggests that maintenance of an energy gradient across the thylakoid membrane as well as adenosine triphosphate homeostasis might be involved in the prevention of irreversible impairment under high-temperature regimes (Sharkey and Zhang 2010).

Altogether, different stressful environments, including salinity, drought, and heat cause generally a considerable reduction in contents of important photosynthetic pigments, particularly chlorophyll. This reduction may occur due to stress-induced impairment in pigment biosynthetic pathways or in pigment degradation. However, the extent of these phenomena depends on the species, variety, duration of plant exposure, and tolerance of the stress. Reduction in photosynthetic pigments, whether through the impairment in pigment biosynthesis or destruction of pigments, may lead to the impairment in electron transport and hence reduced photosynthetic capacity in most plants.

5.11.2. Effects on Photosystems

Photosynthetic pigments present in the photosystems are believed to be damaged by stress factors resulting in a reduced light-absorbing efficiency of both photosystems (PSI and PSII) and hence a reduced photosynthetic capacity (Geissler *et al.* 2009, Zhang *et al.* 2011). Light energy absorbed by chlorophyll is transformed into chlorophyll fluorescence (Maxwell and Johnson 2000). Despite the fact that the extent of chlorophyll fluorescence does not comprise more than 1–2% of total light absorbed by the chlorophyll, its measurement is convenient and noninvasive. It gives a valuable insight into exploitation of the excitation energy by PSII, and indirectly by the other protein complexes of the thylakoid membranes (Roháèek 2002), particularly in plants exposed to stressful conditions. Xanthophylls are carotenoids located in light-harvesting antenna complexes (LHC) of almost all photosynthetic organisms; they play an important role in light harvesting, photoprotection, and assemblage of light harvesting complex (Latowski *et al.* 2004, Misra *et al.* 2006). Zeaxanthin (one of xanthophylls) is formed from violaxanthin by violaxanthin de-epoxidase; it plays a key role in minimizing the overexcitation in higher plants (Kuczyñska *et al.* 2012). In the xanthophyll cycle (Fig. 3), the interconversion of two Carotinoids, violaxanthin and zeaxanthin, takes place and it has a substantial role in photoprotection of plants. Due to its considerable importance, it is a promising target for genetic engineering to enhance stress tolerance in plants. The over expression of the *chyB* gene, which is responsible for encoding β-carotene hydroxylase one of enzymes in zeaxanthin biosynthetic pathway), induced two-fold enhancement in the pool size of the xanthophylls cycle in *Arabidopsis thaliana* (Davison *et al.* 2002). The dissipation of excess light energy as heat within light harvesting complex protects photosynthetic apparatus against the oxidative damage. The xanthophyll cycle, particularly the de-epoxidation of violaxanthin to zeaxanthin through antheraxanthin, was reported to play an important role in energy dissipation at high light intensity (Demmig- Adams and Adams 1992, Misra *et al.* 2006). The dissipation of excitation energy is determined as nonphotochemical quenching (NPQ) of chlorophyll fluorescence during photosynthetic electron transport, which significantly correlates with contents of zeaxanthin and antheraxanthin produced during the xanthophyll cycle (Niyogi *et al.* 1997).

Although the photochemical efficiency of PSII was similar in both cabbage and kidney bean seedlings, the de-epoxidation state of violaxanthin increased six-fold in kidney bean, while no significant change was observed in cabbage under saline treatments. Similarly, nonphotochemical quenching increased in kidney bean, but decreased in cabbage. This showed that pigments involved in the xanthophyll cycle influenced nonphotochemical quenching in both cabbage and kidney. Thus, the increase in the de-epoxidation state of violaxanthin in salt-stressed kidney plants may be a signal to shield the pigment-protein complexes from salt-induced photodamage (Misra *et al.* 2006). During nonphotochemical quenching, the light harvesting complex of PSII undergo conformational changes, thereby generating a modification in pigment interactions causing the development of energy traps. Thus, nonphotochemical quenching plays a key role in the protection of PSII from photodamage. Nonphotochemical quenching is considered as an indicator of excess excitation energy (Joshi *et al.* 1995, Ruban *et al.* 2002, Parida *et al.* 2007). Overall, the fast fluorescence transients following OJIP curve indicate the size of plastoquinone pool within a leaf tissue. However, the curve undergoes changes in response to different stressful environments, such as drought, temperature, salinity, heavy metals, light intensity, *etc.* (Haldimann and Strasser 1999, Popovic *et al.* 2003, Jafarinia and Shariati 2012).

It was widely reported that the suppression in photosynthetic rate can occur due to a number of biotic and abiotic factors, which can substantially alter fluorescence emission kinetic characteristics of plants (Baker and Rosenqvist 2004, Baker 2008). Moreover, the fluorescence induction parameters, such as the minimal fluorescence, Fi – the fluorescence at transient inflection level, the maximal fluorescence, the variable fluorescence, the fluorescence at peak level, and in particular their ratios, are commonly used to determine a number of metabolic disorders in the leaves of many species subjected to a variety of stresses (Baker and Rosenqvist 2004, Baker 2008, B czek-Kwinta *et al.* 2011). The variable fluorescence / the maximal fluorescence ratio is an important parameter, which determines the maximum quantum efficiency of PSII. It provides a measure of the rate of linear electron transport, hence, an indication of overall photosynthetic capacity (Jamil *et al.*, 2007, Tang *et al.*, 2007, Balouchi 2010). In healthy leaves, The variable fluorescence/the maximal fluorescence value is usually close to 0.8 in most plant species, therefore a lower value indicates that a proportion of PSII reaction centers is damaged or inactivated, a phenomenon, termed as photoinhibition, commonly observed in plants under stress (Baker and Rosenqvist, 2004, Zlatev, 2009, Vaz and Sharma, 2011). Salt-stress-induced inhibition in plant is often ascribed to the reduced photosynthetic performance (Wu *et al.*, 2010, Akram and Ashraf, 2011, da Silva *et al.*, 2011), but the underlying mechanisms are still not fully elucidated. However, since PSII is known to play a major role in photosynthetic response to environmental adversity (Han *et al.*, 2010, Liu and Shi 2010, Xu *et al.*, 2010), the salt-induced effect on PSII has been studied thoroughly with a number of contradictory reports appearing in the literature. For example, some studies have shown a significant inhibitory effect of salinity on PSII activity (Everard *et al.*, 1994, Akram and Ashraf, 2011, Saleem *et al.*, 2011), whereas other reports found no significant effect on the structure and function of PSII (Al-Taweel *et al.*, 2007, Abdeshahian *et al.*, 2010). Recently, Mehta *et al.,* (2010) have reported that the donor side of the PSII was damaged more than the acceptor side due to salt stress (0.1–0.5 M NaCl) in wheat (*Triticum aestivum).* Furthermore, the salt-induced damage to PSII was reversible, because 100% recovery of the acceptor side and about 85% of the donor side has been reported (Mehta *et al.*, 2010).

Similarly to others, drought stress is known to alter the chlorophyll *a* fluorescence kinetics and hence to damage the PSII reaction center (Zhang *et al.*, 2011). A number of studies conducted *in vivo* have shown that drought stress causes considerable damage to the oxygen evolving center (OEC) coupled with PSII (Skotnica *et al.*, 2000, Kawakami *et al.* 2009) as well as degradation of D1 polypeptide leading to the inactivation of the PSII reaction center (He *et al.*, 1995, Liu *et al.*, 2006, Zlatev 2009). The changes lead to the generation of reactive oxygen species (ROS), which ultimately cause the photoinhibition and oxidative damage (Ashraf, 2009, Gill and Tuteja, 2010, Anjum *et al.*, 2011). Plants have evolved a variety of protective mechanisms against the ROS induced damage to cellular components, such as the dissipation of excess excitation energy and the synthesis of protective pigments, such as carotenoids and anthocyanins (Efeoglu *et al.*, 2009, Huang *et al.*, 2010). Chlorophyll *a* fluorescence is considered as one of the important indicators of drought tolerance in different species and cultivars/ genotypes, *e.g.*, durum wheat cultivars (Havaux *et al.*, 1988, Flagella *et al.*, 1996, Araus *et al.*, 1998), bread wheat cultivars (Havaux *et al.*, 1988), and tobacco cultivars (Van Rensburg *et al.*, 1996). All these reports suggest that drought-resistant and drought sensitive cultivars can be easily screened at the level of PSII (Guoth *et al.* 2009, da Graça *et al.*, 2010).

A number of earlier studies have shown that drought stress adversely affected the functionality of both PSII and PSI, particularly PSII. This led to decreased electron transport through these two systems (Liu *et al.* 2006, Zlatev 2009). The amounts of PSII proteins, such as D1, D2, and light harvesting complexII as well as mRNA corresponding to genes of *psbA*, *psbD,* and *cab*, also declined markedly due to water deficit; this was ascribed to decreased rates of transcription and translation as well as fast deterioration of proteins and mRNAs (Duan *et al.,* 2006, Liu *et al.*, 2009). Phosphorylation of proteins is known as a key molecular mechanism that plays a vital role in an adaptation of living organisms to unfavorable growth conditions. In chloroplasts, an exceptional, redox-regulated, protein phosphorylation has been found (Wang and Portis 2007, Dutta *et al.*, 2009); it can phosphorylate about 20 thylakoid membrane proteins. Most prominent of these phosphoproteins are those of the light harvesting complexII as well as of PSII reaction center such as D1, D2, CP43, and a 9-kD (PsbH) polypeptide. Phosphorylation of PSII proteins has been reported to regulate the stability, degradation, and turnover of the reaction center proteins (Lundin *et al.*, 2007, Fristedt *et al.*, 2009). However, dephosphorylation of these proteins can also take place under stressful environments and it is catalyzed by phosphatases (Vener *et al.*, 1999, Liu *et al.*, 2009). Phosphorylation and dephosphorylation of PSII are the main regulatory factors and they play a major role in PSII repair. Liu *et al.* (2009) have shown that water stress caused the rapid dephosphorylation of PSII proteins, coupled with the phosphorylation of light harvesting complexII b4 and CP29 in barley (*Hordeum vulgare*). The accelerated dephosphorylation is brought about by both intrinsic and extrinsic membrane protein phosphatases. However, it was also reported that the reduction in dephosphory-lation exacerbated stress induced damages and inhibited the recovery of the photosystems, when the stress was relieved by re-watering. Furthermore, it was also found that the thylakoid structure remained almost intact under water stress except that CP29 migrated slightly from granal thylakoid to stroma thylakoid, whereas the rest of PSII proteins remained unaffected and intact. However, drought stress activated chloroplast proteins and caused the release of TLP40, a potential inhibitor of the membrane phosphatases. It was suggested that phosphorylation of CP29 may cause uncoupling light harvesting complexII from the PSII complex and splitting up the light harvesting complexII trimer and thus causing its degradation. In contrast, dephosphory-lation of PSII proteins may play a role in the repair process of PSII proteins and signal transduction in response to stress (Liu *et al.*, 2009, Los *et al.,* 2010).

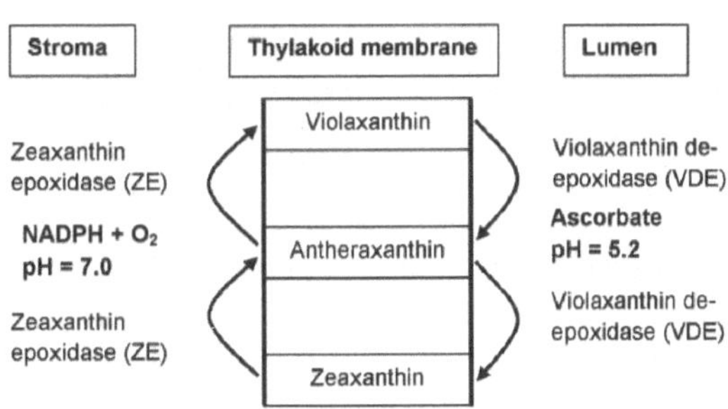

Fig. 3: The xanthophyll cycle in higher plants (modified from Hieber *et al.* 2004).

Photosynthesis is sensitive to heat stress and PSII has been reported to be highly sensitive (Fig. 3) (Allakhverdiev *et al.* 2007, Yan *et al.* 2011). It has been also reported that plants exposed to heat stress for even a short time experience the inhibition of the oxygen evolving complex and the reaction center of PSII (Wang *et al.* 2010, Hamdani *et al.* 2011), although the former is more sensitive to heat than the latter. Smith and Low (1989) reported that the PSII core complex undergoes denaturation at about 60 °C, whereas the denaturation of light harvesting complexII proteins occurs at about 74°C (Smith *et al.* 1989). The major reason considered for heat-induced inactivation of oxygen evolving complex was the release of 33 kDa extrinsic protein from the complex (Zhang *et al.*, 2011). The heat-induced deterioration of PSII leads to considerable perturbance in electron transport mediated by PSII. From a number of studies, it is evident that changes occurring in the ultrastructure of thylakoid membranes above 40°C cause dissociation of the light harvesting complex chlorophyll *a*/*b*-proteins from the PSII core complex (Tang *et al.*, 2007, Iwai *et al.*, 2010). These proteins normally take part in mem-brane stacking (Várkonyi *et al.* 2009). However, heat-induced alteration in these proteins leads to destacking of the appressed membranes of the granum (Fristedt *et al.*, 2009, Lemeille and Rochaix 2010), which is mediated by the separation of nonbilayer- forming lipids. Dobrikova *et al.* (2002) found that the electric dipole moments of the thylakoids and the membranes enriched with PSII were significantly temperature dependent. They also found that the reduction in the electric dipole moments of thylakoids and PSII-enriched membranes correlated well with heat induced nonfunctionality in the PSII photochemical activity and restructuring of the macro assemblies in thylakoid membranes.

As in the case of drought stress, phosphorylation and dephosphorylation of thylakoid proteins has been suggested to play a critical role in responses of plants to elevated temperature (Krishnan and Pueppke 1987, Conde *et al.* 2011). For example, Rokka *et al.* (2000) have shown that rapid dephosphorylation of PSII core proteins took place in isolated spinach (*Spinacia oleracea*) thylakoids at high temperatures. They also found that an increase in temperature from 22°C to 42°C caused a more than 10-fold increase in the dephosphorylation rates of D1 and D2 and of CP43 (chlorophyll *a* binding protein). In contrast, the dephosphorylation rates of light harvesting complex and the 9-kD protein of the PSII (PsbH) were accelerated only 2- to 3-fold. This rapid dephosphorylation is catalyzed by a PSII-specific membrane protein phosphatase, the activity of which is regulated by the temperature.

Thus, the activation of the membrane protein phosphatase was considered to initiate a fast repair of photo damaged PSII and to act as an early signal for other responses to heat stress in chloroplasts. It is evident that drought, salinity, and high-temperature stress adversely affect the functionality of both photosystems and reduce electron transport through them. This results in a low production of ATP and NADPH, the two main products of the light reactions (electron transport), which are essential for CO_2 fixation in the dark reactions of the photosynthesis. Thus, stress-induced impairment of the photosystems ultimately limits the CO_2 reduction process.

5.11.3. Effects on Gas-exchange Characteristics

CO_2-exchange characteristics have been regarded an important indicator of the growth of plants, because of their direct link to net productivity (Ashraf 2004, Piao *et al.*, 2008). However, the effect of any stress on photosynthesis could be caused by stomatal, nonstomatal or both factors (Athar and Ashraf 2005, Saibo *et al.*, 2009), nevertheless, the extent of stress-induced stomatal or nonstomatal regulation depends on the species. It is known that salinity stress, both stomatal and nonstomatal regulation of photosynthesis (Ashraf, 2004, Saibo *et al.*, 2009). Muranaka *et al.* (2002) observed that under salt (NaCl) stress, the rate of photosynthesis in wheat plants decreased in two stages. In the first stage, photosynthetic reduction was slow without any discernible photochemical changes. However, in the second stage, photosynthetic decline was prompt and coupled with a degeneration in the energy conversion efficiency in PSII. They suggested that the salt-induced osmotic effect may induce a gradual decline in photosynthesis due to stomata closure under saline regimes. However, an uptake and accumulation of excessive amounts of Na^+ may directly affect the electron transport and cause a marked reduction in photosynthetic capacity. A reduced activity of the Hill reaction was also observed in salt-stressed chloroplasts (El-Shintinawy, 2000, Zeid, 2009). Salt stress imposed at the reproductive stage was reported to decrease the net CO_2 assimilation rate and stomatal conductance of intact leaves in various wheat genotypes (Shahbaz and Ashraf, 2007, Perveen *et al.*, 2010). The inhibitory effect of salt-induced osmotic stress (water deficit) on the rate of photosynthesis was related to decreased production of ATP due to the impaired electron transport (Moud and Maghsoudi, 2008, Curtiss *et al.*, 2011). Down-regulation of various gas-exchange characteristics to a varying extent has been observed in different plant species exposed to saline stress in a number of studies (Raza *et al.*, 2007, Ali *et al.*, 2008, Ashraf and Akram 2011, Noreen *et al.*, 2012). The salinity-induced osmotic effect on plants causes a substantial accumulation of abscisic acid (ABA), particularly in the guard cells of stomata, which consequently leads to a partial stomata closure thereby lowering the stomatal conductance as well as substomatal CO_2 concentration (Zhao *et al.*, 2009b).

The association of the growth and yield with gas exchange characteristics in several species is summarized in Table 1. It is evident that photosynthetic capacity has a positive association with a biomass production or a seed yield in plants under saline stress, including the crops, *Triticum aestivum* (James *et al.*, 2002), *Oryza sativa* (Moradi and Ismail, 2007), *Phaseolus vulgaris* (Seemann and Critchley, 1985), *Zea mays* (Crosbie and Pearce, 1982), *Vigna mungo* (Chandra Babu *et al.*, 1985), *Gossypium hirsutum* (Pettigrew and Meredith 1994), *Gossypium barbadense* (Cornish *et al.*, 1991), *Spinacia oleracea* (Robinson *et al.*, 1983), *Asparagus officinalis* (Faville *et al.*, 1999), the grass species, *Panicum hemitomon*, *Spartina patens,* and *Spartina alterniflora* (Hester *et al.*, 2001), and six *Brassica* diploid and amphiploid species (Ashraf 2001). In contrast, there are other studies (Table 1), which show no or little association of photosynthetic capacity with the growth in various plant species, *e.g.*, *Hordeum vulgare* (Rawson *et al.*, 1988), *Triticum aestivum* (Ashraf and O'Leary 1996), *Hibiscus cannabinus* (Curtis and Läuchli, 1986), *Olea europea* (Loreto *et al.*, 2003), *Trifolium repens* (Rogers and Noble, 1992), and castor bean (*Ricinus communis*) (Pinheiro *et al.*, 2008). In view of all these reports, it is evident that photosynthetic capacity cannot be used as a general indicator for salt tolerance.

Table 1: Association of growth and yield of different plant species with gas-exchange characteristics under stress conditions

Abiotic Stress	Plant Species	Relatonship of growth /yield with gas exchange	References
Salinity	Wheat (*Triticum aestivum* L.)	A: +ve, biomass and yield	James *et al.*, 2002.
	Common beans (*Phaseolus vulgaris* L.)	A; +ve, Biomass and yield	Seemann and Critchley, 1985.
	Cotton (*Gossypium hirsutam* L.)	A; +ve, biomass & yield	Pettigrew and Meredith, 1994
	Cotton (*Gossypium barbadense* L.)	A; +ve, biomass & yield	Cornish *et al.*, 1991.
	Spinach (*Spinicia oleracea* L.).	A; Biomass & yield +ve	Robinson *et al.*, 1983.
	Asparagus (*Asparagus officinalis* L.).	A; biomass & yield +ve	Faville *et al.*, 1999.
	Grass (*Panicum hemilton*, *Spartina patense and Spartina alterniflora*).	A; biomass & yield +ve.	Hester *et al.*, 2001.
	Six *Brassica* species	A; biomass & yield +ve	Ashraf, 2001.
	Barley (*Hordeum vulgare* L.(.	A; growth –ve	Rawson *et al.*, 1988.
	Wheat (*Triticum aestivum* L.).	A; growth –ve	Ashraf and O'Leary, 1996.
	Kenef (*Hibiscus cannabinus L.)*	A; growth –ve	Curtis and Lauchli, 1986.
	Olive (*Olea europa* L.).	A; growth –ve	Loreto *et al.*, 2003.
	White clover (*Trifolium repens* L.).	A; growth –ve	Rogers and Noble, 1992.
	Caster bean (*Ricunus communis* L.).	A; growth –ve	Pinheiro *et al.*, 2008.
Drought	Wheat (*Triticum aestivum* L.).	A and gs; yield +ve	Fischer *et al.*, 1998.
	Maize (*Zea mays* L.).	A; growth +ve	Ashraf *et al.*, 2007.
	Rice (*Oryza sativa* L.).	A; growth +ve	Garg *et al.*, 2002.
	Cotton (*Gossypium hirsutam* L.)	A; yield +ve	Yan *et al.*, 2004.
	Tobacco (*Nicotiana tabacum* L.)	A; +ve, Growth	Shou *et al.*, 2004.

Contd.

Abiotic Stress	Plant Species	Relatonship of growth /yield with gas exchange	References
	Cotton (*Gossypium barbadense* L.)	A; +ve, yield	Levi *et al.*, 2009.
	Blackgram [*Vigna mungo* (L) Hepper]	A; +ve, Yield	Chandra Babu *et al.*, 1985.
	Cotton (*Gossypium hirsutam* L.)	A; +ve, seed yield	Pettigrew & Meredith, 1994
	Sunflower (*Helianthus annuus* L.)	A: +ve, seed yield	Kiani *et al.*, 2007
	Wheat (*Triticum aestivum* L.)	A: ns, yield and gs	Hura *et al.*, 2006
	Grasses (*Cynodon dactylon* L. & *Cenchrus ciliaris*)	A; ns, grwth and gs	Akram *et al.*, 2007
	Cotton (*Gossypium hirsutam* L.)	A: -ve, yield	Levi *et al.*, 2009
	Tomato (*Lycopersicum esculentum* L.)	A; +ve, growth	Camejo *et al.*, 2005; Singh *et al., 2005*
	Melon (*Cucumis melo*)	A; +ve, growth	Kitroongruang *et al.* 1992
	Creeping bentgrass (*Agrstis palustris* Hudd)	A; +ve, growth	Xu and Huang, 2001.
	Potato (*Solanum tuberosum* L.)	A; -ve, growth	Thornton *et al.*, 1996

A – Photosynthetis capacity; gs – stomatal conductance; +ve – positive association, -ve – negative association, ns – insignificant.

Drought stress is also known to depress gas-exchange characteristics to a varying extent thereby affecting overall photosynthetic capacity of most plants. For example, Lawlor and Cornic (2002) reported that the leaf net CO_2 assimilation rate (PN) of higher plants decreased substantially as the leaf water potential and relative water content (RWC) decreased. However, there are contrasting opinions, whether drought impairs photosynthesis primarily through stomatal or nonstomatal (metabolic) limitations (Saibo *et al.,* 2009, Dias and Brüggemann, 2010, Mafakheri *et al.*, 2010). The control of water loss through stomatal regulation has been recognized as an early plant response to drought (Jia and Zhang, 2008, Harb *et al.*, 2010). As drought continues, the stomata closure occurs for longer periods during the day. This, in turn, leads to the reduced carbon assimilation rate and water loss, resulting in maintenance of the carbon assimilation at the cost of low water availability (Brock and Galen, 2005, Sausen and Rosa, 2010, Pan *et al.*, 2011). Stomatal limitation was generally considered to be the major factor of reduced photosynthesis under water deficit conditions (Galmes *et al.*, 2007, Bousba *et al.*, 2009). This has been ascribed to a decline in both *net photosynthetic rate* and substomatal CO_2 concentration (*Ci*) that consequently inhibit overall photosynthesis.

Several researchers have proposed the use of stomatal conductance (*gs*) as an indicator to assess the difference between stomatal and nonstomatal limitations to photosynthesis under water-limited environments (Xu and Zhou 2008, Yu *et al.,* 2008). For example, Flexas *et al.* (2002) showed that net photosynthetic rate and *Ci* had a strong correlation with stomatal conductance in both field-grown and potted grape wine plants. Such a strong relationship led to the proposal that the down-regulation of photosynthesis depends more on the availability of CO_2 in the chloroplast than on leaf water content or water potential (Flexas *et al.*, 2002, Saibo *et al.*, 2009, Galmes *et al.,* 2011). However, a relationship of stomatal conductance with leaf water potential or relative water content was not observed, *i.e.*, reduced photosynthesis caused by water-deficit conditions occurred at different leaf water levels in different species, even though at similar stomatal conductance (Athar and Ashraf 2005, Peri *et al.* 2009). Thus, it is likely that either water deficit has no effect on photosynthesis until a threshold is reached, beyond which it is impaired or a consistent suppression in metabolism is caused (Athar and Ashraf 2005, Lawlor and Tezara 2009). It has also been shown that, leaves that survive drought often show higher rates of photosynthesis (Akram *et al.*, 2007, Bayramov *et al.*, 2010) and of ribulose-1,5-bisphosphate carboxylase/oxygenase content per unit leaf area (David *et al.*, 1998, Kumar and Singh, 2009).

In several studies, attempts have been made to identify morpho-physiological and molecular processes involved in controlling the yield under water-deficit conditions. Attempts have also been made to show the relationship between gas-exchange parameters, particularly net photosynthetic rate, and drought tolerance of crop species or cultivars (Table 1). For example, a positive association of net photosynthetic rate with drought tolerance has been reported in different crops, *e.g.*, semi-dwarf spring wheat (Fischer *et al.*, 1998), maize (Ashraf *et al.*, 2007), transgenic lines of rice (Garg *et al.*, 2002), and transgenic cotton lines (Yan *et al.*, 2004). Shou *et al.*, (2004) expressed constitutively a tobacco mitogenactivated protein kinase kinase kinase (MAPKKK or NPK1) in maize. It was reported that the NPK1 expression considerably improved drought tolerance in transgenic maize, which was associated with the enhanced rate of photosynthesis, suggesting the effective role of NPK1 in protecting the photosynthetic machinery from the drought-induced injurious effects. In another study (Yu *et al.*, 2008), a mutant, *enhanced drought tolerance1*, of *Arabidopsis* with improved drought tolerance was developed. The mutant showed significantly reduced leaf stomata density, but enhanced photosynthetic capacity. In addition, the positive correlation between the rate of photosynthesis and the crop yield has been reported earlier in several studies (Fischer *et al.* 1998, Ashraf *et al.* 2007, Makino 2011).

However, such a positive relationship of photosynthetic capacity to drought tolerance is not a general rule and in other studies the reverse results were obtained. For example, Hura *et al.* (2006) did not find a significant relationship between net photosynthetic rate and stomatal conductance and the drought tolerance of some winter triticale (× *Triticosecale*) genotypes (Table 1). Similarly, Akram *et al.* (2007) did not find a significant relationship between the photosynthetic capacity and drought tolerance of some genetically diverse populations of two grasses, *Cynodon dactylon* (L.) Pers. and *Cenchrus ciliaris* L. Recently, Levi *et al.* (2009) have developed near-isogenic lines of two species of cotton (*Gossypium hirsutum*, GH, and *G. barbadense*, GB) using marker-assisted selection. It is interesting to note that the GB near-isogenic lines showed a positive association with net photosynthetic rate contrary to the GH near-isogenic lines, even if in the latter case, the parent line and the near-isogenic lines showed almost similar values of net photosynthetic rate. All the reports, showing the positive, null, or negative association of photosynthetic capacity with drought tolerance in different plant species, can be explained in the view of Guo *et al.* (2002) by the fact that also other factors such as assimilate partitioning and utilization may affect the crop yield.

In view of a number of reports, it is evident now that CO_2 uptake and photosynthetic yield decline considerably or even cease, when temperature exceeds an optimum (Aniszewski *et al.,* 2001, Islam 2011, Xie *et al.*, 2011). High temperature can damage the ultrastructure of the thylakoid membrane, thereby modulating its biochemical properties (Wang *et al.*, 2010). Substantially high temperature has been reported to inactivate the oxygen evolving complex of PSII (Zhang *et al.* 2001, Toth *et al.*, 2011). For example, a 15-min treatment (at 38ºC) to leaves previously grown at 23ºC caused a marked inactivation of PSII in potato (Havaux 1996). The heat induced damage to the oxygen evolving complex may occur due to the release of Mn ions from the complex (Nash *et al.*, 1985). Photosynthetic capacity (rate of photosynthesis) has also been considered as one of the potential indicators of heat-stress tolerance in different plant species, because the gas-exchange characteristic showed a positive association with the drought tolerance in different species or cultivars within the species in several studies, *e.g.*, tomato (Camejo *et al.,* 2005, Singh *et al.*, 2005) and melon (Kitroongruang *et al.*, 1992). Although the above reports (Table 1) suggest that photosynthetic rate could be used as a potential indicator of heat tolerance, there are other studies reporting little or no association of this physiological trait with heat tolerance, *e.g.*, in potato (*Solanum tuberosum*) (Thornton *et al.*, 1996) and creeping bentgrass (*Agrostis palustris*) (Xu and Huang 2001), no positive association of *net photosynthetic rate* was observed with heat tolerance of cultivars within each species.

From these reports, it is amply clear that genetic differences in photosynthetic capacity exist at intraspecific and interspecific levels. Since a varying association of photosynthetic capacity with the degree of stress tolerance exists in different species, the use of the photosynthetic capacity as a selection criterion will be plausible and effective only in those species, where a positive association between the photosynthetic capacity and the growth exists under stress conditions.

5.11.4. Effects on Activities of key Photosynthetic Enzymes

One of the most prominent effects of various stresses is the stomata closure, which leads to a lower concentration of intercellular CO_2, which in turn causes deactivation of ribulose-1,5-bisphosphate carboxylase/oxygenase as well as other enzymes such as sucrosephosphate synthase (SPS) and nitrate reductase (Chaves *et al.*, 2009, Mumm *et al.*, 2011). Stress-induced increase and decrease in the activities of enzymes involved in photosynthesis are summarized in Table 2. Increased levels of Na^+ and Cl^- (above 250 mM) in the leaf tissue may substantially perturb the metabolic processes of photosynthesis (Tavakkoli *et al.* 2009, Biswal *et al.*, 2011). Furthermore, the salt-induced, osmotic effect can adversely affect the activities of a number of stroma enzymes involved in CO_2 reduction (Kaiser and Heber, 1981, Xue *et al.*, 2008).

Flowers *et al.* (1977) reported that ribulose-1,5-bisphosphate carboxylase/ oxygenase activity is inhibited *in vitro* by high levels of salt, while Aragao *et al.*

(2005) suggested that ribulose-1,5-bisphosphate carboxylase/oxygenase can also be affected by salt *in vivo* (Table 2). The increase in the amount of ribulose-1,5-bisphosphate carboxylase/oxygenase can be beneficial for the survival of plants under harsh environmental conditions, because a positive relationship of leaf ribulose-1,5-bisphosphate carboxylase/oxygenase contents with *net photosynthetic rate* has been reported in most C3 plants (Taub 2010, Makino 2011). However, in contrast, such a relationship between leaf ribulose-1,5-bisphosphate carboxylase/oxygenase and *net photosynthetic rate* was not found in a glycophytic species, *Vigna unguiculata,* where ribulose-1,5-bisphosphate carboxylase/oxygenase activity decreased in a salt-tolerant cultivar Vita 3, whereas it increased markedly in a salt-sensitive cultivar Vita 5 (Aragao *et al.* 2005). Thus, in view of such contrasting reports, further research is needed to affirm whether the association between these two traits is positive or negative. In another study with two species of rice, Ghosh *et al.* (2001) found that not only ribulose-1,5-bisphosphate carboxylase/oxygenase is affected, but also the enzymes involved in regeneration of the ribulose-1,5-bisphosphate carboxylase / oxygenase substrate, ribulose-1,5-bisphophate (RuBP) are regulated by salt stress and they can play a key role in the regulation of the Calvin cycle (Table 2). Fructose-1,6-bisphosphatase has been considered as one of the potential enzymes that can cause the decline of photosynthetic activity under stressful conditions, because it is involved in the regeneration of RuBP. Earlier, Seemann and Sharkey (1982) observed that salt stress caused a substantial reduction in the RuBP pool size in *Phaseolus vulgaris*, which was primarily ascribed to the salt-induced effect on the RuBP regeneration potential (Table 2). Similar findings have been also reported in *Phaseolus vulgaris* by Caemmerer and Farquhar (1984) and in sunflower (*Helianthus annuus*) (Gimenez *et al.* 1992). Ghosh *et al.* (2001) reported that the activity of fructose-1,6-bisphosphatase was positively associated with the different salt tolerance of the two rice species, because it was inhibited in the salt-sensitive species (*Oryza sativa*), but it remained unaffected in the wild, salt-tolerant species (*Porteresia coarctata*) under saline regimes (Table 2). Similarly, in another study with potato, a transgenic line expressing the chloroplastic fructose-1,6-bisphosphatase at less than 15% compared with the wild-type plants showed considerably reduced net photosynthetic rate and growth; it shows that the enzyme has a key role in the regulation of the Calvin cycle (Kossman *et al.* 1994). Abdel-Latif (2008) observed that although the activities of phosphoenolpyruvate carboxylase (PEPC) isolated from wheat (C3 plant) and that from maize (C4 plant) were inhibited under saline regimes, the enzyme from wheat was less sensitive to salt than that from maize (Table 2).

Table 2: Stress-induced increase/decrease in the activities of enzymes involved in photosynthesis.

Stress	Plant species	Enzyme	Enzyme activity	Reference
Salt	Blackeyed pea (*Vigna unguiculata* L.)	Rubisco	Decreased	Aragao *et al.*, 2005
	Common bean (*Phaseolus vulgaris* L.)	FBP	Decreased	Seemann and Sharkey 1982
	Sunflower (*Helianthus annuus* L.)	FBP	Decreased	Gimenez *et al.*, 1992
	Common bean (*Phaseolus vulgaris* L.)	FBP	Decreased	Caemmerer and Farquhar, 1984
	Potato (*Solanum tuberosum* L.)	FBP	Increased	Kossman *et al.*, 1994
	Rice (*Oryza sativa* L.)	FBP	Increased	Ghosh *et al.*, 2001.
	Wheat (Triticum aestivum L.) and Maize (*Zea mays* L.)	PEPC	Decreased	Abdel-latif, 2008
Drought	Subterranean clover (*Medicago sativa* L.)	Rubisco	Not affected	Medrano *et al.*, 1997
	Setaria (*Setaria sphacelota)*	Rubisco	Decreased	Marques and Arrabica, 1995
	Tobacco (*Nicotiana tabacum* L.)	Rubisco	Not affected	Gunsekera and Berkowitz, 1993
	Maize (*Zea mays* L.)	PEPC	Increased	Jeanneau *et al.*, 2002
	Sugar cane (*Saccharum officinarum* L.)	PEPC	Decreased	Du *et al.*, 1998
Heat	Maize (*Zea mays* L.)	Rubisco	Decreased	Salvucci and Crafts-Brandner 2004
	Cotton (*Gossypium hirsutam* L.) and Tobacco (*Nicotiana tabacum* L.)	Rubisco	Decreased	Crafts-Brandner and Salvucci 2000
	Pea (*Pisum sativum* L.) and *Amaranthus hypochondriacus*	PEPC	Decreased	Chinthapalli *et al.*, 2003
	Wheat (*Triticum aestivum* L.)	Rubisco	Decreased	Xu *et al.*, 2003

Rubisco – ribulose-1,5-bisphosphate carboxylase/oxygenase; FBP – fructose-1,6-bisphosphatase; PEPC – phospho*enol*pyruvate carboxylase; NADP-ME – NADP malic enzyme; PPDK – pyruvate orthophosphate dikinase.

Similarly as the salt stress, drought stress can also adversely affect levels and activities of different enzymes involved in the mechanism of photosynthesis, thereby impairing the synthesis of carbohydrates as well as water use efficiency of plants, which in turn adversely affects the overall plant yield (Kumar and Singh 2009, Gill *et al*., 2011). An appraisal of ribulose-1,5-bisphosphate carboxylase/oxygenase content and activity has been considered as one of the useful indicators for breeding programmes aiming to enhance water use efficiency as well as the yield (Hirel *et al*., 2007). For example, in subterranean clover (*Trifolium subterraneum*), water deficit conditions reduced the initial and total ribulose-1,5-bisphosphate carboxylase/oxygenase activity, but it did not decrease the overall amount of ribulose-1,5-bisphosphate carboxylase/ oxygenase protein per unit of leaf area (Medrano *et al*., 1997). In another study, Marques and Arrabica (1995) found that ribulose-1,5-bisphosphate carboxylase/oxygenase activity in *Setaria sphacelota* declined slightly under moderate drought, but substantially under severe drought. The decline in the rate of ribulose-1,5-bisphosphate carboxylase/oxygenase-catalyzed reaction was reported to be due to alteration in CO_2 availability at the chloroplast level or the availability of the substrate, ribulose-1,5-bisphosphate , or the deactivation of ribulose-1,5-bisphosphate carboxylase/oxygenase. The main reason for the inhibition of photosynthesis under both mild and severe drought was suggested to be a low supply of CO_2 to ribulose-1,5-bisphosphate carboxylase/oxygenase, as the substomatal CO_2 is generally known to be reduced by stress conditions. However, in order to assess the extent of the decline in photosynthesis due to reduced ribulose-1,5-bisphosphate carboxylase/oxygenase activity under drought conditions, Gunasekera and Berkowitz (1993) used transgenic tobacco plants transformed with antisense ribulose-1,5-bisphosphate carboxylase/oxygenase. They showed that a 68% decrease in ribulose-1,5-bisphosphate carboxylase/ oxygenase activity did not hamper net photosynthetic rate under water-limited regimes, suggesting that drought stress may affect any of the steps involved in the regeneration of RuBP rather than ribulose-1,5-bisphosphate carboxylase/ oxygenase itself. Using the transgenic approach, Jeanneau *et al*. (2002) developed a maize line with enhanced expression of C4-phosphoenolpyruvate carboxylase. This transgenic line showed a 30% increase in water use efficiency and a 20% increase in dry biomass under moderate water deficit conditions.

In another study concerning nonstomatal limitations of photosynthesis, Du *et al*. (1998) reported that the activities of the various C4 photosynthesis enzymes, such as PEPcase, NADPmalic enzyme (NADP-ME), ribulose-1,5-bisphosphate carboxylase/oxygenase, and fructose-1,6- bisphosphatase, decreased 2 to 4 times under drought conditions, whereas the activity of phosphopyruvate dikinase (PPDK) declined 9.1 times in C4 plant, sugar cane. This suggests that phosphopyruvate dikinase is potentially the limiting enzyme to photosynthesis under water-deficit conditions. High temperature has been reported to reduce the solubility of CO_2 relative to O_2 within the leaf tissue resulting in reduced availability of CO_2 as a substrate for CO_2 concentrating mechanisms (Salvucci and Crafts-Brandner 2004, Li *et al.*, 2007). This in turn significantly affects the activities of key photosynthetic enzymes. At temperatures above the thermal optimum, ribulose-1,5-bisphosphate carboxylase/oxygenase activase, the key regulatory enzyme of ribulose-1,5-bisphosphate carboxylase/ oxygenase, is reported to dissociate in both C3 and C4 species, causing a reduction in the photosynthetic capacity (Raines, 2011, Sage and Zhu, 2011). C3 plants require a stabilization of ribulose-1,5-bisphosphate carboxylase/ oxygenase activase to acclimate themselves to high temperatures (Kurek *et al.*, 2007, Kumar *et al.*, 2009). Crafts-Brandner and Salvucci (2000) reported that ribulose-1,5-bisphosphate carboxylase/oxygenase activity was markedly inhibited in both cotton and tobacco, when leaf temperatures exceeded 35°C. They suggested that the decrease in ribulose-1,5-bisphosphate carboxylase/ oxygenase activation could happen, when the rate of enhanced ribulose-1,5-bisphosphate carboxylase/oxygenase activation was insufficient to overcome the accelerated rate of ribulose-1,5-bisphosphate carboxylase/oxygenase deactivation at higher temperatures. Furthermore, they also observed that in the absence of activase, the rate of deactivation of isolated ribulose-1,5-bisphosphate carboxylase/oxygenase increased substantially with rising temperature. The activase capability to maintain or enhance ribulose-1,5-bisphosphate carboxylase/oxygenase activation *in vitro* was also reported to decline with temperature. A mechanism of an acclimation to elevated temperatures has also been found in C4 maize, wherein the acclimation to high temperature was associated with manifestation of a larger subunit of ribulose-1,5-bisphosphate carboxylase/oxygenase and limited recovery of the ribulose-1,5-bisphosphate carboxylase/oxygenase activation state (Crafts-Brandner and Salvucci, 2002). C4 plants are generally more tolerant to elevated temperatures than C3 plants. However, photosynthesis in C4 plants is also perturbed by high temperatures. For example, as well as high-temperature induced impairment in ribulose-1,5-bisphosphate carboxylase/oxygenase activase functioning and the activation state in C4 plants, the inhibition occurs also in electron transport, PEP carboxylation, and PEP regeneration at elevated temperatures (Raines 2006, Kelly *et al.* 2006). Chinthapalli *et al.* (2003) showed that the phosphoenolpyruvate carboxylasease from the C4 plant (*Amaranthus hypochondriacus*) was less sensitive to supraoptimal temperature, but more sensitive to suboptimal temperature than the enzyme from the C3 species *(Pisum sativum*). This shows that the key photosynthetic enzymes in C3 and C4 plants are unlikely sensitive to temperature. Furthermore, various enzymes differ in their response to temperature stress. For example, in wheat (*Triticum aestivum*), heat stress imposed for 12 days caused a sharp decline in ribulose-1,5-bisphosphate carboxylase/oxygenase activity, whereas the activity of phosphoenolpyruvate carboxylase increased first, but later decreased, which caused a marked increase in phosphoenolpyruvate carboxylase/ribulose-1,5-bisphosphate carboxylase/oxygenase ratio (Xu *et al.* 2003).

From the preceding discussion, it is evident that the activities of enzymes involved in C3 and C4 photosynthetic pathways are altered to a varying extent under stressful environments. However, it depends on the type of species, stomatal and nonstomatal factors, as well as their interaction, how far the changes in the activities of more, a number of enzymes of both CO_2 concentrating mechanisms are regulated by light (Taiz and Zeiger 2010, Doubnerová and Ryšlavá 2011), so the light intensity and its duration in addition to the intensity of additional stress play also a significant role in the regulation of these enzymes. It has also been observed that under most stresses the balance between different enzymes is also perturbed. However, an optimal balance of different enzymes is necessary for the normal functioning of the photosynthetic pathways.

5.12. Improvement in Photosynthetic Capacity Under Stress by Engineering Photosynthesis-related Genes or Transcription Factors

Two approaches are currently used to improve the photosynthetic capacity in different plant species. Firstly, an effort is devoted to developing transgenic C3 plants with over-expression of ribulose-1,5-bisphosphate carboxylase/ oxygenase genes or other enzymes of the C3 pathway. Secondly, attempts are underway to transfer the genes of a key C4 photosynthetic pathway to C3 plants. Considerable progress has been made during the last two decades in engineering the photosynthetic genes using advanced protocols of the recombinant DNA technology. Using the first strategy, Feng *et al.* (2007) produced a transgenic line of japonica rice by overexpressing *OsSbp* cDNA from an indica rice cultivar. This transgenic line showed beside an enhanced activity of the C3 enzyme sedoheptulose-l,7-bisphosphatase (SBPase, EC 3.1.3.37), which is involved in the regeneration of RuBP, also the enhanced photosynthetic capacity as well as the increased tolerance to salt stress. They suggested that the enhanced rate of photosynthesis under salt stress might have been due to the enhanced supply of RuBP to ribulose-1,5-bisphosphate carboxylase/oxygenase by the activation of SBPase. In another study, Tanaka *et al.* (2005) assessed the role of the overexpression of chloroplastic glutamine synthetase (a key enzyme of photorespiratory pathway involved in the utilization of NH_3 released as a result of conversion of two molecules of glycine into serine in mitochondrion), in salt tolerance of rice. They showed that the overexpression of glutamine synthetase significantly reduced the tissue Na+

content under saline regimes, suggesting the putative role of this enzyme in salt tolerance of rice. The enhanced salt tolerance in this transgenic rice was suggested to be associated with increased reassimilation of NH_3 in the photorespiratory process. Similarly to this study, Kozaki and Takeba (1996) showed enhanced protection from photoinhibition of transgenic tobacco plants overexpressing glutamine synthetase. Karaba *et al.* (2007) found a significant improvement in water use efficiency and in biomass production due to improved photosynthetic assimilation and decreased transpiration rate in transgenic rice plants under salt- and drought stress conditions by overexpressing *Arabidopsis* HARDY (*HRD*) gene. The expression of the pea *ABR17* (ABA-responsive17) cDNA, a member of pathogenesis related proteins (PR10), in *Arabidopsis thaliana* showed the improved rate of germination under salt, cold temperature, or both conditions. Furthermore, the transgenic *Arabidopsis* plants showed the enhanced tolerance to freezing temperature, suggesting the potential utility of the *ABR17* gene to engineer multiple stress tolerance due to *ABR17*-mediated increased photosynthesis (Srivastava *et al.* 2004).

Water deficits and heat stress or both together adversely affected photosynthesis of transgenic *Arabidopsis thaliana* plants constitutively expressing *ABP9*, a bZIP transcription factor (Zhang *et al.* 2008) by regulating photosynthetic carbon- and light-use efficiencies together with a leaf ABA content, pigment composition, and an expression of stress- and light harvesting-responsive genes. net photosynthetic rate and stomatal conductance of the transgenic plants were less reduced by a single stress, although net photosynthetic rate and the electron transport rate declined more under well watered conditions in the transgenic plants in comparison with those under water-deficit conditions. The results revealed that *ABP9* transgenic plants were less susceptible to the stress than the wild type plants. In addition, the increased ABA contents in both transgenic and wild type plants in response to water stress and/or heat stresses suggest that declines in net photosynthetic rate and stomatal conductance might have been due to ABA-induced stomata closure. Recently, Patra *et al.* (2010) have shown that expression of *PcINO1* and *McIMTI* genes in chloroplasts as well as in cytosol caused higher accumulation of total inositol compared with nontransgenic plants under saline regimes. The transgenic plants showed higher photosynthetic activity and the growth, while lesser oxidative damage was found as compared with the wild type plants under saline conditions.

5.12.1. Transfer of C4 genes/traits to C3 plants

Several attempts have been made during the past three decades to improve capacity of C3 plants by transferring C4 traits to C3 plants. Initially, classical hybridization of C4 plants with those of C3 was done, but most of the resultant C3-C4 hybrids were infertile (Brown and Bouton 1993). However, with the advent of recombinant DNA technology, it is possible now to transfer and express substantially the enzymes of C4 pathways in appropriate places within the leaves of C3 plants (Miyao 2003, Begonia and Begonia 2007, Kajala *et al.* 2012). For example, Häusler *et al.* (2002) reported that the first cDNAs of the C4-phosphoenolpyruvate carboxylase sequences for cloning were derived from maize (Izui *et al.* 1986) and *Flaveria trinervia* (Poetsch *et al.* 1991). Later on, a tobacco transgenic line was developed overexpressing maize phosphoenol-pyruvate carboxylase gene under the control of its own promoter (Hudspeth *et al.* 1992) or by using the constitutively expressing *Cauliflower mosaic virus* (CaMV) 35S promoter (Benfey and Chua 1990, Kogami *et al.* 1994). However, in all these transgenic lines, although the activity of C4-phosphoenolpyruvate carboxylase increased two-fold compared with that in nontransformed plants, the transgenic and nontransformed lines did not differ significantly in the net CO_2 assimilation rate. In another study, Ku *et al.* (1999) transferred the complete maize phosphoenolpyruvate carboxylase gene to rice plants; the activity of phosphoenolpyruvate carboxylase in the transgenic rice increased about 110-fold compared with the nontransgenic plants and three-fold the maize activity. The incorporation of the intact maize C4-*Pdk* gene into rice was also reported to be effective in overproducing phosphopyruvate dikinase about 40-fold over that of wild-type rice plants (Fukayama *et al.* 2001).

Similarly, the introduction of the maize C4-specific NADP-malic enzyme cDNA to rice plants enhanced the activity of NADP-malic enzyme to 30- or 70-fold in rice leaves compared with the non-transgenic rice (Takeuchi *et al.* 2000, Tsuchida *et al.* 2001). Although Matsuoka *et al.* (2001) have reported that a number of C3 plants over-expressing C4 phosphoenolpyruvate carboxylase have shown enhanced photosynthetic capacity, the success is not so significant, when the phosphoenolpyruvate carboxylase gene is transferred to a phylogenetically distant plant species. For example, the intact, maize C4-specific phosphoenolpyruvate carboxylase gene was transferred to tobacco plants, but it was not fully expressed in the tobacco leaves (Hudspeth *et al.* 1992). This was ascribed to incorrect transcription initiation. In addition, incorrect splicing may take place, when genes from monocots are transferred to dicots (Goodall and Filipowicz 1991, Ruan *et al.* 2012). Furthermore, the conversion of C3 to C4 plant requires a rearrangement or *de novo* formation of anatomical structures within the leaves to localize the photosynthetic enzymes in appropriate leaf tissues. Such promising anatomical changes, parallel to C4 leaf anatomy, have not been achieved so far. Therefore, the manipulation of genes responsible for the development of C4 specific structures in C3 plants is essential rather than only engineering the genes of key C4 enzymes. Thus, this is one of the major causes of the little success in achieving the desired goal in terms of developing C3 transgenic lines with C4 trait. In addition, although photosynthesis-related genes have been transferred from C4 to C3 plants and their photosynthetic capacity has been examined under nonstress conditions, no attempts have been reported to test the effectiveness of this approach under stressful conditions.

5.12.2. Role of Transcription Factors in Regulation of Genes Involved in Photosynthesis

Transcription factors are found in all organisms, because they are essential for the regulation of the gene expression. Different types of transcription factors exist and an organism with a larger genome usually contains more transcription factors than one with a smaller genome. Elucidation of the mechanism of the gene expression for a particular trait is a major focus of molecular biologists to find out how different types of transcription factors are involved in the gene expression. Recently, Saibo *et al.* (2009) have described the role of a number of transcription factors involved directly or indirectly in the regulation of genes involved in photosynthesis. For example, a transcription factor LONG HYPOCOTYL 5 (HY5), a bZIP-type, was reported to be mainly involved in the regulation of *CAB* gene expression by light, although it may also exhibit a significant role in abiotic stress tolerance (Maxwell *et al.*, 2003, Saibo *et al.*, 2009). This transcription factor, despite controlling the expression of chlorophyll *a/b* binding protein 2 (CAB2) (Maxwell *et al.*, 2003), regulates the expression of the gene for the ribulose-1,5-bisphosphate carboxylase/oxygenase small subunit (RbcS1A) (Chattopadhyay *et al.*, 1998, Lee *et al.*, 2007). There is another transcription factor, OsMYB4, the overexpression of which has been reported to be involved in high accumulation of glycine betaine, which in turn increases stress tolerance in *Arabidopsis thaliana* (Mattana *et al.*, 2005), because glycine betaine can stabilize ribulose-1,5-bisphosphate carboxylase/ oxygenase structure under high-saline regimes (Sakamoto and Murata 2002, Yang *et al.*, 2005, Khafagy *et al.*, 2009). Thus, the overexpression of this transcription factor has an indirect effect on the regulation of photosynthetic genes under stressful environments. In maize, the expression of photosynthetic genes has been reported to be partly controlled by two factors DOF1 and DOF2. From the expression studies, it was evident that DOF1 is an activator of transcription, whereas DOF2 is a repressor (Yanagisawa and Sheen 1998). However, DOF1 was found to enhance the expression of the maize C4-phosphoenolpyruvate carboxylase gene.

Saibo *et al.* (2009) have reported that enhanced expression of crassulacean acid metabolism-specific genes in plants under drought or saline conditions is mediated through transcriptional regulation, in which both *cis*-acting DNA sequences and *trans*-acting factors play a vital role. Schaeffer *et al.* (1995) showed enhanced transcription of the *Ppcl* gene encoding a crassulacean acid metabolism-specific isozyme of phosphoenolpyruvate carboxylase and *Gapl* encoding NAD-dependent glyceraldehyde-3-phosphate dehydrogenase under salt stress. Since several of the sequences of the promoters of both genes resemble consensus binding sites for the MYB class of transcription factors, it was suggested that the salt-induced up-regulation of these specific photosynthesis-related genes may be mediated by MYB-type transcription factors (Schaeffer *et al.*, 1995).

5.13 Role of Mitogen-activated Protein Kinases (MAPKs) in Photosynthesis

Mitogen-activated protein kinases respond to a variety of extracellular stimuli including mitogens, proinflammatory cytokines, osmotic stress, and heat shock and they pass on information from sensors to cellular processes in all eukaryotes (Nakagami *et al.*, 2005, Diédhiou *et al.*, 2008, Inagaki *et al.*, 2008). They are known to regulate a multitude of cellular processes, such as mitosis, differentiation, gene expression, and cell death or survival (Zhang and Klessig 2001). MAP kinases are widespread in eukaryotes. They are effectively involved in the signal transduction of various metabolic processes. MAPK cascades comprise minimally three protein kinases, a MAP kinase kinase kinase (MAPKKK), a MAP kinase kinase (MAPKK), and a MAP kinase (MAPK), which function in concert. Activation of each MAPK occurs when it undergoes phosphorylation. Following the activation, mitogen-activated protein kinases move from the cytoplasm into the nucleus, wherein they phosphorylate many transcription factors leading to alterations in the gene expression (Whitmarsh and Davis 2000, Janknecht 2003, Zhang *et al.*, 2011). There is strong evidence now that MAP kinases are involved in the mechanisms of tolerance to a variety of biotic and abiotic stresses in plants (Huang *et al.*, 2011, Zhang *et al.*, 2011). For example, increased expression of genes encoding a MAP kinase module is strong evidence of the involvement of this enzyme cascade in plants exposed to cold, salt, or drought stress (Mizoguchi *et al.*, 1996, Dai *et al.*, 2007, Kant *et al.*, 2007). In *Arabidopsis*, the MAPK kinase 2 (MKK2) has been reported to be activated by both cold and salt stress (Teige *et al.*, 2004). Similarly, mitogen-activated protein kinases have been reported to be activated by osmotic stresses in *Medicago sativa* and tobacco (Jonak *et al.*, 2002) and by salt stress in rice (Diédhiou *et al.*, 2008). Shou *et al.*, (2004) have shown that the tobacco MAPKKK (NPK1) expressed constitutively in maize resulted in a maize transgenic line with enhanced drought tolerance trait measured in terms of kernel mass and *net photosynthetic rate*. The enhanced drought tolerance of the transgenic line of maize was shown to be strongly associated with a markedly high rate of photosynthesis, suggesting the active role of NPK1 in the mechanism of protection of photosynthesis machinery from water stress damage.

From the above, it is evident that development of transgenic lines of C3 plants overexpressing C4 photosynthetic enzymes is a meaningful approach to improve photosynthetic capacity of C3 plants and to bring it to the level of C4 plants. However, the extent, to which C4 enzymes transferred to C3 plants play a role in effectively fixing CO_2, depends on a number of factors including the localization of introduced C4 enzymes within the leaf tissues and coordination of C4 enzymes with already CO_2 fixing pathways or other allied pathways operative in C3 plants. It has been observed that although overexpression of a single C4 enzyme can modulate the photosynthetic metabolism in C3 plants, in most cases, it does not have significant effects on photosynthesis. Furthermore, little success of attaining photosynthetic capacity in C3 plants equivalent to that of C4 plants by transferring C4 genes to C3 plants could have been due to non-transformation of C3 leaf anatomical structure to that of C4 leaf. This may be the main reason of a low effectiveness of the attempts to bring C3 photosynthetic capacity at par with that of C4 through genetic engineering. Thus, alternatively, efforts should be made to improve the efficiency of CO_2 concentrating processes in C3 plants by enhancing the activities of key enzymes through genetic manipulation. However, transgenic C3 plants overexpressing multiple C4 enzymes are now in the focus of most scientists to improve photosynthetic capacity in C3 plants (Miyao 2003, Begonia and Begonia 2007, Kajala *et al.* 2012). Little information is available in the literature on the components involved in either the perception or signaling involved in a stress response. This necessitates a comprehensive elucidation of the signal transduction pathways induced by different stresses so that appropriate programmes can be devised to improve plant tolerance to a variety of abiotic stresses including the functioning of the photosynthetic system. Thus, identification of signaling components involved in the stress adaptation in plants is a meaningful approach to identify transcriptional activators of adaptive mechanisms to stressful environments that are promising for improvement of crop tolerance.

5.14 Conclusion and Future Prospects

It is now evident that stresses lead to the considerable reduction in photosynthetic performance mediated through stress-induced stomatal or nonstomatal limitations (Athar and Ashraf 2005, Rahnama *et al.* 2010, Taiz and Zeiger 2010). However, it is not easy to discriminate between the effects of these limitations on overall photosynthetic capacity of a plant. Certainly, it depends on the species to what extent the process of photosynthesis under stress conditions is controlled by stomatal or nonstomatal factors. Nevertheless, knowledge of the proportion of the stomatal and nonstomatal factors controlling the process of photosynthesis is vital for appointing the future research on photosynthesis. Salinity, drought, or high temperatures modulate gas exchange characteristics, such as net photosynthetic rate, stomatal conductance, *E*, and *Ci*. Although genetic differences in photosynthetic capacity exist at intraspecific and interspecific levels, net photosynthetic rate is considered as one of the potential, physiological, selection criteria for stress tolerance (Ashraf 2004). Since an unreliable association of the photosynthetic capacity with the degree of stress tolerance exists in many species, use of the photosynthetic capacity as a physiological marker will be realistic only in those species for which a positive relationship between the photosynthetic performance and the growth under stress conditions has been confirmed. Improvement in the photosynthetic performance of C3 plants to bring it at par with that of C4 plants has been an important goal of plant biologists. This could be achieved either by overexpressing the key C3 enzymes or by transferring the C4 enzymes to C3 plants. The production of transgenic lines overexpressing C3 or C4 enzymes has had limited success in enhancing photosynthetic perform ance in most transformed lines. It is not clear yet up to what extent C4 enzymes transferred to C3 plants take part in effective assimilation of CO_2 in C3 engineered plants.

It depends undoubtedly on a multitude of factors including the partitioning of introduced C4 enzymes within the leaf tissues and the interlinking of C4 enzymes with photosynthesis-related pathways functioning in C3 plants. In most of the studies on generations of transgenic lines made for enhanced photosynthetic performance that were published so far, a single C4 enzyme gene has been transferred, but in most cases this did not result in a major success in terms of enhanced photosynthetic capacity in transgenic C3 lines. Thus, a generation of transgenic C3 plants overexpressing multiple C4 enzymes is one of the premier themes of future research.

The elucidation of stress-induced signal transduction pathways is vital to enhance plant tolerance to different stresses. Undoubtedly, by integrating the advances in molecular genetics and cell biology, our knowledge of the signaling pathways has increased greatly over recent years, although it is still far from a full understanding to perception and signaling of environmental cues in plants. There is little information on the components involved in either the perception or signaling involved in stress responses. One of several promising advances made recently in stress biology, knowledge of long- and shortdistance signaling is contemplated as of vital importance for understanding up- or down-regulation of photosynthesis under various stresses, because signaling pathways involved in plant stress responses are interconnected at several points (Chaves *et al.* 2009). This calls for ample understanding of the signal transduction pathways induced by different stresses to improve plant tolerance to various, stressful conditions. Efforts get currently underway in different laboratories all over the world to identify signaling components, such as transcription factors and protein kinases, particularly mitogen-activated protein kinases, involved in stress adaptation in plants. Thus, understanding the prospective association between the stress-induced protein kinase pathways and the genetic and epigenetic regulation of the gene expression in plants remains the important area of the future research. Photosynthetic response to different types of stresses is quite intricate, because it entails the interaction of several restrictions occurring at different locations of the cell or the leaf, and at different phases of the plant growth and development. Furthermore, the duration and intensity of the stress can also significantly affect the photosynthetic capacity. Thus, the complete elucidation of factors involved in the regulation of photosynthetic capacity under stressful environments would help devise appropriate strategies to grow plants successfully in stress-prone areas.

It is evident that stressful factors depending on their intensity and duration can differently down- or upregulate the genes involved in the mechanism of photosynthesis in plants. Thus it could be useful to know expression patterns of such genes for understanding plant photosynthetic or other metabolic responses to various stresses and to develop transgenic lines of different crops with enhanced photosynthetic capacity under stressful conditions.

References

Abdel-Latif, A. 2008. Phosphoenolpyruvate carboxylase activity of wheat and maize seedlings subjected to salt stress. *Aust. J. Basic Appl. Sci.*, 2: 37-41.

Abdeshahian, M., Nabipour, M. and Meskarbashee, M. 2010. Chlorophyll fluorescence as criterion for the diagnosis salt stress in wheat (*Triticum aestivum*) plants. *Int. J. Chem. Biol. Eng.* 4: 184-186.

Adam, S. and Murthy, S.D.S. 2014. Effect of cold stress on photosynthesis of plants and possible protection mechanisms, Springer India Ltd.

Ahmad, T., Sablok, G., Tatarinova, T.V., Xu, Q., Deng, X.X. and Guo, W.W., 2013. Evaluation of codon biology in citrus and Poncirus trifoliata based on genomic features and frame corrected expressed sequence tags. DNA Res. *Int. J. Rapid Publ. Rep. Genes Genomes*, 20: 135–150.

Ain-Lhout, F., Zunzunegui, M., Diaz Barradas, M.C., Tirado, R., Clavijo, A. and Gracia Novo, F. 2001. Comparison of proline accumulation in two Mediterranean shrubs subjected to natural and experimental water deficit. *Plant Soil*, 230: 175–80

Akram, M.S. and Ashraf, M. 2011. Exogenous application of potassium dihydrogen phosphate can alleviate the adverse effects of salt stress on sunflower (*Helianthus annuus* L.). *J. Plant Nutr.* 34: 1041-1057.

Akram, M.S., Athar, H.U.R. and Ashraf, M. 2007. Improving growth and yield of sunflower (*Helianthus annuus* L.) by foliar application of potassium hydroxide (KOH) under salt stress. *Pak. J. Bot.* 39:769-776.

Akram, N.A. and Ashraf, M. 2011. Improvement in growth, chlorophyll pigments and photosynthetic performance in salt-stressed plants of sunflower (*Helianthus annuus* L.) by foliar application of 5-aminolevulinic acid. *Agrochimica*, 55: 94-104.

Allakhverdieva, Y.M., Mamedov, M.D. and Gasanov, R.A., 2001. The effect of glycinebetaine on the heat stability of photosynthetic membranes. *Turk. J. Bot.*, 25: 11–17.

Ali, Q., Athar, H.R. and Ashraf, M. 2008. Modulation of growth, photosynthetic capacity and water relations in salt stressed wheat plants by exogenously applied 24-epibrassinolide. *Plant Growth Regul.* 56: 107-116.

Allakhverdiev, S.I., Los, D.A., Mohanty, P., Nishiyama, Y. and Murata, N. 2007. Glycinebetaine alleviates the inhibitory effect of moderate heat stress on the repair of photosystem II during photoinhibition. *Biochim. Biophys. Acta*, 1767: 1363-1371.

Al-Taweel, K., Iwaki, T., Yabuta, Y., Shigeoka, S., Murata, N. and Wadano, A. 2007. A bacterial transgene for catalase protects translation of D1 protein during exposure of salt-stressed tobacco leaves to strong light. *Plant Physiol.*, 145: 258-265.

Alexieva, V., Sergiev, I., Mapelli, S. and Karanov, E. 2001. The effect of drought and ultraviolet radiation on growth and stress markers in pea and wheat. *Plant Cell Environ.*, 24: 1337–1344.

Allen, D.J. and Ort, D.R. 2001. Impact of chilling temperatures on photosynthesis in warm climate plants. *Trends Plant Sci.*, 6: 36-42.

Alonso, R., Elvira, S., Castillo, F.J. and Gimeno, B.S. 2001. Interactive effects of ozone and drought stress on pigments and activities of antioxidative enzymes in *Pinis halpensis. Plant Cell Environ,* 24: 905–916.

Amiard, V., Morvan-Bertrand, A., Billard, J.P., Huault, C., Keller, F. and Prud'homme, M.P. 2003.

Aniszewski, T., Drozdov, S.N., Kholoptseva, E.S., Kurets, V.K., Obshatko, L.A., Popov, E.G. and Talanov, A.V.2001. Effects of light and temperature parameters on net photosynthetic carbon dioxide fixation by whole plants of five lupin species (*Lupinus albus* L., *Lupinus angustifolius* L., *Lupinus luteus* L., *Lupinus mutabilis* Sweet. and *Lupinus polyphyllus* Lindl.). *Acta Agr. Scand., Sect. B, Soil Plant Sci.*, 51: 17-27.

Anjum, S.A., Wang, L.C, Farooq, M., Hussain, M., Xue, L.L. and Zou, C.M. 2011a. Brassinolide application improves the drought tolerance in maize through modulation of enzymatic antioxidants and leaf gas exchange. *J. Agron. Crop Sci.*, 197(3): 177-185.

Anjum, S.A., Xie, X., Wang. L.L., Saleem, M.F., Man, C. and Lei, W. 2011b. Morphological, physiological and biochemical responses of plants to drought stress. *Afr. J. Agric. Res.*, 6(9): 2026-2032.

Aragao, M.E.F., Guedes, M.M., Otoch, M.L.O., Guedes, M.I.F., Melo, D.F. and Lima, M.G.S. 2005. Differential responses of ribulose-1,5-bisphosphate carboxylase/oxygenase activities of two *Vigna unguiculata* cultivars to salt stress. *Braz. J. Plant Physiol.*, 17: 207-212.

Araus, J.L., Amaro, T., Voltas, J., Nakkoul, H., and Nachit, M.M. 1998. Chlorophyll fluorescence as a selection criterion for grain yield in durum wheat under Mediterranean conditions. *Field Crops Res*, 55: 209- 223.

Arfan, M., Athar, H. R., Ashraf, M. 2007. Does exogenous application of salicylic acid through the rooting medium modulate growth and photosynthetic capacity in differently adapted spring wheat cultivars under salt stress? *J. Plant Physiol.*, 6: 685-694.

Asada, K. 2000. The water-water cycle as alternative photon and electron sinks. *Philos. Trans. R. Soc. Lond. B Biol. Sci.,* 355: 1419–1430.

Ashraf, M. 1994. Breeding for salinity tolerance in plants. *Crit. Rev. Plant Sci.*, 13: 17-42.

Ashraf, M. 2001. Relationships between growth and gas exchange characteristics in some salt-tolerant amphidiploid *Brassica* species in relation to their diploid parents. *Environ. Exp. Bot.*, 45: 155-163.

Ashraf, M. 2004. Some important physiological selection criteria for salt tolerance in plants. *Flora*, 199: 361-376.

Ashraf, M. 2009. Biotechnological approach of improving plant salt tolerance using antioxidants as markers. *Biotechnol. Adv.*, 27: 84-93.

Ashraf, M., Karim, F. 1991. Screening of some cultivars/lines of black gram (*Vigna mungo* I., Hepper) for resistance to water stress. *Trop. Agr.*, 68: 57-62.

Ashraf, M., Mehmood, S. 1990. Response of four *Brassica* species to drought stress. *Environ.Exp. Bot.*, 30: 93-100.

Ashraf, M., Nawazish, S., Athar, H.R. 2007. Are chlorophyll fluorescence and photosynthetic capacity potential physiological determinants of drought tolerance in maize (*Zea mays* L.). *Pak. J. Bot.*, 39: 1123-1131.

Ashraf, M., O'Leary, J.W. 1996. Responses of some newly developed salt-tolerant genotypes of spring wheat to salt stress, II. Water relations and photosynthetic capacity. *Acta Bot. Neerl.*, 45: 29-39.

Ashraf, M., Sultana, R. 2000. Combination effect of NaCl salinity and N-form on mineral composition of sunflower plants.*Biol. Plant.*, 43: 615-619.

Ashraf, M.Y., Azmi, A.R., Khan, A.H., Ala, S.A. 1994. Effect of water stress on total phenol, peroxidase activity and chlorophyll contents in wheat (*Triticum aestivum* L.). *Acta Physiol. Plant.*, 16: 185-191.

Athar, H.R. and M. Ashraf. 2005. Photosynthesis under drought stress. In "*Handbook of Photosynthesis*", Ed.M. Pessarakli,CRC Press, Taylor and Francis Group, NY, pp 793-804.

B¹czek-Kwinta, R., Kozie³, A., Seidler-£o¿ykowska, K. 2011. Are the fluorescence parameters of German chamomile leaves the first indicators of the anthodia yield in drought conditions? *Photosynthetica*, 49: 87-97.

Bailey, S., Walters, R.G., Jansson, S. and Horton, P. 2001. Acclimation of Arabidopsis thaliana to the light environment: the existence of separate low-light and high-light responses. *Planta*, 293: 794–801.

Baker, N.R. 1991. A possible role for photosystem II in environmental perturbations of photosynthesis. *Physiol. Plant*, 81: 563–570.

Baker, N.R. 2008. Chlorophyll fluorescence: A probe of photosynthesis *in vivo*. *Annu. Rev. Plant Biol.*, 59: 89-113.

Baker, N.R., Rosenqvist, E. 2004. Applications of chlorophyll fluorescence can improve crop production strategies: an examination of future possibilities. *J. Exp. Bot.*, 55: 1607-1621.

Balouchi, H.R. 2010. Screening wheat parents of mapping population for heat and drought tolerance, detection of wheat genetic variation. *Int. J. Biol. Life Sci.*, 6: 56-66.

Bayramov, S.M., Babayev, H.G., Khaligzade, M.N. Guliyev, N.M., and Raines, C.A. 2010. Effect of water stress on protein content of some Calvin cycle enzymes in different wheat genotypes. *Proceedings of ANAS (Biological Sciences)*, 65(5-6): 106-111.

Begonia, G.B., Begonia, M.T.: Plant photosynthetic production as controlled by leaf growth, phenology, and behavior. *Photosynthetica* 45: 321-333, 2007.

Benfey, P.N., Chua, N.H. 1990. The cauliflower mosaic virus 35S promoter: combinatorial regulation of transcription in plants. *Science*, 25: 959-966.

Bernstein, N., Shoresh, M., Xu, Y. and Huang, B. 2010. Involvement of the plant antioxidative response in the differential growth sensitivity to salinity of leaves vs roots during cell development. *Free Radical Bio Med.*, 49: 1161–1171.

Bethke, P.C. and Drew, M.C. 1992. Stomatal and nonstomatal components to inhibition of photosynthesis in leaves of capsicum annuum during progressive exposure to NaCl salinity. *Plant Physiol.*, 99: 219–226.

Bijanzadeh, E., Emam, Y. 2010. Effect of defoliation and drought stress on yield components and chlorophyll content of wheat. *Pak. J. Biol. Sci.*, 13: 699-705.

Biswal, B., Joshi, P.N., Raval, M.K., Biswal, U.C. 2011 Photosynthesis, a global sensor of environmental stress in green plants: stress signalling and adaptation. *Curr. Sci.* 101:47-56.

Biswal, B., Raval, M.K., Biswal, U.C., Joshi, P. 2008. Response of photosynthetic organelles to abiotic stress: modulation by sulfur metabolism. In "*Sulfur Assimilation and Abiotic Stress in Plants*, Ed. N.A. Khan, S. Singh, S. Umar, pp. 167-191. Springer-Verlag, Berlin – Heidelberg.

Bläsing, O.E., Gibon, Y., Günther, M., Höhne, M., Morcuende, R., Osuna, D., Thimm, O., Usadel, B., Scheible, W-R. and Stitt, M. 2005. Sugars and circadian regulation make major contributions to the global regulation of diurnal gene expression in *Arabidopsis*. *Plant Cell.*, 17: 3257–3281.

Blumwald, E., Aharon, G.S. and Apse, M.P., 2000. Sodium transport in plant cells. Biochim. Biophys. Acta BBA - Biomembr. 1465: 140–151.

Bohnert, H.J. and Jenson, R.G. 1996. Plant stress adaptations making metabolism move. *Trends Biotech.*, 14: 267–274.

Bor, M., Ozdemir, F. and Turkan, I. 2003. The effects of salt stress on lipid peroxidation and antioxidants in leaves of sugar beet *Beta vulgaris* L. and wild beet *Beta maritima* L. *Plant Sci.*, 164: 77–84.

Bousba, R., Ykhlef, N., Djekoun, A. 2009. Water use efficiency and flag leaf photosynthetic in response to water deficit of durum wheat (*Triticum durum* Desf.). *World J. Agr. Sci.*, 5: 609-616.

Boyer, J.S. 1970. Leaf enlargement and metabolic rates in corn, soybean, and sunflower at various leaf water potentials. *Plant Physiol.*, 46: 233-235.

Boyer, J.S., Wong, S.C. and Farquhar, G.D. 1997. CO_2 and water vapor exchange across leaf cuticle (epidermis) at various water potentials. *Plant Physiol.*, 114: 185–191.

Brinker, M., Brosche, M., Vinocur, B., Abo-Ogiala, A., Fayyaz, P., Janz, D., Ottow, E.A., Cullmann, A.D., Saborowski, J., Kangasjarvi, J., Altman, A. and Polle, A., 2010. Linking the salt transcriptome with physiological responses of a salt-resistant *Populus* species as a strategy to identify genes important for stress acclimation. *Plant Physiol.*, 154: 1697–1709.

Brock, M.T., Galen, C. 2005. Drought tolerance in the alpine dandelion, *Taraxacum ceratophorum* (Asteraceae), its exotic congenter *T. officinale* and interspecific hybrids under natural and experimental conditions. *Amer. J. Bot.*, 92: 1311-1321.

Brown, R.H., Bouton, J.H. 1993. Physiology and genetics of interspecific hybrids between phytosynthetic type. *Annu. Rev. Plant Physiol. Plant Mol. Biol.*, 44: 435-456.

Brugnoli, E. and Lauteri, M. 1991. Effects of salinity on stomatal conductance, photosynthetic capacity, and carbon isotope discrimination of salt-tolerant (*Gossypium hirsutum* L.) and salt-sensitive (*Phaseolus vulgaris* L.) C_3 non-halophytes. *Plant Physiol.*, 95: 628–635.

Brugnoli, E., Scartazza, A., De Tullio, M.C., Monteverdi, M.C., Lauteri, M. and Augusti, A. 1998. Zeaxanthin and non-photochemical quenching in sun and shade leaves of C3 and C4 plants. *Physiol. Plant.*, 104: 727-734.

Bukhov, N.G., Wiese, C., Neimanis, S., Heber, U., 1999. Heat sensitivity of chloroplasts and leaves: leakage of protons from thylakoids and reversible activation of cyclic electron transport. *Photosyn. Res.*, 59: 81–93.

Caires, A.R.L., Scherer, M.D., Santos, T.S.B., Pontim, B.C.A., Gavassoni, W.L., Oliveira, S.L. 2010. Water stress response of conventional and transgenic soybean plants monitored by chlorophyll *a* fluorescence. *J. Fluorescence*, 20: 645-649.

Camejo D, Rodríguez P, Morales MA, Dell'Amico JM, Torrecillas A, Alarcón JJ. 2005. High temperature effects on photosynthetic activity of two tomato cultivars with different heat susceptibility. *J. Plant Physiol.*, 162: 281-289.

Camejo, D., Jim´enez, A., Alarc´on, J.J., Torres, W., G´omez, J.M., Sevilla, F., 2006. Changes in photosynthetic parameters and antioxidant activities following heat-shock treatment in tomato plants. *Funct. Plant Biol.*, 33: 177–187.

Carbone, M.S., Czimczik, C.I., McDuffee, K.E. and Trumbore, S.E. 2007. Allocation and residence time of photosynthetic products in a boreal forest using a low-level ^{14}C pulse-chase labeling technique. *Global Change Biology*, 13: 466–477.

Chaitanya, K.V., Sundar, D., Reddy, A.R., 2001. Mulberry leaf metabolism under high temperature stress. *Biol. Plant.*, 44: 379–384.

Chaitanya, K.V., Masilamani, S., Jutur, P.P. and Ramachandra Reddy, A. 2002. Variation in photosynthetic rates and biomass productivity among four mulberry cultivars. *Photosynthetica*, 40:305–308.

Chaitanya, K.V., Sundar, D., Jutur, P.P. and Ramachandra Reddy, A. 2003. Water stress effects on photosynthesis in different mulberry cultivars. *Plant Growth Regul.,* 40: 75–80.

Chandra Babu, R., Srinivasan, P., Natarajaratnam, N., Rangasamy, S. 1985. Relationship between leaf photosynthetic rate and yield in blackgram (*Vigna mungo* L. Hepper) genotypes. *Photosynthetica*, 19: 159-163.

Chandra, S. 2003.Effects of leaf age on transpiration and energy exchange of *Ficus glomerata,* a multipurpose tree species of central Himalayas. *Physiol. Mol. Biol. Plants*, 9: 255-260.

Chattopadhyay S, Ang LH, Puente P, Deng XW, Wei N. 1998. *Arabidopsis* bZIP protein HY5 directly interacts with light-responsive promoters in mediating light control of gene expression. *Plant Cell*, 10(5): 673–683.

Chaves, M.M., Pereira, J.S., Maroco, J., Rodrigues, M.L., Ricardo, C.P.P., Osorio, M.L., Carvalho, I., Faria, T. and Pinheiro, C. 2002. How plants cope with water stress in the field. Photosynthesis and growth. *Ann. Bot.*, 89: 907–916.

Chaves, M.M., Flexas, J., Pinheiro, C. 2009. Photosynthesis under drought and salt stress: regulation mechanisms from whole plant to cell. *Ann. Bot.*, 103: 551–560.

Chinthapalli, B., Murmu, J., Raghavendra, A.S. 2003. Dramatic difference in the responses of phosphoenolpyruvate carboxylase to temperature in leaves of C3 and C4 plants. *J. Exp. Bot.*, 54: 707-714.

Chourey, K., Ramani, S. and Apte, S.K. 2003. Accumulation of LEA proteins in salt (NaCl) stressed young seedlings of rice (*Oryza sativa* L.) cultivar Bura Rata and their degradation during recovery from salinity stress. *J. Plant Physiol.*, 160: 1165–1174.

Chow, W.S. and Anderson, J.M. 1987. Photosynthetic responses of Pisum sativum to an increase in irradiance during growth. I. Photosynthetic activities. *Aus. J. Pl. Physiol.*, 14: 1–8.

Ciais, P., M. Reichstein, M., Viovy, N., Granier, A., Ogée, J., Allard, V., Aubinet, M., Buchmann, N., Bernhofer, C., Carrara, A., Chevallier, F., De Noblet, N., Friend, A.D., Friedlingstein, P., T. Grünwald, T., Heinesch, B., Keronen, P., A. Knohl, A., Krinner, G., Loustau, D., Manca, G., Matteucci, G., Miglietta, F., Ourcival, J.M., Papale, D., Pilegaard, K., Rambal, S., Seufert, G., Soussana, J.F., Sanz, M.J., Schulze, E.D., Vesala, T. and Valentini, R. 2005. Europe-wide reduc- tion in primary productivity caused by the heat and drought in 2003. *Nature*, 437: 529–533.

Cornic, G. 2000. Drought stress inhibits photosynthesis by decreasing stomatal aperture – not by affecting ATP synthesis. *Trends Plant Sci.*, 5: 187–188.

Conde, A., Chaves, M.M. and Geros, H., 2011. Membrane transport, sensing and signalling in plant adaptation to environmental stress. *Plant Cell Physiol.*, 52: 1583–1602.

Cornic, G. 2005. Drought stress inhibits photosynthesis by decreasing stomatal aperture not by affecting ATP synthesis. *Trends Plant Sci.*, 5: 187–188.

Conde, A., Chaves, M.M., Gerós, H. 2011. Membrane transport, sensing and signaling in plant adaptation to environmental stress. *Plant Cell Physiol.*, 52: 1583-1602.

Conklin, P.I. 2001. Recent advances in the role and biosynthesis of ascorbic acid in plants. *Plant Cell Environ.*,24: 383–394.

Cornic, G. and Briantais, J.M., 1991. Partitioning of photosynthetic electron flow between CO_2 and O_2 reduction in a C3 leaf (*Phaseolus vulgaris* L.) at different CO_2 concentrations and during drought stress. *Planta*, 183: 178–184.

Cornish, K., Radin, J.W., Turcotte, E.L., Luand, Z., Zeiger, E. 1991. Enhanced photosynthesis and stomatal conductance of Pima cotton (*Gossypium barbadense* L.) bred for increased yield. *Plant Physiol.*, 97: 484-489.

Cossins, A.R. 1994. Homeoviscous adaptation of biological membranes and its functional significance. In "*Temperature Adaptation of Biological Membranes*" Ed. A.R. Cossins, Portland Press, London, 1994, pp. 63–76.

Crafts-Brandner, S.J., Salvucci, M.E. 2000. Rubisco activase constrains the photosynthetic potential of leaves at high temperature and CO_2. *Proc. Natl. Acad. Sci. USA*, 97: 13430-13435.

Crafts-Brander, C., Salvucci, M.E., 2002. Sensitivity to photosynthesis in the C4 plant, maize to heat stress. *Plant Cell*, 12: 54–68.

Crosbie, T.M., Pearce, R.B. 1982. Effects of recurrent phenotypic selection for high and low photosynthesis on agronomic traits in two maize populations. *Crop Sci.*, 22: 809-813.

Curtis, P.S., Läuchli, A. 1986. The role of leaf area development and photosynthetic capacity in determining growth of kenaf under moderate salt stress. *Aust. J. Plant Physiol.*, 13: 353-365.

Curtiss, J., Rodriguez-Uribe, L., Stewart, J.M., Zhang, J. 2011. Identification of differentially expressed genes associated with semigamy in pima cotton (*Gossypium barbadense* L.) through comparative microarray analysis. *BMC Plant Biol.*, 11: 49-.

da Graça, J.P., Rodrigues, F.A., Farias, J.R.B., de Oliveira, M.C.N., Hoffmann-Campo, C.B., Zingaretti, S.M. 2010. Physiological parameters in sugarcane cultivars submitted to water deficit. *Braz. J. Plant Physiol.*, 22: 189-197.

da Silva, E.N., Ribeiro, R.V., Ferreira-Silva, S.L., Viégas, R.A., Silveira, J.A.G. 2007. Salt stress induced damages on the photosynthesis of physic nut young plants. *Sci. Agr.*, 68: 62-68.

Dai, X., Xu, Y., Ma, Q., Xu, W., Wang, T. and Xue, Y. 2007. Overexpression of an R1R2R3 MYB Gene, *OsMYB3R-2*, increases tolerance to freezing, drought, and salt stress in transgenic *Arabidopsis*. *Plant Physiol.*, 143: 1739-1751.

Damayanthi, M.M.N., Mohotti, A.J., Nissanka, S.P.: Comparison of tolerant ability of mature field grown tea (*Camellia sinensis* L.) cultivars exposed to a drought stress in passara area. *Trop. Agr. Res.*, 22: 66-75.

David, M.M., Coelho, D., Barrote, I., Correia, M. J. 1998. Leaf age effects on photosynthetic activity and sugar accumulation in droughted and rewatered *Lupinus albus* plants. *Aust. J. Plant Physiol.*, 25: 299-306.

Davison, P.A., Hunter, C.N., Horton, P. 2002. Overexpression of â-carotene hydroxylase enhances s tress tolerance in *Arabidopsis*. *Nature*, 418: 203-206.

De Las Rivas, J., Barber, J., 1997. Structure and thermal stability of photosystem II reaction centers studied by infrared spectroscopy. *Biochemistry*, 36: 8897–8903.

Del Blanco, I.A., Rajaram, S., Kronstad, W.E. and Reynolds, M.P. 2000. Physiological performance of synthetic hexaploid wheat–derived populations. *Crop Sci.*, 40: 1257-1263.

Delfine, S., Alvino, A., Villani, M.C. and Loreto, F. 1999. Restrictions to carbon dioxide conductance and photosynthesis in spinach leaves recovering from salt stress. *Plant Physiol.*, 119: 1101–1106.

Demmig-Adams, B., Adams, W.W. 1992. III: Carotenoid composition in sun and shade leaves of plants with different life forms. *Plant Cell Environ.*, 15: 411-419.

Deng, X., Joly, R. and Hahn, D. 1990 . The influence of plant water deficit on photosynthesis and translocation of ^{14}C-labeled assimilates in cacao seedlings. *Physiol. Plantarum*, 78: 623–627.

De Ronde, J.A.D., Cress, W.A., Kruger, G.H.J., Strasser, R.J., Staden, J.V., 2004. Photosynthetic response of transgenic soybean plants containing an *Arabidopsis P5CR* gene, during heat and drought stress. *J. Plant Physiol.*, 61: 1211–1244.

Dias, M.C., Brüggemann, W. 2010a. Limitations of photosynthesis in *Phaseolus vulgaris* under drought stress: gas exchange, chlorophyll fluorescence and Calvin cycle enzymes. *Photosynthetica*, 48: 96-102.

Dias, M.C., Brüggemann, W. 2010b. Water-use efficiency in *Flaveria* species under drought-stress conditions. *Photosynthetica*, 48: 469-473.

Diédhiou, C.J., Popova, O.V., Dietz, K.J., Golldack, D. 2008. The SNF1-type serine-threonine protein kinase SAPK4 regulates stress-responsive gene expression in rice. *BMC Plant Biol.*, 8: 49.

Din, J., Khan, S.U., Ali, I., Gurmani, A.R. 2011. Physiological and agronomic response of canola varieties to drought stress. *J. Anim. Plant Sci.*, 21: 78-82.

Dionisio-Sese, M. and Tobita, S. 2000. Effects of salinity on sodium content and photosynthetic responses of rice seedlings differing in salt tolerance. *J Plant Physiol.*, 157: 54–58.

Dobrikova, A., Petkanchin, I., Taneva, S.G. 2002. Temperature induced changes in the surface electric properties of thylakoids and photosystem II membrane fragments. *Colloid. Surface. A*, 209: 185-192.

Dodd, I.C. 2003. Hormonal interactions and stomatal responses. *J. Plant Growth Regul.*, 22: 32-46.

Doubnerová, V., Ryšlavá, H. 2011. What can enzymes of C4 photosynthesis do for C3 plants under stress? *Plant Sci.*, 180: 575-583.

Du, Y.C., Nose, A., Wasano, K., Uchida, Y. 1998. Responses to water stress of enzyme activities and metabolite levels in relation to sucrose and starch synthesis, the Calvin cycle and the C4 pathway in sugarcane (*Saccharum* sp.) leaves. *Aust. J. Plant Physiol.*, 25: 253-260.

Duan, H.G., Yuan, S., Liu, W.J., Xi, D.H., Qing, D.H., Liang, H.G., Lin, H.H. 2006. Effects of exogenous spermidine on photosystem II of wheat seedlings under water stress. *J Integr Plant Biol*, 48(8): 920–927

Dulai, S., Molnár, I., Molnár-Láng, M. 2011. Changes of photosynthetic parameters in wheat / barley introgression lines during salt stress. *Acta Biol. Szeged*, 55: 73-75.

Dutta, S., Mohanty, S., Tripathy, B.C. 2009. Role of temperature stress on chloroplast biogenesis and protein import in pea. *Plant Physiol.*, 150: 1050-1061.

Eckardt, N.A. 2009. A new chlorophyll degradation pathway. *Plant Cell*, 21: 700.

Efeoglu, B., Ekmekçi, Y., Çiçek, N. 2009. Physiological responses of three maize cultivars to drought stress and recovery. *S. Afr. J. Bot.*, 75: 34-42.

Efeoglu, B., Terzioglu, S. 2009. Photosynthetic responses of two wheat varieties to high temperature. *EurAsia J. Bio. Sci.*, 3: 97-106.

El-Far, I.A. and A.Y. Allan. 1995. Responses of some wheat cultivars to sowing methods and drought at different stages of growth. *Assuit J. Agric. Sci.*, 26(1): 267-277.

El-Shintinawy, F. 2000. Photosynthesis in two wheat cultivars differing in salt susceptibility. *Photosynthetica*, 38: 615-620.

Evans, J.R. and Poorter, H. 2001. Photosynthetic acclimation of plants to growth irradiance: the relative importance of specific leaf area and nitrogen partitioning in maximizing carbon gain. *Pl. Cell Environ.*, 24:,755–767.

Everard, J.D., Gucci, R, Kann, S.C., Flore, J.A., Loescher, W.H. 1994. Gas exchange and carbon partitioning in the leaves of celery (*Apium graveolens* L.) at various levels of root zone salinity. *Plant Physiol.*, 106: 281-292.

Fructans, but not the sucrosy-galactosides, raffinose and loliose, are affected by drought stress in perennial ryegrass. *Plant Physiol.*, 132: 2218–2229.

Anderson, J.M., Chow, W.S. and Park, Y.I. 1995. The grand design of photosynthesis: acclimation of the photosynthetic apparatus to environmental cues. *Photosynthesis Res.*, 46:129-139.

Farooq, M., Wahid, A., Kobayashi, N., Fujita, D. and Basra, S.M.A. 2009. Plant drought stress: effects, mechanisms and management. *Agron. Sustain. Dev.*, 29: 185-212.

Farquhar, G.D., Von Caemmerer, S, and Berry, J.A. 2001. Models of photosynthesis. *Plant Physiol.*, 125: 42–45.

Faville, M.J., Silvester, W.B., Allan Green, T.G., Jermyn, W.A. 1999. Photosynthetic characteristics of three asparagus cultivars differing in yield. *Crop Sci.*, 39: 1070-1077.

Fischer, R.A., Rees, D., Sayre, K.D. Lu, J.M., Condon, A.G. and Larque-Saavedra, A. 1998. Wheat yield progress is associated with higher stomatal conductance, higher photosynthetic rate and cooler canopies. *Crop Sci.*, 38: 1467- 1475.

Flagella, Z., Campanile, R.G., Ronga, G., Stoppelli, M.C., Pastore, D., De Caro, A., and Di Fonzo, N. 1996. The maintenance of photosynthetic electron transport in relation to osmotic adjustment in durum wheat cultivars differing in drought resistance. *Plant Sci.*, 118:127-133

Flexas, J., Bota, J., Escalona, J.M., Sampol, B. and Medrano, H. 2002. Effects of drought on photosynthesis in grapevines under field conditions: an evaluation of stomatal and mesophyll limitations. *Funct. Plant Biol.*, 29: 461-471.

Flexas, J., Bota, J., Loreto, F., Cornic, G. and Sharkey, T.D. 2004. Diffusive and metabolic limitations to photosynthesis under drought and salinity in C3 plants. *Plant Biology*, 6: 1–11.

Flexas, J., Bota, J., Galmes, J., Medrano, H. and Ribas-Carbo, ìM. 2006 . Keeping a positive carbon balance under adverse conditions: responses of photosyn- thesis and respiration to water stress. *Physiol. Plantarum*, 127: 343–352

Flexas, J., Diaz-Espejo, A., Galmes, J., Kaldenhoff, R., Medrano, H., and Ribas-Carbo, M., 2007. Rapid variations of mesophyll conductance in response to changes in CO_2 concentration around leaves. *Plant Cell Environ.*, 30: 1284–1298.

Flexas, J., Ribas-Carbo, M., Diaz-Espejo, A., Galmes, J. and Medrano, H., 2008. Mesophyll conductance to CO_2: current knowledge and future prospects. *Plant Cell Environ.*, 31: 602–621.

Flowers, T.J., Troke, P.F., Yeo, A.R. 1977. The mechanism of salt tolerance in halophytes. *Annu. Rev. Plant Physiol.*, 28: 89-121.

Flowers, T.J. and Colmer, T.D. 2008. Salinity tolerance in halophytes. *New Phytol.*, 179: 945–963.

Foyer, C.H. and Noctor, G. 2000. Oxygen processing in photosynthesis: regulation and signaling. *New Phytol.,* 146: 359–388.

Foyer, C.H. 2001. Prospects for enhancement the soluble antioxidants ascorbate and glutathione. *Biol. Fac.*, 15: 75–78.

Foyer, C.H., Vanacker, H., Gomez, L.D. and Harbinson, J. 2002. Regulation of photosynthesis and antioxidant metabolism in maize leaves at optimal and chilling temperatures. *Review Plant Physiol Biochem.*, 40: 659–668.

Freschi, L., Mercier, H. 2012. Connecting environmental stimuli and crassulacean acid metabolism expression: Phytohormones and other signaling molecules. *Prog. Bot.*, 73: 231-255.

Fricke, W., Akhiyarova, G., Veselov, D. and Kudoyarova, G., 2004. Rapid and tissue-specific changes in ABA and in growth rate in response to salinity in barley leaves. *J. Exp. Bot.*, 55: 1115–1123.

Fristedt R, Willig A, Granath P, Crèvecoeur M, Rochaix JD, Vener AV. 2009. Phosphorylation of photosystem II controls functional macroscopic folding of photosynthetic membranes in *Arabidopsis. Plant Cell*, 21: 3950-3964.

Fu, J., Huang, B. and Fry, J. 2010. Osmotic potential, sucrose level, and activity of sucrose metabolic enzymes in tall fescue in response to deficit irrigation. *J. Amer. Soc. Hort. Sci.*, 135: 506–510.

Fukayama, H., Tsuchida, H., Agarie, S. 2001. Significant accumulation of C4-specific pyruvate, orthophosphate dikinase in a C3 plant, rice. *Plant Physiol.*, 127: 1136-1146.

Fukushima, E., Arata, Y., Endo, T. Sonnewald, U., and sato, F. 2001. Improved salt tolerance of transgenic tobacco expressing apoplastic yeastderived invertase. *Plant Cell Physiol.*, 42: 245-249.

Galmés, J., Medrano, H., Flexas, J. 2007. Photosynthetic limitations in response to water stress and recovery in Mediterranean plants with different growth forms. *New Phytol.*, 175: 81-93.

Galmés, J., Ribas-Carbó, M., Medrano, H., Flexas, J. 2011. Rubisco activity in Mediterranean species is regulated by the chloroplastic CO_2 concentration under water stress. *J. Exp. Bot.*, 62: 653-665.

Garcia-Sanchez, F., Jifon, J.L., Carvaial. M. and Syvertsen, J.P. 2002. Gas exchange, chlorophyll and nutrient contents in relation to Na^+ and Cl2 accumulation in 'Sunburst' Mandarin grafted on different rootstocks. *Plant Sci.*, 162: 705–712.

Garg AK, Kim JK, Owens TG, Ranwala AP, Choi YD, Kochian LV, Wu RJ. 2002. Trehalose accumulation in rice plants confers high tolerance levels to different abiotic stresses. *Proc. Natl. Acad. Sci. USA*, 99: 15898- 15903.

Geissler, N., Hussin, S., Koyro, H.W. 2009. Interactive effects of NaCl salinity and elevated atmospheric CO_2 concentration on growth, photosynthesis, water relations and chemical composition of the potential cash crop halophyte *Aster tripolium* L. *Environ. Exp. Bot.*, 65: 220-231.

Ghosh, S., Bagchi, S., Majumder, A.L. 2001 Chloroplast fructose- 1,6-bisphosphatase from *Oryza* differs in salt tolerance property from the Porteresia enzyme and is protected by osmolytes. *Plant Sci.*, 160: 1171-1181.

Gill, S.S, Tuteja, N. 2010. Reactive oxygen species and antioxidant machinery in abiotic stress tolerance in crop plants. *Plant Physiol. Biochem.*, 48: 909-930.

Gill, S.S., Khan, N.A., Tuteja, N. 2011. Differential cadmium stress tolerance in five indian mustard (*Brassica juncea* L.) cultivars: An evaluation of the role of antioxidant machinery. *Plant Signal Behav.*, 6: 293-300.

Gillon, J.S. and Yakir, D., 2000. Internal conductance to CO_2 diffusion and C18OO discrimination in C3 leaves. *Plant Physiol.*, 123: 201–214.

Gilmour, S.J., Sebolt, A.M., Salazar, M.P., Everard, J.D. and Thomashow, M.F. 2000. Over expression of the Arabidopsis CBF3 transcriptional activator mimics multiple biochemical changes associated with coldacclimation. *Plant Physiol.*, 124: 1854–186.

Gimenez, C., Mitchell, V.J., Lawlor, D.W. 1992. Regulation of photosynthesis rate of two sunflower hybrids under water stress. *Plant Physiol.*, 98: 516-524.

Gomathi, R., Rakkiyapan, P. 2011. Comparative lipid peroxidation, leaf membrane thermostability, and antioxidant system in four sugarcane genotypes differing in salt tolerance. *Int. J. Plant Physiol. Biochem.*, 3: 67-74.

Gombos, Z., Wada, H., Hideg, E., Murata, N. 1994. The unsaturation of membrane lipids stabilizes photosynthesis against heat stress. *Plant Physiol.*, 104: 563-567.

Goodall, G.J., Filipowicz, W. 1991. Different effects of intron nucleotide composition and secondary structure on premRNA splicing in monocot and dicot plants. *EMBO J.*, 10: 2635-2644.

Gorissen, A,, Tietema, A., Joosten, N., Estiarte, M., Penuelas, J., Sowerby, A., Emmett, B. and Beier, C. 2004. Climate change affects carbon allocation to the soil in shrublands. *Ecosystems*, 7: 650–661.

Grassmann, J., Hippeli, S. and Elstner, E.F. 2002. Plant's defence and its benefits for animals and medicine: role of phenolics and terpenoids in avoiding oxygen stress. *Plant Physiol. Biochem.*, 40: 471–478.

Guerfel, M., Baccouri, O., Boujnah, D., Chaibi, W. and Zarrouk, M. 2009. Impacts of water stress on gas exchange, water relations, chlorophyll content and leaf structure in the two main Tunisian olive (*Olea europaea* L.) cultivars. *Sci. Horticult.*, 119: 257-263.

Guidi, L., Nali, C., Ciompi, S., Lorenzini, G. and Soldatini, G.F. 1997. The use of chlorophyll fluorescenceand leaf gas exchange as methods for studying the different responses to ozone of two bean cultivars. *J. Exp. Bot.*, 48: 173-179.

Gunasekera, D., Berkowitz, G.A 1993. Use of transgenic plants with ribulose-1,5-bisphosphate carboxylase / oxygenase antisense DNA to evaluate the rate limitation of photosynthesis under water stress. *Plant Physiol.*, 103: 629-635.

Guo, B.Z., Butrón, A., Li, H., Widstrom, N.W., Lynch, R.E 2002. Restriction fragment length polymorphism assessment of the heterogeneous nature of maize population GT-MAS:gk and field evaluation of resistance to aflatoxin production by *Aspergillus flavus*. *J. Food Prot.*, 65: 167-171.

Guo, Y.-P., Zhou, H.-F., Zhang, L.-C., 2006. Photosynthetic characteristics and protective mechanisms against photooxidation during high temperature stress in two citrus species. *Sci. Hort.*, 108: 260–267.

Guóth, A., Tari, I., Gallé, Á., Csiszár, J., Horváth, F., Pécsváradi, A., Cseuz, L. and Erdei, L. 2009. Chlorophyll *a* fluorescence induction parameters of flag leaves characterize genotypes and not the drought tolerance of wheat during grain filling under water deficit. *Acta Biol. Szeged.*, 53:1-7.

Guy, L. 1990. Cold acclimation and freezing stress tolerance: role of protein metabolism, *Annu. Rev. Plant. Physiol.*, 41: 187–223.

Guy, C.L. and Li, Q.B. 1998. The organization and evolution of the spinach stress 70 molecular chaperone gene family, *Plant Cell*, 10: 539–556.

Guy, C., Haskell, D. and Li, Q.B. 1998. Association of proteins with the stress 70 molecular chaperones at low temperature evidence for the existence of cold labile proteins in spinach. *Cryobiology*, 36: 301–314.

Haldimann, P., Strasser, R.J. 1999. Effects of anaerobiosis as probed by the polyphasic chlorophyll fluorescence rise kinetics in pea (*Pisum sativum* L.). *Photosynth. Res.*, 62: 67-83.

Hall, A.E., 1992. Breeding for heat tolerance. *Plant Breed. Rev.*, 10: 129–168.

Hamada, A.M., El-Enany, A.E. 1994. ffect of NaCl salinity on growth, pigment and mineral element contents, and gas exchange of broad bean and pea plants. *Biol. Plant.*, 36: 75-81.

Hamdani, S., Gauthier, A., Msilini, N., Carpentier, R. 2011. Positive charges of polyamines protect PSII in isolated thylakoid membranes during photoinhibitory conditions. *Plant Cell Physiol.*, 52: 866-873.

Hamilton, D.W.A., Hills, A., Kohler, B., Blatt, M.R. 2000. Ca2+ channels at the plasma membrane of stomatal guard cells are activated by hyperpolarization and abscisic acid. *Proc. Natl. Acad. Sci. USA*, 97: 4967-4972.

Han, W., Xu, X.W., Li, L., Lei, J.Q., and Li, S.Y. 2010. Chlorophyll *a* fluorescence responses of *Haloxylon ammodendron* seedlings subjected to progressive saline stress in the Tarim desert highway ecological shelterbelt. *Photosynthetica*, 48: 635-640.

Harb, A., Krishnan, A., Madana, M.R. 2010. Molecular and physiological analysis of drought stress in *Arabidopsis* reveals early responses leading to acclimation in plant growth. *Plant Physiol.*, 154: 1254-1271.

Harpaz-Saad S, Azoulay T, Arazi T, Ben-Yaakov E, Mett A, Shiboleth YM, Hörtensteiner S, Gidoni D, Gal-On A, Goldschmidt EE, Eyal Y. 2007. Chlorophyllase is a rate-limiting enzyme in chlorophyll catabolism and is posttranslationally regulated. *Plant Cell*, 19: 1007-1022.

Hasegawa, P.M., Bressan, R.A., Zhu, J.K. and Bohnert, H.J. 2000. Plant cellular and molecular responses to high salinity. *Annu. Rev. Plant Physiol. Plant Mol. Biol.*, 51: 463–499.

Hatzimanikatis, V. and Lee, K.H., 1999. Dynamical analysis of gene networks requires both mRNA and protein expression information. *Metab. Eng.*, 1: 275–281.

Haupt-Herting, S. and Fock, H.P. 2002. Oxygen exchange in relation to carbon assimilation in water-stressed leaves during photosynthesis. *Ann. Bot.*, 89: 851–859.

Häusler, R.E., Hirsch, H.J., Kreuzaler, F., Peterhansel, C. 2002. Overexpression of C4-cycle enzymes in transgenic C3 plants: a biotechnological approach to improve C3-photosynthesis. *J. Exp. Bot.*, 53: 591-607.

Havaux, M. 1992. Stress tolerance of photosystem II in vivo: antagonistic effects of water, heat, and photoinhibition stresses. *Plant Physiol.*, 100: 424–432.

Havaux, M. and Tardy, F. 1999. Loss of chlorophyll with limited reduction of photosynthesis as an adaptive response of Syrian barley landraces to high-light and heat stress. *Aust. J. Plant Physiol.*, 26: 569–578.

Havaux, M. 1996. Short-term responses of photosystem I to heat stress - Induction of a PS II-independent electron transport through PS I fed by stromal components. *Photosynth. Res.*, 47: 85-97.

Havaux, M., Ernez, M., Lannoye, R. 1988. Correlation between heat tolerance and drought tolerance in cereals demonstrated by rapid chlorophyll fluorescence tests. *J. Plant Physiol.*, 133: 555-560.

Havaux, M., Tardy, F., Ravenel, J., Chanu, D. and Parot, P. 1996. Thylakoid membrane stability to heat stress studied by flash spectroscopic measurements of the electrochromic shift in intact potato leaves: influence of the xanthophyll content. *Plant Cell Environ.*, 19:1359-1368.

Hawkins, H. J., Lewis, O.A.M. 1993. Combination effect of NaCl salinity, nitrogen form and calcium concentration on the growth and ionic content and gaseous properties of *Triticum aestivum* L. cv. Gamtoos. *New Phytol.*, 124: 161-170.

He, J.X., Wang, J., Liang, H.G. 1995. Effects of water stress on photochemical function and protein metabolism of photosystem II in wheat leaves. *Physiol. Plant.*, 93: 771-777.

Hernan'dez, J.A., Olmos, E., Corpas, F.J., Sevilla, F. and Del Rio, L.A. 1995. Salt-induced oxidative stress in chloroplasts of pea plants. *Plant Sci.*, 105: 151-167.

Herna'ndez, J.A., Ferrer, M.A., Jime'nez, A., Ros-Barcelo', A. and Sevilla, F. 2001. Antioxidant system and O_2/H_2O_2 production in the apoplast of *Pisum sativum* L. leaves: its relation with NaCl- induced necrotic lesions in minor veins. *Plant Physiol.*, 127: 817–831.

Hester, M.W., Mendelsohn, I.A., Mckee, K.L. 2001. Species and population variation to salinity stress in *Panicum hemitomon*, *Spartina patens*, and *Spartina alterniflora*: morphological and physiological constraints. *Environ. Exp. Bot.*, 46: 277-297.

Hieber, A.D., Kawabata, O., Yamamoto, H.Y. 2004. Significance of the lipid phase in the dynamics and functions of the xanthophyll cycle as revealed by PsbS overexpression in tobacco and *in-vitro* de-epoxidation in monogalactosyldiacylglycerol micelles. *Plant Cell Physiol.*, 45: 92-102.

Hirel, B., Le Gouis, J., Ney, B., Gallais, A. 2007. The challenge of improving nitrogen use efficiency in crop plants towards a more central role for genetic variability and quantitative genetics within integrated approaches. *J. Exp. Bot.*, 58: 2369-2387.

Ho gberg, P. and Read, D.J. 2006. Towards a more plant physiological perspec- tive on soil ecology. *Trends in Ecology and Evolution*, 21: 549–554.

Hogberg, P., Hogberg, M.N., Go ttlicher, S.G., Betson, N.R., Keel, S.G., Metcalfe, D.B., Campbell, C., Schindlbacher, A., Hurry, V., Lundmark T, Linder, S. and Näsholm, T. 2008 . High temporal resolution tracing of photosynthate carbon from the tree canopy to forest soil microorganisms. *New Phytologist*, 177: 220–228.

Hopkins, W.G. 1999. The physiology of plants under stress. In "*Introduction to Plant Physiology*" 2nd Edition, Wiley, New York, pp. 451–475.

Hu, T., Li, H., Zhang, X., Luo, H. and Fu, J. 2011. Toxic effect of NaCl on ion metabolism, antioxidative enzymes and gene expression of perennial ryegrass. *Ecotox Environ Safe*, 74: 2050–2056.

Huang, H., Zhang. Q., Zhao, L., Feng, J., and Peng, C. 2010. Does Lutein plays a key role in the rotection of photosynthetic apparatus in *Arabidopsis* under severe oxidative stress? *Pak. J. Bot.*, 42: 2765-2774.

Huang, X.S., Luo, T., Fu, X., Fan, Q., Liu, J. 2011. Cloning and molecular characterization of a mitogen-activated protein kinase gene from *Poncirus trifoliata* whose ectopic expression confers dehydration/drought tolerance in transgenic tobacco. *J. Exp. Bot.*, 62: 5191-5206.

Hudspeth RL, Grula JW, Dai Z, Edwards GE, Ku MS. 1992. Expression of maize phosphoenolpyruvatecarboxylase in transgenic tobacco. Effects on biochemistry and physiology.*Plant Physiol.*, 98: 458-464.

Hubbard, R.M., Ryan, M.G., Stiller, V. and Sperry, J.S. 2001. Stomatal conductance and photosynthesis vary linearly with plant hydraulic conductance in Ponderosa pine. *Plant Cell Environ.*, 24: 113–121.

Humbeck, K. and Krupinska, K. 2003. The abundance of minor Chlorophyll a/b binding proteins CP29 and LHCI of barley (*Hordeum vulgare* L.) during leaf senescence is controlled by light. *J. Expt. Bot.,* 54: 372–383.

Hury, V.M., Malmberg, G., Gardestro¨m, P. and O¨quist, G. 1994. Effect of short-term shift to low temperature and of long-term cold hardening on photosynthesis and Ribulose-1,5-Bisphosphate carboxylase/oxygenase and sucrose phosphate synthase activity in leaves of winter rye (*Secale cereal* L.). *Plant Physiol.*, 106: 983-990.

Hurry, V., Strand, A., Furbank, R. and Stitt, M. 2000. The role of inorganic phosphate in the development of freezing tolerance and the acclimatization of photosynthesis to low temperature is revealed by the pho nutants of *Arabidopsis thaliana. Plant J.*, 24: 383–396

Huseynova, M., Suleymanov, S.Y., Rustamova S.M., Aliyev. J.A. 2009. Drought-induced changes in photosynthetic membranes of two wheat (*Triticum aestivum* L.) cultivars. *Russ. Biokhimiya*, 74: 1109-1116.

Imai, T., Kingston-Smith, A.H. and Foyer, C.H. 1999. Inhibition of endogenous ascorbate synthesis in potato leaves supplied with exogenous ascorbate. *Free Rad. Res.*, 31: 171–179.

Inagaki, M., Omori, E., Kim, J.Y., Komatsu, Y., Scott, G., Ray, M.K., Yamada, G., Matsumoto, K., Mishina, Y., Ninomiya-Tsuji, J. 2008. TAK1-binding protein 1, TAB1, mediates osmotic stress-induced TAK1 activation but is dispensable for TAK1-mediated cytokine signaling. *J. Biol. Chem.*, 283: 33080-33086.

Inan, G., Zhang, Q., Li, P.H., Wang, Z.L., Cao, Z.Y., Zhang, H., Zhang, C.Q., Quist, T.M., Goodwin, S.M., Zhu, J.H., Shi, H., Damsz, B., Charbaji, T., Gong, Q., Ma, S., Fredricksen, M., Galbraith, D.W., Jenk, M.A., Rhodes, D., Hasegawa, P.M., Bohnert, H.J., Joly, R.J., Bressan, R.A. and Zhu, J.K. 2004. Salt cress: a halophyte and cryophyte Arabidopsis relative model system and its applicability to molecular genetic analyses of growth and development of extremophiles. *Plant Physiol.*, 135: 1718–1737.

IPCC 2007. Climate Change 2007: The physical science basis. Contribution of working group I to the fourth assessment report of the intergovernmental panel on climate change . Cambridge, UK and NY, USA: Cambridge University Press.

Irrigation salinity, 2011. NSW Environment & Heritage, URL http://www.environment.nsw.gov.au/salinity/solutions/irrigation.htm .

Isaksson, C., Andersson, S. 2008. Oxidative stress does not influence carotenoid mobilization and plumage pigmentation. *Proc. R Soc. Biol. Sci. Ser. B*, 275: 309-314.

Islam, M.T. 2011. Effect of temperature on photosynthesis, yield attributes and yield of aromatic rice genotypes. *Int. J. Sustain. Crop Prod.*, 6: 14-16.

Iturbe-Ormaetxe, I., Escuredo, P.R., Arrese-Igor, C. and Becana, M. 1998. Oxidative damage in pea plants exposed to water deficit or paraquat. *Plant Physiol.*, 116: 173–181.

Iwaia, M., Yokonoa, M., Inadab, N., Minagawa, J. 2010. Live-cell imaging of photosystem II antenna dissociation during state transitions. *Proc. Natl. Acad. Sci. USA,* 107: 2337-2342.

Izui, K., Ishijima, S., Yamaguchi, Y., Murata, T., Shigesada, K., Sugiyama, T. and Katsuki, H. 1986. Changing and sequence analysis of cDNA encoding active phosphor enol pyruvate carboxylase of the C4-pathway from maize. *Nuclic acids Res.,* 14: 1615-1628.

Jafarinia, M., Shariati, M. 2010. Effects of salt stress on photosystem II of canola plant (*Brassica napus* L.) probing by chlorophyll a fluorescence measurements. *Iran. J. Sci. Technol.*, A1: 71-76.

Jain, M., Tiwary, S., Gadre, R. 2010. Sorbitol-induced changes in various growth and biochemical parameters in maize. *Plant Soil Environ.*, 56: 263-267.

Jaleel, C.A., Somasundaram, R., Farooq, M. Manivannan, P., Al-Juburi, H.J. Vam, R.P. Wahid, A., 2009. Drought stress in plants: a review on morphological characteristics and pigments composition. *Int. J. Agr. Biol.*, 11: 100-105.

James, R.A., Rivelli, A.R., Munns, R., von Caemmerer, S. 2002. Factors affecting CO_2 assimilation, leaf injury and growth in salt-stressed durum wheat. *Funct. Plant Biol.*, 29: 1393-1403.

Jamil, M., Rehman, S., Lee, K.J., Kim, J.M., Kim, H., Rha, E.S. 2007. Salinity reduced growth PS2 photochemistry and chlorophyll content in radish. *Sci. Agr.*, 64: 111-118.

Janknecht, R. 2003. Regulation of the ER81 transcription factor and its coactivators by mitogen- and stress-activated protein kinase 1 (MSK1). *Oncogene*, 22: 746-755.

Jeanneau M, Gerentes D, Foueillassar X, Zivy M, Vidal J, Toppan A, and Perez P. 2002. Improvement of drought tolerance in maize: towards the functional validation of the Zm-Asr1 gene and increase of water use efficiency by over-expressing C4–PEPC. *Biochimie*, 84: 1127-1135.

Jia, W., Zhang, J. 2008. Stomatal movements and long-distance signaling in plants. *Plant Signal Behav.,* 3: 772-777.

Jiang, Q.W., Kiyoharu, O. and Ryozo, I. 2002. Two novel mitogen-activated protein signaling components, OsMEK1 and OsMAP1, are involved in a moderate low-temperature signaling pathway in Rice1. *Plant Physiol.*, 129: 1880–1891.

Jin, M-X, Li, D-Y, Mi, H. 2002. Effects of high temperature on chlorophyll fluorescence induction and the kinetics of far red radiation-induced relaxation of apparent Fo in maize leaves. – *Photosynthetica*, 40: 581-586.

Johnson, G.N. 2005. Cyclic electron transport in C3 plants: fact or artefact? *J Exp Bot.*, 56: 407–416.

Jonak, C., Ökrész, L., Bögre, L., Hirt, H. 2002. Complexity, cross talk and integration of plant MAP kinase signalling. *Curr. Opin. Plant Biol.* 5: 415-424.

Jones, P.G. and Inouye, M. 1994. The cold shock response—a hot topic, *Mol. Microbiol.*, 11: 811–818.

Joshi, M.K., Desai, T.S., Mohanty, P. 1995. Temperature-dependent alterations in the pattern of photochemical and nonphotochemical quenching and associated changes in the photosystem II conditions of the leaves. *Plant Cell Physiol.* 36: 1221-1227.

Juan, M., Rivero, R.M., Romero, L., Ruiz, J.M. 2005. Evaluation of some nutritional and biochemical indicators in selecting saltresistant tomato cultivars. *Environ. Exp. Bot.*, 54: 193-201.

Jun, H., Hariji, R., Adam, L., Rozwadowski, K.L., Hammerlindl, J.L., Keller, W.A. and Selvaraj, G. 2000. Genetic engineering of glycine betaine production towards enhancing stress tolerance in plants. *Plant Physiol.*, 122: 747–756.

Kacperska, A. 1989. Metabolic consequences of low temperature stress in chilling-insensitive plants. In "*Low Temperature Stress Physiology in Crops*" Ed. P.H. Li, CRC Press, Boca Raton, FL, 1989, pp. 27–40.

Kaewsuksaeng, S. 2011. Chlorophyll degradation in horticultural crops. *Walailak J. Sci. Technol.*, 8: 9-19.

Kaiser, W.M., Heber, U. 1981. Photosynthesis under osmotic stress. Effect of high solute concentration on the permeability of the chloroplast envelope and on the activity of stroma enzymes. *Planta*, 153: 423-429.

Kaiser, W.M., Kaiser, G., Schöner, S. and Neimanis, S. 1981. Photosynthesis under osmotic stress. Differential recovery of photosynthetic activities of stroma enzymes, intact chloroplasts and leaf slices after exposure to high solute concentrations. *Planta*, 153: 430-435.

Kajala, K,, Brown, N.J., Williams, B.P., Borrill, P., Taylor, L.E., Hibberd, J.M. 2012. Multiple *Arabidopsis* genes primed for recruitment into C4 photosynthesis. *Plant J.*, 69: 47-56.

Kannan, N.D., Kulandaivelu, G. 2011. Drought induced changes in physiological, biochemical and phytochem icalproperties of *Withania somnifera* Dun. *J. Med. Plants Res.*, 5: 3929-3935.

Kant, S., Kant, P., Raveh, E. and Barak, S. 2006. Evidence that differential gene expression between the halophyte, *Thellungiella halophila*, and *Arabidopsis thaliana* is responsible for higher levels of thecompatible osmolyte proline and tight control of Na^+ uptake in *T. halophila*. *Plant Cell Environ.*, 29: 1220–1234.

Kant, P., Kant, S., Gordon, M., Shaked, R., Barak, S. 2007. *STRS1* and *STRS2*, two DEAD-box RNA helicases that attenuate *Arabidopsis* responses to multiple abiotic stresses. *Plant Physiol.*, 145: 814-830.

Kao, W.Y., Tsai, H. and Tsai, T.T., 2001. Effect of NaCl and nitrogen availability on growth andphotosynthesis of seedlings of a mangrove species, *Kandelia candel* (L.) Druce. *J. Plant Physiol.,* 158: 841–846.

Karaba, A., Dixit, S., Greco, R., Aharoni, A., Trijatmiko, K.R., Marsch-Martinez, N., Krishnan, A., Nataraja, K.N., Udayakumar, M., Pereira, A. 2007. Improvement of water use efficiency in rice by expression of *HARDY*, an Arabidopsis drought and salt tolerance gene. *Proc. Natl. Acad. Sci. USA*, 104: 15270–15275.

Karim, M.A., Fracheboud, Y., Stamp, P., 1997. Heat tolerance of maize with reference of some physiological characteristics. *Ann. Bangladesh Agri.*, 7: 27–33.

Karim, M.A., Fracheboud, Y., Stamp, P., 1999. Photosynthetic activity of developing leaves of maize (*Zea mays* L.) is less affected by heat stress than that of developed leaves. *Physiol. Plant.*, 105: 685–693.

Kawakami, K., Umenab, Y., Kamiyab, N., Shen, J. 2009. Location of chloride and its possible functions in oxygen-evolving photosystem II revealed by X-ray crystallography. *Proc. Natl. Acad. Sci. USA*, 106: 8567-8572.

Kawamitsu, Y., Driscoll, T. and Boyer, J.S. 2000. Photosynthesis during desiccation in an Intertidal alga and a land plant. *Plant Cell Physiol.*, 41(3): 344-353.

Kelly, G.J. 2006. Photosynthesis, carbon metabolism: The Calvin cycle's golden jubilee. In: "*Thirty Years of Photosynthesis 1974–2004" Ed. G.J.* Kelly, and E. Latzko. Springer, Heidelberg pp. 382-410.

Kempa S, Krasensky J, Dal Santo S, Kopka J, Jonak C. 2008. A central role of abscisic acid in stress-regulated carbohydrate metabolism. *PLoS ONE*, 3: 3935.

Kepova, K.D., Holzer, R., Stoilova, L.S., Feller, U., 2005. Heat stress effects on ribulose-1,5-bisphosphate carboxylase/oxygenase, Rubisco bindind protein and *Rubisco activase* in wheat leaves. *Biol. Plant.*, 49: 521–525.

Khafagy, M.A., Arafa, A.A., El-Banna, M.F. 2009. Glycinebetaine and ascorbic acid can alleviate the harmful effects of NaCl salinity in sweet pepper. *Aust. J. Crop Sci.*, 3: 257-267.

Khan, M.A., Shirazi, M.U., Khan, M.A., Mujtaba, S.M., Islam, E., Mumtaz, S., Shereen, A. Ansari, R.U. and Ashraf, M.Y. 2009. Role of proline, K/Na ratio and chlorophyll content in salt tolerance of wheat (*Triticum aestivum* L.). *Pak. J. Bot.*, 41: 633-638.

Kiani, S.P., Grieu, P., Maury, P., Hewezi, T., Gentzbittel, L., Sarrafi, A. 2007. Genetic variability for physiological traits under drought conditions and differential expression of water stress associated genes in sunflower (*Helianthus annuus* L.). *Biomed. – Life Sci.,* 114: 193-207.

Kitroongruang, N., Jodo, S., Hisai, J., Kato, M. 1992. Photosynthesis characteristics of melons grown under high temperatures. *J. Japan. Soc. Hort. Sci.*, 61: 107-114.

Knight, H., Trewavas, A.J. and Knight, M.R. 1996. Cold calcium signaling in *Arabidopsis* involves two cellular pools and a change in calcium signature after acclimation. *Plant Cell,* 8: 489–503.

Knight, H., Brandt, S. and Knight, M.R. 1998. A history of drought stress alters calcium signaling pathways in *Arabidopsis. Plant J.,* 16: 681-387.

Koch, K. 2004. Sucrose metabolism: regulatory mechanisms and pivotal roles in sugar sensing and plant development. *Curr. Opin. Plant Biol.*, 7: 235–246.

Koch, K.E., Nolte, K.D., Duke, E.R., McCarty, D.R. and Avigne, W.T. 1992. Sugar levels modulate differential expression of maize sucrose synthase genes. *Plant Cell*, 4: 59–69.

Kogami, H., Shono, M., Koike, T., Yanagiswa, S., Izui, K., Sentoku, S., Tanifugi, S. and chimiya, H. 1994. Molecular and physiological evaluation of transgenic tobacco plants expressing a maize phosphoenolpyruvate carboxylase gene under the control of the cauliflower mosaic virus 35S promoter. *Transgenic Res.*, 3: 287-296.

Kohler, B., Blatt, M.R. 2002. Protein phosphorylation activates the guard cell Ca^{2+} channel and is a prerequisite for gating by abscisic acid. *Plant J.*, 32: 185-194.

Kossman, J., Sonnewald, U., Willmitzer, L. 1994. Reduction of the chloroplastic fructose-1,6-bisphosphatase in transgenic potato plants impairs photosynthesis and plant growth. *Plant J.*, 6: 637-650.

Kozaki, A., Takeba, G. 1996. Photorespiration protects C3 plants from photooxidation. *Nature*, 384: 557-560.

Krishnan, H.B., Pueppke, S.G. 1987. Heat shock triggers rapid protein phosphorylation in soybean seedlings. *Biochem. Biophys. Res. Commun.*, 148: 762-767.

Ku, M. S. B., S. Agarie, M. Nomura, H. Fukuyama, H. Tsuchida, K. Ono, S. Hirose, S. Toki, M. Miyao-Tokutomi, and M. Matsuoka. 1999. High-level expression of maize phosphoenolpyruvatecarboxylase in transgenic rice plants. *Nature Biotech*, 17: 72-80.

Kuczyñska, P., Latowski, D., Niczyporuk, S., Olchawa-Pajor, M., Jahns, P., Gruszecki, W.I., and Strza³ka, K. 2012. Zeaxanthin epoxidation–an in vitro approach. *Acta Biochim. Polonica*, 59: 105-107.

Kulshrehtha, S., Mishra, D.P., Gupta, R.K. 1987. Changes in contents of chlorophyll, proteins and lipids in whole chloroplasts and chloroplast membrane fractions at different water potential in drought resistant and sensitive genotypes of wheat. *Photosynthetica*, 21: 65-70.

Kumar, A., Li, C., Portis, A.R.J. 2009. *Arabidopsis thaliana* expressing a thermostable chimeric Rubisco activase exhibits enhanced growth and higher rates of photosynthesis at moderately high temperatures. *Photosynth. Res.*, 100: 143-153.

Kumar, S., Singh, B. 2009. Effect of water stress on carbon isotope discrimination and Rubisco activity in bread and durum wheat genotypes. *Physiol. Mol. Biol. Plants*, 15: 281-286.

Kurek, I., Chang, T.K., Bertain, S.M., Madrigal, A., Liu, L., Lassner, M.W. and Zhu, G. 2007. Enhanced thermostability of *Arabidopsis* rubisco activase improves photosynthesis and growth rates under moderate heat stress. *Plant Cell*, 19: 3230-3241.

Kyparissis, A., Petropoulun, Y. and Manetas, Y. 1995. Summer survival of leaves in a soft-leaved shrub (*Phlomis fruticosa* L., Labiatae) under Mediterranean field conditions: avoidance of photoinhibitory damage through decreased chlorophyll contents. *J. Exp. Bot.*, 46: 1825-1831.

Latowski, D., Åkerlund, H.E., Strzalka, K. 2004. Violaxanthin deepoxidase, the xanthophylls cycle enzyme, requires lipid inverted hexagonal structures for its activity. *Biochemistry*, 43: 4417-4420.

Lawlor, D.W. 2001. Photosynthesis. 3rd Ed. – Scientific Publishers Limited, Oxford.

Lawlor, D.W. 2002. Limitation to photosynthesis in water stressed leaves: stomata versus metabolism and the role of ATP. *Ann. Bot.*, 89: 1-15.

Lawlor, D.W. and Cornic, G. 2002. Photosynthetic carbon assimilation and associated metabolism in relation to water deficits in higher plants. *Plant Cell Environ.*, 25: 275–294.

Lawlor, D.W. 2009. Musings about the effects of environment on photosynthesis. *Ann. Bot.*, 103: 543-549.

Lawlor, D.W., Cornic, G. 2002. Photosynthetic carbon assimilation and associated metabolism in relation to water deficits in higher plants. *Plant Cell Environ.*, 25: 275-294.

Lawson, T., Oxborough, K., Morison, J.I.L. and Baker, N.R. 2003. The responses of guard and mesophyll cell photosynthesis to CO_2, O_2, light, and water stress in a range of species are similar. *J Exp Bot.*, 54: 1743–52.

Lee, Y.P., Kim, S.H., Bang, J.W., Lee, H.S., Kwak, S.S. and Kwon, S.Y. 2007. Enhanced tolerance to oxidative stress in transgenic tobacco plants expressing three antioxidant enzymes in chloroplasts. *Plant Cell Rep.* 26: 591-598.

Leegood, R.C. and Edwards, G.E. 1996. Carbon metabolism and photorespiration: temperature dependence in relation to otherenvironmental factors. In "*Photosynthesis and the environment*" Ed. N.R. Baker, Kluwer, pp. 191-221.

Lemeille, S., Rochaix, J.D. 2010. State transitions at the crossroad of thylakoid signaling pathways. *Photosynth. Res.*, 106: 33-46.

Levi, A., Ovnat, L., Paterson, A.H., Saranga, Y. 2009. Photosynthesis of cotton near-isogenic lines introgressed with QTLs for productivity and drought related traits. *Plant Sci.*, 177: 88-96.

Li, T., Li, S., Wang, J. nd Yu, K. 2003 . Effects of water stress at different deficit intensities on transport and distribution of 14C-assimilates in micropropagated apple plants. *Eu. J. Hort. Sci.*, 68:227–233.

Li F, Vallabhaneni R, Yu J, Rocheford T, Wurtzel ET. 2008. The maize phytoene synthase gene family: overlapping roles for carotenogenesis in endosperm, photomorphogenesis and thermal stress tolerance. *Plant Physiol.*, 147(3): 1334-1346.

Li M., Liu H., Li L., Yi X., Zhu X. 2007. Carbon isotope composition of plants along an altitudinal gradient and its relationship to environmental factors on the Qinghal-Tibet Plateau. *Polish J. Ecol.* 55: 67-78.

Li, T., Zhang, Y., Liu, H., Wu, Y.T., Li, W.B. and Zhang H.X. 2010.Stable expression of *Arabidopsis* vacuolar Na^+/H^+ antiporter gene *AtNHX1* and salt tolerance in transgenic soybean for over six generations. *Chinese Sci. Bull.*, 55: 1127-1134.

Liang, Y., Chen, Q., Liu, Q., Zhang, W. and Ding, R. 2003. Exogenous silicon (Si) increases antioxidant enzyme activity and reduces lipid peroxidation in roots of salt-stressed barley (*Hordeum vulgare* L.). *J. Plant Physiol.*, 160: 1157–1164.

Lichtenthaler, H.K. 1996. Vegetation stress: an introduction to the stress concept in plants. *J. Plant Physiol.*, 148: 4–14.

Lichtenthaler, H.K. 1997. Fluorescence imaging as a diagnostic tool for plant stress. *Trends Plant Sci.*, 2: 316–320.

Lichtenthaler, H.K. 1998. The stress concept in plants: an introduction. In "*Stress of life: from molecules to man*" Ed. P. Csermely, Annals of New York Academy of Sciences, 851. New York, NY, USA, pp. 187–98.

Lingling Feng, L.L., Yujun Han, Y., Liu, G., An, B.,Yang, J., Yang, G., Li, Y. And Zhu, Y. 2007. Overexpression of sedoheptulose-1, 7-bisphosphatase enhances photosynthesis and growth under salt stress in transgenic rice plants. *Funct. Plant Biol.*, 34: 822-834.

Liska, A.J., Shevchenko, A., Pick, U. and Katz, A. 2004. Enhanced photosynthesis and redox energy production contribute to salinity tolerance in Dunaliella as revealed by homology-based proteomics. *Plant Physiol.*, 136: 2806–2817.

Liu, J., Shi, D.C. 2010. Photosynthesis, chlorophyll fluorescence, inorganic ion and organic acid accumulations of sunflower in responses to salt and salt-alkaline mixed stress. *Photosynthetica*, 48: 127-134.

Liu, N., Ko, S., Yeh, K.C., Charng, Y. 2006. Isolation and characterization of tomato Hsa32 encoding a novel heat-shock protein. *Plant Sci.*, 170: 976–985.

Liu, H., Wang, Q., Yu, M., Zhang, Y., Wu, Y., and Xhang, H. 2008. Transgenic salt-tolerant sugar beet (*Beta vulgaris* L.) constitutively expressing an *Arabidopsis thaliana* vacuolar Na^+/H^+ antiporter gene, AtNHX3, accumulates more soluble sugar but less salt in storage roots. *Plant Cell Environ.*, 31: 1325–1334.

Lu, X., Wang, Z., Wang, L., Wu, R., Pholips, J. and Deng, X. 2009. LEA 4 group genes from the resurrection plant *Boea hygrometrica* confer dehydration tolerance in transgenic tobacco. *Plant Sci.*, 176: 90-98.

Loreto, F., Centritto, M., Chartzoulakis, K. 2003. Photosynthetic limitations in olive cultivars with different sensitivity to salt stress. *Plant Cell Environ.*, 26: 595-604.

Los, D.A., Zonia, A., Sinetova, M., Kryazhov, S., Mironov, K., Valdislay, V. and Zinchenko, V.V. 2010. Stress sensors and signal transducers in cyanobacteria. *Sensors*, 10: 2386-2415.

Lundin, B., Thuswaldner, S., Shutova, T., Eshaghi, S., Samuelsson, G., Barbar, J., Anderson, B.. and Spetea, C. 2007. Subsequent events to GTP binding by the plant PsbO protein: structural changes, GTP hydrolysis and dissociation from the photosystem II complex. *Biochim. Biophys. Acta*, 1767: 500-508.

Lynch, D.V. and Thompson Jr., G.A. 1982. Low temperature-induced alterations in the chloroplast and microsomal membrane of *Dunaliella salina, Plant Physiol.*, 69: 1369–1375.

Lynch, D.V. 1990. Chilling injury in plants: the relevance of membrane lipids. In "*Environmental Injury to Plants*" Ed. F. Katterman, Academic press, New York, pp. 17–34

Mafakheri, A., Siosemardeh, A., Bahramnejad, B., Struik, P.C. and Sohrabi, Y. 2010.Effect of drought stress on yield, proline and chlorophyll contents in three chickpea cultivars. *Aust. J. Crop Sci.*, 4: 580-585.

Maggio, A., Miyazaki, S., Veronese, P., Fujita, T., Ibeas, J.I., Damsz, B., Narasimhan, M.L., Hasegawa, P.M., Joly, R.J. and Bressan, R.A. 2002. Does proline accumulation play an active role in stress-induced growth reduction. *Plant J.*, 31: 699–712.

Mahajan, S. and Tuteja, N. 2005. Cold, salinity and drought stress: An overview. *Arch. Biochem Biophys.*, 444: 139-158.

Makino, A., Mae, T. and Ohira, K .1985. Photosynthesis and ribulose-1,5-bisphosphate carboxylase /oxygenase in rice leaves from emergence through senescence. Quantitative analysis by carboxylation/ oxygenation and regeneration of ribulose-1,5- bis phosphate. *Planta*, 166: 414–420.

Makino, A. 2011. Photosynthesis, grain yield, and nitrogen utilization in rice and wheat. *Plant Physiol.*, 155: 125-129.

Manchanda, G. and Garg, N., 2008. Salinity and its effects on the functional biology of legumes. *Acta Physiol. Plant.*, 30: 595–618.

Marchand, F.L., Mertens, S., Kockelbergh, F., Beyens, L., Nijs, I., 2005. Performance of high arctic tundra plants improved during but deteriorated after exposure to a simulated extreme temperature event. *Global Change Biol.* 11: 2078–2089.

Marques da Silva, J., Arrabica, M.C. 1995. Effect of water stress on Rubisco activity of *Setaria sphacelota*. In "*Photosynthesis: From Light to Biosphere.*" Ed. P. Mathis, Vol. IV. Kluwer, Dordrect – Boston – London, pp. 545-548.

Matsuoka, M., Furbank, R., Fukayama, H., Miyao, M. 2001. Molecular engineering of C4 photosynthesis. *Annu. Rev. Plant Physiol. Plant Mol. Biol.*, 52: 297-314.

Mattana, M., Biazzi, E., Consonni, R., Locatelli, F., Vannini, C., Provera, S., and Coraggio, I. 2005. Overexpression of *Osmyb4* enhances compatible solute accumulation and increases stress tolerance of *Arabidopsis thaliana*. *Physiol. Plant.*, 125: 212-223.

Maxwell BB, *Andersson CR*, *Poole DS*, Kay SA, Chory J. 2003. HY5, Circadian clock-associated 1, and a cis-element, DET1 dark response element, mediate DET1 regulation of chlorophyll a/b-binding protein 2 expression. *Plant Physiol.*, 133: 1565-1577.

Maxwell, K., Marrison, J.L., Leech, R.M., Griffiths, H. and Horton, P. 1999. Chloroplast acclimation in leaves of *Guzmania monostachia* in response to high light. *Plant Physiology*, 121: 89–95.

Maxwell, K., Johnson, G.N. 2000. Chlorophyll fluorescence—a practical guide. *J. Exp. Bot.*, 50: 659-668.

McKersie, B.D. and Bowley, S.R. 1997. Active Oxygen and Freezing Tolerance in Transgenic Plants,Plant Cold Hardiness, Molecular Biology, Biochemistry and Physiology, Plenum, New York, pp. 203–214.

Medici, L..O., Azevedo, R..A., Canellas, L.P., Machado, A.T., Pimentel, C. *2007*. Stomatal conductance of maize under water and nitrogen deficits. *Pesq. Agropec. Bras.*, 42: 599-601.

Medrano, H., Parry, M.A., Socias, X. and Lawlor, D.W. 1997. Long term water stress inactivates Rubisco in subterranean clover. *Ann Appl Biol.*, 131: 491–501.

Medrano, H., Escalona, J.M., Bota, J., Gulías, J., Flexas, J. 2002. Regulation of photosynthesis of C3 plants in response to progressive drought: the stomatal conductance as a reference parameter. *Ann. Bot.*, 89: 895-905.

Medrano, H., Parry, M.A.J., Socias, X., Lawlor, D.W. 1997. Longterm water stress inactivates rubisco in subterranean clover. *Ann. Appl. Biol.*, 131: 491-501.

Mehta, P., Jajoo, A., Mathur, S., Bharti, S. 2010. Chlorophyll a fluorescence study revealing effects of high salt stress on photosystem II in wheat leaves. *Plant Physiol. Biochem.*, 48: 16-20.

Melcher, K., Ng, L.M., Zhou, X.E., Soon, F.F., Xu, Y., Suino-Powell, K.M., Park, S.Y., Weiner, J.J., Fujii, H., Chinnusamy, V., Kovach, A., Li, J., Wang, Y., Li, J., Peterson, F.C., Jensen, D.R., Yong, E.L., Volkman, B.F., Cutler, S.R., Zhu, J.K., Xu, H.E. 2009. A gate-latch-lock mechanism for hormone signalling by abscisic acid receptors. *Nature*, 462: 602-608.

Meloni, D.A., Oliva, M.A., Martinez, C.A. and Cambraia, J. 2003. Photosynthesis and activity of superoxide dismutase, peroxidase and glutathione reductase in cotton under salt stress. *Environ.Exp Bot.,* 49: 69–76.

Meyer, S. and Genty, B. 1999. Heterogenous inhibition of photosynthesis over the leaf surface of *Rosa rubinosa* L. during water stress and abscisic acid treatment: induction of a metabolic component by limitation of CO_2 diffusion. *Planta* , 210: 126–131.

Mittova, V., Guy, M., Tal, M, and Volokita, M. 2004. Salinity upregulates the antioxidative system in root mitochondria and peroxisomes of the wild salt-tolerant tomato species *Lycopersicon pennellii. J. Exp. Bot.,* 55: 1105–1113.

Misra, A.N., Latowski, D., Strzalka, K. 2006. The xanthophyll cycle activity in kidney bean and cabbage leaves under salinity stress. *Russ. J. Plant Physiol.*, 53: 113-121.

Miyao, M. 2003. Molecular evolution and genetic engineering of C4 photosynthetic enzymes. *J. Exp. Bot.*, 54: 179-189.

Mizoguchi, T., Irie, K., Hirayama, T., Hayashida, N., Yamaguchi-Shinozaki, K., Katsumoto, K. and Shinozaki, K. 1996.A gene encoding a mitogen-activated protein kinase kinase kinase is induced simultaneously with genes for a mitogen-activated protein kinase and an S6 ribosomal protein kinase by touch, cold, and water stress in *Arabidopsis thaliana*. *Proc. Natl. Acad. Sci. USA*, 93: 765-769.

Moing, A., Carbonne, F., Rashad, M.H. and Gaudill, J.P. 1992. Carbon fluxes in mature peach leaves. *Plant Physiol.*, 100: 1878–1884.

Moghaieb, R.E.A., Tanaka, N., Saneoka, H., Murooka, Y., Ono, H. and Monikawa, H. 2006. Characterization of salt tolerance in ectoine-transformed tobacco plants (*Nicotiana tabacum*): photosynthesis, osmotic adjustment, and nitrogen partitioning. *Plant Cell Environ.*, 29: 173-182.

Mohanty, S., Baishna, B.G., Tripathy, C. 2006. Light and dark modulation of chlorophyll biosynthetic genes in response to temperature. *Planta*, 224: 692-699.

Moradi, F., Ismail, A.M. 2007. Responses of photosynthesis, chlorophyll fluorescence and ROS-scavenging systems to salt stress during seedling and reproductive stages in rice. *Ann. Bot.*, 99: 1161-1173.

Morales, D., Rodr´ýguez, P., Dell'amico, J., Nicol´as, E., Torrecillas, A., S´anchez-Blanco, M.J., 2003. High-temperature preconditioning and thermal shock imposition affects water relations, gas exchange and root hydraulic conductivity in tomato. *Biol. Plant.* 47: 203–208.

Morot-Guadry, J.-F., Job, D. and Lea, P.J. 2001. Amino acid metabolism. In "*Plant nitrogen*" Eds. P.J. Lea, J.-F. Morot-Guadry, Berlin: Springer; pp. 167–211.

Moud, A.M., Maghsoudi, K. 2008. Salt stress effects on respiration and growth of germinated seeds of different wheat (*Triticum aestivum* L.) cultivars. *World J. Agr. Sci.*, 4: 351-358.

Mumm, P., Wolf, T., Fromm, J., Roelfsema, M.R., Marten, I. 2011. Cell type-specific regulation of ion channels within the maize stomatal complex. *Plant Cell Physiol.*, 52: 1365-1375.

Munne-Bosch, S. and Alegre, L. 2002. The function of tocopherols and tocotrienols in plants. *Crit. Rev. Plant Sci.*, 21: 31–57.

Munne-Bosch, S. and Algere, L. 2003. Drought-induced changes in the redox state of a-tocopherol, ascorbate and the diterpene cornosic acid in chloroplasts of labiatae species differing in carnosic acid contents. *Plant Physiol.*, 131: 1816–1825.

Munns, R. 2005. Genes and salt tolerance: bringing them together. *New Phytol.*, 167: 645–663.

Muranaka, S., Shimizu, K., Kato, M. 2002. A salt-tolerant cultivar of wheat maintains photosynthetic activity by suppressing sodium uptake. *Photosynthetica*, 40: 509-515.

Murchie, E.H. and Horton, P. 1998. Contrasting patterns of photosynthetic acclimation to the light environment are dependent on the differential expression of the responses to altered irradiance and spectral quality. *Pl. Cell Enviro.*, 21: 139–148.

Murchie, E.H., Hubbart, S., Peng, S. and Horton, P. 2005. Acclimation of photosynthesis to high irradiance in rice: gene expression and interactions with leaf development. *J. Expt. Bot.*, 56(411): 449-460.

Murchie, E.H., Hubbart, S., Chen, Y.Z., Peng, S.B. and Horton, P. 2002. Acclimation of rice photosynthesis to irradiance under field conditions. *Pl. Physiol.*, 130: 1999–2010.

Nakagami, H, Pitzschke A, Hirt H. 2005. Emerging MAP kinase pathways in plant stress signalling. *Trends Plant Sci.*, 10: 339-346.

Nakamura, T., Nomura, M., Mori. H., Jagendroff, A.T., Ueda, A. and Takabe, T. 2001. An isozyme of betaine aldehyde dehydrogenase in barley. *Plant cell Physiol.*, 42: 1088–1092.

Nakamoto, H., Hiyama, T., 1999. Heat-shock proteins and temperature stress.In "*Handbook of Plant and Crop Stress*" Ed. M. Pessarakli, Marcel Dekker, New York, pp. 399–416.

Nash, D., Miyao, M., Murata, N. 1985. Heat inactivation of oxygen evolution in photosystem II particles and its acceleration by chloride depletion and exogenous manganese. *Biochim. Biophys. Acta*, 807: 127-133.

Negi, J., Hashimoto-Sugimoto, M., Kusumi, K. and Iba, K., 2014. New approaches to the biology of stomatal guard cells. *Plant Cell Physiol.*, 55: 241–250.

Neta-Sharir, I., Isaacson, T., Lurie, S., Weissa, D. 2005. Dual role for tomato heat shock protein 21: protecting photosystem II from oxidative stress and promoting color changes during fruit maturation. *Plant Cell*, 17: 1829-1838.

Nishida, I. and Murata, N. 1996. Chilling sensitivity in plants and Cyanobacteria: the crucial contribution of membrane lipid, *Annu. Rev. Plant Physiol. Plant Mol. Biol.*, 47: 541–568.

Niyogi, K.K., Björkman, O., Grossman, A.R. 1997. Chlamydomonas xanthophyll cycle mutants identified by video imaging of chlorophyll fluorescence quenching. *Plant Cell*, 9: 1369-1380.

Nizam, I. 2011. Effects of salinity stress on water uptake, germination and early seedling growth of perennial ryegrass. *Afr. J. Biotechnol.*, 10: 10418–10424.

Noctor, G. and Foyer, C.H. 1998. Ascorbate and glutathione: keeping active oxygen under control. *Ann.Rev. Plant Physio.l Plant Mo.l Biol.*, 49: 249–279.

Noctor, G., Gomez, L., Vanacker, H. and Foyer, C.H. 2002. Interactions between biosynthesis, compartmentation and transport in the control of glutathione homeostasis and signaling. *J. Exp. Bot.*, 53: 1283–1304.

Noreen, Z., Ashraf, M., Akram, N.A. 2010. Salt-induced modulation in some key gas exchange characteristics and ionic relations in pea (*Pisum sativum* L.) and their use as selection criteria. *Crop Pasture Sci.*, 61: 369-378.

Noreen, Z., Ashraf, M., Akram, N.A. 2012. Salt-induced regulation of photosynthetic capacity and ion accumulation in some genetically diverse cultivars of radish (*Raphanus sativus* L.). *J. Appl. Bot. Food Qual.*, 85: 91-96.

Oguchi, R., Hikosaka, K. and Hirose, T. 2003. Does the change in light acclimation need leaf anatomy? *Pl. Cell Environ.*, 26: 505–512.

Olien, C.R. and Smith, M.N. 1997. Ice adhesions in relation to freeze stress, *Plant Physiol.*, 60: 499–503.

Ommen, O.E., Donnelly, A., Vanhoutvin, S., van Oijen, M. and Manderscheid, R. 1999. Chlorophyll content of spring wheat flag leaves grown under elevated CO_2 concentrations and other environmental stresses within the ESPACE-wheat project. *Eur. J. Agron.*, 10: 197-203.

Oren, R., Sperry, J.S., Katul, G.G., Pataki, D.E., Ewers, B.E., Phillips, N. and Schaffer, K.V.R. 1999. Survey and synthesis of intra and inter specific variation of stomatal sensitivity to vapour pressure deficit. *Plant Cell Environ.*, 22: 1515–1526.

Osmond, B., Schwartz, O. and Gunning, B. 1999. Photoinhibitory printing on leaves, visualized by chlorophyll fluorescence imaging and confocal microscopy, is due to the diminished fluorescence from grana. *Aust. J. Plant Physiol.*, 26: 171–724.

Palta, J, and Gregory, P. 1997. Drought affects the fluxes of carbon to roots and soil in ^{13}C pulse-labelled plants of wheat. Soil Biology & Biochemistry 29:1395–1403

Pan, J., Lin, S., Woodbury, N.W. 2012. Bacteriochlorophyll excitedstate quenching pathways in bacterial reaction centers with the primary donor oxidized. *J. Phys. Chem. B.*, 116: 2014-2022.

Pan, X., Lada, R.R., Caldwell, C.D., Falk, K.C. 2011. Water-stress and N-nutrition effects on photosynthesis and growth of *Brassica carinata. Photosynthetica*, 49: 309-315.

Pankoviæ, D., Sakaè, Z., Kevrešan, S. and Plesnièar, M., 1999. Acclimation to long-term water deficit in the leaves of two sunflower hybrids: photosynthesis, electron transport and carbon metabolism. *J. Exp. Bot.*, 50: 128–138.

Papageorgiou, G.C. and Morata, N. 1995. The usually strong stabilizing effects of glycine betaine on the structure and function in the oxygen evolving photosystem-II complex. *Photosynth Res.*, 44: 243–252.

Parida, A.K. and Das, A.B., 2005. Salt tolerance and salinity effects on plants: a review. *Ecotoxicol.Environ. Saf.,* 60: 324–349.

Parida AK, Dasgaonkar VS, Phalak MS, Umalkar GV, Aurangabadkar LP. 2007. Alterations in photosyn-thetic pigments, protein, and osmotic components in cotton genotypes subjected to short-term drought stress followed by recovery. *Plant Biotechnology Reports*, 1: 37-48.

Parry, M.A.J., Andralojic, P.J., Khan, S., Lea, P.J. and Keys, A.J. 2002. Rubisco activity: effects of drought stress. *Ann. Bot.*, 89: 833–839.

Parida, A.K., Mittra, B., Das, A.B., Das, T.K., and Mohanty, P. 2005. High salinity reduces the content of a highly abundant 23-kDa protein of the mangrove *Bruguiera parviflora. Planta*, 221: 135-140.

Patra, B., Roy, S., Richter, A., Majumder, A.L. 2010. Enhanced salt tolerance of transgenic tobacco plants by co-expression of *PcINO1* and *McIMT1* is accompanied by increased level of myoinositol and methlated inositol. *Protoplasma*, 245: 143-152.

Percival, G.C. 2005. The use of chlorophyll fluorescence to identify chemical and environmental stress in leaf tissue of three oak (*Quercus*) species. *J. Arboric.*, 31: 215-227.

Peri, P., Martinez, P.G., Lencinas, M.V. 2009. Photosynthetic response to different light intensities and water status of two main Nothofagus species of southern Patagonian forest, Argentina. *J. Forest. Sci.*, 55: 101-111.

Perveen, S., Shahbaz, M., Ashraf, M. 2010. Regulation in gas exchange and quantum yield of photosystem II (PSII) in saltstressed and non-stressed wheat plants raised from seed treated with triacontanol. *Pak. J. Bot.*, 42: 3073-3081.

Pettigrew, W., Meredith, J.W. 1994. Leaf gas exchange parameters vary among cotton genotypes. *Crop Sci.*, 34: 700-705.

Piao, S., Ciais, P., Friedlingstein, P., Peylin, P., Reichstein, M., Luyssaert, S., Margolis, H., Fang, J., Barr, A., Chen, A., Grelle, A., Hollinger, D.Y., Laurila, T., Lindroth, A., Richardson, A.D., Vesala, T. 2008. Net carbon dioxide losses of northern ecosystems in response to autumn warming, *Nature*, 451(7174): 49-52.

Pinhero, R.G., Rao, M.V., Palyath, G., Murr, D.P. and Fletcher, R.A. 2001. Changes in the activities of antioxidant enzymes and their relationship to genetic and paclobutrazol-induced chilling tolerance of maize seedlings. *Plant Physiol.*, 114: 695–704.

Pinheiro, H.A., Silva, J.V., Endres, L., Ferreira, V.M., Ca^mara, C.A., Cabral, F.F., Oliveira, J.F., Carvalho, L.W.T., Santos, J.M., Santos Filho, B.G., 2008. Leaf gas exchange,chloroplastic pigments and dry matter accumulation in castor bean (*Ricinus communis* L.) seedlings subjected to salt stress conditions. *Industrial Crops and Products,* 27, 385–392.

Pirzad, A., Shakiba, M. R., Zehtab-Salmasi, S., Mohammadi, S. A., Sharifi, R. S., and Hassani, 2011. Effect of water stress on leaf relative water content, chlorophyll, proline and soluble carbohydrates in *Matricaria chamomilla* L. *J. Med. Plants Res.*, 5: 2483-2488.

Pitman, M., and Lauchli, A., 2002. Global impact of salinity and agricultural ecosystems, in "*Salinity: Environment – Plants – Molecules*" Eds. A. Lauchl, U. Luttge, Kluwer Academic Publishers, Dordrecht, pp. 3–20.

Plaut, Z., Carmi, A., and Grava, A. 1996. Cotton root and shoot response to subsurface drip irrigation and partial wetting ofthe upper soil profile. *Irrig. Sci.,*16(3):107-113.

Poetsch, W., Hermans, J., Westhoff, P. 1991. Multiple cDNAs of phosphoenolpyruvate carboxylase in the C4 dicot *Flaveria trinervia*. *FEBS Lett.*, 292: 133-136.

Pollock, C.J. and Lloyed, E.J. 1987. The effect of low growth temperature upon stach, sucrose and fructose synthesis in leaves. *Ann. Bot.*, 60: 231-235.

Popovic, R., Dewez, D., Juneau, P. 2003. Application of chlorophyll fluorescence in ecotoxicology: heavy metals, herbicides, and air pollutants. In: "*Practical Applications of Chlorophyll Fluorescence in Plant Biology.*" Ed. J.R. DeEll, P.M.A. Toivonen, Kluwer Academic Publishers, Boston: pp. 151-184.

Prado, F.E., Boero, C., Gallardo, M. and Gonzalez, J.A. 2000. Effect of NaCl on germination, growth, and soluble sugar content in *Chenopodium quinoa* Willd. seeds. *Bot. Bull. Acad. Sinica.*, 41: 27–34.

Pyke, K. and Leech, R.M. 1991. A rapid image analysis screening procedure for identifying chloroplast number mutants in mesophyll cells of *Arabidopsis thaliana*. *Plant Physiol.*, 99: 1005–1008.

Quartin, V.L., Ramalho, J.C., Campos, P.S. and Nunes, M.A. 2004. A importa^ncia da investigacao na cafercultura : o problema do frio. In "Livro de Actas do 1^0 Coloquio Angola – Agricultura," Eds. Tembo, PJL, Neto, JFC, e Pomball AM, Sociedade e Desenvolvimento Rural, Instituto Superior de Agronomia Press, Lisboa, pp. 64-72.

Rahnama, A., Poustini, K., Tavakkol-Afshari, R., Tavakoli, A. 2010. Growth and stomatal responses of bread wheat genotypes in tolerance to salt stress. *Int. J. Biol. Life Sci.*, 6: 216-221.

Raines, C.A. 2006. Transgenic approaches to manipulate the environmental responses of the C3 carbon fixation cycle. *Plant Cell Environ.*, 29: 331-339.

Raines, C.A. 2011. Increasing photosynthetic carbon assimilation in C3 plants to improve crop yield: current and future strategies. *Plant Physiol.*, 155: 36-42.
Raison, J.K. and Orr, G.R. 1986. Phase transition in liposomes formed from polar lipids of mitochondria from chilling sensitive plants. *Plant Physiol.* 81: 807–811.
Ramalho, J.C., Quartin, V., Fahl, J.L., Carelli, M.L. Leitao, A.E. and Nunes, M.A. 2003. Cold acclimation ability of photosynthesis among species of the tropical *Coffea* genus. *Plant Biol.*, 5: 631-641.
Rawson, H.M., Richards, R.A., Munns, R. 1988. An examination of selection criteria for salt tolerance in wheat, barley and triticale genotypes. *Aust. J. Agr. Res.*, 39: 759-772.
Raza, S.H., Athar, H.R., Ashraf, M. 2006. Influence of exogenously applied glycinebetaine on the photosynthetic capacity of two differently adapted wheat cultivars under salt stress. *Pak. J. Bot.*, 38: 341-351.
Raza, S.H., Athar, H.R., Ashraf, M., Hameed, A. 2007. GB-induced modulation of antioxidant enzymes activities and ion accumulation in two wheat cultivars differing in salt tolerance. *Environ. Exp.Bot.*, 60: 368-378.
Reda, F., Mandoura, H.M.H. 2011. Response of enzymes activities, photosynthetic pigments, proline to low or high temperature stressed wheat plant (*Triticum aestivum* L.) in the presence or absence of exogenous proline or cysteine. *Int. J. Acad. Res.,* 3: 108-115.
Rhodes, D., and Samaras, T. 1994. Genetic control of osmoregulation in plants. In "*Cellular and molecular physiology of cell volume regulation*" Ed. S.K. Strange, Boca Raton, FL, USA: CRC Press; pp. 347–361.
Robinson, S.P., Downton, W.J.S., Millhouse, J.A. 1983. Photosynthesis and ion content of leaves and isolated chloroplasts of salt-stressed spinach. *Plant Physiol.*, 73: 238-242.
Robinson, M. and Bunce, J.A. 2000. Influence of drought-induced water stress on soybean and spinach leaf ascorbate dehydroascorbate level and redox status. *Int. J. Plant Sci.,* 161: 271–279.
Rogers, M.E., Noble, C.L. 1992. Variation in growth and ion accumulation between two selected populations of *Trifolium repens* L. differing in salt tolerance. *Plant Soil,* 146: 131-136.
Roháèek, K. 2002. Chlorophyll fluorescence parameters: the definitions, photosynthetic meaning, and mutual relationships. *Photosynthetica*, 40: 13-29.
Rokka, A., Aro, E.M., Herrmann, R.G., Andersson, B., Vener, A.V. 2000. Dephosphorylation of photosystem II reaction center proteins in plant photosynthetic membranes as an immediate response to abrupt elevation of temperature. *Plant Physiol.*, 123: 1525-1536.
Rolland, F., Baena-Gonzalez, E. and Sheen, J. 2006. Sugar sensing and signaling in plants: conserved and novel mechanisms. *Annu. Rev. Plant Biol.*, 57: 675–709.
Rontein, D., Bassat, G. and Hanson, H.D.2002. Metabolic engineering of osmoprotectant accumulation in plants. *Metab. Eng.*, 4: 49–56.
Routley, D.G. 1966. Proline accumulation in wilted ladino clover leaves. *Crop Sci.*, 6: 358-361.
Ruan, C., Shao, H., da Silva, J.A.T. 2012. A critical review on the improvement of photosynthetic carbon assimilation in C3 plants using genetic engineering. *Crit. Rev. Biotechnol.*, 32: 1-21.
Ruban, A.V., Horton, P. 1995. Regulation of non-photochemical quenching of chlorophyll luorescence in plants. *Aust. J. Plant Physiol.*, 22: 221-230.
Ruban, A.V., Pascal, A.A., Robert, B., Horton, P. 2002. Activation of zeaxanthin is an obligatory event in the regulation of photosynthetic light harvesting. *J. Biol. Chem.*, 277: 7785-7789.
Ruehr, N.K., Offermann, C.A., Gessler, A., Winkler, J.B., Ferrio, J.P., Buchmann, N. and Barnard, R.L. 2009. Drought effects on allocation of recent carbon: from beech leaves to soil CO_2 efflux. *New Phytologist*, 184: 950-961.
Ruelland, E. and Zachowski, A. 2010. How plants sense temperature. *Environ Exp Bot.*, 69: 225–232.

Sabir, P., Ashraf, M., Hussain, M., Jamil, A. 2009. Relationship of photosynthetic pigments and water relations with salt tolerance of proso millet (*Panicum miliaceum* L.) accessions. *Pak. J. Bot.*, 41: 2957-2964.

Sage, R.F., Zhu, X.L. 2011. Exploiting the engine of C4 photosynthesis. *J. Exp. Bot.*, 62: 2989-3000.

Saibo, N.J.M., Lourenço, T., Oliveira, M.M. 2009. Transcription factors and regulation of photosynthetic and related metabolism under environmental stresses. *Ann. Bot.*, 103: 609-623.

Sairam, R.K., Rao, V.K., Srivastava, G.C. 2002. Differential response of wheat genotype to long term salinity stress in relation to oxidative stress, antioxidant activity and osmolyte concentration. *Plant Sci.*, 163: 1037-1046.

Sakamoto, A., Murata, N. 2002. The role of glycinebetaine in the protection of plants from stress: clues from transgenic plants. *Plant Cell Environ.*, 25: 163-171.

Saleem, A., Ashraf, M., Akram, N.A. 2011. Salt (NaCl)-induced modulation in some key physio-biochemical attributes in okra (*Abelmoschus esculentus* L.). *J. Agron. Crop Sci.*, 197: 202-213.

Salinity Factsheet, 2011. CSIRO Land and Water, URL http://www.csiro.au/resources/Salinity-Factsheet

Salvucci, M.E., Crafts-Brandner, S.J. 2004. Relationship between the heat tolerance of photosynthesis and the thermal stability of rubisco activase in plants from contrasting thermal environments. *Plant Physiol.*, 134: 1460-1470.

Salvucci, M.E., Crafts-Brandner, S.J., 2004b. Inhibition of photosynthesis by heat stress: the activation state of Rubisco as a limiting factor in photosynthesis. *Physiol. Plant.* 120: 179–186.

Samarah, N.H., Alqudah, A.M., Amayreh, J.A. and McAndrews, G.M. 2009. The effect of late-terminal drought stress on yield components of four barley cultivars. *J. Agron. Crop Sci.*, 195: 427-441.

Sanchez, F.J., Manzanares, M., de Andres, E.F., Tenorio, J.L. and Ayerbe, L. 1998. Turgor maintenance, osmotic adjustment and soluble sugar and praline accumulation in 49 pea cultivars in response to water stress. *Field Crops Res.*, 59: 225–235.

Santos, C., Azevedo H., Caldeira, G. 2001. *In situ* and *in vitro* senescence induced by KCI stress: nutritional imbalance, lipid peroxidatin and antioxidant metabolism. *J. Exp. Bot.* 52: 351-360.

Santos, C.V. 2004. Regulation of chlorophyll biosynthesis and degradation by salt stress in sunflower leaves. *Sci. Hort.*, 103: 93-99.

Saravanavel, R., Ranganathan, R., Anantharaman, P. 2011. Effect of sodium chloride on photosynthetic pigments and photosynthetic characteristics of *Avicennia officinalis* seedlings. *Recent Res. Sci. Technol.*, 3: 177-180.

Sausen, T.L., Rosa, L.M.G. 2010. Growth and carbon assimilation limitations in *Ricinus communis* (Euphorbiaceae) under soil water stress conditions. *Acta Bot. Bras.*, 24: 648-654.

Schaeffer, H.J., Forsthoefel, N.R., Cushman, J.C. 1995. Identification of enhancer and silencer regions involved in salt-responsive expression of crassulacean acid metabolism (CAM) genes in the facultative halophyte *Mesembryanthemum crystallinum*. *Plant Mol. Biol.*, 28: 205-218.

Scandalios, J.G., Guan, L. and Polidoros, A.N. 1997. Catalases in plants: gene structure, properties, regulation and expression. In "*Oxidative stress and the molecular biology of antioxidant defenses*" Ed. J.G. Scandalios, Cold Spring Harbor, NY, USA: Cold Spring arbor Laboratory; pp. 343–406.

Schimel, D. 1995 . Terrestrial ecosystems and the carbon-cycle. *Global Change Biology*, 1: 77–91.

Schrader, S.M., Wise, R.R., Wacholtz, W.F., Ort, D.R. and Sharkey, T.D. 2004. Thylakoid membrane responses to moderately high leaf temperature in Pima cotton. *Plant Cell Environ.*, 27: 725-735.

Seckin, B., Turkan, I., Sekmen, A.H. and Ozfidan, C. 2010. The role of antioxidant defense systems at differential salt tolerance of *Hordeum marinum* Huds. (Sea barleygrass) and *Hordeum vulgare* L. (cultivated barley). *Environm Exp Bot.*, 69: 76–85.

Seemann, J.R., Sharkey, T.D. 1982. The effect of abscisic acid and other inhibitors on photosynthetic capacity and the biochemistry of CO_2 assimilation. *Plant Physiol.*, 84: 696-700.

Seemann, J.R., Critchley, C. 1985. Effects of salt stress on the growth, ion content, stomatal behaviour and photosynthetic capacity of a salt sensitive species. *Phaseolus vulgaris* L. *Plant Physiol.*, 82: 555-560.

Seemann, J.R. and Sharkey, T.D. 1986. Salinity and nitrogen effects on photosynthesis, ribulose-1, 5-bisphosphate carboxylase and metabolite pool sizes in *Phaseolus vulgaris* L. *Plant Physiol.*, 82: 555–560.

Seo, P.J., Kim, M.J., Park, J.Y., Kim, S.Y., Jeon, J., Lee, Y.H., Kim, J. and Park, C.M. 2010. Cold activation of a plasma membrane-tethered NAC transcription factor induces a pathogen resistance response in *Arabidopsis. Plant J.*, 61: 661–671.

Serraj, R., Krishnamurthy, L., Kashiwagi, J., Kumar, J., Chandra, S. and Crouch, J.H. 2004. Variation in root traits of chickpea (*Cicer arietinum* L.) grown under terminal drought. *Field Crops Res.*, 88:115–127.

Shabala, S. and Cuin, T.A. 2007. Potassium transport and plant salt tolerance. *Physiol Plant*, 133: 651–669.

Shahbaz, M., Ashraf, M. 2007. Influence of exogenous application of brassinosteroid on growth and mineral nutrients of wheat (*Triticum aestivum* L.) under saline conditions. *Pak. J. Bot.*, 39: 513-522.

Shao, H.B., Liang, Z.S. and Shao, M.A. 2006. Osmotic regulation of 10 wheat (*Triticum* aestivum L.) genotypes at soil water deficits. *Biointerfaces.*, 47: 132–139.

Sharkey, T.D. 2005. Effects of moderate heat stress on photosynthesis: importance of thylakoid reactions, Rubisco deactivation, reactive oxygen species, and thermotolerance provided by isoprene. *Plant Cell Environ.*, 28: 269-277.

Sharkey, T.D., Zhang, R. 2010. High temperature effects on electron and proton circuits of photosynthesis. *J. Integr. Plant Biol.*, 52: 712-722.

Sharkova, V.E., 2001. The effect of heat shock on the capacity of wheat plants to restore their photosynthetic electron transport after photoinhibition or repeated heating. *Russ. J. Plant Physiol.*, 48: 793–797.

Shou, H., Bordallo, P., Wang, K. 2004. Expression of the *Nicotiana* protein kinase (NPK1) enhanced drought tolerance in transgenic maize. *J. Exp. Bot.*, 55: 1013-1019.

Sikuku, P.A., Netondo, G.W., Onyango, J.C., Musyimi, D.M. 2010. Chlorophyll fluorescence, protein and chlorophyll content of three NERICA rainfed rice varieties under varying irrigation regimes. *ARPN J. Agr. Biol. Sci.*, 5: 19-25.

Sims, D.A. and Pearcy, R.W. 1992. Response of leaf anatomy and photosynthetic capacity in *Alocasia macrorrhiza* (Araceae) to a transfer from low to high light. *Am. J. Bot.* 79: 449–455.

Sims, D.A. and Pearcy, R.W. 1994. Scaling sun and shade photosynthetic acclimation of *Alocasia macrorrhiza* to whole plant performance. I. Carbon balance and allocation at different daily photon flux densities. *Pl. Cell Environ.*, 17:, 881–887

Singh, C.R., Curtis, C., Yamamoto, Y., Hall, N.S., Kruse, D.S., He, H., Hannig, E.M., Asano, K. 2005. Eukaryotic translation initiation factor 5 is critical for integrity of the scanning preinitiation complex and accurate control of GCN4 translation. *Mol. Cell Biol.,* 25: 5480-5491.

Skotnica, J., Matoušková, M., Nauš, J., Lazár, D., Dvoøák, L. 2000. Thermoluminescence and fluorescence study of changes in Photosystem II photochemistry in disiccating barley leaves. *Photosynthesis Research*, 65: 29-40.

Smeekens, S. 2000, Sugar-induced signal transduction in plants. *Annu. Rev. Plant Biol.*, 51: 49–81.

Smirnoff, N. 1995. Antioxidant systems and plant response to the environment. In "*Environment and Plant Metabolism: Flexibility and Acclimation*" Ed. V. Smirnoff, BIOS Scientific Publishers, Oxford, UK.

Smith, K.A., Ardelt, B.K., Huner, N.P.A., Krol, M., Myscich, E. and Low, P.S. 1989. Identification and partial characterization of the denaturation transition of the light harvesting complex-II of spinach chloroplast membranes. *Plant Physiol.*, 90: 492-499.

Smith, K.A., Low, P.S. 1989. Identification and partial characterization of the denaturation transition of the photosystem-II reaction centre of spinach chloroplast membranes. *Plant Physiol.*, 90: 575-581.

Socias, F.X., Correia, M.J., Chaves, M.M. and Medrano, H. 1997. The role of abscisic acid and water relations in drought responses of subterranean clover. *J. Exp. Bot.*, 48: 1281–1288.

Sohn, S.O., Back, K. 2007. Transgenic rice tolerant to high temperature with elevated contents of dienoic fatty acids. *Biol. Plant.*, 51: 340-342.

Srivastava, S., Fristensky, B., Kav, N.N.V. 2004. Constitutive expression of a PR10 protein enhances the germination of Brassica napus under saline conditions. *Plant Cell Physiol.*, 45: 1320-1324.

Stepien, P. and Klobus, G. 2005. Antioxidant defense in the leaves of C3 and C4 plants under salinity stress. *Physiol Plant*, 125: 31–40.

Stepien, P. and Klobus, G. 2006. Water relations and photosynthesis in Cucumis sativus L. leaves under salt stress. *Biol. Plant.*, 50: 610–616.

Stepien, P. and Johnson, G.N., 2009. Contrasting responses of photosynthesis to salt stress in the glycophyte *Arabidopsis* and the *Halophyte Thellungiella*: Role of the plastid terminal oxidase as an alternative electron sink. *Plant Physiol.*, 149: 1154–1165.

Steponkus, P.L., Uemura, M. and Webb, M.S. 1993. A contrast of the cryostability of the plasma membrane of winter rye and spring oat-two species that widely differ in their freezing tolerance and plasma membrane lipid composition. In "*Advances in Low-Temperature Biology*" Ed. P.L. Steponkus, vol. 2, JAI Press, London, pp. 211–312.

Stiles, K.A. and Van Volkenburgh, E.. 2002. light regulated leaf expansion in two Populus species: dependence on developmentally controlled ion transport. *J. Expt. Bot.*, 53: 1651–1657.

Stitt, M. and Hurry, V. 2002. A plant foe all seasons: alteration in photosynthetic carbon metabolism during cold acclimation in *Arabidopsis*. *Curr. Opin. Plant Biol.*, 5: 199-206.

Stone, P. 2001. The effects of heat stress on cereal yield and quality. In "*Responses and Adaptation to Temperature Stress*" Ed. A.S. Basra, Crop Food Products Press, Binghamton, NY, pp. 243–291.

Stewart, C.R. 1981. Proline accumulation: Biochemical aspects. In "*Physiology and Biochemistry of drought resistance in plants*" Eds. L.G. Paleg, and D. Aspinall, pp. 243-251.

Taiz, L., Zeiger, E. 2010. Plant Physiology. 5th Ed. Sinauer Associates, Sunderland.

Takeuchi, Y., Akagi, H., Kamasawa, N., Osumi, M., Honda, H. 2000. Aberrant chloroplasts in transgenic rice plants expressing a high level of maize NADP-dependent malic enzyme. *Planta*, 211: 65-274.

Tamura, T., Hara, K., Yamaguchi, Y., Koizumi, N. and Sano, H. 2003. Osmotic stress tolerance of transgenic tobacco expressing a gene encoding a membrane-located receptor-like protein from tobacco plants. *Plant Physiol.*, 131: 454–462.

Tanaka, Y., Sano, T., Tamaoki, M., Nakajima, N., Kondo, N., Hasezawa, S. 2005. Ethylene inhibits abscisic acid-induced stomatal closure in *Arabidopsis. Plant Physiol.*, 138: 2337-2343.

Tang, D., Qian, H., Zhao, L., Huang, D., Tang, K. 2005. Transgenic tobacco plants expressing *BoRS1* gene from *Brassica oleracea* var. *acephala* show enhanced tolerance to water stress. *J. Biosci.*, 30: 647- 655.

Tang, Y., Wen, X., Lu, Q., Yang, Z., Cheng, Z., Lu, C. 2007. Heat stress induces an aggregation of the light-harvesting complex of photosystem II in spinach plants. *Plant Physiol.*, 143: 629-638.

Taub, D. 2010. Effects of rising atmospheric concentrations of carbon dioxide on plants. *Nature Educ.Knowl.*, 1: 21.

Tavakoli, H., Mohtasebi, S.S.M., Jafari, A., Galedar, M.N. 2009. Some engineering properties of barley straw. *Appl. Eng. Agr.*, 25: 627-633.

Tayefi-Nasrabadi, H., Dehghan, G., Daeihassani, B. 2011. Some biochemical properties of guaiacol peroxidases as modified by salt stress in leaves of salt-tolerant and saltsensitive safflower (*Carthamus tinctorius* L.) cultivars. *Afr. J. Biotechnol.*, 10: 751-763.

Teige, M., Scheikl, E., Eulgem, T., Doczi, R., Ichimura, K., Shinozaki, K., Dangl, J., Hirt, H. 2004. The MKK2 pathway mediates cold and salt stress signaling in *Arabidopsis*. *Mol. Cell,* 15: 141-152.

Tewari, A.K., Tripathy, B.C. 1998. Temperature-stress-induced impairment of chlorophyll biosynthetic reactions in cucumber and wheat. *Plant Physiol.*, 117: 851-858.

Tewari, A.K., Tripathy, B.C. 1999. Acclimation of chlorophyll biosynthetic reactions to temperature stress in cucumber (*Cucumis sativus* L.). *Planta*, 208: 431-437.

Tezara, W., Mitchell, V.J., Driscoll, S.D. and Lawlor, D.W. 1999. Water stress inhibits plant photosynthe-sis by decreasing coupling factor and ATP. *Nature*, 1401: 914–917.

Thomas, R. 1999. Source-sink regulation by sugar and stress. *Curr. Opin. Plant Biol.*, 2: 198–206.

Thornton, M.K., Malik, N.J., Dwelle, R.B. 1996. Relationship between leaf gas exchange characteristics and productivity of potato clones grown at different temperatures. *Amer. Potato J.*, 73: 63-77.

Todorov, D.T., Karanov, E.N., Smith, A.R., Hall, M.A., 2003. Chlorophyllase activity and chlorophyll content in wild type and *eti 5* mutant of *Arabidopsis thaliana* subjected to low and high temperatures. *Biol. Plant.*, 46: 633–636.

Toth, S.Z., Schansker, G., Kissimon, J., Kovacs, L., Garab, G., Strasser, R.J., 2005. Biophysical studies of photosystem II-related recovery processes after a heat pulse in barley seedlings (*Hordeum vulgare* L.). *J. Plant Physiol.*, 162: 181–194.

Tóth, S.Z., Nagy, V.,Puthur, J.T., Kovács, L., Garab, G. 2011. The physiological role of ascorbate as photosystem II electron donor: protection against photoinactivation in heat-stressed leaves. *Plant Physiol.*, 156: 382-392.

Tomasz Hura. T., Grzesiak, S., Hura, K., Grzesiak, M. and Rzepka, A. 2006. Differences in the physiological state between triticale and maize plants during drought stress and followed rehydration expressed by the leaf gas exchange and spectrofluorimetric methods. *Acta Physiol. Plant.*, 28: 433-443.

Trebst, A., Depka, B. and Hollander-Czytko, H. 2002. A specific role for tocopherol and of chemical singlet oxygen quenchers in the maintenance of photosystem II structure and function in *Chlamydomonas reinhardtii*. *FEBS Lett.*, 516: 156–160.

Tsuchida, H., Tamai, T., Fukayama, H. Agarie, S., Nomura, M., Onodera, H., Ono, K., Nishizawa, Y., Lee, B.H., Hirose, H., Toki, S., Ku, M.S.B. and Miyao, M. 2001. High level expression of C4-specific NADP-malic enzyme in leaves and impairment of photoautotrophic growth in a C3 plant, rice. *Plant Cell Physiol.*, 42: 138-145.

Tuteja, N., 2007. Abscisic Acid and Abiotic Stress Signaling. *Plant Signal. Behav*, 2: 135–138.

Uemura, M. and Steponkus, P.L. 1997. Effect of Cold Acclimation on Membrane Lipid Composition and Freeze Induced Membrane Destabilization, Plant Cold Hardiness. Molecular Biology, Biochemistry and Physiology, Plenum, New York, pp. 171–79.

Van Camp, W., Capiau, K., Van Montagu, M., Inzé, D., Slooten, L. 1996. Enhancement of oxidative stress tolerance in transgenic tobacco plants overproducing Fe-superoxide dismutase in chloroplasts. *Plant Physiol.*, 112: 1703-1714.

Van Rensburg, L., Krüger, G.H.J., Eggenberg, P., Strasser, R.J. 1996. Can screening criteria for drought resistance in *Nicotiana tabacum* L. be derived from the polyphasic rise of the chlorophyll a fluorescence transient (OJIP)? *S. Afr. J. Bot.*, 62: 337-341.

Van Volkenburgh, E. 1999. Leaf expansion—an integrating plant behaviour. *Pl. Cell and Enviro.*, 22:, 1463–1473

Várkonyi, Z., Nagy, G., Lambrev, P., Kiss, A.Z., Székely, N., Rosta, L., Garab, *G*. 2009. Effect of phosphorylation on the thermal and light stability of the thylakoid membranes. *Photosynth. Res.*, 99: 161-171.

Vaz, J., Sharma, P.K. 2011. Relationship between xanthophy cycle and non-photochemical quenching in rice (*Oryza sativa* L.) plants in response to light stress. *Indian J. Exp. Bot.*, 49: 60-67.

Velikova, V., Sharkey, T.D., Loreto, F. 2012. Stabilization of thylakoid membranes in isoprene-emitting plants reduces formation of reactive oxygen species. *Plant Signal. Behav.*, 7:139-141.

Vener, A.V., Rokka, A., Fulgosi, H., Andersson, B., Herrmann, R.G. 1999. A cyclophilin regulated PP2A-like protein phosphatase in thylakoid membranes of plant chloroplasts. *Biochemistry*, 38: 14955-14965.

Verbruggen, N. and Hermans, C. 2008. Proline accumulation in plants: a review. *Amino Acids*, 35: 753 759.

Verma, S., Mishra, S.N. 2005. Putrescine alleviation of growth in salt stressed *Brassica juncea* by inducing antioxidative defense system. *J. Plant Physiol.*, 162: 669-677.

von Caemmerer, S., Farquhar, G.D. 1984. Effects of partial defoliation, changes of irradiance during growth at enhanced p(CO_2) on the photosynthetic capacity of leaves of *Phaseolus vulgaris* L. *Planta*, 160: 320-329.

Vu, J.C.V., Gesch, R.W., Allen, L.H., Boote, K.J. and Bowes, G.. 1999. CO_2 enrichment delays a rapid, drought induced decrease in Rubisco small subunit transcript abundance. *J. Plant. Physiol.,* 155: 139.

Vu, J.C.V., Gesch, R.W., Pennanen, A.H., Allen, L.H.J., Boote, K.J., Bowes, G., 2001. Soybean photosynthesis, Rubisco and carbohydrate enzymes function at supra-optimal temperatures in elevated CO_2. *J. Plant Physiol.*, 158: 295–307.

Wahid, A., Shabbir, A., 2005. Induction of heat stress tolerance in barley seedlings by pre-sowing seed treatment with glycinebetaine. *Plant Growth Reg.*, 46: 133–141.

Wahid, A., Ghazanfar, A., 2006. Possible involvement of some secondary metabolites in salt tolerance of sugarcane. *J. Plant Physiol.*, 163: 723–730.

Walters, R.G., Shephard, F., Rogers, J.J.M., Rolfe, S.A. and Horton, P. 2003. Identification of mutants of *Arabidopsis* defective in acclimation of photosynthesis to the light environment *Pl. Physiol.*, 131: 472–481

Wanner, L.A. and Junttila, O. 1999. Cold induced freezing in *Arabidopsis. Plant Physiiol.,* 120: 391-400.

Wang, D., Portis, A.R.Jr. 2007. A novel nucleus-encoded chloroplast protein, PIFI, is involved in NAD(P) H dehydrogenase complex-mediated chlororespiratory electron transport in *Arabidopsis. Plant Physiol.*, 144: 1742-1752.

Wang, W.H., Yi, X.Q., Han, A.D., Liu, T.W., Chen, J., Wu, F.H., Dong, X.J., He, J.X., Pei, Z.M., Zheng, H.L.. 2012. Calcium-sensing receptor regulates stomatal closure through hydrogen peroxide and nitric oxide in response to extracellular calcium in *Arabidopsis. J. Exp. Bot.*, 63: 177-190.

Wardlaw, I.. 1969. Effect of water stress on translocation in relation to photosynthesis and growth II. Effect during leaf development in lolium temulentum l. *Aus. J. Biol. Sci.*, 22: 1–

16. Whitmarsh, A.J., Davis, R.J. 2000. Regulation of transcription factor function by phosphorylation. *Cell. Mol. Life Sci.*, 57: 1172-1183.

Wingler, A., Quick, W.P., Bungard, R.A., Bailey, K.J., Lea, P.J. and Leegood, R.C. 1999. The role of photorespiration during drought stress: an analysis utilizing barley mutants with reduced activities of photorespiratory enzymes. *Plant cell Environ.*, 22: 361–373.

Williams, J.P., Kahn, M.U., Mitchell, K. and Johnson, G. 1988. The effect of temperature on the level and biosynthesis of unsaturated fatty acids in diglycerols of *Brassica napus* leaves. *Plant Physiol.*, 87: 904–910.

Wise, R.R., Olson, A.J., Schrader, S.M., Sharkey, T.D., 2004. Electron transport is the functional limitation of photosynthesis in field-grown Pima cotton plants at high temperature. *Plant Cel Environ.*, 27: 717–724.

Wong, C.E., Li, Y., Labbe, A., Guevara, D., Nuin, P., Whitty, B., Diaz, C., Golding, G.B., Gray, G.R., Weretilnyk, E.A., Griffith, M. and Moffatt, B.A. 2006. Transcriptional profiling implicates novel interactions between abiotic stress and hormonal responses in *Thellungiella*, a close relative of *Arabidopsis*. *Plant Physiol.*, 140: 1437–1450.

Wu, W, Zheng Y, Chen L, Wei Y, Yang R, and Yan, Z.H. 2005. Evaluation of genetic relationships in the genus *Thunb.* in China based on RAPD and ISSR markers. *Biochem Syst Ecol.*, 33: 1141–1157.

Wu, X.X., Ding, H., Chen, J., Zhang, H. and Zhu, W. 2010. Attenuation of salt-induced changes in photosynthesis by exogenous nitric oxide in tomato (*Lycopersicon esculentum* Mill. L.) seedlings. *Afr. J. Biotechnol.*, 9: 7837-7846.

Xie, X.J., Shen, *S.H.H., Li*, Y.X., Xao, X.Y. *Li, B.B.,* and Xu, D.F. *2011.* Effect of photosynthetic characteristic and dry matter accumulation of rice under high temperature at heading stage. *Afr. J. Agr. Res.*, 6: 1931- 1940.

Xin, Z. and Browse, J. 2001. Cold comfort farm: the acclimation of plants to freezing temperatures. *Plant Cell Environ.*, 23: 893-902.

Xiong, Y., Fei, S., Arora, R., Brummer, E.C., Barker, R.E., Jung, G. and Warnke, S.E. 2007. Identification of quantitative trait loci controlling winter hardiness in an annual perennial ryegrass interspecific hybrid population. *Mol. Breeding*, 19: 125–136.

Xu, Q.Z., Huang, B.R. 2001. Morphological and physiological characteristics associated with heat tolerance in creeping bentgrass. *Crop Sci.*, 41: 127-133.

Xu, W., Zhou, Y., Chollet, R. 2003. Identification and expression of a soybean nodule-enhanced PEP-carboxylase kinase gene (*NEPpcK*) that shows striking up-/down-regulation *in vivo*. *Plant J.*, 34: 441-452.

Xue-Xuan, X., Hong-Bo, S., Yuan-Yuan, M., Gang, X., Jun-Na, S., Dong-Gang, G. and Cheng-Jiang, R. 2010. Biotechnological implications from abscisic acid (ABA) roles in cold stress and leaf senescence as an important signal for improving plant sustainable survival under abiotic-stressed conditions. *Crit. Rev. Biotechnol.*, 30: 222-230.

Xu, Z., Zhou, G. 2008. Responses of leaf stomatal density to water status and its relationship with photosynthesis in a grass. *J. Exp. Bot.*, 59: 3317-3325.

Xue, G.P., McIntyre, C.L., Glassop, D., Shorter, R. 2008. Use of expression analysis to dissect alterations in carbohydrate metabolism in wheat leaves during drought stress. *Plant Mol. Biol.*, 67: 197-214.

Yamada, M., Hidaka, T., Fukamachi, H., 1996. Heat tolerance in leaves of tropical fruit crops as measured by chlorophyll fluorescence. *Sci. Hortic.*, 67: 39–48.

Yamaguchi-Shinozaki, K. and Shinozaki, K. 2006. Transcriptional regulatory networks in cellular responses and tolerance to dehydration and cold stresses. *Annu. Rev. Plant Biol.*, 57: 781–803.

Yamane, Y., Kashino, Y., Koike, H., Satoh, K., 1998. Effects of high temperatures on the photosynthetic systems in spinach: oxygen-evolving activities, fluorescence characteristics and the denaturation process. *Photosynth. Res.*, 57: 51–59.

Yamazaki, J.-Y., Kamimura, Y., Okada, M. and Sugimara, Y. 1999. Changes in photosynthetic characteristics and photosystem stoichiometries in the lower leaves in rice seedlings. *Pl. Sc.* 148: 155–163

Yan, J., He, C., Wang, J., Mao, J.H., Holaday, S.A., Allen, R.D. and Zhang, H. 2004. Overexpression of the *Arabidopsis* 14-3-3 protein GF14 in cotton leads to a "staygreen" phenotype and improves stress tolerance undermoderate drought conditions. *Plant Cell Physiol.*, 45: 1007- 1014.

Yan, K., Chen, P., Shao, H., Zhang, L. and Xu, G. 2011. Effects of short-term high temperature on photosynthesis and photosystem II performance in *Sorghum. J. Agron. Crop Sci.*, 97: 400-408.

Yanagisawa, S., Sheen, J. 1998. Involvement of maize Dof zinc finger proteins in tissue-specific and light-regulated gene expression. *Plant Cell*, 10: 75-89.

Yang, X., Liang, Z., Lu, C. 2005. Genetic engineering of the biosynthesis of glycinebetaine enhances photosynthesis against high temperature stress in transgenic tobacco plants. *Plant Physiol.*, 138: 2299-2309.

Yang, J.Y., Zheng, W., Tian, Y., Wu, Y. and Zhou, D.W. 2011. Effects of various mixed salt-alkaline stresses on growth, photosynthesis, and photosynthetic pigment concentrations of *Medicago ruthenica* seedlings. *Photosynthetica*, 49: 275-284.

Yano, S. and Terashima, I. 2004. Developmental process of sun and shade leaves in *Chenopodium album* L. *Pl. Cell and Enviro.*, 27: 781–793.

Yildiz, M., Terzi, H. 2008. Small heat shock protein responses in leaf tissues of wheat cultivars with different heat susceptibility. *Biologia*, 63: 521-525.

Yu, H., Chen, X., Hong, Y.Y., Wang, Y., Xu, P.. Ke, S.D., Liu, H.Y., Zhu, J.K., Liner, D.J. and Xiang, C.B. 2008. Activated expression of an Arabidopsis HD-START protein confers drought tolerance with improved root system and reduced stomatal density. *Plant Cell*, 20: 1134-1151.

Zeid, I.M. 2009. Trehalose as osmoprotectant for maize under salinity-induced stress research. *J.Agr. Biol. Sci.*, 5: 613-622.

Zhang, J. and Kirkham, M.B. 1996. Antioxidant response to drought in sunflower and sorghum seedlings. *New Phytol.*, 132: 361-373.

Zhang, H.X., Hodson, J.N., Williams, J.P., Blumwald, E. 2001. Engineering salt-tolerant *Brassica* plants: characterization of yield and seed oil quality in transgenic plants with increased vacuolar sodium accumulation. *Proc. Natl. Acad. Sci. USA*, 98: 12832-12836.

Zhang, J., Jia, W., Yang, J., Ismail, A.M. 2006. Role of ABA in integrating plant responses to drought and salt stresses. *Field Crops Res.*, 97: 111-119.

Zhang, L.T., Zhang, S.S., Gao, H.Y., Xue, Z.C., Yang, C., Meng, X.L. and Meng, Q.W. 2011. Mitochondrial alternative oxidase pathway protects plants against photoinhibition by alleviating inhibition of the repair of photodamaged PSII through preventing formation of reactive oxygen species in *Rumex* K-1 leaves. *Physiol. Plant.*, 143: 396-407.

Zhang, S., Klessig, D.F. 2001. MAPK cascades in plant defense signaling. *Trends Plant Sci.*, 6: 520-527.

Zhang, X., Wollenweber, B., Jiang, D., Liu, F. and Zhao, J. 2008.. Water deficits and heat shock effects on photosynthesis of a transgenic *Arabidopsis thaliana* constitutively expressing *ABP9*, a bZIP transcription factor. *J. Exp. Bot.*, 59: 839-848.

Zhao, J., Guo, S., Chen, S., Zhang, H. and Zhao, Y. 2009a.: Expression of yeast *YAP1* in transgenic *Arabidopsis* results in increased salt tolerance. *J. Plant Biol.*, 52: 56-64.

Zhao, X., Tan, H.J., Liu, Y.B., Li, X.R. and Chen, G.X. 2009b. Effect of salt stress on growth and osmotic regulation in *Thellungiella* and *Arabidopsis* callus. *Plant Cell Tiss. Organ Cult.*, 98: 97-103.

Zhu, J.K. 2001. Plant salt tolerance. *Trends Plant Sci.*, 6: 66–71.

Zhu, X.G., de Sturier, E. and Long S.P. 2007. Optimizing the distribution of resources between enzymes of carbon metabolism can dramatically increase photosynthetic rate: a numerical simulation using an evolutionary algorithim. *Plant Physiol.*, 145: 513-526.

Zhu, B., Xiong, A.S., Peng, R.H., Xu, J., Zhou, J., Xu, J.T., Jin, X.F., Zhang, Y., Hou, X.L. and Yao, Q.H. 2008. Heat stress protection in Aspen spl transgenic *Arabidopsis thaliana. BMB Rep.*, 41: 382-387.

Ziaf, K., Amjad, M., Pervez, M.A. and Rajwana, I.A. 2009. Evaluation of different growth and physiological traits as indices of salt tolerance in hot pepper *(Capsicum annuum L.). Pak. J. Bot.,* 41: 1797-1809.

Zlatev, Z. 2009. Drought-induced changes in chlorophyll fluorescence of young wheat plant. *Biotechnology*, 23: 437-441.

6

Abiotic Stress and Mineral Nutrition in Plants

A. Bhattacharya

Drought stress is one of the major limitations to the agricultural productivity worldwide. The management of plant nutrients is very helpful to develop plant tolerance to drought. Better plant nutrition can effectively alleviate the adverse effects of drought by a number of mechanisms. Drought results in increased generation of the reactive oxygen species (ROS) due to energy accumulation in stressed plants which increases the photo-oxidative effect and damage the chloroplast membrane. Application of macro-nutrients like N, K and Ca reduce the toxicity of ROS by increasing the concentration of antioxidants like superoxide dismutase (SOD); Catalase (CAT) and peroxidise (POD) in the plant cells. These antioxidants scavenge the ROS and reduce the photo-oxidation and maintain the integrity of chloroplast membrane and increase the photosynthetic rate in the crop plants. Similarly, the application of some micro- nutrients like Zn, Si and Mg also increase antioxidants concentration and improves drought tolerance in plants. In other mechanism, nutrients like P, K, Mg and Zn improve the root growth which in turn increases the intake of water which helps in stomatal regulation and enhances the drought tolerance. Application of nutrients like potassium and calcium help to maintain high tissue water potential under drought condition and improve drought tolerance by osmotic adjustment. The micronutrients like Cu and B alleviate the adverse effects of drought indirectly by activating the physiological, biochemical and metabolic processes in the plants.

Potassium status increased cell membrane stability, root growth, leaf area and total dry mass for plants living under drought conditions and also improved water uptake and water conservation. Maintaining an adequate potassium nutritional status is critical for plant osmotic adjustment and for mitigating ROS damage as induced by drought stress.

Plants can be damaged by infectious microbes such as fungi, bacteria, viruses, and nematodes. They can also be damaged by noninfectious factors, causing problems that can collectively be termed "abiotic diseases" or "abiotic disorders". Unfavorable soil properties, fertility imbalances, moisture extremes, temperature extremes, chemical toxicity, physical injuries, and other problems are examples of abiotic disorders that can reduce plant health and even kill plants. Furthermore, many of these abiotic disorders can predispose plants to diseases caused by infectious microbes. Biotic and abiotic plant diseases can be compared to infectious and non-infectious human diseases. In humans, the flu virus and streptococcal bacteria cause infectious diseases, and they spread from person to person. In contrast, non-infectious (abiotic) disorders in humans include health conditions which are not transmitted from person to person or caused by an infectious agent. For example, vitamin A deficiency can cause blindness, vitamin C deficiency can cause scurvy, excessive dietary cholesterol can cause many health problems, and exposure to mercury or lead is toxic.

Soil structure determines the soil's ability to hold water, nutrients, and oxygen and make them available to plants. The most common issue related to soil structure is compaction, which results in inadequate pore space for root growth. Clay soils, with their smaller particle size, have naturally smaller pore space and are at high risk for becoming severely compacted. Compaction can occur from a variety of sources including traffic (particularly heavy farming or construction equipment), raindrop impact, tilling operations (plow layer), and minimal crop rotation. Reduced water availability is an obvious consequence of compaction as runoff occurs more frequently in a compacted soil and available pore space to hold water is limited. However, low oxygen availability for root respiration can also be a serious consequence of restricted soil pore spaces. Although some plant species are more tolerant of the consequences of compacted soils than others, if the problem is severe, all plants in the compacted area will be affected. If compaction is suspected, use a spade to check for resistance of underlying soil. Many compacted soils are 2:1 shrink-swell clays (smectites), so cracking at the soil surface during dry periods can be an indication of compaction. Conversely, water pooling in low spots after a precipitation event could indicate reduced infiltration resulting from an underlying compacted soil.

Soil pH is the measure of the H^+ ion activity in the soil solution. A high amount of H^+ activity results in an acidic soil condition, while low activity results in a predominance of OH^- activity, leading to alkaline soil. Although some plant species have preferences for more extreme acidic or alkaline soil conditions, it is generally regarded that a slightly acidic pH range of 6-7 is most favorable for plant growth. Soil pH outside of this range can have a dramatic impact on the solubility and therefore availability of plant nutrients. Soil pH below 5.5 generally results in low availability of calcium (Ca), magnesium (Mg), and phosphorus (P), and increased solubility of aluminum (Al), iron (Fe), and boron (B). High levels of these three nutrients in low soil pH are common, and can induce toxicity symptoms in plants. Soils with pH levels above 7.8 have a high availability of Ca and Mg at the expense of P, B, Fe, manganese (Mn), zinc (Zn), and copper (Cu). Plants grown in these alkaline soils often have deficiency symptoms to these nutrients.

In most cases, low soil pH can be adjusted to a more alkaline condition prior to planting with the application of calcium carbonate (limestone). Due to the immobility of the material, tilling limestone into the top 4-6 inches will result in greater pH adjustment. If soil pH is too alkaline, sulfur (S) can be incorporated in the same manner for pH reduction. Nitrogen (N) fertilizer sources can also impact soil pH due to the role of ion exchange at the root surface. Ammonium forms of N tend to reduce soil pH while nitrate forms will cause a soil pH increase. Similarly, ammonium sulfate fertilizer is commonly used to reduce soil pH, particularly for crops that prefer these conditions. Soil borne pathogens, as well as plants, prefer certain pH ranges. Changing fertilizer N source can sometimes suppress soil borne disease as it changes the pH of the soil near the root surface, thereby changing the environment where infection occurs. Although this strategy can be successful under some circumstances, changing the bulk soil pH using fertilizer alone can be difficult. If a large change in soil pH is desired, soil tests should be conducted to determine how much limestone or S should be incorporated to reach the desired soil pH.

Plant nutrition is the study of the chemical elements and compounds that are necessary for plant growth, and also of their external supply and internal metabolism. The criteria for an element to be essential for plant growth:

- In its absence the plant is unable to complete a normal life cycle; or
- That the element is part of some essential plant constituent or metabolite.

This is in accordance with Liebig's Law of the Minimum. There are 14 essential plant nutrients. Carbon and oxygen are absorbed from the air, while other nutrients including water are obtained from the soil. Plants must obtain the following mineral nutrients from the growing media:

The primary macronutrients: nitrogen (N), phosphorus (P), potassium (K), the three secondary macronutrients: calcium (Ca), sulphur (S), magnesium (Mg), the micronutrients/trace minerals: boron (B), chlorine (Cl), manganese (Mn), iron (Fe), zinc (Zn), copper (Cu), molybdenum (Mo), nickel (Ni). The macronutrients are consumed in larger quantities and are present in plant tissue in quantities from 0.2% to 4.0% (on a dry matter weight basis). Micro nutrients are present in plant tissue in quantities measured in parts per million, ranging from 5 to 200 parts per million (ppm), or less than 0.02% dry weight.

Most soil conditions across the world can provide plants with adequate nutrition and do not require fertilizer for a complete life cycle. However, humans can artificially modify soil through the addition of fertilizer to promote vigorous growth and increase yield. The plants are able to obtain their required nutrients from the fertilizer added to the soil. A colloidal carbonaceous residue, known as humus, can serve as a nutrient reservoir. Even with adequate water and sunshine, nutrient deficiency can limit growth. Nutrient uptake in the soil is achieved by cation exchange, where root hairs pump hydrogen ion (H^+) into the soil through proton pump. These hydrogen ions displace cations attached to negatively charged soil particles so that the cations are available for uptake by the root.

Root-zone temperature is an important factor affecting plant growth and uptake of water and nutrient (Marschner *et al.,* 1996; Bode Stoltzfus *et al.* 1998). Actually, plant growth is controlled by various factors, especially root temperature and nutrient supply. In particular, nutrient uptake is affected by soil temperature (Xu and Huang, 2006). Soil temperature may influence the physico-chemical and biological processes which affect nutrient availability in soils, and in turn affect plant nutrient uptake (Hussain and Maqsood, 2011). Clarkson *et al.* (1992) reported that specific absorption rates of plant nutrients depended on soil temperature to a large extent; even a small raise in soil temperature could induce large changes in plant growth and nutrient absorption. This indicates that root zone temperature is crucially important in plant nutrient uptake and utilization.

In cold seasons, air temperature in greenhouses can rise suddenly on sunny days from a low night temperature of 12°C to high temperatures of 30°C (Miao *et al*., 2009), while the soil or solution temperature may change slowly and stay around 10°C. So, root physiology is usually limited by low root zone temperature, even as shoots are at a suitable temperature. Lee *et al.* (2004) reported root pressure, hydraulic conductivity and nutrients active transport were seriously reduced when roots were exposed to low temperature. So, nutrient uptake could be inhibited by low root zone temperature (Peng and Dang, 2003). In another example, tomato (*Lycopersicon esculentum* Mill.) shoot mineral element uptake was shown to be significantly slowed at root zone temperature cooler than 15°C (Cornillon, 1974). On the other hand, elevated root temperature promoted plant nutrient uptake by (1) increasing new root formation (Domisch *et al*., 2002), (2) changing root physiology and improving nutrient uptake (Kozlowski and Pallardy, 1997), and (3) accelerating nutrient mineralization in soil (Domisch *et al.* 2002).

Over a wide range of nutrient concentrations, plant performance has been shown to be affected by different root temperatures (Niedziela, Jr. *et al.* 2008; Ambebe *et al.* 2009). Engels (1993) reported different potassium and phosphorus uptake rates between maize (*Zea mays* L. cv. Bastion) and wheat (*Triticum aestivum* L. cv. Star) at different root zone temperatures. Maize shoot demand for nutrients seemed to control nitrogen, and potassium uptake and translocation, but not phosphorus at low solution temperatures (Engels *et al.* 1992). Moreover, phosphorus uptake was usually more suppressed by low soil temperatures than the uptake of other nutrients (Bravo-F and Uribe, 1981). The effect of temperature on nutrient uptake is difficult to generalize because these effects vary with different physiological processes and plant organs. In many studies, the nutrient solution was varied by increasing or decreasing the whole NPK concentration, while neglecting changes in single element concentrations and their effects at different root zone temperature. Yan *et al.* (2012) reported that plant response to different root zone temperature is affected by varying single nitrogen, phosphorus or potassium nutrient concentration. Quantification of the effects of root zone temperature on nutrient uptake should be investigated separately for each nutrient element.

Significant research has focused on the relationship between root zone temperature and plant utilization of nitrogen; such as the effect of root zone temperature on selective absorption of nitrate and ammonium by soybean (*Glycine max* L. Merr. cv. Wells) (Duke *et al.*, 1979), ryegrass (*Lolium multiflorum* and *Lolium perenne*) (Clarkson and Warner, 1979), arctic plant species (Atkin and Cummins, 1994), *Eucalyptus nitens* (Garnett and Smethurst, 1999), and rose (*Rosa* × *hybrida* cv. Grand Gala) plants (Calatayud *et al.*, 2008). Mean global temperature is also predicted to rise by 1.5–4.5°C, with local variations, within the next 100 years (Houghton *et al.*, 1992). It is well known that low temperature can limit shoot and root growth, and nutrient and water uptake, especially below 10 °C which is a common temperature for early crop growth in temperate regions (Bowen, 1991). In this temperature range, specific absorption rates of nutrients can show marked temperature dependency, with large changes in plant growth and nutrient absorption after a relatively small increase in soil temperature (Clarkson *et al.*, 1992; Engels and Marschner, 1992). Temperatures below the plant's optimum range usually result in increased relative investment of biomass in roots, whether temperature for the entire plant or only root temperature is considered (Clarkson *et al.*, 1992). Low soil temperatures reduce overall growth and tend to increase carbon allocation to roots because nutrient and water uptake are reduced (Lambers *et al.*, 1995). At low soil temperatures, shoot demand seems to control uptake and translocation of nitrogen, potassium and calcium but not phosphorus (Engels *et al.*, 1992). Phosphorus uptake is usually more depressed by low soil temperature than is the uptake of other nutrients (Engels *et al.*, 1992). This may be mainly due to reduced root growth because at low temperatures phosphate uptake is dominated by root production rather than by chemical nutrient availability in soil or by physiological uptake capacity (Mackay and Barber, 1984). However, direct effects of low temperature on roots or on functioning of active transport systems cannot be ruled out (Engels *et al.*, 1992; Engels and Marschner, 1992).

Three major growth limiting factors of temperate regions are thus likely to be changed in the coming century: temperature, atmospheric CO_2 concentration, and nutrient requirements. Moreover, their effects are known to be interactive and are therefore of particular interest in understanding and predicting climate change effects on agriculture. Experimental soil temperature studies are few since soil temperature is difficult to manipulate, yet it is much more stable than air temperature and changes slowly making its effects easier to model and predict than those of air temperature. Several authors have reported interactive effects of atmospheric CO_2 and air temperature on wheat biomass (Mitchell *et al.*, 1993, 1995; Kimball *et al.*, 1995; Rawson, 1995; Batts *et al.*, 1997), but the direction of the effects was not always the same or varied from year to year. However, most of these studies have been conducted in the field, with no or inadequate monitoring of soil temperature. Also, their emphasis was on plant development and crop yield measurements whereas much less attention was paid to the underlying physiological mechanisms.

Plant nutrition is a difficult subject to understand completely, partly because of the variation between different plants and even between different species or individuals of a given clone. An element present at a low level may cause deficiency symptoms, while the same element at a higher level may cause toxicity. Further, deficiency of one element may present as symptoms of toxicity from another element. An abundance of one nutrient may cause a deficiency of another nutrient. For example, lower availability of a given nutrient such as SO_2^{4-} can affect the uptake of another nutrient, such as NO_3^-. As another example, K^+ uptake can be influenced by the amount of NH_4^+ available.

The root, especially the root hair, is the most essential organ for the uptake of nutrients. The structure and architecture of the root can alter the rate of nutrient uptake. Nutrient ions are transported to the center of the root, the stele in order for the nutrients to reach the conducting tissues, xylem and phloem. The Casparian strip, a cell wall outside of the stele but within the root, prevents passive flow of water and nutrients, helping to regulate the uptake of nutrients and water. Xylem moves water and inorganic molecules within the plant and phloem accounts for organic molecule transportation. Water potential plays a key role in a plants nutrient uptake. If the water potential is more negative within the plant than the surrounding soils, the nutrients will move from the region of higher solute concentration—in the soil—to the area of lower solute concentration: in the plant.

There are three fundamental ways plants uptake nutrients through the root: 1) simple diffusion, occurs when a non-polar molecule, such as O_2, CO_2, and NH_3 follows a concentration gradient, moving passively through the cell lipid bilayer membrane without the use of transport proteins. 2) facilitated diffusion, is the rapid movement of solutes or ions following a concentration gradient, facilitated by transport proteins. 3) Active transport, is the uptake by cells of ions or molecules against a concentration gradient; this requires an energy source, usually ATP, to power molecular pumps that move the ions or molecules through the membrane.

Nutrients are moved inside a plant to where they are most needed. For example, a plant will try to supply more nutrients to its younger leaves than to its older ones. When nutrients are mobile, symptoms of any deficiency become apparent first on the older leaves. However, not all nutrients are equally mobile. Nitrogen, phosphorus, and potassium are mobile nutrients, while the others have varying degrees of mobility. When a less mobile nutrient is deficient, the younger leaves suffer because the nutrient does not move up to them but stays in the older leaves. This phenomenon is helpful in determining which nutrients a plant may be lacking. Though nitrogen is plentiful in the Earth's atmosphere, relatively few plants harbor nitrogen fixing bacteria, so most plants rely on nitrogen compounds present in the soil to support their growth. These can either be supplied by decaying matter, nitrogen fixing bacteria, animal waste, or through the agricultural application of purpose-made fertilizers.

6.1. Macronutrients (Primary)

Phosphorus

Phosphorus is important in plant bioenergetics. As a component of ATP, phosphorus is needed for the conversion of light energy to chemical energy (ATP) during photosynthesis. Phosphorus can also be used to modify the activity of various enzymes by phosphorylation, and can be used for cell signaling. Since ATP can be used for the biosynthesis of many plant biomolecules, phosphorus is important for plant growth and flower / seed formation. Phosphate esters make up DNA, RNA, and phospholipids. Most common in the form of polyprotic phosphoric acid (H_3PO_4) in soil, but it is taken up most readily in the form of H_2PO_4. Phosphorus is limited in most soils because it is released very slowly from insoluble phosphates. Under most environmental conditions it is the limiting element because of its small concentration in soil and high demand by plants and microorganisms. Plants can increase phosphorus uptake by a mutualism with mycorrhiza. A phosphorus deficiency in plants is characterized by an intense green coloration in leaves. If the plant is experiencing high phosphorus deficiencies the leaves may become denatured and show signs of necrosis. Occasionally the leaves may appear purple from an accumulation of anthocyanin. Because phosphorus is a mobile nutrient, older leaves will show the first signs of deficiency. It is useful to apply a high phosphorus content fertilizer, such as bone meal, to perennials to help with successful root formation. In phosphorus deficiency, leaves remain small, erect, unusually dark green with greenish red in sweet potato (O'Sullivan *et al.*, 1993), bluish green in chili (Balakrishnan, 1999), brown in birds foot trefoil (Russelle and McGraw, 1986) or purplish tinge in sugar maple (Bernier and Brazeau, 1988); blueberry (Tamada,1989) and sugarcane (Nautiyal *et al.*, 2000). The underside develops bronzy appearance. The root growth is also restricted under phosphorus stress in black pepper (Nybe and Nair, 1986). Anthocyanin pigment increases in leaves of barley (Hamy, 1983) and *Arabidopsis thaliana* (Trull *et al.*, 1997) under phosphorus stress,

Potassium

Potassium regulates the opening and closing of the stomata by a potassium ion pump. Since stomata are important in water regulation, potassium reduces water loss from the leaves and increases drought tolerance. Potassium deficiency may cause necrosis or interveinal chlorosis. K^+ is highly mobile and can aid in balancing the anion charges within the plant. Potassium helps in fruit colouration, shape and also increases its brix. Hence, quality fruits are produced in Potassium rich soils. It also has high solubility in water and leaches out of rocky or sandy soils. This water solubility can result in potassium deficiency. Potassium serves as an activator of enzymes used in photosynthesis and respiration Potassium is used to build cellulose and aids in photosynthesis by the formation of a chlorophyll precursor potassium deficiency may result in higher risk of pathogens, wilting, chlorosis, brown spotting, and higher chances of damage from frost and heat. Under potassium stress condition, yellowing of leaves starts from the tips or margins of leaves extending towards the center of leaf base. The yellowing is interveinal and irregular in the leaves of tomato (Besford, 1978) and blueberry (Tamada, 1989). These yellow parts become necrotic (dead spots) with leaf curling in tobacco (Arnold *et al.*, 1986); sugar maple (Bernier and Brazeau, 1988); sapota (Nachegowda *et al.*,1992) and sugarcane (Nautiyal *et al.*, 2000). There is a sharp difference between green, yellow and necrotic parts.

Nitrogen

Nitrogen is an essential component of all proteins. Most of the nitrogen taken up by plants is from the soil in the forms of NO_3^-, although in acid environments such as boreal forests where nitrification is less likely to occur, ammonium NH_4^+ is more likely to be the dominating source of nitrogen. Amino acids and proteins can only be built from NH_4^+ SO NO_3^- must be reduced. Under many agricultural settings, nitrogen is the limiting nutrient of high growth. Some plants require more nitrogen than others, such as corn (*Zea mays*). Because nitrogen is mobile, the older leaves exhibit chlorosis and necrosis earlier than the younger leaves. Soluble forms of nitrogen are transported as amines and amides. The characteristic deficiency symptom of nitrogen is the appearance of uniform yellowing of leaves including the veins, this being more pronounced on older leaves as expressed in rabbit-eye and blueberries (Tamada, 1989); Fescue (Razmjoo, 1997); *Ailanthus triphysa* (Anoop *et al.*, 1998); chili (Balakrishnan 1999) and sugarcane (Nautiyal *et al.*, 2000). The leaves become stiff and erect. In dicotyledonous crops the leaves detach easily under extreme deficiency condition. Cereal crops show characteristics 'V' shaped yellowing at the tip of lower leaves. O'Sullivan *et al.* (1993) observed relatively small and pale green leaves with dull appearance in sweet potato. If such condition of nitrogen stress do persist, the result is a decreased foliage growth and shoot growth.

6.2. Macronutrients (Secondary and Tertiary)

Sulphur

Sulphur is a structural component of some amino acids and vitamins, and is essential in the manufacturing of chloroplasts. Sulphur is also found in the iron sulphur complexes of the electron transport chains in photosynthesis. It is immobile and deficiency therefore affects younger tissues first. Symptoms of deficiency include yellowing of leaves and stunted growth. Sulfur deficiency cause leaves to become yellowish in black pepper (Nybe and Nair, 1987); potato (Gupta and Sanderson, 1993) and *Brassica oleracea* (Stuiver *et al.*, 1997) and it appears similar to nitrogen deficiency, but the symptoms are first visible on younger leaves (Russelle and McGraw, 1986). The affected leaves are narrow and the veins are paler and chlorotic than interveinal portion, especially towards the base with marginal necrosis in sugarcanc (Nautiyal *et al.*, 2000).

Calcium

Calcium regulates transport of other nutrients into the plant and is also involved in the activation of certain plant enzymes. Calcium deficiency results in stunting. This nutrient is involved in photosynthesis and plant structure. Blossom end rot is also a result of inadequate calcium. Calcium stress in plants results in chlorosis of young leaves along the veins of birds foot trefoil (Russelle and McGraw, 1986) and blueberry (Tamada, 1989), if deficiency persist longer, bleaching of upper half leaf followed by leaf tip curling do occur in black pepper (Nybe and Nair, 1987) and sugarcane (Nautiyal *et al.*, 2000). The growing bud and leaf becomes chlorotic white with base remaining green, the distortion of the tips of shoots *i.e.*, dieback was observed by Edwards and Hortan, (1979) in peach seedlings. Similarly, Spehar and Galway, (1997) found brown spots on leaves, reduced expansion and premature leaf senescence under calcium stress in soybean crop. Stress during fruiting in tomato increases susceptibility to blossom end rot (Sonneveld and Voogt, 1991 and Ho *et al.*, 1999). Calcium stress is also responsible for other disorders such as bitter pit in apple (Monge *et al.*, 1995 and Silva and Rodriguez, 1996); leaf tip burn in cabbage (Miao *et al.*, 1997) and lettuce; black heart of celery; cavity spot of carrots (Scaife and Clarkson 1978); vitrescence in melons (Jean-Baptist *et al.*, 1999).

Magnesium

Magnesium is an important part of chlorophyll, a critical plant pigment important in photosynthesis. It is important in the production of ATP through its role as an enzyme cofactor. Magnesium deficiency causes yellowing, but differs from that of nitrogen. The yellowing takes place in between veins of older leaves (Makkanen, 1995) of *Picea abies* and veins remain green, this is followed by necrosis of tissues in birdsfoot trefoil (Russelle and McGraw, 1986), melons (Simon *et al.*, 1986), black pepper (Nybe and Nair, 1987) and blueberry (Tamada, 1989). Mg deficiency may be induced in tomatoes by high levels of ammonium in the nutrient solution (Kafkafi *et al.*, 1971).

Silicon

In plants, silicon strengthens cell walls, improving plant strength, health, and productivity. Other benefits of silicon to plants include improved drought and frost resistance, decreased lodging potential and boosting the plant's natural pest and disease fighting systems. Silicon has also been shown to improve plant vigor and physiology by improving root mass and density, and increasing above ground plant biomass and crop yields. Although not considered an essential element for plant growth and development (except for specific plant species - sugarcane and members of the horsetail family), silicon is considered a beneficial element in many countries throughout the world due to its many benefits to numerous plant species when under abiotic or biotic stress. Silicon is the second most abundant element in earth's crust. Higher plants differ characteristically in their capacity to take up silicon. Depending on their SiO_2content they can be divided into three major groups:

- Wetland graminae-wetland rice, horsetail (10-15%)
- Dryland graminae-sugar cane, most of the cereal species and few dicotyledons species (1-3%)
- Most of dicotyledons especially legumes (<0.5%)
- The long distance transport of Si in plants is confined to the xylem. Its distribution within the shoot organ is therefore determined by transpiration rate in the organs
- The epidermal cell walls are impregnated with a film layer of silicon and effective barrier against water loss, cuticular transpiration rate in the organs.

Silicon can stimulate growth and yield by several indirect actions. These include decreasing mutual shading by improving leaf erectness, decreasing susceptibility to lodging, preventing manganese and iron toxicity.

6.3. Micronutrients

Some elements are directly involved in plant metabolism however, this principle does not account for the so-called beneficial elements, whose presence, while not required, has clear positive effects on plant growth. Mineral elements those either stimulates growth but are not essential, or that are essential only for certain plant species, or under given conditions, are usually defined as beneficial elements.

Iron

Iron is necessary for photosynthesis and is present as an enzyme cofactor in plants. Iron deficiency can result in interveinal chlorosis and necrosis. Iron is not the structural part of chlorophyll but very much essential for its synthesis. Copper deficiency can be responsible for promoting an iron deficiency. The principal veins remain conspicuously green and surrounding portion of the younger leaves turn yellow tending towards whiteness in chickpea (Mehrotra and Gupta, 1990; Saxena *et al.*, 1990); groundnut (Reddy *et al.*, 1993); radish, cauliflower, cabbage and sorghum (Preeti *et al.*, 1994); lentil (Zaiter and Ghalayini, 1994) and soybean (Fonts and Cox, 1998). Under sever iron deficiency, most part of the leaf becomes white (Russelle and McGraw, 1986).

Molybdenum

Molybdenum is a cofactor to enzymes important in building amino acids. Involved in nitrogen metabolism. It is part of nitrate reductase enzyme. The common symptoms of molybdenum deficiency in plants include a general yellowing, marginal and interveinal chlorosis, marginal necrosis, rolling, scorching and downward curling of margins in poinsettia cultivars (Cox and Bartley, 1987; Cox, 1992) and in various field, horticulture and forage crops (Gupta, 1997). The deficiency of molybdenum in cauliflower causes the disorder described as 'Whiptail' (Duval *et al.*, 1991).

Boron

Boron is important for binding of pectins in the RGII region of the primary cell wall, secondary roles may be in sugar transport, cell division, and synthesizing certain enzymes. Boron deficiency causes necrosis in young leaves and stunting. Boron deficiency causes yellowing or chlorosis of youngest leaves and stems (Yu *et al.*, 1998) which starts from the base to the tip. Rosetting of terminal shoots of potato (Roberts and Rhee, 1990). Leaf tip burn, elongate and become whitish brown in rice (Yu *et al.*, 1998). Death of terminal bud occurs in extreme cases. Boron deficiency causes brown heart in radish (Shelp *et al.*, 1987) and crown choking in coconut (Baranwal *et al.*, 1989).

Copper

Copper is important for photosynthesis. Symptoms for copper deficiency include chlorosis. Involved in many enzyme processes. Copper is necessary for proper photosynthesis and is involved in the manufacture of lignin (cell walls). Copper is also involved in grain production. It is also hard to find in some conditions. In copper deficiency, visible foliar symptoms appear on young leaves as chlorosis changing to necrosis (Del, 1994); rolling, wilting and twisting of leaves in wheat (Owuoche, 1995). The later affected leaves appear papery and twisted in rice (Nautiyal *et al.*, 1999).

Manganese

Manganese is necessary for photosynthesis, including the building of chloroplasts. Manganese deficiency may result in coloration abnormalities, such as discolored spots on the foliage. The principal veins as well as smaller veins are green, the interveinal portion become chlorotic in *Ailanthus triphysa* (Anoop *et al.*, 1998) followed by necrosis and browning of interveinal tissue in melons (Simon *et al.*, 1986). The affected young leaves remain small and abscise before older leaves in birds foot trefoil (Russelle and McGraw, 1986).

Sodium

In C4 plants, sodium is a micronutrient that aids in metabolism, specifically in regeneration of phosphoenolpyruvate and synthesis of chlorophyll (Kening, 2008). In others, it substitutes for potassium in several roles, such as maintaining turgor pressure and aiding in the opening and closing of stomata (Subbarao *et al.*, 2003). Excess sodium in the soil limits the uptake of water due to decreased water potential, which may result in wilting; similar concentrations in the cytoplasm can lead to enzyme inhibition, which in turn causes necrosis and chlorosis (Zhu, 2001). To avoid these problems, plants developed mechanisms that limit sodium uptake by roots, store them in cell vacuoles and control them over long distances excess sodium may also be stored in old plant tissue, limiting the damage to new growth.

Zinc

Zinc is required in a large number of enzymes and plays an essential role in DNA transcription. A typical symptom of zinc deficiency is the stunted growth of leaves, commonly known as "little leaf" and is caused by the oxidative degradation of the growth hormone auxin. The leaves become narrow and small in chili (Balakrishnan, 1999), the lamina becomes chlorotic in sweet potato (O'Sullivan *et al.*, 1993), sour orange seedlings (Swietlik, 1995) and chickpea (Khan *et al.*, 1998), while veins remain green. Subsequently, dead spots develop all over the leaf including veins, tips and margins under sever deficiency, shoot growth is reduced (O'Sullivan *et al.*, 1993; Swietlik, 1995; Yu and Rengel, 1999). Khaira disease in rice results due to zinc deficiency (Sahi *et al.*, 1992). Shoot elongation is reduced and a tuft or rosette of distinctly narrow leaves is produced at the shoot terminal in apple and pear. The symptoms are termed 'little leaf' or 'rosette' (Hanson, 1993).

Nickel

In higher plants, Nickel is absorbed by plants in the form of Ni^{2+} ion. Nickel is essential for activation of urease, an enzyme involved with nitrogen metabolism that is required to process urea. Without nickel, toxic levels of urea accumulate, leading to the formation of necrotic lesions. In lower plants, nickel activates several enzymes involved in a variety of processes, and can substitute for Zinc and Iron as a cofactor in some enzymes. Plant growth is reduced and older leaves turn chlorotic giving plants a nitrogen deficient phenotype, when grown on urea-based nutrient solutions not supplemented with Ni in tomato and soybean (Krogmeier *et al.*, 1991). Similar results were obtained in oilseed-rape, zucchini and soybean by Gerendas and Sattelmacher (1997).

Chlorine

Chlorine, as compounded chloride, is necessary for osmosis and ionic balance; it also plays a role in photosynthesis. The symptoms of chlorine deficiency develop first on the older leaves. Discrete patches of pale green chlorotic tissue appear between the main vein near the tip of the leaf, downward cupping of some of the older leaves of Kiwifruit was observed by Smith *et al.* (1987). The leaflets of youngest leaves shrivel completely, older leaflets develop a brown necrosis which start near the tip and extend backwards particularly at the margins of red clover (Whitehead, 1985).

Cobalt

Cobalt has been proven to be beneficial to at least some plants, but is essential in others, such as legumes where it is required for nitrogen fixation for the symbiotic relationship it has with nitrogen-fixing bacteria. Vanadium may be required by some plants, but at very low concentrations. It may also be substituting for molybdenum. Selenium and sodium may also be beneficial. Sodium can replace potassium's regulation of stomatal opening and closing.

- The requirement of Co for N_2 fixation in legumes and non-legumes have been documented clearly
- Protein synthesis of Rhizobium is impaired due to Co deficiency
- It is still not clear whether Co has direct effect on higher plant

Aluminium

Tea has a high tolerance for aluminium toxicity and the growth is stimulated by aluminium application. The possible reason is the prevention of copper, manganese or phosphorus toxicity effects. There have been reports that aluminium may serve as fungicide against certain types of root rot. Aluminum is the major component of soil clay minerals. In acidic soils (pH, efficiency, especially of phosphorus and calcium, tolerance to Al toxicity is the major trait that allows plants to adapt to acidic soils (Baligar and Fageria, 1997).

It has been suggested that aluminium toxicity invariably affects biochemical and physiological processes, such as: (1) cellular and ultrastructural modifications in leaves (Vitorello *et al.*, 2005), (2) changes in chloroplasts form and arrangement of the granum (Moustakas *et al.*, 1997), (3) reduction of stomatal opening and decreased photosynthetic activity (Vitorello *et al.*, 2005), (4) reduction of photochemical efficiency of photosystem II (Moustakas and Ouzounidou, 1994), (5) chlorosis and leaf necrosis (Vitorello *et al.*, 2005), (6) decrease of ATP concentration in leaves (Loren-Plucinska and Ziegler, 1996),

(7) damage of the outer membrane of chloroplasts (Hampp and Schnabi, 1975), (8) cell membrane lipid peroxidation (Yamamoto *et al.*, 2001), (9) imbalance in cell wall assembly (Teroaka *et al.*, 2002), (10) interaction with cytosolic calcium (Jones *et al.*, 1999), and (11) intermediates of the phosphoinositide signaling cascades (Mart´ýnez-Estevez *et al.*, 2003). The effects of aluminium on photosynthesis are probably indirect (Mihailovic *et al.*, 2008). The first symptom of aluminium toxicity is the inhibition of root growth, which can occur after a brief exposure to aluminium (Dipierro *et al.*, 2005). The root meristem is considered the primary site of aluminium toxicity, suggesting that aluminium interacts actively with the cell division and elongation (Ciamporova, 2002). In addition, aluminium induces disturbances in the trans-membrane transport of ions [nitrogen, phosphorus, potassium, calcium and magnesium in plant roots (Kochian, 1995), becoming indirectly responsible for the impairment of root-shoot transport and metabolic processes in shoots (Mihailovic *et al.*, 2008). Aluminium inhibits the net uptake of Ca^{2+}, Mg^{2+}, and nitrogen by roots, whereas phosphorus uptake appears to be unaffected (Moustakas *et al.*, 1995). Aluminium inhibits Ca^{2+} uptake more than that of other ions (Moustakas and Ouzounidou, 1994). Witches Broom disease (*Monililiophthroa pernicosa*) severely affects the production of cacao and genotypes resistant to this fungal disease are needed to overcome the effects of this disease (Gesteira *et al.*, 2007).

Damage from excessive macronutrient levels can occur in crop and ornamental plants as the result of over-application of fertilizers or manures. Nitrogen toxicity is most typical under hot, dry conditions and plants turn an overly-deep shade of green. Lesions often occur on the stems of annual seedlings and these can be confused with canker diseases. Similarly, twisting and distorting of mature tomato plants that experience ammonium toxicity may appear similar to symptoms caused by viruses. Ammonium toxicity can be a problem in greenhouse soils because of the lack of specific microorganisms that convert ammonium to nitrite and then to nitrate.

Micronutrient toxicities are common in many production systems. Symptoms often include chlorosis or necrosis on leaf margins or tips, but leaf spotting, flecking, and other symptoms can occur. Many nutrient toxicities are triggered by excessively low or high soil pH. Micronutrient toxicities are particularly common in greenhouse floriculture. For example, iron and manganese toxicity often occurs in greenhouse crops when the growing medium has a low pH. Excessive micronutrients can also occur when irrigation water or soil has significant concentrations of micronutrients. Other elements that are not plant nutrients, *e.g.*, lead and arsenic, and heavy metals can also be toxic to plants.

6.4. Inter-relationships Between Nutrient Elements

Just like humans require a balanced diet with appropriate amounts of carbohydrates, proteins, vitamins, minerals, fats and water, plants too require conditions of balanced plant nutrition. Nutrient balancing in micronutrients is as important and yet more difficult than balancing between macronutrients. There is a pre-determined ratio of nutrients that is required by the plant system, depending on its life cycle, environment and its genotypic characteristics to realize its maximum genetic potential. This ratio of elements is more critical than the actual concentration of the individual elements.

There are basically two kinds of interactions between nutrients. SYNERGISM is a positive effect between nutrients and ANTAGONISM is a negative effect between nutrients. Two or more elements working together to create an overall improved physiological state in the plant is called physiological synergism while, excess of one nutrient reducing the uptake of another nutrient is called physiological antagonism. These interactions depend on soil type, physical properties, pH, ambient temperatures and proportion of participating nutrients. There is a highly controlled selectivity process involved in uptake of nutrients by plants and that is the reason why the plant does not contain the same ratio of nutrients inside the plant as found in the soil. For example, alkaline soils normally contain higher calcium levels than potassium but, when a plant growing in this field is analyzed, it contains higher potassium level than calcium. Synergism and antagonism between two mineral nutrients become even more important when the contents of both elements are near deficiency range.

6.4.1. Synergism

Many soil scientists, plant physiologists and plant biochemists have tried to clarify the much complicated relationships between nutrients. Some of these relationships are straight forward but, most are not. A few examples from agricultural laboratory research and field based experiments have shown that an:

- Optimum supply of nitrogen ensures optimum uptake of potassium as well as phosphorus, magnesium, iron, manganese and zinc from the soils.
- Optimal levels of copper and boron improve nitrogen uptake by plant.
- Optimal levels of molybdenum improve utilization of nitrogen as well as increases uptake of phosphorus.
- Optimal levels of calcium and zinc improve uptake of phosphorus and potassium.

- Optimal levels of sulphur increases the uptake of manganese and zinc.
- Optimal levels of manganese increases uptake of copper.

6.4.2. Antagonism

- Excessive amounts of nitrogen reduce the uptake of phosphorus, potassium, iron and almost all secondary and micronutrients like calcium and magnesium iron, manganese, zinc and copper.
- Excessive amounts of phosphorus reduces uptake of cationic micronutrients like iron, manganese, zinc and copper.
- Excessive amounts of potassium reduce uptake of magnesium to a greater extent and calcium to a lesser extent.
- Excessive amounts of calcium reduces uptake of iron.
- Excessive Iron reduces zinc uptake.
- Excessive zinc reduces manganese uptake.

The examples above show that the interrelationships between nutrients in the plant system are quite complicated and interdependent. More research will need to be done on the molecular levels to elucidate the actual relationships if possible.

6.5. Potassium Inter-relationships with Macronutrients

The interaction between nitrogen and potassium is historically well documented by experiments started in 1852 at Rothamsted Station. There is a strong interaction between these two nutrients in crop growth. It was observed that the crop response to applied nitrogen fertilizers decreases when the exchangeable potassium content of a soil is below the optimal level. The "reading" of the genetic code in plant cells to produce proteins and enzymes would be impossible without adequate potassium. Although, nitrogen is fundamental in production of proteins, plants deficient in potassium will not produce proteins despite an abundance of available nitrogen. Instead, incomplete protein such as amino acids, amides and nitrate accumulate in the cell. This is because; the enzyme nitrate reductase which catalyzes the formation of proteins is activated by potassium. Similarly, the enzyme responsible for synthesis of starch, starch synthetase, is also activated by potassium. Under inadequate potassium levels, the level of starch declines, while soluble carbohydrates and nitrogen-based compounds accumulate. Lower amounts of starch means less amounts of it moving from source to sink and hence, poor quality of end product. In addition, leaves that harbor excessive amounts of soluble carbohydrates turn soft and fleshy and are often perfect candidates for a pest/fungus attack. Practical implication of this interaction is that there is little point in applying large amounts of nitrogen when, there is too little exchangeable potassium in the soil because the nitrogen is used inefficiently and represents a financial cost to the grower. Sucking pest attack also seems to be an after effect of imbalance between nitrogen and potassium.

Potassium also plays a major role in the transport of water and nutrients throughout the plant in the xylem. Maintaining the water content of the vacuole requires a sufficient concentration of salts in the water to sustain the osmotic concentration and potassium is the salt most plants seem to prefer to do this. When potassium supply is reduced, translocation of nitrates, phosphates, calcium magnesium and amino acids is depressed. As with phloem transport systems, the role of potassium in xylem transport is often in conjunction with specific enzymes and plant growth hormones. An ample supply of potassium is essential to efficient operation of these systems. Optimal levels of zinc and copper increase uptake of phosphorus. Zinc is a component of dehydrogenases, proteinase and peptidase enzyme. It promotes growth hormones and starch formation and promotes seed maturation. Copper also has very important roles to play in plant biochemistry. It is a component of lactose and other oxidase enzymes. It also has a role to play in phosynthesis and protein and carbohydrate metabolism. Plants that have excessive amounts of phosphorus and potassium often produce "bitter" fruit/end product.

6.5.1. Potassium Inter-relationships with Secondary Nutrients

Agronomists and plant nutritionists have long been aware of the greatest competition that occurs between ions with similar size, valency and ion charge. Calcium, magnesium and potassium ions are quite similar in size and charge and hence, exchange sites cannot distinguish the difference between the ions. Often times, they indiscriminately accept either ion regardless of which ion is meant for that site. Generally, the binding strengths of potassium and calcium are much stronger than magnesium and they easily out-compete magnesium at exchange sites. Applications of potassium fertilizers reduce a plant's ability to absorb magnesium. Similarly, very high rates of magnesium fertilizers will indeed depress potassium absorption by plants, but this antagonism is not nearly as strong as the inverse relation of potassium on magnesium. The rate of potassium uptake is also influenced by soil pH, which further aggravates potassium magnesium antagonism in high pH soils. Reiterating, calcium, magnesium, and potassium compete with each other and the addition of any one of them will reduce the uptake rate of the other two.

6.5.1.1. Potassium and Sodium

A huge issue that is stifling cereal crop production is dry land salinity worldwide. The problems of salinity are twofold: lack of water and excessive sodium. The bread bowls of the world like the plains of Canada, Australia, India, Central Asia, Russia and California are all severely affected by salinization of soils. This means that once productive lands have been taken out of agricultural productivity. Today approximately 20% of the world's cultivated land and nearly half of all irrigated lands are salt-affected. Alkaline soils are difficult to take into agricultural production. Rain water stagnates on the soil easily and in dry periods, irrigation is hardly possible. Agriculture is limited to crops tolerant to surface water logging (*e.g.* rice, grasses) and the productivity is low.

The only element that affects sodium levels in the soil is potassium. When sodium percentage is higher than potassium then there is often trouble on the horizon from soil health and crop productivity point of view. At higher levels of sodium, plants will preferentially take up sodium in place of potassium. Plants use both low and high affinity systems for potassium uptake. Under sodium stress, it is necessary for plants to operate the more selective high-affinity potassium uptake system in order to maintain adequate potassium nutrition. Potassium deficiency inevitably leads to growth inhibition because potassium plays a critical role in maintaining cell turgor, membrane potential and enzyme activities. Once sodium gets into the cytoplasm, it inhibits the activities of many enzymes. This inhibition is also dependent on how much potassium is present: higher sodium/potassium ratio more the damage. Even in the case of halophytes that accumulate large quantities of sodium inside the cell, their cytosolic enzymes are just as sensitive to sodium as enzymes of glycophytes. Sodium when taken up by the cells, starts expanding especially under hot temperatures by absorption of cell moisture. This often leads to cell wall rupturing and oozing of cell components and slow programmed death of the plant.

As a domino effect, when sodium and potassium ratio is excessive (ratio >10) then plant will not get enough manganese. It isn't that manganese is absent; it is just tied up and hence, unavailable to the plant. Now, manganese is essential to grain formation and hence, there will be a dip in yields, high degree of unfilled grain and impoverished stands. An important factor in the battle between sodium and potassium ions is calcium. Increased calcium supply has a protective effect on plants under sodium stress. Adequate levels of calcium assist in potassium/sodium selectivity. This beneficial effect of calcium is mediated through a signaling pathway that regulates activity of potassium and sodium transporters. Calcium may also directly suppress sodium import mediated by nonselective cation channels.

6.5.2. Potassium Inter-relationships with Micronutrients

Potassium has direct synergistic relationships with two micronutrients namely: iron and manganese. Manganese is a very important component of photosynthesis, nitrogen metabolism and nitrogen assimilation; it activates decarboxylase, dehydrogenase and oxidase enzymes. Iron plays a very important part in chlorophyll formation. It is a component of ferrodoxin which is responsible for oxidation/reduction reactions in the plant system like - nitrate and sulphate reduction and nitrogen fixation. Iron is also a constituent of peroxidase and catalase, which are defense enzymes of the plant.

Domino effects of potassium are very common in the plant system due to the complex relationship between potassium and other nutrients. Boron and potassium have overlapping roles to play in plant physiology and hence, are synergistic. Like potassium, boron is also involved in some aspects of flowering and fruiting processes, pollen germination, cell division, nitrogen metabolism, carbohydrate metabolism, active salt absorption, hormone movement and action, water metabolism and the water relations in plants. They both serve in acting as a buffer and are necessary in the maintenance of conducting tissues and to exert a regulatory effect on other elements. It has been show that an optimal level of boron increases potassium permeability in the cell membrane.

An inverse relationship is seen between potassium and molybdenum. Low supplies of potassium reduce the uptake of molybdenum from soil, thereby resulting in inefficient nitrogen utilization by the plant. If this occurs during the reproductive phase, this can often lead to a higher proportion of sterile female flowers. Often times along with potassium deficiency symptoms the plant will show classic nitrogen deficiency symptoms in spite of good levels of nitrogen in the soils. Zinc and copper also have indirect effects on potassium. The tendency of plants to maintain a constant amount of total cations, on a chemically equivalent basis, however, leads to some rather complex relationships. To ascribe the effect of a particular manifestation on the concentration of one particular element becomes difficult.

6.6. Nutrient Deficiency

Nutrient deficiencies often result from a lack of plant nutrients in the soil (Table 1). However, as noted above, certain nutrient deficiencies, such as Ca, P, and Fe, occur due to poor soil conditions (such as inappropriate soil pH) for nutrient uptake and utilization by the plant. Two of the easiest ways to recognize nutrient deficiencies are the lack of visible pathogen signs (infectious microbe parts, such as mycelium) and the relatively uniform distribution pattern of symptoms in the field as compared to many diseases caused by plant pathogens.

However, plant nutrient deficiencies are best diagnosed using plant tissue analysis. As opposed to soil nutrient analysis, plant tissue analysis allows one to determine plant nutrient uptake rather than plant nutrient availability. Because nutrient deficiencies lack visible signs, they are often mistaken for virus diseases. One of the best ways to diagnose nutrient disorders is the distribution of symptoms on the plant. Mobile nutrients are readily transferred within the plant to the growing points so symptoms appear on the lower (older) leaves of the plant. Conversely, immobile nutrients display symptoms on the meristem of the plant during nutrient deficiencies.

Table 1: A list of some essential plant nutrients (elements) and their relative importance and mobility in plants

Plant Nutrient	Macro / Micro	Mobility	Comments
Nitrogen (N)	Primary Macro	Mobile	Easily leached
Phosphorus (P)	Primary Macro	Mobile	pH strongly affects
Potassium (K)	Primary Macro	Mobile	Fruit quality
Calcium (Ca)	Secondary Macro	Non mobile	Excess can limit others
Magnesium (Mg)	Secondary Macro	Mobile	Leach if Ca not present
Sulphur (S)	Secondary Macro	Non mobile	Sulphur may acidify soil
Boron (B)	Micronutrient	Non mobile	Remedy with borax
Copper (Cu)	Micronutrient	Non mobile	Deficiency rare
Iron (Fe)	Micronutrient	Non mobile	pH strongly affect
Manganese (Mn)	Micronutrient	Non mobile	Absorb via leaves
Molybdenum (Mo)	Micronutrient	Non mobile	Important for legumes
Zinc (Zn)	Micronutrient	Non mobile	High pH leads to deficiency

Nitrogen (N) deficiency is a major limitation for non-leguminous agricultural plants, and use of N fertilizer for crop and ornamental plants is higher than any other single macro- or micronutrient. Nitrogen is important for the production of chlorophyll, the pigment that makes plant tissues green. Plants deficient in N typically have a pale yellow color (chlorosis) as a result of reduced chlorophyll production. Nitrogen is also a vital element for many other plant physiological processes as a component of proteins. Therefore, plants that are deficient in N may also appear stunted and display poor vigor. Because N is highly mobile in the plant, N deficiency is usually observed first in the older leaves of the plant. These older leaves may senesce while younger shoots near the apical meristem remain healthy. Nitrogen deficiencies are common in non-legume crops because N is quickly leached out of the soil once it is converted into NO_3^- by soil microbes. Plants can absorb N in two ionic forms, NO_3 and NH_4. Soil amendments that are commonly used to provide N to plants include a variety of synthetic fertilizers in addition to cover crops and compost applications. Nitrogen deficiencies may also result from infection by root pathogens such as root-knot nematodes (*Meloidogyne* spp.). Nitrogen deficiencies can cause increased susceptibility to certain leaf pathogens such as *Alternaria solani*, while excessive plant N levels may result in increased susceptibility to other pathogens such as *Botrytis cinerea*, or *Rhizoctonia solani*.

Phosphorus (P) deficiency can be a serious problem in some plants because it is typically not very mobile in soils. In addition, certain environmental conditions can make it difficult for plants to absorb and translocate this macronutrient. Phosphorus is utilized in the plant for a number of activities including photosynthesis, the transport of energy (in the form of ATP) throughout the plant, and as an important component of DNA. Phosphorus is also important for flowering and seed production. Plants deficient in phosphorus have weaker stems, which can result in problems such as lodging in grain crops. Phosphorus deficiency can also result in poor growth and stunting, a blue/green hue to the leaves, and/or purple-colorations to stems and undersides of the leaves. Soil type and biological characteristics greatly affect P availability in soil. Plants grown in acid and clay soils are particularly prone to P deficiency. Cool conditions or poor oxygen availability to the roots can lead to P deficiency. Phosphorus can be supplied to soils as commercial fertilizers, natural rock materials or animal residues.

Iron (Fe) deficiency (chlorosis) is a significant problem for crop and ornamental plants, particularly in calcareous and high pH soils. Iron chlorosis is prevalent in some areas of the US on tree and landscape species as well as turfgrasses. Iron is a key component in the production of chlorophyll within the leaf. Therefore, plants with an Fe deficiency typically have similar leaf size/shape compared to normal plants, but will display interveinal chlorosis. Symptoms first develop in the new growth which appears as yellow-green leaves, often with a striped appearance. Most soils have adequate supplies of Fe, but availability to the plant decreases as soil pH increases. Maintaining pH <7 is critical to optimize availability of Fe for plants. Iron deficiencies can also occur when soil oxygen availability is low, such as heavily compacted areas. Iron availability can also be affected by soil microbial activity and can therefore be reduced during low temperatures, low light conditions, and wet or excessively dry soils.

Potassium (K) deficiency arises in the older leaves of the plant. Potassium deficiency can be particularly important in certain fruit and vegetable production systems for its role in fruit quality. Potassium plays a key role in cellular signaling and growth regulation in plants. It also functions in other processes such as photosynthesis. Symptoms of K deficiency include necrosis (tissue death) on leaf margins, leaf curling and browning, and interveinal chlorosis. Plants that are deficient in K can be more prone to frost damage as well as certain diseases. Potassium is found in large quantities throughout soils. However, due to soil parental material and the effect of weathering, soil K levels vary significantly.

Potassium availability is reduced by the presence of competing cations such as Ca^{2+} and NH_4^+. Potassium can also be readily leached from sandy soils. Plant uptake of K may be reduced by certain environmental conditions including temperature, soil moisture, and oxygen availability. Potassium can be applied in conventional fertilizers or with rock phosphate (potash).

Calcium (Ca) deficiency occurs in many fruiting vegetables and can be a severe problem in acid soils. Calcium deficiency can be a result of inadequate levels of Ca within the soil or growing medium, but it is often the result of poor transpiration or fluctuations in soil moisture levels. Calcium is necessary for plant growth and in particular is an important component in the production of cell walls. Calcium is also important as a signal regulator and serves to strengthen cell membranes. Blossom end rot is a common symptom of Ca deficiency on fruits. Other symptoms manifest as plant stunting, localized tissue necrosis, and leaf marginal chlorosis. Blossom end rot and other fruit disorders caused by Ca deficiency (*e.g.,* bitter pit of apples) can often lead to secondary colonization by fungi. Death of terminal buds and root tips can occur and root growth is often inhibited. Because Ca^{2+} is highly immobile, mature older leaves are typically unaffected and symptoms are pronounced in the growing tissues. Calcium deficiency can be overcome by raising pH (in acid soils) and applying Ca fertilizer. Calcium is also available via additions of calcitic lime and gypsum. Calcium has long been studied in its relationship to plant disease and plant pathogenesis. Due to its role in strengthening cell walls, Ca deficiencies may lead to increased susceptibility to certain plant diseases. Similarly, Ca is a component of host response proteins to pathogen toxins, such as oxalic acid, which is utilized by some fungi such as *Sclerotium rolfsii* during infection.

Magnesium (Mg) deficiency can occur in alkaline soils, but is most prevalent in strongly acidic, sandy soils where Mg can be easily leached away. Magnesium is an important component of the chlorophyll molecule and is an important cofactor in the production of ATP. Plants that are Mg-deficient display interveinal chlorosis and have reduced photosynthesis. Magnesium deficiency is often mistaken for K deficiency due to similar plant symptoms. Magnesium deficiency results in chlorotic and necrotic tissues with an orange, red, or brownish color. Yellowing of the leaf margin is also common on many plant species. Early leaf senescence may also occur, particularly on older leaves as Mg is easily translocated through the plant. Over-application of K and/or Ca, which competes at soil cation exchange sites, can also lead to Mg deficiency. Similarly, over-application of Mg can lead to Ca deficiency; it is important to maintain a suitable Ca:Mg ratio in agricultural soils. Magnesium deficiencies are often confused with viruses and other nutrient problems. However, symptoms of viruses are typically manifested in the young, growing part of the plant. Foliar and granular applications of Epsom salt can be used to remedy Mg deficiency in crop plants.

6.7. Role of Mineral Nutrition Against Environmental Stress

Most of the yield losses caused by abiotic stresses are attributed to drought, salinity, extreme temperatures, acidity, and impairment of the mineral nutrition status of plants. Cakmak (2002) reported that at least 60% of cultivated soils worldwide have growth limiting problem arising from mineral nutrition deficiencies and toxicities. Survival and productivity of crop plants exposed to environmental stresses are dependent on their ability to develop adaptive mechanisms to avoid or tolerate stress.

Low temperatures have an effect on mineral nutrition of plants. Absorption of ions by roots is difficult, as well as their movement in the above ground parts of plants (Lukatkin *et al.*, 2012). The distribution of nutrients between the plant organs is disrupted, with general decrease in the nutrient content in the plant (Лукаткин, 2002). Chilling of plants leads to a decrease in the activity of nitrate reductase, reduction in the nitrogen incorporation in the amino acids and proteins, and a drop in the proportion of organic phosphorus and an increase in inorganic phosphorus content (Zia *et al.*, 1994), which is a consequence of a breach of phosphorylation and enhanced decomposition of organic phosphorus compounds. Mechanisms to reduce the absorption of nutrients by chilling temperatures include depression of respiration and/or oxidative phosphorylation, impair enzymatic transport systems associated with conformational proteins changes in membranes, changes in membrane potential, reducing the supply of ATP to H^+-transporting ATPase, as well as lowering the permeability coefficients for ions (Clarkson *et al.*, 1988). Special attention of researchers has been drawn to two hypotheses to explain the induction of chilling damage to a rapid increase in the concentration of free cytosolic Ca^{2+} ($[Ca^{2+}]_{cyt}$) (Minorsky, 1985) and the occurrence of oxidative stress upon chilling of chilling-sensitive plants (Hariyadi and Parkin, 1993). Minorsky (1985) proposed a hypothesis to explain most of the secondary effects of chilling shock, which suddenly increases (by 1–2 orders) in the concentration of $[Ca^{2+}]_{cyt}$. It is assumed that the rapid increase in $[Ca^{2+}]_{cyt}$ due to chilling, may serve as the primary physiological signal of cold exposure. It was shown that changes in intracellular calcium compartmentation in chilled plants, leading to an increase in $[Ca^{2+}]_{cyt}$, stop cytoplasmic streaming and affect the sub-cellular structures (Woods *et al.*, 1984). There is evidence that input of $^{45}Ca^{2+}$ in maize root cells increased by 20–25% at a temperature of 2°C (Zocchi and Hanson, 1982). Changes in $[Ca^{2+}]_{cyt}$ trigger cascade reactions in the cell, which leads to numerous disturbances at all levels of an organization. It has been shown that chilling induces abrupt reduction of Ca^{2+}-ATPase activity, which pumps out Ca^{2+} in apoplast and/or in intracellular depots. So, this enhances the $[Ca^{2+}]_{cyt}$ level in cytoplasm. During the growth of maize seedlings on nutrient media with different calcium status more intense chilling injury was observed at reduced or enhanced Ca^{2+} doses in comparison with optimal dose (Lukatkin and Isaikina, 1997). In recent years, the calcium hypothesis has been further developed in view of oxidative stress that occurs when cooling the chilling-sensitive plants. Oxidative stress that occurs during cooling of chilling-sensitive plants plays a leading role in the transduction of chilling injury (Hu *et al.*, 2008). The reason why production of free radicals and reactive oxygen species (ROS) increased is singlet oxygen, superoxide anion,hydroxyl radical, hydrogen peroxide (Suzuki and Mittler, 2006). These ROS cause considerable damage to membrane lipids and other cellular components (Lukatkin *et al.*, 1995; Lukatkin, 2003). It was shown that $[Ca^{2+}]_{cyt}$ changes are intimately connected to an oxidative stress. Oxidative stress causes an immediate increase in cytosolic calcium (Price *et al.*, 1994), acting the same as chilling shock (Knight *et al.*, 1996). This reaction is transient, and finishes within 1–2 minutes. In turn, $[Ca^{2+}]_{cyt}$ influences a level of free radicals, inhibiting activity of SOD (Price *et al.*, 1994). So, increasing the concentration of ionized calcium causes increased oxidative stress (Price *et al.*, 1994; Lock and Price, 1994), *i.e.* is the signal amplification cascade that causes chilling damage.

The major types of stresses, which potentially affect nutrient uptake, are flooding, salinity and drought. Other stresses, which also at the same time affect the normal growth of plants and the nutrient uptake circumstances include, seedbed preparation, plant population, planting time, types of soil, soil N, P, K, Ca, Mg, S, Fe, Mn, Zn, Cu, B, Mo, Co, Si, Cd, Cr, Se and Al, soil organisms, diseases, weeds, toxic metals, air pollution, growth regulators, wind, hailstorms, water-table and allelopathy (Alam, 1999). Salinity and drought can differentially affect the mineral nutrition of plants. Due to the competition of Na^{+} and Cl^{-} with nutrients such as Ca^{2+}, K^{+} and NO^{-} salinity may cause nutrient deficiencies. On the other hand drought can affect nutrient uptake and translocation of some nutrients (Hu, and Schmidhalter, 2003). Potassium is a most important nutrients element that has an important role for endurance of plants in stress condition (Baque *et al.*, 2006). The availability of K^{+} to the plant, due to the decreasing mobility of K^{+} under stress environment, decreases with decreasing soil water content (Arjenaki *et al.*, 2012). Macro-nutrients like K;N and Ca and micro-nutrients like Zn, Si and Mg reduce the toxicity of ROS by increasing the concentration of antioxidants like superoxide dismutase (SOD) and improve drought tolerance in plants. The micronutrients like B and Cu lighten the adverse effects of drought in biochemical, physiological and metabolic processes in the plants (Ahmad *et al.*, 2011). In the case of water logging, even in tolerant plants the growth rate, nutrient uptake and root-shoot ratio reduced (Ferreira *et al.*, 2008).

Of the mineral nutrients, nitrogen plays a major role in utilization of light energy and photosynthetic carbon metabolism (Kato *et al.*, 2003; Huang *et al.*, 2004). An excess of non-utilized light energy can be expected to occur in N-deficient leaves, which leads to a high risk of photo-oxidative damage. In rice plants under high light intensity, N-deficiency is associated with enhanced lipid per-oxidation (Huang *et al.*, 2004), and Kato *et al.* (2003) showed that plants grown under high light intensity and with high nitrogen supply had greater tolerance to photo-oxidative damage and higher photosynthetic capacity than those grown under similar high light intensity and low nitrogen supply.

In plants suffering from N deficiency, the conversion of xanthophylls cycle pigments and formation of zeaxanthin were enhanced, and were accompanied by chlorophyll bleaching, particularly under high light intensity (Kato *et al.*, 2003). In spinach, N-deficient plants dissipate a greater fraction of absorbed light energy than N-adequate ones: up to 64% and only 36%, respectively. This difference was associated with corresponding changes in xanthophylls cycle pigments (Verhoeven *et al.*, 1997). The reduction in utilization of light energy and the consequently elevated need for protection against photo-oxidative damage in N-deficient plants can be more marked when the N-deficiency stress is combined with an environmental stress.

The form in which N is applied affects plants tolerance to photo-damage. The light induced conversion of violaxanthin to zeaxanthin, as a means to dissipate excess light energy was found to be stronger in bean leaves supplied with nitrate than in those supplied with ammonium (Bendixen *et al.*, 2001, Zhu *et al.*, 2000).

Deficiency of K, Mg and Zn also enhance the sensitivity of plants to photo-oxidative damage. When supplies of these nutrients are low, leaf symptoms of chlorosis and necrosis and disturbances of plant growth become more severe when plants exposed to high light intensity (Cakmak *et al.*, 1995; Polle, 1996). Deficiencies of potassium and/or magnesium cause marked decreases in photosynthetic carbon metabolism and utilization of fixed carbon (Cakmak and Engels, 1999; Mengel and Kirkby, 2001). Consequently their deficiencies cause massive accumulation of carbohydrates in source leaves, with consequent inhibition of photosynthetic carbon reduction. Consistent with these changes in photosynthetic carbon metabolism, an excess of non-utilized light energy and photoelectrons is expected in K and Mg deficient plants. This is the main reason why magnesium and potassium deficient leaves are highly light sensitive. In contrast to magnesium and potassium deficiency, phosphorus deficiency had no effect on sucrose transport. Leaf chlorosis is not typical of phosphorus deficient plants (Cakmak, 1994a). Because of the distinct effects of magnesium and potassium on photosynthetic carbon metabolism and on ROS formation in chloroplasts, photo-oxidative damage in plants grown under marginal conditions, such as drought, chilling and salinity can be exacerbated when the soil supply of magnesium or potassium is low.

Zinc ions are known to be strong inhibitors of NADPH oxidase. In bean and cotton root cells of Zn deficiency caused a significant increase in activity oh NADPH-dependent O_2 production, and a resumed supply of Zn to Zn-deficient plants for 12 or 24 hours caused a distinct reduction in the activity of O_2 -generating enzymes (Cakmak and Marschner, 1988a; Pinton *et al.*, 1994). Similarly in tobacco cell cultures salt-induced O_2 generation by NADPH oxidase was strongly inhibited by Zn (Kawano *et al.*, 2002). It has often been said that improving the Zn nutritional status of plants growing in saline conditions was critical for protection of plant against toxicity. This protective role of Zn was ascribed to its role in maintenance of the structural integrity of the plasma membrane and thus controlling the uptake of Na and other toxic ions (Cakmak and Marschner, 1988b). In light of the protective roles of Zn against ROS it can be said that Zn ions protect salt-stressed plants not only from uptake of toxic ions across plasma membranes but also from damaging attack of ROS.

Like Zn, K too is a critical nutrient that protects plant cells from salt-induced cell damage. Impairment of the K nutritional status of plants by increased Na uptake is a well known phenomenon (Zhao *et al.*, 2001; Jiang and Zhang, 2002a, b). It appears that in addition to ROS formation by photosynthetic electron transport, ROS production by NADPH oxidase activity is involved in cell damage and plant growth depression under drought stress. It has been suggested that the protective role of Zn and K against drought stress seems also to be related to their inhibitory effects on NADPH-dependent O_2 generation. Thus, in case of deficiency of these nutrients, plants become more sensitive to drought stress.

There are several examples from field experiments that demonstrate a role of K and Zn in protection of plants under low-temperature conditions, frost damage and related decrease in potato plant yields were alleviated by application of large doses of K (Gerewal and Singh, 1980). Citrus trees are found to be vulnerable to low temperature and peroxidative damage when grown under Zn-deficient condition (Cakmak *et al.*, 1995). N too is involved in protection of plants against chilling stress (Close *et al.*, 2003). Like low N supply, also excess N results in high sensitivity to environmental stress. Marschner (1995) found that a very high supply of N often led to a reduced root-to shoot ratio and that in turn impaired the support of shoot biomass and mineral nutrient and water. Also, in plants receiving a high N supply, most of the roots may grow near to the soil surface, with consequently higher sensitivity to frost and drought damage (Saebo *et al*., 2001).

Generally, plant genotypes that tolerate low-temperature stress are able to maintain high water potential by closing their stomata and preventing transpirations loss (Wilkinson *et al*., 2001). Calcium has been shown to be an essential requirement for chilling-induced stomatal closure in chilling-tolerant genotypes. Increasing the Ca supply induces stomatal closure, and this evvect in most distinct in plants grown at low temperature. It is also believed that ABA-induced stomatal closure is partially mediated by Ca released from internal guard cell stores or the apoplasts (Wilkinsons *et al*., 2001) and this function seems to make Ca a major contributing factor to chilling tolerance and protection of leaves from dehydration.

In plants, water and nutritional deficiency, high salinity and extreme temperature are some of the most studied stress factors (Lambers *et al*., 1998). Water deficit, heat stress, chilling and freezing, salinity, oxygen and nutrient are major stress factors restricting plant growth and devel-opment (Bray, 2000). In addition, stress plays a major role in determining how soil and climate limit the distribution of plant species (Shinozaki *et al*., 2000). The physiological process that underlies stress injury and the adaptation and acclimation mechanisms of plants to environmental stress is of huge importance to both agriculture and the environment (Hong *et al.,* 2000). Nutrient stress can result either from the form in which the nutrient exist, the process by which they become available to the plant; content of soil solution and soil pH (Evans, 1989). The demand by a plant for a given nutrient changes with time because it is influenced by changes in all other environmental factors that control plant growth which include other nutrients, water, radiation, temperature and age among others (Har-tung *et al*., 1998). Nutrient stress can be evaluated as the proportion by which the growth rate of the plant or crop is limited by that nutrient under the prevailing conditions. It affects all aspects of growth and so should be quantified in terms of growth parameters such as dry weight and biomass accumulation (Pollock and Cairns, 1991). The present study should serve to provide more information on the effect of nitrogen nutritional stress on some aspects of the primary metabolic activities of *Zea mays* and *Vigna unguiculata*.

6.8. Role of Mineral Nutrition Against Drought Stress

Water stress is one of the major limitations to the agricultural productivity worldwide, particularly in warm, arid and semi arid parts of the world (Boyer, 1982). Recent trends indicate that productivity and fertility of soils are globally declining due to degradation and intensive use of soils without consideration of proper soil management practices (Gruhn *et al*., 2000; Cakmak, 2002).

Environmental problems (*e.g.,* water deficiency and salinity) are increasing as a result of burgeoning population of world and intensive use of natural resources. These environmental stresses contribute significantly in reduction of crop yields well below the potential maximum yields. Bray *et al.* (2000), reported that the relative decreases in potential maximum crop yields (*i.e.,* yields under ideal conditions) associated with abiotic stress factors including drought, vary between 54% and 82%. Therefore, for sustaining food security, a high priority should be given to minimizing the detrimental effects of drought.

Drought results in the increased generation of reactive oxygen species (ROS) due to energy accumulation in stressed plants which consume less light energy through photosynthetic carbon fixation (Asada, 2006). Drought inhibits or slows down photosynthetic carbon fixation mainly through limiting the entry of CO_2 into the leaf or directly inhibiting metabolism (Loggini *et al* 1999; Apel and Hirt, 2004). Plants have developed a wide range of adaptive and/or resistance mechanisms to maintain productivity and ensure survival under drought stress condition. To reduce the toxicity of ROS, plant cells have developed an anti oxidative system, consisting of low molecular-weight antioxidants like ascorbate, a-tocopherol, glutathione, and carotenoids, as well as protective enzymes. Superoxide radicals are scavenged by superoxide dismutase (SOD), while the resulting H_2O_2 is reduced to H_2O by catalase (CAT) and peroxidase (POD) (Apel and Hirt, 2004). Despite the internal resistance of the plants to drought stress, the detrimental effects of drought can be minimized by adequate and balanced supply of mineral nutrients. Increasing evidence suggests that mineral-nutrient status of plants plays a critical role in increasing plant resistance to drought stress (Marschner, 1995). Optimal nutrition and most favourable soil tillage greatly affect water circulation within plants, which is a highly effective method of combating drought. Under low nutrient concentrations in soil, plants have to absorb more water to be able to take up the same amount of mineral nutrients for their metabolism than they would from soil with satisfactory fertility. On the other hand, in conditions of lacking soil moisture, plants are unable to get optimal amounts of nutrients, which has negative effects on the overall condition of plants, especially their growth and fruit quality.

Proper nutrition is the basic need of every living organism. There are now 17 elements which are considered essential for plants to complete their life cycle (Waraich *et al.,* 2011). These essential plant nutrients are divided into two categories; macronutrients and micronutrients. Macronutrients include carbon, hydrogen, oxygen, nitrogen, phosphorus, potassium, calcium) magnesium and sulfur. Micronutrients are zinc, copper, iron, manganese, boron, molybdenum, chlorine and nickel (Waraich *et al.,* 2011). Although silicon is not essential, it is considered as a beneficial plant nutrient. These plant nutrients are not only required for better plant growth and development, but also helpful to alleviate different kinds of abiotic stresses like drought stress. Plants have developed a wide range of adaptive and/or resistance mechanisms to maintain productivity and ensure survival under a variety of environmental stress conditions. Increasing evidence suggests that mineral-nutrient status of plants plays a critical role in increasing plant resistance to environmental stress factors (Marschner, 1995).

Drought-induced nitrogen deficiency largely contributes to growth inhibition under water deficit (Heckathorn *et al.*, 1997) mainly affecting the leaf size through decreasing the cell number and cell size (MacAdam *et al*, 1989). Toth *et al.* (2002) reported reduction in leaf production, individual leaf area and total leaf area under N deficient conditions. Increased leaf area index, leaf area duration, crop photosynthetic rate, radiation interception and radiation use efficiency have been reported by enhanced nitrogen supply (Muchow, 1988). Change in leaf photosynthesis in response to variations in plant nitrogen supply has also been observed (Just *et al.*, 1989). Lower rates of photosynthesis under conditions of nitrogen limitation are often attributed to reduction in chlorophyll contents and rubisco activity (Verhoeven *et al*, 1997; Toth *et al.*, 2002). Lawlor (2002) reported that plant metabolic processes, based on proteins, leading to increase in vegetative and reproductive growth and yield are totally dependent upon the adequate supply of nitrogen. Nitrate reductase (NR), the first enzyme in the pathway of nitrogen assimilation has been shown to decrease in water-stressed leaves of sunflower (Azedo-Silva *et al.*, 2004) and it increased with nitrogen application (Kathju *et al.*, 1990). Dehydration adversely affected the activity of nitrate reductase in roots of sunflower (Azedo-Silva *et al.* 2004), whereas contrasting results were observed in roots of maize and no effect of dehydration on nitrate reductase activity was recorded (Abd-El Baki *et al.*, 2000). Possible mechanisms to minimize the detrimental effects of drought by improving water use efficiency with N nutrition were described by Waraich *et al* (2011). Inorganic fertilization has been reported to mitigate the adverse effects of water stress on crop growth and development (Payne *et al.*, 1995; Raun and Johnson 1999). Water stress at seedling stage might lead to higher dry root weights, longer roots, coleoptiles and higher root/shoot ratios which could be exploited as selection criteria for stress tolerance in crop plants at very early stage of growth (Dhanda *et al.*, 2004; Kashiwagi *et al.*, 2004). Whereas, at reproductive stage, flag leaf area (Ali *et al.*, 2010), specific leaf weight, leaf dry matter (Aggarwal and Sinha, 1984), excised leaf weight loss (Bhutta, 2007), relative dry weight (Jones *et al.*, 1980), relative water content (Colom and Vazzana, 2003), residual transpiration (Sabour *et al.*, 1997) and cell membrane stability are the characters of interest and had been widely exploited as reliable morph-physiological markers contributing towards drought tolerance for various crop plants.

Phosphorus is found in less quantity in soils as compared to nitrogen and potassium. Total phosphorus concentration in surface soils varies from 0.005 to 0.15% (Havlin *et al.*, 2007). Phosphate is the principal element involved in plant energy processes. The relative leaf-growth rate is one of the most sensitive parameter to phosphorus deficiency (Kirschbaum and Tompkins, 1990), and it affects the photosynthetic rate per unit area. Phosphorous deficiency induced decline in leaf growth and photosynthetic rate. The reported accumulation of starch indicates that photosynthates cannot be used for plant growth under phosphorus limited condition (Fredeen *et al.*, 1989). Phosphorus deficiency is also known to reduce the uptake rate of nitrates and its assimilation by the nitrate reductase (Pilbeam *et al.*, 1993). It is generally accepted that the uptake of phosphorus by crop plants is reduced in dry-soil conditions (Pinkerton and Simpson, 1986). Application of phosphorus fertilizer can improve plant growth considerably under drought conditions (Garg *et al.*, 2004). The positive effects of phosphorus on plant growth under drought have been attributed to an increase in stomatal conductance (Brck *et al.*, 2000), photosynthesis (Ackerson, 1985), higher cell-membrane stability, water relations (Sawwan *et al.*, 2000) and drought tolerance. Phosphorus improves the root growth and maintains high leaf water potential. Phosphorus also maintains the cell turgidity by maintaining the high leaf water potential which in turn increases the stomatal conductance and increases the photosynthetic rate under drought.

Potassium plays an important role in survival of plants under environmental stress conditions. It is essential for many physiological processes, such as photosynthesis, translocation of photosynthates into sink organs, maintenance of turgescence, activation of enzymes, and reducing excess uptake of ions such as sodium and iron in saline and flooded soils (Marschner, 1995; Mengel and Kirkby, 2001). Plants suffering from environmental stresses like drought have a larger internal requirement for potassium (Cakmak and Engels, 1999). Environmental stress factors that enhance the requirement for potassium also cause oxidative damage to cells by inducing formation of ROS, especially during photosynthesis (Elstner and Osswald, 1994; Foyer *et al.*, 1994). Increase in ROS production in drought-stressed plants are well known and related to impairment in photosynthesis and associated disturbances in carbohydrate metabolism (Quartacci *et al.*, 1994; Jiang and Zhang, 2002). Decrease in photosynthesis caused by drought stress is particularly high in plants supplied with low potassium and are minimal when potassium is sufficient (Sen Gupta *et al.*, 1989). Alleviation of detrimental effects of drought stress, especially on photosynthesis, by sufficient potassium supply has also been shown in legumes (Sangakkara *et al.*, 2000). Under water-deficit conditions, potassium nutrition increases crop tolerance to water stress by utilizing the soil moisture more efficiently than in K-deficient plants. Potassium maintains the osmotic potential and turgor of the cells (Lindhauer, 1995) and regulates the stomatal functioning under water stress conditions (Kant & Kafkafi, 2002), It enhances photosynthetic rate, plant growth and yield under stress conditions (Egila *et al.*, 2001). The protective role of K in plants suffering from drought stress by maintenance of a high pH in stroma and against the photooxidative damage to chloroplasts was also reported by Cakmak (1997).

Calcium is not just a macronutrient but also a major controller of plant metabolism and development (Poovaiah and Reddy, 2000). Calcium plays a role in mediating stress response during injury, recovery from injury, and acclimation to stress (Palta, 2000). It has been said that calcium is necessary for recovery from drought by activating the plasma membrane enzyme ATPase which is required to pump back the nutrients that were lost in cell damage (Palta, 2000). Since dehydration is the common denominator, calcium also has a role to play in freeze injury tolerance. Calcium has a very prominent role in the maintenance of cell structure. Its activates the plasma membrane enzyme ATPase which pumps back the nutrients lost during cell membrane damage due to calcium deficiency and recover the plant from injury. Calcium also plays a role as calmodulin which controls the plant metabolic activities and enhances the plant growth under drought condition.

Magnesium is involved in several physiological and biochemical processes in plants affecting growth and development. Epstein and Bloom (2004) reported that magnesium is exceptional in activating more enzymes than any other mineral nutrient. Mg-activated enzymes are ATPases, ribulose-1, 5-bisphosphate (RuBP) carboxylase, RNA polymerase and protein kinases (Marschner, 1995; Shaul, 2002). As the central atom of the chlorophyll molecule is perhaps the best-known function of magnesium in plants. Magnesium deficiency can be induced not only by a direct lack of magnesium but also by the presence of competing cations that prevent magnesium uptake, such as Ca^{+2} in calcareous soils; H^{+}, NH_4^{+} and Al^{3+} in acidic soils and Na^{+} in saline soils (Mengel and Kirkby, 2001; Shaul, 2002). It has been suggested that magnesium plays a fundamental role in phloem export of photosynthates from the source to the sink organs, and its deficiency results in dramatic increases in accumulation of carbohydrates in the source leaves (Cakmak *et al.*, 1994b, 1994c; Marschner *et al.*, 1996). Asada (2006) reported that reduced transport and hence accumulation of carbohydrates in Mg-deficient leaves causes alterations in photosynthetic carbon metabolism and restrict CO_2 fixation. Impairment of the photosynthetic electron transport to CO_2 through photosynthetic membranes may cause an accumulation of non-utilized electrons and absorbed energy. Under such conditions, the electrons and excitation energy not used in photosynthetic CO_2 fixation is channeled to molecular O_2, leading to the generation of highly reactive O_2 species (ROS) and consequently to damage of chloroplast constituents such as chlorophyll and membrane lipids (Mittler, 2002). Magnesium increases the root growth and root surface area which helps to increase uptake of water and nutrients by root and transport of sucrose from leaves to roots. Magnesium improves CHO translocation by increasing phloem export and reduces ROS generation and photo-oxidative damage to chloroplast under drought conditions.

Zinc is an important micronutrient essential for plant growth and development. The soil in dry regions is often poor in plant-available zinc associated with high calcium carbonate content and alkaline pH (Liu, 1996). Drought stress reduces the net photosynthetic rate of the plants. This decline is related to a reduction in light interception due to lower leaf area, to reduction in carbon fixation per unit leaf area or to damage of the photosynthetic apparatus (Castrillo *et al*., 2001; Bruce *et al*., 2002). In cauliflower, a reduction in photosynthesis induced by zinc deficiency was associated with a decrease in stomatal conductance and intercellular CO_2 concentration (Sharma *et al*., 1994). A decrease of carbonic anhydrase activity due to zinc deficiency also contributed to the reduced photosynthetic rate (Cakmak and Engels, 1999; Hacisalihoglu *et al*., 2003). In cabbage, zinc deficiency lowered osmotic potential and increased water saturation deficit (Sharma *et al*., 1994). The transpiration rate of pecan plants declined under zinc deficiency (Hu and Sparks, 1991). Khan *et al*., (2003) reported that applying zinc increased chickpea grain yields when the plants were well-watered, but not under water stress, except for the Zn-efficient and drought-resistant genotype. It has been reported that zinc is important for its ability to influence auxin levels and has long been known to be a co-enzyme for production of tryptophane (Bennett and Skoog, 2002; Waraich *et al*, 2011). Increase in auxin levels due to zinc application enhances the root growth which in turn improves the drought tolerance in plants. Zinc application reduces the activity of membrane-bound NADPH oxidase which in turn decreases the generation of ROS (Waraich *et al*, 2011) and reduces photoxidation damage while the activities of SOD, POD, and CAT are enhanced indicating that zinc lowers the ROS generation and protect cells against ROS attack under water stress (Waraich *et al*, 2011).

A primary function of boron is related to cell wall formation in plants. The plants suffering from drought stress may be stunted. Sugar transport in plants, flower retention, pollen formation, seed germination and grain production are reduced with drought stress. By improving the boron nutrition, the detrimental effects of drought can be corrected. Boron improves the drought tolerance in plants by improving sugar transport, flower retention, pollen formation and seed germination. Seed and grain production are also increased with proper boron supply. Boron nutrition under drought condition results in reduction in stunted appearance (rosetting), barren ears due to poor pollination, hollow stems and fruit (hollow heart) and brittle, discolored leaves and loss of fruiting bodies.

Copper is an important micronutrient essential for carbohydrate and nitrogen metabolism. Copper is also required for lignin synthesis which is needed for cell wall strength and prevention of wilting. Drought stress adversely affects all these processes in plants. Proper copper nutrition alleviates the adverse affects of drought by reducing dieback of stems and twigs, yellowing of leaves, stunted growth, pale green leaves that wither easily, and improves CHO and nitrogen metabolism which in turn improves the growth of plants.

Silicon is the second most abundant element in soil after oxygen. It occurs in two major forms: silica and oxides of silicon, and both types exist in crystalline and/or amorphous forms such as quartz, flint, sand-stone, opal and diatomaceous earth's silicates. In soil solution silicon occurs as silicic acid at concentration ranging from 0.1-0.6 mM (Epstein, 1999). Despite silicon being ubiquitous and prominent of constituent of plants, it is still widely not recognized as an essential nutrient for plants. However, it is proved to be beneficial for better plant growth and development, especially in plants of gramineae family (Shi *et al*, 2005). Silicon can improve plant growth and tolerance to biotic and abiotic stresses (Liang *et al*., 2007; Neumann and Niede, 2001). The possible mechanisms to alleviate detrimental effects of drought in crop plants by improving silicon nutrition were described by Waraich *et al*. (2011). Silicon has a positive effect on plants under drought stress. Gao *et al* (2004, 2006) reported that the addition of silicon increased water use efficiency by reducing leaf transpiration and the water flow rate in the xylem vessel in maize. Silicon could facilitate water uptake and transport in *Sorghum bicolor* in drought conditions (Hattori *et al*., 2005, 2007). Silicon alleviated oxidative stress by regulating the activities of antioxidant enzymes under drought in potted wheat, (Gong *et al*, 2005). In addition to antioxidant defense, plants can also adapt to water stress by changing solute levels so that turgor and hence physiological activity are maintained at low leaf water potentials (Zhu *et al*, 2005). It has been suggested that accumulation of solutes in the stressed leaves contributes to dehydration tolerance (Wood *et al*, 1996; Smienoff, 1998). Silicon nutrition increases the antioxidants production and reduces ROS generation which in turn reduces the photo-oxidative damage and maintain the integrity of chloroplast membrane and enhances the drought tolerance in plants (Waraich *et al*, 2011).

6.9. Role of Potassium in Abiotic Stress Resistance

Potassium is an essential nutrient and is also the most abundant cation in plants. The concentration of K^+ in the cytoplasm has consistently been found to be between 100 and 200 mM (Shabala and Pottosin, 2010), and apoplastic K^+ concentration may vary between 10 and 200 or even reach up to 500 mM (White *et al.*, 2010). Potassium plays essential roles in enzyme activation, protein synthesis, photosynthesis, osmoregulation, stomatal movement, energy transfer, phloem transport, cation-anion balance and stress resistance (Marschner, 2012).

6.9.1. Potassium and Drought Stress

The major limitation for plant growth and crop production in arid and semi-arid regions is soil water availability. Plants that are continuously exposed to drought stress can form ROS, which leads to leaf damage (Cakmak, 2005; Oerke and Dehne, 2004; Foyer *et al.*, 2002) and, ultimately, decreases crop yield. During drought stress, root growth and the rates of K^+ diffusion in the soil towards the roots were both restricted, thus limiting potassium acquisition. The resulting lower potassium concentrations can further depress the plant resistance to drought stress, as well as potassium absorption. Maintaining adequate plant potassium is, therefore, critical for plant drought resistance. A close relationship between potassium nutritional status and plant drought resistance has been demonstrated. The roles of potassium in physiological and molecular mechanisms of plant drought resistance have been explored.

6.9.2. Cell Elongation and Cell Membrane Stability

Crop tolerance to low soil moisture stress can be improved by inducing deeper rooting, larger absorption surfaces and greater water retention in plant tissues (Wang *et al.*, 2013). Deeper rooting could be achieved by deep placement of potassium fertilizer that is associated with other mineral nutrients, such as phosphorus and nitrogen, which both have root signaling functions (Kirkby *et al.*, 2009). Adequate amounts of potassium can enhance the total dry mass accumulation of crop plants under drought stress in comparison to lower potassium concentration (Egilla *et al.*, 2001). Furthermore, potassium is also essential for the translocation of photoassimilates in root growth (Romheld and Kirkby, 2010). Root growth promotion by increased appropriate potassium supply under K-deficient soil was found to increase the root surface that was exposed to soil as a result of increased root water uptake (Romheld and Kirkby, 2010). Lindhauer (1985) reported that fine potassium nutrition not only increased plant total dry mass and leaf area, but also improved the water retention in plant tissues under drought stress.

It had been reported that the maintenance of membrane integrity and stability under drought stress is also essential for plant drought tolerance (Baiji *et al.*, 2002). Cell membrane stability was significantly declined under drought stress (Wang and Huang, 2004). Higher potassium applications showed greater adaptation to water stress in maize (Premachandra *et al. 1991)* and was due to the role of potassium in improving cell membrane stability and osmotic adjustment ability. An adequate potassium supply is essential to enhancing drought resistance by increasing root elongation and maintaining cell membrane stability.

6.9.3. Aquaporins and Water Uptake

Aquaporins are channel proteins that are present in the plasma and intracellular membranes of plant cells. They play a crucial role in plant water relations by regulating the osmotic potential and hydraulic conductivity of membranes and make changes in plant water permeability (Heinen and Ye, 2009; Maurel *et al.*, 2001). Under drought stress conditions, aquaporin gene expression can be regulated (Lian *et al.*, 20044; Tyerman *et al.*, 2002) to help plants maintain their water balance (Tyerman *et al.*, 2002; Galmes *et al.*, 2007; Kaldenhoff *et al.*, 2008).

Under drought, roots regulated their water and ion uptake capacities by modifying *PIPs* (plasma membrane intrinsic proteins) and K^+ channel at the transcription level to cope with the water deficiency (Galmes *et al.*, 2007; Smart *et al.*, 2001; Alexandersson *et al.*, 2005; Liu *et al.*, 2006; Cuéllar *et al.*, 2010). Kanai *et al. (2011)* observed close coupling between aquaporin activities and K-channel transporters.

6.9.4. Osmotic Adjustment

The maintenance of a favorable water status is critical for plant survival under drought stress. Osmotic adjustment is a major trait that is associated with maintaining high cellular turgor potential and water retention in response to drought stress. Many studies have shown that osmotic adjustment of leaves is positively correlated with drought tolerance in various plant species (DaCosta and Huang, 2006). As one of the most prominent inorganic osmotica in plants, K^+ plays a key role in formation of the osmotic adjustment ability, even under drought conditions (Marschner, 2012). Cell turgor recovery in osmotically-generated stress was regulated by increasing K^+, Cl^- and Na^+ uptake by root cells, which was partly mediated by voltage-gated K^+ transporters at the cellular plasma membrane (Shabala and Lew, 2002). Furthermore, sufficient potassium induces solute accumulation, thus lowering osmotic potential and helping to maintain plant cell turgor under osmotic stress. An adequate potassium status may facilitate osmotic adjustment, which maintains higher turgor pressure, relative water content and lower osmotic potential, thus improving the ability of plants to tolerate drought stress (Kant and Kafkafi, 2002; Egilla *et al.*, 2005).

6.9.5. Stomatal Regulation

One of the major functions of the stomata is to control plant water loss via transpiration. During drought stress, quick stomatal closure and internal moisture preservation are essential for plant adaptation to drought conditions. potassium plays a crucial role in turgor regulation within the guard cells during stomatal movement (Marschner, 2012). As stomatal closure is preceded by a rapid release of K^+ from the guard cells into the leaf apoplast, it is reasonable to think that stomata would be difficult to remain open under K-deficient conditions. Some studies also stated that potassium deficiency may induce stomatal closure and inhibit photosynthetic rates in several crop plants (Jin *et al.*, 2011; Tomemori and Hamamura, 2002). Conversely, many studies suggest that potassium had no effect on stomatal conductance and photosynthetic rates under well-watered conditions, but potassium starvation could favor stomatal opening and promote transpiration, compared with potassium sufficiency in several plants under drought stress (Pervez *et al.*, 2004; Benlloch-Gonzalez *et al.*, 2008, 2010). Furthermore, photosynthetic rate was decreased under drought stress in K-deficient plants (Egilla *et al.*, 2005; Pervez *et al.*, 2004; Tsonev *et al.*, 2011). This discrepancy may be related to the plant species, experimental system and environmental factors within the experimental field or interspecific differences.

The effects of drought stress on stomata closure in olive trees and sunflower plants were found to be dependent on the K^+ nutrient status (Benlloch-Gonzalez *et al.*, 2008, 2010). When plants were supplied with different K^+ concentrations and then subjected to drought stress, their stomatal conductance was more markedly reduced in normal potassium plants than in low potassium plants. Benlloch-Gonzalez *et al.* (2010) explained that the low plant potassium status could inhibit water-stress-induced stomatal closure via ethylene synthesis, and stomatal conductance could be significantly reduced in K^+-starved plants after the adding of an ethylene synthesis inhibitor (cobalt). K^+ starvation increases the transcription of genes involved in ethylene production and signaling and stimulates ethylene production (Benlloch-Gonzalez *et al.*, 2008; Shin and Schachtman, 2004). Then, the increased ethylene could inhibit the action of abscisic acid (ABA) on stomata and delay stomata closure (Tanaka *et al.*, 2005, 2006). During drought stress, the stomata cannot function properly in K^+-deficient plants, resulting in greater water loss. Drought stress did not decrease water use efficiency (WUE), whereas it did increase WUE by rapid stomata closing during water deficit (Kant and Kafkafi, 2002). Adequate levels of potassium nutrition enhanced plant drought resistance, water relations, WUE and plant growth under drought conditions (Kant and Kafkafi, 2002).

6.9.6. Detoxification of Reactive Oxygen Species

Stomatal closing in response to drought stress leads to a reduction in photosynthetic efficiency as a consequence of chloroplast dehydration (Cakmak, 2005). Photosynthesis inhibition can further disturb the balance between ROS production and antioxidant defense (Cruz de Carvalho, 2008; Fu and Huang, 2001; Reddy *et al.*, 2004), resulting in ROS accumulation. The ROS have a dual action in biotic and abiotic stresses that depends on their cellular concentration (Dat *et al.*, 2000). Low levels of ROS could be involved in the stress-signaling pathway by triggering stress defense/acclimation responses (Dat *et al.*, 2000; Vranova *et al.,* 2002). However, ROS became extremely injurious to cellular membranes and other cellular components when its concentrations reached the point of phytotoxicity, resulting in oxidative stress and, eventually, cell death (Dat *et al.*, 2000; Mittler, 2002).

Drought stress-induced ROS production can additionally be enhanced in K-deficient plants (Cakmak, 2005). Under drought stress, photosynthetic CO_2 fixation in K-deficient plants is substantially limited by impairment in stomata regulation, conversion of light energy into chemical energy and phloem export of photosynthates from source leaves into sink organs (Egilla *et al.*, 2005). As the impairment in photosynthetic CO_2 fixation occurs, molecular O_2 is activated, leading to extensive generation of ROS (Cakmak, 2000) and, thereby, oxidative degradation of chlorophyll and membranes. The maintenance of adequate potassium nutrition is critical for mitigating or preventing damage by drought stress and controlling the water balance (Abdel Wahab and Abd-Alla, 1995). Egilla *et al.* (2005) suggested that increasing extra chloroplastic K^+ concentrations in plant cells with an excess K^+ supply could prevent photosynthesis inhibition under drought stress. An adaptive potassium requirement for drought-stressed plants could be related to the role of potassium in enhancing photosynthetic CO_2 fixation and transport of photosynthates into sink organs and inhibiting the transfer of photosynthetic electrons to O_2, thus reducing ROS production (Cakmak, 2005).

Beside the photosynthetic electron transport, nicotinamide adenine dinucleotide phosphate (NADPH)-dependent oxidase activation represents another major source for production of ROS in plant cells by a number of biotic and abiotic stress factors (Vranova *et al.*, 2002). NADPH-oxidizing enzymes catalyze one-electron reduction of O_2 to $O_2^{\bullet-}$ by using NADPH as an electron donor (Cakmak, 2005). Cakmak (2005) reported that activity of NADPH oxidase was increased in cytosolic fractions of bean roots with increasing severity of potassium deficiency, resulting in an increase in NADPH-dependent $O_2^{\bullet-}$ generation. The reason for the increase of NADPH oxidase by potassium deficiency is probably that potassium deficiency induced ABA accumulation

(Peuke *et al.*, 2002). Furthermore, ABA has also been shown to be effective in increasing H_2O_2 and $O_2^{\bullet -}$ accumulations in roots or leaves (Jiang and Zhang, 2001; Lin and Kao, 2001), but this point needs to be clarified in future studies (Wang *et al.*, 2013). An improvement in the plant potassium supply can inhibit ROS production under drought stress by reducing NADPH oxidase activity and maintaining photosynthetic electron transport (Cakmak, 2005). In addition to potassium, various micronutrients, including Zn, B, Cu and Mn, have also been shown to be involved in detoxifying oxygen radicals (Marschner and Cakmak, 1989). The potassium supply is thus associated with other mineral nutrients and is essential for the detoxification of active oxygen under drought stress.

6.9.7. Potassium and Salt Stress

Salt-stressed root growth is restricted by osmotic effects and toxic effects of ions, which results in lower nutrient uptake and inhibits the translocation of mineral nutrients, especially K^+. As a result of the similarities in physicochemical properties between Na^+ and K^+, Na^+ could compete with K^+ for major binding sites in key metabolic processes, including both low-affinity (*e.g.*, non-selective cation channels (NSCC)) and high-affinity (*e.g.*, KUP and high-affinity K^+ transporter (HKT)) transporters and could also disturb plant metabolism (Marschner, 2012; Shabala and Cuin, 2008). K^+ deficiency can usually be observed under salinity stress. First, high levels of Na^+ inhibit K^+ activity in the soil solution, resulting in a reduction of K^+ availability. Second, Na^+ not only interferes with K^+ translocation from root to shoot (especially in low K^+ status) (Botella *et al.*, 1997), but also competes with K^+ for uptake sites at the plasma membrane, resulting in lower K^+ uptake. Third, salinity stress leads to plasma membrane disintegrity and favors K^+ leaking, resulting in a rapid decline in cytosolic K^+ (Coskun *et al.*, 2010). Also, salinity induces significant membrane depolarization and favors K^+ leaking through depolarization-activated outward-rectifying (KOR) K^+ channels (Shabala and Cuin, 2008). Therefore, keeping cellular K^+ content above a certain threshold and maintaining a high cytosolic K^+/Na^+ ratio (either by retaining K^+ or preventing Na^+ from accumulating in the leaves) is critical for plant growth and salt tolerance. An increasing potassium supply corresponded with higher K^+ accumulation in plant tissue, which reduced the Na^+ concentration and resulted in a higher K^+/Na^+ ratio. Members of the HKT transporter (high-affinity K^+ transporter) family that mediate Na^+-specific transport or Na^+-K^+ co-transport play a key role in plant Na^+ tolerance mechanisms (Mian *et al.*, 2011; Platten *et al.*, 2006). HKT represents a primary mechanism in the regulation of Na^+ and K^+ homeostasis, as well as Na^+ exclusion (Byrt *et al.*, 2007; Horie *et al.*, 2009).

Plant growth and salt tolerance were sharply reduced when exposed to a combination of salt stress and K-deficiency stress. K^+ deficiency significantly increased the negative effects that were induced by salt in the photosynthesis of barley and was accompanied by an increase in salt sensitivity (Degl'Innocenti *et al.*, 2009). Similar results were found by Qu *et al.* (2011, 2012) in which K^+ deficiency significantly inhibited nitrogen and photosynthetic carbon assimilation and also impaired the light reaction pathways of PS I and PS II in maize under salt stress. In a study by Chen *et al.* (2007) it was reported that K^+ flux from barley root in response to NaCl treatment was highly positively correlated with net CO_2 assimilation, plant growth, survival rate, relative grain yield and tolerance to salt stress (Wang *et al.*, 2013).

Increased evidence has shown that potassium can involve osmotic adjustment of salt-stressed plants. During salt stress conditions, increased Na^+ concentrations were accumulated in the vacuole and a substantial osmotic potential gradient was established between the vacuole and the cytosol by depressing the cytosol's water activity. This change requires a coordinated increase in compatible solutes in the cytosol to balance out the osmotic pressure. Munns and Tester (2008) reviewed that plants have a Na^+ exclusion mechanism that maintains a low level of Na^+ in the leaves during salt stress; thus, the major osmoticum in leaves was K^+. K^+ plays an important role in maintaining cell turgor and osmotic adjustment. The vacuole and the cytosol are the two major pools of potassium in plant cells. Cytosolic K^+ concentrations are maintained at a constant level and are essential for plant metabolism, while vacuolar K^+ concentrations may vary dramatically. Under K^+-deficient conditions, a constant cytosolic K^+ concentration was attributed to the consumption of vacuolar potassium (Walker *et al.*, 1996).

Low K^+ status might induce the formation of ROS and related cell damage under saline conditions, which was attributed to the effects of K^+ deficiency and/or Na^+ toxicity on stomatal closing and the inhibition of photosynthetic activity and ultimately inhibits plant growth and reduces crop production (Gong *et al.*, 2011). It has been shown that salinity-induced ROS formation can lead to programmed cell death (PCD), and a high cytosolic K^+/ Na^+ ratio is essential for triggering salinity-induced PCD (Shabala, 2009). A decrease in the cytosolic K^+ pool would activate caspase-like proteases and lead to PCD under saline conditions. The ability of plants to satisfy their metabolic requirements for K^+ in the presence of salinity by using higher K^+ fluxes and lower Na^+ fluxes that result in a higher K^+/ Na^+ selectivity ratio is essential for salt tolerance. The addition of K^+ to a saline culture solution has been found to increase K^+ concentrations in plant tissue that corresponds with a decrease in Na^+ content, with a further increase in plant growth and salt tolerance. Increased evidence shows that it is not the absolute quantity of Na^+ *per se* that influences salt resistance, but rather the cytosolic K^+ / Na^+ ratio that determines plant salt tolerance (Shabala and Potosin, 2010; Shabala and Cuin, 2008).

6.9.8. Potassium and Low-Temperature Stress

Cold stress inhibits plant growth and development, which results in limited crop productivity. It affects plants by directly inhibiting metabolic reactions and indirectly influencing cold-induced osmotic, oxidative and other stresses. The effect of increasing K^+ applications on yield and cold tolerance studied by Devi *et al.* (2012) in *Panax ginseng* showed that a high K^+ concentration activated the plant's antioxidant system and increased levels of ginsenoside-related secondary metabolite transcripts, which are associated with cold tolerance. Cold stress may destroy photosynthetic processes and reduce the effectiveness of antioxidant enzymes, resulting in ROS accumulation (Suzuki and Mittler, 2006; Xiong *et al.*, 2002; Mittler, 2002). Potassium improved plant survival under cold stress by increasing antioxidant levels and reducing ROS production (Cakmak, 2005; Devi *et al.*, 2012).

Greater frost damage in K-deficient plants is related to water deficiency from the chilling-induced inhibition of water uptake and freezing-induced cellular dehydration (Zhu, 2001a). A significant negative correlation was found between frost damage and leaf potassium concentration, and an adequate potassium supply can effectively increase frost resistance (Romheld and Kirkby, 2010; Kant and Kafkafi, 2002). Bogdevitch (2000) found that oats that were supplied with sufficient potassium could survive late frost without obvious damage, whereas much of the crop that was grown on K-deficient soil did not survive. This finding could be attributed to a regulation of osmotic and water potential and a reduction of electrolyte leakage caused by cold stress (Kant and Kafkafi, 2002; Webster and Ebdon, 2005). High concentrations of K^+ protected against freezing by lowering the freezing point of the plant's cell solution. Furthermore, an adapted cytosol K^+ concentration is also essential for enzyme activities that are involved in regulating frost resistance (Kant and Kafkafi, 2002).

Because the plasma membrane is the primary site for perceiving changes in temperature, membrane fluidity can be decreased by cold stress as a result of changes in fatty acid unsaturation and the lipid-protein composition of the cell membrane (Wang *et al.*, 2006). The ratio of unsaturated/saturated fatty acids in the cell membrane was essential for plant cold tolerance, and the higher the ratio in the cell membrane, the more tolerant the tissue is to cold stress (McKersie and Leshem, 1994). A decrease in membrane fluidity could further affect the transport of ions, water and metabolites. The effects of nitrogen and potassium on spikelet sterility induced by low temperature at the reproductive stage of rice were studied by Haque (1988). The spikelet sterility induced by low temperature was decreased with the increase of K^+ supply and the increase of the K / N ratio in the rice leaves. Increasing plant frost resistance by the addition of potassium is associated with the increase in phospholipids, membrane permeability and improvement in the biophysical and biochemical properties of cell (Hakerlerler *et al.*, 1997). Higher potassium tissue concentrations reduced chilling damage and increased cold resistance, ultimately increasing yield production (Kant and Kafkafi, 2002; Mengel, 2001). Frost damage was inversely related to potassium concentration and was significantly reduced by potassium fertilization.

6.9.9. Potassium and Waterlogging Stress

Waterlogging affects approximately 10% of the global land area (Mugnai *et al.*, 2011) and is a serious impediment for sustainable agriculture development. Yield losses due to waterlogging may vary between 15% and 80%, depending on the crop species and growth stage, soil type and duration of the stress (Zhu, 2010), resulting in severe economic penalties in some area. The important biological consequence of waterlogging is that the respiration of roots and micro-organisms depletes the residual oxygen and the environment becomes hypoxic (*i.e.*, oxygen levels limit mitochondrial respiration) and, later, anoxic (*i.e.*, respiration is completely inhibited) (Bailey-Serres *et al.*, 2008; Wegner, 2010). The low energy status under oxygen deficient conditions results in a substantial depolarization of plasma membrane potential (Shabala, 2011), subsequent impairment of ion transport processes through voltage regulated uptake channels and a decrease of the uptake of most essential cations (*e.g.*, K^+, NH_4^+ or Mg^{2+}) (Colmer and Greenway, 2011; Kirmizi and Bell, 2012). Pang *et al.* (2006) reported that hypoxia-induced K^+ flux responses are mediated by both inwardly rectifying potassium (KIR) and NSCC channels in the elongation zone, while KOR channels in the mature zone are likely to play a critical role. Avoiding K^+ loss during hypoxia or anoxia stress is the key mechanism responsible for waterlogging resistance in plants (Pang *et al.*, 2006; Teakle *et al.*, 2013; Mancuso and Marras, 2006; Mugnai *et al.*, 2011).

Furthermore, as flooding time increases, potentially toxic compounds, such as sulfides, soluble iron and manganese, ethanol, CO_2, ethylene, lactic acid, acetaldehyde and acetic and formic acid, were accumulated as the result of the reduced soil redox potential (Shabala, 2011; Fiedler *et al.*, 2007). Those compounds acted on cellular membranes, leading to phospholipid oxidation and a subsequent change in membrane integrity and membrane transport (Pang *etal.*, 2006; Erlejman *et al.*, 2004). Rapid changes in net K^+ were measured in response to the application of secondary metabolites (various monocarboxylic acids and phenolic acids) produced by waterlogged soils (Pang *et al.*, 2007). Shabala (2011) assumed that organic acid uptake across the plasma membrane results in a net H^+ influx and causes a substantial membrane depolarization. Such a depolarization will significantly affect intracellular K^+ homeostasis by reducing K^+ uptake via KIR, as well as enhancing K^+ efflux via KOR.

Waterlogging is known to block the oxygen supply to the roots, thus inhibiting root respiration, resulting in a severe decline in energy status of root cells, affecting important metabolic processes of plants. Under waterlogged conditions, the stomata conductance, photosynthesis rate and root hydraulic conductivity of plant were hampered (Else *et al.*, 2001). The oxidative damage induced by the generation of reactive oxygen species affects the integrity of membranes and induces damage to the efficiency of photosystem II, thereby, causing a considerable decrease in net photosynthetic rates (Ashraf, 2012). Exogenous application of potassium could effectively ameliorate the adverse effects of waterlogging on plants. potassium supplement under waterlogging not only increased plant growth, photosynthetic pigments and photosynthetic capacity, but also improved plant nutrient uptake as a result of higher K^+, Ca^{2+}, N, Mn^{2+} and Fe^{2+} accumulation (Ashraf *et al.*, 2011). Ashraf *et al.* (2011) also reported that exogenous application of potassium in soil and as foliar spray alleviated the adverse effects of waterlogging on cotton plants.

6.10. Partition of Photosynthates under Nutrient Deficiency

Nutrient deficiency may not only affect the provision of photosynthates by decreasing source capacity (leaf area index, leaf area duration), but also by altering photosynthate partitioning between the source leaves and various sinks. Under nitrogen deficiency, the increase in root-shoot dry weight ratio (Engels and Marschner 1995; Marschner, 1995) is not only caused by preferential phloem export of sucrose to the roots, but also by export of nitrogen, which can exceed the xylem import of nitrogen from roots to the shoot (Peuke *et al.,* 1994). Similarly, under phosphorus deficiency root-shoot dry weight ratio increases, in soybean for example, from 0.24 in sufficient to 1.0 in deficient plants (Fredeen *et al.,* 1989). This increase in root-shoot dry weight ratio might also be achieved in part by enhanced net re-translocation of phosphorus from shoot to roots in deficient plants (Smith *et al.,* 1990). These results suggest that under nitrogen or phosphorus deficiency not only is a higher proportion of the nutrients taken up from the substrate retained in the roots but also that cycling from shoot to roots provides additional nitrogen and phosphorus to the roots. This cycled fraction of these nutrients may not only contribute to, but may even cause, the shift in sink strength for photosynthates of the roots at the expense of the shoot apex. Under phosphorus deficiency, despite strongly impaired shoot growth, the chlorophyll concentrations and photosynthetic activity of the source leaves are usually at least maintained (Fredeen *et al.,* 1989; Cakmak, 1994a), ensuring continuous photosynthate export to the roots as a dominant sink in the deficient plants.

During phosphorus deficiency shoot dry weight is much lower particularly because of the strongly inhibited growth of the shoot apex (Cakmak, 1994). However, in the primary leaves (source leaves) sucrose concentration and sucrose export in phloem exudate are maintained and, therefore, root growth is similar in the deficient and phosphorus sufficient control plants.

Plants deficient in either potassium or magnesium behave differently in respect to photosynthate partitioning between shoot and roots as compared with plants deficient in nitrogen (Anderson, 1988) or phosphorus. Although phloem mobility as well as cycling of potassium and magnesium are similarly high as for phosphorus and nitrogen, root-shoot dry weight ratio decreases rather than increases under potassium or magnesium deficiency. For plants deficient is magnesium, a decrease in root-shoot dry weight ratio is often observed (Ericsson and Kahr, 1995; Marschner, 1995). Despite the requirement of both these nutrients in various steps in photosynthesis , this decrease in magnesium- and potassium-deficient plants is not caused by impaired photosynthesis *per se,* but results from impaired photosynthate export in the phloem, leading to the accumulation of carbohydrates such as sucrose in the source leaves (Fischer and Bremer, 1993), despite their much lower chlorophyll concentrations. Partitioning of nonstructural carbohydrates between shoot and roots is strongly altered in the phosphorus-, potassium- and magnesium-deficient plants, and closely correlated with the effect of the respective deficiency on dry matter partitioning between shoot and roots.

Cycling of potassium and magnesium through source leaves is high (Jeschke and Pate, 1991b) and essential for photosynthate export in the phloem. In potassium-deficient source leaves, both phloem loading of photosynthates and subsequent solute volume flow in the sieve tubes were depressed and sucrose accumulated in these leaves. In magnesium-deficient source leaves, accumulation of sucrose was particularly high, suggesting specific inhibition of phloem loading of sucrose, probably as the result of low activity of the proton pumping ATPase at the sieve tube membranes. In magnesium-deficient leaves, not only is the export of sucrose impaired but also that of other solutes such as potassium and particularly amino acids. Resupplying magnesium to the roots of deficient plants for only 12 h in the dark period strongly enhanced phloem export of magnesium, sucrose, amino acids, and to some extent potassium. This interdependence of phloem loading and export of sucrose and amino acids is in agreement with results of Winter *et al.* (1992). It is further stated that this interdependence includes the mineral nutrients potassium and magnesium. Export of photosynthates could thus rapidly deplete potassium and magnesium at the phloem loading sites unless continuously replenished by the bulk leaf tissue. In source leaves, the phloem loading step might, therefore, be particularly sensitive to magnesium and potassium deficiency.

Accumulation of photosynthates in source leaves, as a result of either impaired phloem loading or export under magnesium and potassium deficiency, or impaired sink activity under zinc deficiency (Marschner and Cakmak, 1989), enhances the formation of toxic oxygen species such as superoxide radicals (O^{2-}) and hydrogen peroxide. Accordingly, the activity of detoxifying enzymes in source leaves increases under either magnesium or potassium deficiency, but not under phosphorus deficiency where less photosynthates are accumulated.

Under conditions of high light intensity, however, the production of toxic oxygen species often exceeds the capacity of the detoxifying enzymes leading to photooxidation of chloroplast pigments and severe symptoms of chlorosis and necrosis in potassium- and magnesium deficient leaves. The particularly high light sensitivity of potassium- and magnesium- deficient source leaves can be readily demonstrated by partial shading of the leaf blades, where the shaded plants remain green and the light parts become chlorotic and necrotic (Marschner and Cakmak, 1989). In agreement with this observation, in tobacco and tomato plants with genetically manipulated inhibition of phloem loading of sucrose, accumulation rates of carbohydrates in the source leaves was associated with severe chlorosis and necrosis of these leaves (Riesmeier *et al.,* 1994).

Zinc deficiency combined with high light intensity, leads to photo-oxidation of chloroplast pigments and, thus, to a particularly evident rapid destruction of functioning of source leaves. In zinc- deficient plants, although growth and sink activity of the shoot apex are strongly depressed and carbohydrates accumulate in the source leaves, root growth is not enhanced model (Cakmak *et al.,* 1989). Besides the lower export of photosynthates, the activity of Cu/Zn superoxide dismutase (SOD) is impaired in zinc deficient leaves and raises the level of O^{2-} in the cells accordingly (Cakmak and Marschner, 1988). The source function of the leaf is thus destroyed (Marschner *et al.* 1996).

References

Abd El-Hadi, A.H., Ismail. K.M. and El-Akahawy, M.A. 1997. Effect of potassium on the drought resistance of crops in Egyptian conditions, In, "*Food Security in the WANA Region, the Essential Need for Balanced Fertilization,*" Ed. A.E. Johnston, Int Potash Inst Basel: 328-336.

Abdel Wahab, A.M. and Abd-Alla, M.H. 1995. The role of potassium fertilizer in nodulation and nitrogen fixation of faba bean (*Vicia faba* L.) plants under drought stress. *Biol. Fert. Soils.,* 20:147–150.

Abd-El Baki, G.K., Siefriz, F., Man, H.M., Weiner, H., Kaldenhoff, R. and Kaiser, W.M. 2000. Nitrate reductase in *Zea mays* L. under salinity. *Plant Cell Environ.*, 23: 515-521.

Ackerson, R.C. 1985. Osmo-regulation in cotton in response to water-stress. 3. Effects of phosphorus fertility. *Plant Physiol.*, 77: 309-312.

Aggarwal, P.K. and Sinha, S.K. 1984. Differences in water relations and physiological characteristics in leaves of wheat associated with leaf position on the plant. *Plant Physiol.,* 74: 1041-1045.

Ahmad, E.. Rashid, A., Saifullah, A., Ashraf, M.Y. and Ehsanullah, 2011. Role of mineral nutrition in alleviation of drought stress in plants. *A.J.C.S.,* 5(6): 764-777.

Alam, S.M., 1999. Nutrient Uptake by Plants Under Stress Conditions. Marcel Dekker, Inc.

Alexandersson, E., Fraysse, L., Sjovall-Larsen, S., Gustavsson, S., Fellert, M., Karlsson, M., Johanson, U. and Kjellbom, P. 2005. Whole gene family expression and drought stress regulation of aquaporins. *Plant Mol. Biol.,* 59: 469–484.

Ali, M.A., Hussain, M., Khan, M.L. Ali, Z., Zulkiffal, M., Anwar, J., Sabir, W. and Zeeshan, M. 2010. Source-sink relationship between photosynthetic organs and grain yield attributes during grain filling stage in spring wheat (*Triticum aestivum*). *Int J Agrie Biol.*, 12:509-515.

Ambebe, T.F., Dang, Q.L. and Li, J.L. 2009. Low soil temperature inhibits the effect of high nutrient supply on photosynthetic response to elevated carbon dioxide concentration in white birch seedlings. *Tree Physiol.*, 30: 234–243.

Anderson, E.L. 1988. Tillage and N fertilization effects on maize root growth and root:shoot ratio. *Plant and Soil,* 108: 245-251.

Anoop, E.V.; Gopikumar, K. and Babu, L.C. 1998. Visual symptoms and chlorophyll production of *Ailanthrus triphysa* seedlings in response to nutrient deficiency. *J. Tropical Forest Sci.* 10(3): 304-311.

Apel, K. and Hirt, H. 2004. Reactive oxygen species: Metabolism, oxidative stress, and signal Transductioa. *Ann Rev Plant Bio.*, 55: 373-399.

Arjenaki, F.G., Jabbari, R. and Morshedi, A. 2012. Evaluation of drought stress on relative water content, chlorophyll content and mineral elements of wheat (*Triticum aestivum L.)* varieties. *Int. J. Agric. Crop Sci.,* 4(11): 726-729.

Arnold, N., Chong, C. and Binns, M. 1986. Seasonal nutrient contents in grey and non-grey flue-ured tobacco. *Beitrage zur Tabakforschung Int.*, 13(5): 265-269.

Asada, K. 2006. Production and scavenging of reactive oxygen species in chloroplasts and their functions. *Plant Physiol.,* 141: 391-396.

Ashraf, M.A., Ahmad, M.S.A., Ashraf, M., Al-Qurainy, F. and Ashraf, M.Y. 2011. Alleviation of waterlogging stress in upland cotton (*Gossypium hirsutum* L.) by exogenous application of potassium in soil and as a foliar spray. *Crop Pasture Sci.,* 62: 25–38.

Ashraf, M.A. 2012. Waterlogging stress in plants: A review. *Afr. J. Agri. Res.,* 7: 1976–1981.

Atkin, O.K. and Cummins, W.R. 1994. The effect of root temperature on the induction of nitrate reductase activities and nitrogen uptake rates in arctic plant species. *Plant Soil*, 159: 187–197.

Azedo-Silva, J.J., Osorio, F.F. and Correia, M.J. 2004. Effects of soil drying and subsequent re-watering on the activity of nitrate reductase in root and leaves of *Helianthus annum. Funct. Plant Biol.*, 31: 611-621.

Bajji, M., Kinet, J.M. and Lutts, S. 2002. The use of the electrolyte leakage method for assessing cell membrane stability as a water stress tolerance test in durum wheat. *Plant Growth Regul.,* 36: 61–70.

Bailey-Serres, J. and Voesenek, L.A.C.J. 2008. Flooding stress: Acclimations and genetic diversity. *Annu. Rev. Plant Biol.,* 59: 313–339.

Balakrishnan, K. 1999. Studies on nutrients deficiency symptoms in chilli (*Capsicum* annum) *Indian J. Plant Physiol.* 4(3): 229-231.

Baligar, V.C., and Fageria, N.K. 1997. Nutrient use efficiency in acid soils: Nutrient management and plant use efficiency. In, "*Plant-Soil Interactions at Low pH: Sustainable and Forestry Production,*" Eds. A.C. Moniz, A.M.C. Furlani, R.E. Schaffert, N.K. Fageria, C.A. Rosolem, and H. Cantarella, Campinas, Brazil: Brazilian Soil Science Society, pp. 75–95..

Baque, M.A., Karim, M.A., Hamid, A. and Tetsushi, H. 2006. Effects of fertilizer potassium on growth, yeild and nutrient uptake of wheat (*Triticum aestivum*) under water stress condition. *South pacific Studies*, 27: 1.

Baranwal, V.K.; Manikandan, P.; Ray, A.K. 1989. Crown choking disorder of coconut: a case of boron deficiency. *J. Plantation Crops*. 17(2): 114-120.

Batts, G.R., Morison, J.I.L., Ellis, R.H., Hadley, P. and Wheeler, T.R. 1997. Effects of CO_2 and temperature on growth and yield of crops of winter wheat over several seasons. *European Journal of Agronomy*, 7: 43–52.

Bendixen, R., Gerendas, J., Schinner, K., Sattelmacher, B. and Hansen, U.P. 2001. Differences in zeaxanthin formation in nitrate and ammonium grown *Phaseolus vuklgaris. Physiologia Plantarum*, 111: 255-261.

Bennett, J.P. and Skcog, F. 2002. Preliminary experiments on the relation of growth-promoting substances to the rest period in fruit trees. *Plant Physiol.*, 13: 219-225.

Bernier, B. and Brazeau, M. 1988. Nutrient deficiency symptoms associated with sugar maple dieback and decline in the Quebec Appalachians. *Canadian J. Forest Res.* 18(6): 762-767.

Benlloch-Gonzalez, M., Arquero, O., Fournier, J.M., Barranco, D. and Benlloch M. 2008. K^+ starvation inhibits water-stress-induced stomatal closure. *J. Plant Physiol.,* 165: 623-630.

Benlloch-Gonzalez, M., Romera, J., Cristescu, S., Harren, F., Fournier, J.M. and Benlloch M. 2010. K^+ starvation inhibits water-stress-induced stomatal closure via ethylene synthesis in sunflower plants. *J. Exp. Bot..*, 61: 1139–1145.

Bhutta, W.M. 2007. The effect of cultivar on the variation of spring wheat grain quality under drought conditions. *Cereal Res Commun.*, 35: 1609-1619.

Bode Stoltzfus, R.M., Taber, H.G. and Aiello, A.S. 1998. Effect of increasing root-zone temperature on growth and nutrient uptake by 'Gold Star' muskmelon. *J. Plant Nutr.*, 21: 321–328.

Bogdevitch, I. 2000. IPI Internal Report. International Potash Institute; Basel, Switzerland: 2000.

Botella, M.A., Martinez, V., Pardines, J. and Cerda, A. 1997. Salinity induced potassium deficiency in maize plants. *J. Plant Physiol.*, 150: 200–205.

Bowen, G.D. 1991. *Soil temperature, root growth and plant function. In, "Plant roots: the hidden half,*" Eds. Y. Waisel, A. Eshel, U. Kafkaki, New York: Plenum Press, 309–*330.*

Boyer, J.S. 1982. Plant prod, and environ. *Sci.* 218: 443-448.

Bruce, W.B., Edmeades, G.O. and Barker, T.C. 2002. Molecular and physiological approaches to maize improvement for drought tolerance. *J Exp Bot.,* 53: 13-25.

Bravo-F, P. and Uribe, E.G. 1981. Temperature dependence of the concentration kinetics of absorption of phosphate and potassium in corn roots. *Plant Physiol.*, 67: 815–819.

Bray, E.A., Bailey-Serres, J. and Werelnyk, E 2000. Responses to abiotic stresses. In,"*Biochemistry and Molecular Biology of Plants,*" Eds. B. Buchanan, W. Gruissem, and R. Jones, American Society of Plant Physiology, pp. 1158-1203.

Brck, H., Payne, W.A. and Sattelmacher, B. 2000. Effects of phosphorus and water supply on yield, ranspirational water-use efficiency, and carbon isotope discrimination of pearl millet. *Crop Sci.*,40: 120-125.

Byrt, C.S., Platten, J.D., Spielmeyer, W., James, R.A., Lagudah, E.S., Dennis, E.S., Tester, M. and Munns, R. 2007. HKT1;5-like cation transporters linked to Na^+ exclusion loci in wheat, *Nax2 and Knat1. Plant Physiol.,* 143: 1918–1928.

Cakmak, I. and Marschner H. 1988. Enhanced superoxide radical production in roots of zinc-deficient plants. *Journal of Experimental Botany* 39: 1449-1460.

Cakmak, I, Marschner, H. and Bangerth F. 1989. Effect of zinc nutritional status on growth, protein metabolism and levels of indole-3-acetic acid and other phytohormones in bean *(Phaseolus vulgaris* L.). *Journal of Experimental Botany,* 40: 405-412.

Cakmak, I. and Marschner, H. 1988a. Increase in membrane permeability and exudation in roots of zinc deficient pants. *Journal of plant Physiiology,* 132: 356-361.

Cakmak, I. and Marschner, H. 1988b. Enhanced superoxide redical production in roots of zinc deficient plants. *Journal of Experimental Botany,* 39: 1449-1460.

Cakmak, I. 1994a. Activity of ascorbate dependent H_2O_2 scavenging enzymes and leaf chlorosis are enhanced in magnesium and potassium deficient leaves, but not in phosphorus deficieant leaves. *Journal of Experimental Botany,* 45: 259-1266.

Cakmak, I,, Hengelerm C, and Marschner, H. 1994b. Partitioning of shoot and root dry matter and carbohydrates in bean plants and suffering from phosphorus, potassium and magnesium deficiency. *J Exp Bot.*, 45: 1245-1250.

Cakmak, I., Hengeler, C. and Marschner, H. 1994c. Changes in phloem export of sucrose in leaves in response to phosphorus, potassium and magnesium deficiency in bean plants. *J Exp Bot.*, 45: 1251-1257.

Cakmak, I., Atli, M., Kaya, R., Evliya, H. and Marschner, H. 1995. Association of high light and zinc dieficiency and cold induced leaf chlorosis in grapefruit and mandarin trees. *Journal of Plant Physiol.*, 146: 355-360.

Cakmak, I. 1997. Role of potassium in protecting higher plants against photo-oxidative damage. In, "*Food security in the WANA region, the essential need for balanced fertilization,*" Ed. A.E. Johnston, International Potash Institute, Basel Switzerland, pp. 345-352.

Cakmak, I. and Engels, C. 1999. Role of mineral nutrients and photosynthesis and formation. In. "*Mineral nutrition of crops: Mechanism and implecations,*" Ed. Z. Rengel, The Howorth Press, New Yorlk, pp. 141-168.

Cakmak, I. 2000. Possible roles of zinc in protecting plant cells from damage by reactive oxygen species. *New Phytol.* 146: 185–205.

Cakmak, I. 2002. Plant nutrion research: Priorities to meet human needs for food in sustainable ways. *Plant and Soil*, 247: 3-24.

Cakmak, I. 2005. The role of potassium in alleviating detrimental effects of abiotic stresses in plants. *J. Plant Nutr. Soil Sci.,* 168: 521–530.

Calatayud, Á., Gorbe, E., Roca, D. and Martínez, P.F. 2008. Effect of two nutrient solution temperatures on nitrate uptake, nitrate reductase activity concentration and chlorophyll a fluorescence in rose plants. *Enviro. Exp. Bot.*, 64: 65–74.

Castrillo, M., Fernandez, D., Calcagno, A.M., Trujillo, I. and Guenni, L. 2001. Responses of ribulose-l,5- bisphosphate carboxylase, protein content, and stomatal conductance to water deficit in maize, tomato, and bean. *Photosyn.,* 39: 221-226.

Chen, Z.H., Zhou, M.X., Newman, I.A., Mendham, N.J., Zhang, G.P. and Shabala, S. 2007. Potassium and sodium relations in salinised barley tissues as a basis of differential salt tolerance. *Funct. Plant Biol.,* 34: 150–162.

Ciamporova, M. 2002. Morphological and structural responses of plant roots to aluminium at organ, tissue and cellular levels. *Biology of Plants*, 45: 161–171.

Clarkson, D.T. and Warner, A.J. 1979. Relationships between root temperature and the transport of ammonium and nitrate ions by Italian and perennial ryegrass (*Lolium multiflorum* and *Lolium perenne*). *Plant Physiol.*, 64: 557–561.

Clarkson, D.Ò., Earnshaw, M.J., White, P.A. and Cooper, H.D. 1988. Temperature dependent factors influencing nutrient uptake: an analysis of responses at different levels of organization. Plant and Temperature: *Symposium of Society for Experimental Biology.Cambridge, UK*, 1988, pp. 281–309.

Clarkson, D.T., Jones, L.H.P. and Purves, J.V. 1992. Absorption of nitrate and ammonium ions by *Lolium perenne* from flowing solution cultures at low root temperatures. *Plant Cell Environ.*, 15: 99–106.

Close, D.C., Beadle, C.L. and Hovenden, M.J. 2003. Interactive effects of nitrogen and irradiance on sustained xanthophylls cycle engagement in *Euclyptus nitens* leaves during winter. *Oecologia*, 134: 32-36.

Colmer T.D. 2011. Greenway H. Ion transport in seminal and adventitious roots of cereals during O_2 deficiency. *J. Exp. Bot.*, 62: 39–57.

Colom, M.R. and Vazzana, C. 2003. Photosynthesis and PSII functionality of drought-resistant and drought sensitive weeping love grass plants. *Environ Exp Bot*., 49: 135-144.

Cornillon, P. 1974. Behavior of the tomato in terms of substrate temperature. *Ann. Agron.*, 25: 753–777.

Coskun, D., Britto, D.T. and Kronzucker, H.J. 2010. Regulation and mechanism of potassium release from barley roots: an in plant $^{42}K^+$ analysis. *New Phytol.* 188:1028–1038.

Cox, D.A. 1992. Poinsettia cultivars differs in their response to molybdenum deficiency. *Hort Science.* 27(8): 892-893.

Cox, D.A. and Bartley, G.C. 1987. Lime, molybdenum and cultivars effects on molybdenum deficiency of poinsettia. Res. Report Series *Albama Agric. Expt. Station,* Aubum Univ.. Publ. 1988. No. 5, pp. 13.

Cuéllar, T., Pascaud, F., Verdeil, J.L., Torregrosa, L., Adam-Blondon, A.F., Thibaud, J.B., Sentenac, H., and Gaillard, I. 2010. A grapevine Shaker inward K^+ channel activated by the calcineurin B-like calcium sensor 1-protein kinase CIPK23 network is expressed in grape berries under drought stress conditions. *Plant J.*, 61: 58–69.

Cruz de Carvalho M.H. 2008. Drought stress and reactive oxygen species: Production, scavenging and signaling. *Plant Signal. Behav.* 3: 156–165.

DaCosta, M. and Huang, B.R. 2006. Osmotic adjustment associated with variation in bentgrass tolerance to drought stress. *J. Am. Soc. Hortic. Sci.,* 131: 338–344.

Dat, J., Vandenabeele, S., Vranova, E., van Montagu, M., Inze, D., van Breusegem, F. 2000. Dual action of the active oxygen species during plant stress responses. *Cell. Mol. Life Sci.,* 57: 779–795.

Degl'Innocenti, E., Hafsi, C., Guidi, L. and Navari-Izzo, F. 2009. The effect of salinity on photosynthetic activity in potassium-deficient barley species. *J. Plant Physiol.,* 166: 1968–1981.

Dell, B.1994. Copper nutrition of *Eucalyptus maculata* Hook seedlings: requirements for growth, distribution of copper and the diagnosis of copper deficiency. *Plant and Soil.* 167(2): 181-187.

Devi, B.S.R., Kim, Y.J., Selvi, S.K., Gayathri, S., Altanzul, K., Parvin, S., Yang, D.U., Lee, O.R., Lee, S. and Yang D.C. 2012. Influence of potassium nitrate on antioxidant level and secondary metabolite genes under cold stress in *Panax ginseng. Russ. J. Plant Physiol.,* 59: 318–325.

Dhanda, S.S., Sethi, G.S. and Behl, R.K. 2004. Indices of drought tolerance in wheat genotypes at early stages of plant growth. *J Agron Crop Sci.*, 190: 6-12.

Dipierro, N., Mondelli, D., Paciolla, C., Brunetti, G. and Dipierro, S. 2005. Changes in the ascorbate system in the response of pumpkin (*Cucurbita pepo* L.) roots to aluminium stress. *Journal of Plant Physiology*, 162: 529–536.

Domisch, T., Finér, L., Lehto, T. and Smolander, A. 2002. Effects of soil temperature on nutrient allocation and mycorrhizas in Scots pine seedlings. *Plant Soil*, 239: 173–185.

Duke, S.H., Schrader, L.E., Henson, C.A., Servaittes, J.C., Vogelzang, R.D. and Pendleton, A.W. 1979. Low root temperature effects on soybean nitrogen metabolism and hotosynthesis. *Plant Physiol.*, 63: 956–962.

Duval, L , More, E. and Sicot, A. 1991. Observations on molybdenum deficiency in cauliflower in Brittany, *Comtes Rendus de l Academic d Agriculture de France.* 78(1): 27-34.

Edwards, J.H. and Horton, B. D. 1979. Response of peach seedlings to calcium concentration in nutrients solution. *J. American Soc. Hort. Sci.* 104(1): 97-99.

Egila, J.N., Davies, F.T.J. and Drew, M.C. 2001. Effect of potassium on drought resistance of Hibiscus rosa-sinensis cv. Leprechaun: plant growth, leaf macro and micronutrient content and root longevity. *Plant Soil*, 229: 213-224.

Egilla, J.N., Davies, F.T. and Boutton, T.W. 2005. Drought stress influences leaf water content,photosynthesis, and water-use efficiency of hibiscus rosa-sinensis at three potassium concentrations. *Photosynthetica.* 43: 135–140.

Egilla, J.N., Davies, F.T. and Drew, M.C. 2001. Effect of potassium on drought resistance of *Hibiscus rosa-sinensis*cv. Leprechaun: Plant growth, leaf macro- and micronutrient content and root longevity. *Plant Soil.*, 229: 213–224.

Else, M.A., Coupland, D., Dutton, L. and Jackson, M.B. 2001. Decreased root hydraulic conductivity reduces leaf water potential, initiates stomatal closure and slows leaf expansion in flooded plants of castor oil (*Ricinus communis*) despite diminished delivery of ABA from the roots to shoots in xylem sap. *Physiol. Plantarum.,* 111: 46–54.

Elstner, E.F. and Osswald, W. 1994. Mechanism of oxygen activation during plant stress. Proc. Royal Soc. Edinburgh, Section-B 102: 131-154.

Engels, C. and Marschner, H.. 1992. Root to shoot translocation of macronutrients in relation to shoot demand in maize (Zea +99mays L.) grown at different root zone temperatures. Zeitung für *Pflanzenernährung und Bodenkunde*, 155: 121–*128.*

Engels, C., Münkle, L. and Marschner, H. 1992. Effect of root zone temperature and shoot demand on uptake and xylem transport of macronutrients in maize (*Zea mays* L.). *J. Exp. Bot.*, 43: 537–547.

Engels, C. 1993. Differences between maize and wheat in growth-related nutrient demand and uptake of potassium and phosphorus at suboptimal root zone temperature. *Plant Soil*, 150: 129–138.

Engels, C. and Marschner, H. 1995. Plant uptake and utilization of nitrogen. In, "*Nitrogen fertilization in the environment,*" Ed. P.E. Bacon, Marcel Dekker Inc., New York: pp. 41-81.

Epstein, E. 1999. Silicon. *Annu. Rev. Plant Physiol. Plant Mol Biol.*, 50: 641-664.

Epstein, E. and Bloom, A.J. 2004. Mineral Nutrition of Plants: Principles and Perspectives, 2nd Edn. Sinauer Associates, Sunderland, MA.

Ericsson, T. and Kahr, M. 1995. Growth and nutrition of birch seedlings at varied relative addition rates of magnesium. *Tree Physiology* 15: 85-93.

Erlejman, A.G., Verstraeten, S.V., Fraga, C.G. and Oteiza, P.I. 2004. The interaction of flavonoids with membranes: Potential determinant of flavonoid antioxidant effects. *Free Radical Res.,* 38: 1311–1320.

Evans, J.R. 1989. Photosynthesis and nitrogen relationship in leaves of C_3 plants. *Oceologia*, 78(1): 9-19.

Ferreira, J.L., Coelho, C.H.M., Magalhães, P.C. Sant'ana, G.C. and Borém, A. 2008. Evaluation of mineral content in maize under flooding. *Crop Breed. and Applied Biotech.*, 8: 134-140.

Fiedler, S., Vepraskas, M.J. and Richardson, J.L. 2007. Soil redox potential: Importance, field measurements, and observations. *Adv. Agron.,* 94: 1–54.

Fischer ES, Bremer E. 1993. Influence of magnesium deficiency on rates of leaf-expansion, starch and sucrose accumulation, and net assimilation in *Phaseolus vulgaris. Physiologia Plantarum* 89: 271-276.

Fonts, R.L.F. and Cox, F.R. 1998. Zinc toxicity in soybean grown at high iron concentration in nutrient solution. *J. Plant Nut.* 21(8): 1723-1730.

Foyer, C.H., Lelandais, M. and Kunert, K.J. 1994. Photooxidative stress in plants. *Physiol Plant*, 92: 696-717.

Foyer, C.H., Vanacker, H., Gomez, L.D. and Harbinson J. 2002. Regulation of photosynthesis and antioxidant metabolism in maize leaves at optimal and chilling temperatures: Review. *Plant Physiol. Biochem.*, 40: 659–668.

Fredeen, A.L., Rao, I.M. and Terry, N. 1989. Influence of phosphorous nutrition on growth and carbon partitioning in *Glycine max. Plant Physiol.*, 89: 225-230.

Fu, J.M. and Huang, B.R. 2001. Involvement of antioxidants and lipid peroxidation in the adaptation of two cool-season grasses to localized drought stress. *Environ. Exp. Bot.,* 45: 105–114.

Galmes, J., Pou, A., Alsina, M.M., Tomas, M., Medrano, H. and Flexas, J. 2007. Aquaporin expression in response to different water stress intensities and recovery in Richter-110 (*Vitis* sp.): Relationship with eco-physiological status. *Planta.* 226: 671–681.

Gao, X., Zou, C., Wang, L. and Zhang, F. 2004. Silicon improves water use efficiency in maize plants. *J Plant Nutr.*, 27(8): 1457-1470

Gao, X., Zou, C., Wang, L. and Zhang, F. 2006. Silicon decreases transpiration rate and conductance from stornata of maize plants. *J Plant Nutr.*, 29:1637-1647.

Garg, B.K., Burman, U. and Kathju, S. 2004. The influence of phosphorus nutrition on the physiological response of moth bean genotypes to drought. *J Plant Nutr Soil Sci.,* 167:503-508.

Garnett, T.P. and Smethurst, P.J. 1999. Ammonium and nitrate uptake by *Eucalyptus nitens*: effects of pH and temperature. *Plant Soil*, 214: 133–140.

Gong, H., Zhu, X., Chen, K., Wang, S. and Zhang, C. 2005. Silicon alleviates oxidative damage of wheat plants in pots under drought. *Plant Sci.*, 169: 313-321.

Gong, X., Chao, L., Zhou, M., Hong, M., Luo, L., Wang, L., Ying, W., Jingwei, C., Songjie, G. and Fashui, H. 2011. Oxidative damages of maize seedlings caused by exposure to a combination of potassium deficiency and salt stress. *Plant Soil,* 340: 443–452.

Gruhn, P., Goletti, F. and Yudelman, M. 2000. Integrated nutrient management, soil fertility, and sustainable agriculture: current issues and future challenges. Food, Agriculture, and the Environment Discussion Paper 32, International Food Policy Research Institute, Washington, D.C.

Gerendás, J. and Sattelmacher, B. 1997. Significance of Ni supply for growth, urease activity and the concentrations of urea, amino acids and mineral nutrients of urea-grown plants. *Plant and Soil.* 190: 153-162.

Gerewal, J.S. and Singh, S.N. 1980. Effect of potassium nutrition on frost damage and yield of potato plants on alluvial soils of Punjab (India). *Plant and Soil*, 57: 105-110.

Gesteira, A.S., Micheli, F., Carels, N., Silva, A.C., Gramacho, K.P., Schuster, I., Macedo, J.N., Pereira, G.A. and Cascardo, J.C. 2007. Comparative analysis of expressed genes from cacao meristems infected by Moniliophthora perniciosa. *Annuals of Botany*, 100: 129–140.

Gupta, U.C. and Sanderson, J.B. 1993. Effect of S, Ca and B on tissue nutrient concentration and potato yield. *J. Plant Nut.* 16(6): 1013-1023.

Gupta, U.C. 1997. Symptoms of molybdenum deficiency and toxicity in crops. In, "*Molybdenum in Agriculture*," Ed. U,C. Gupta, Cambridge Universiy Press, pp. 160-170.

Hacisalihoglu, G., Hart, J.J., Wang, Y., Cakmak, I. and Kochian, L.V. 2003. Zinc efficiency is correlated with enhanced expression and activity of Cu/Zn superoxide dismutase and carbonic anhydrase in wheat. *Plant Physiol.,* 131: 595-602.

Hakerlerler, H., Oktay, M., Eryuce, N. and Yagmur, B. 1997. Effect of Potassium Sources on the Chilling Tolerance of Some Vegetable Seedlings Grown in Hotbeds. In, "*Food Security in the WANA Region, the Essential Need for Balanced Fertilization*", Ed. A.E. Johnston, International Potash Institute; Basel, Switzerland, pp. 353–359.

Hampp, R. and Schnabi, H. 1975. Effect of aluminium ions on $^{14}CO_2$-fixation and membrane system of isolated spinach chloroplasts. *Zeitschrift fur Pflanzenphysiologie*, 76: 300–306.

Hamy, A. 1983. Effect of phosphorus deficiency on pigments of barley leaves. *Compets Rendus des Seances de l Academie d Agriculture de France.* 69(12): 935-943.

Hanson, E. 1993. Apples and Pears, In: *Nutrient Deficiencies and Toxicities in Crop Plants*. The American Phytopathological Society, St. Paul Minnesota, pp. 1-7.

Hariyadi, P. and Parkin, K.L. 1993. Chilling-induced oxidative stress in cucumber (*Cucumis sativus* L. cv. Calypso) seedlings. *Journal of Plant Physiology*, 141: 733–738.

Hartung, W.S., Wilkinson, Davies. W.J. 1998.. Factors that regu-late absiscic acid concentrations at the primary site of action at the guard cell. *J Mol Microbiol Biotechnol.* 1: 231-242.

Hattori, T., Inanaga, S., Araki, H., An, P., Morita, S., Luxova', M. and Lux, A. 2005. Application of silicon enhanced drought tolerance in *Sorghum bicolour. Physiol Plant*, 123: 459-466.

Hattori, T., Sonobe, K., Inanaga, S., An, P., Tsuji, W., Araki, H., Eneji, A.E. and Morita, S. 2007. Short term stomatal responses to light intensity changes and osmotic stress in sorghum seedlings raised with and without silicon. *Environ Exp Bot.*, 60: 177-182.

Haque, M.Z. 1988. Effect of nitrogen, phosphorus and potassium on spikelet sterility induced by low temperature at the reproductive stage of rice. *Plant and soil.,* 109: 31–36.

Havlin, J.L., Tisdale, S.L., Nelson, W.L. and Beaton, J.D. 2007. Soil Fertility and Fertilizer, An introductionto nutrient management 7th Edition Prentice Hall, Upper Saddle River, NJ, U.S.A.

Heckathorn, S.A., De-Lucia. E.H. and Zielinki, R.E. 1997. The contribution of drought-related decreases in foliar nitrogen concentration to decreases in photosynthetic capacity during and after drought in prairie grasses. *Physiol Plant*., 122: 62-67.

Heinen, R.B., Ye, Q. and Chaumont, F. 2009. Role of aquaporins in leaf physiology. *J. Exp. Bot.,* 60: 2971–2985.

Ho, L.C., Hand, D.J., Fussell, M. and Papadopoulos, A.P. 1999. Improvement of tomato fruit quality by calcium nutrition. *Acta Horticulturae*. 481(2): 417-423.

Hong, S.W. and Vierling, E. 2000. Mutants of *Arabidopsis thaliana* defective in the acquisition of tolerance to high temperature stress. *Proc Natl Acad Sci USA*, 97: 4392-4397.

Horie, T., Hauser, F. and Schroeder J.I. 2009. HKT transporter-mediated salinity resistance mechanisms in *Arabidopsis* and monocot crop plants. *Trends Plant Sci.,* 14: 660–668.

Houghton, J.T., Callander, B.A. and Varney, S.K. 1992. Climate change 1992. *The supplementary report to the IPCC scientific assessment.* Cambridge: Press Syndicate of the University of Cambridge.

Hu, H. and Sparks, D. 1991. Zinc deficiency inhibits chlorophyll synthesis and gas exchange in 'Stuart' pecan. *Hort Sci.*, 26: 267-268.

Hu, Y. and Schmidhalter, U. 2003. Drought and: A comparison of their effects on mineral nutrition of plants. *J. Plant Nutr. Soil Sci.*, 168: 541-549.

Hu, W.H., Song, X.S., Shi, K., Xia, X.J., Zhou, Y.H. and Yu, J.Q. 2008. Changes in electron transport, superoxide dismutase and ascorbate peroxidase isoenzymes in chloroplasts and mitochondria of cucumber leaves as influenced by chilling. *Photosynthetica*, 46(4): 581–588.

Huang, Z.A., Jiang, D.A., Yang, Y., Sun, J.W. and Jin, S.H. 2004. Effect of nitrogen deficiency on gas exchange, chlorophyll flurpsence and antioxidant enzymes in leaves of rice plants. *Photosynthetica,* 42: 357-364.

Hussain, A and Maqsood, M.A. 2011. Root zone temperature influences nutrient accumulation and use in maize. *Pak. J. Bot.*, 43: 1551–1556.

Jeschke, W.D. and Pate, J.S. 1991. Cation and chloride partitioning through xylem and phloem within the whole plant of *Ricinus communis* L. under conditions of salt stress. *Journal of Experimental Botany* 42: 1105-1116.

Jiang, M. and Zhang, J. 2001. Effect of abscisic acid on active oxygen species, antioxidative defence system and oxidative damage. *Plant Cell Physiol.,* 42: 1265–1273.

Jiang, M.Y. and Zhang, J.H. 2002a. Involvement of plasma membrane NADPH oxidase in abscisic acid-and water stress-induced antioxidant defense in leaves of maize seedlings. *Plant*, 215: 1022-1030.

Jiang, M.Y. and Zhang, J.H. 2002b. Water stress-induced abscisic acid accumulation triggers the increased generation of reactive oxygen species and up-regulates the activities of antioxidant enzymes in maize leaves. *Journal of Experimental Botany*, 52: 2401-2410.

Jin, S.H., Huang, J.Q., Li, X.Q., Zheng, B.S., Wu, J.S., Wang, Z.J., Liu, G.H. and Chen M. 2011. Effects of potassium supply on limitations of photosynthesis by mesophyll diffusion conductance in *Carya cathayensis*. *Tree Physiol.* 31: 1142–1151.

Jean-Baptist, I.; Morard, P.; Bernadac, A. and Papadopoulos, A.P. 1999. Effects of temporary calcium deficiency on the incidence of a nutritional disorder in melon. *Acta Horticulturae.* 481(1): 417-423.

Jones, M.M., Osmond, C.B. and Turner, N.C. 1980. Accumulation of solutes in leaves of sorghum and sunflower in response to water deficits. *Aust J Plant Physiol.*, 7: 193-205.

Jones, D.L., Gilroy, S., Larsen, P.B., Howell, S.H. and Kochian, L.V. 1999. Effect of aluminum on cytoplasmic Ca^{2+} homeostasis in root hairs of *Arabidopsis thaliana* (L.). *Planta*, 206: 378–387.

Just, D., Saux, C., Richaud, C. and Andre, M. 1989. Effect of nitrogen stress on sunflower gas exchange. 1. Photorespiration and carbon partitioning. *Plant Physiol Biochem.*, 27: 669-677.

Kafkafi, U., Walerstein, I. and Feigenbaum, S. 1971. Effect of potassium nitrate and ammonium nitrate on the growth, cation uptake and water requirement of tomato grown in sand soil culture. *Israel J. Agric. Res.* 21:13-30.

Kaldenhoff, R., Ribas-Carbo, M., Flexas, J., Lovisolo, C., Heckwolf, M. and Uehlein, N. 2008. Aquaporins and plant water balance. *Plant Cell Environ.,* 31: 658–666.

Kanai, S., Moghaieb, R.E., El-Shemy, H.A., Panigrahi, R., Mohapatra, P.K., Ito, J., Nguyen, N.T.,Saneoka, H. and Fujita, K. 2011. Potassium deficiency affects water status and photosynthetic rate of the vegetative sink in green house tomato prior to its effects on source activity. *Plant Sci.,* 180: 368–374.

Kant, S., and Kafkafi, U. 2002. Potassium and abiotic stresses in plants. In, "*Role of Potassium in Nutrient Management for Sustainable Crop Production in India*", Eds. N.S. Pasricha and S.K. Bansal, Potash Institute of India; Gurgaon, India: 2002. pp. 233–251.

Kashiwagi, J., Krishnamurthy, L., Upadhyaya, H.D., Krishna, H., Chandra, S., Vadez, V. and Serraj, R. 2004. Genetic variability of drought avoidance root traits in the mini-core germplasm collection of chickpea (*Cicer arietinum* L.). *Euphytica*, 146: 213-222.

Kathju, S., Vyas, S.P., Garg, B.K. and Lahiri, A.N. 1990. Fertility Induced Improvement in Performance and Metabolism of Wheat under Different Intensities of Water Stress. *Proceedings of the Int Cong of Plant Physiology* 88: New Delhi, India, pp. 854-858.

Kato, M.C., Hikosaka, K., Hirotsu, N., Makino, A. and Hirose, T. 2003. The excess light energy that is neither utilized in photosynthesis nor dissipated by photoprotective mechanisms determines the rate of photoinactivation in photosystem II. *Plant Cell Physiol.*, 44:31-325.

Kawano, T., Kawano, N., Muto, S. and Lapeyrie, F. 2002. Retardation and inhibition of carbon-induced superoxide generation in BY-2 tobacco cell suspension culture by Zn^{2+} and Mn^{2+}. *Physiologia Plantarum*, 114: 395-404.

Kening, M.K. 2008. Manganese nutition and photosynthesis in NAD-malic enzyme C4 plants. Ph.D. Desertation, University of Missouri-Columbia.

Khan, H.R., McDonald, G.K. and Rengel, Z. 1998. Assessment of the Zn status of chickpea by plant analysis. *Plant & Soil.* 198(1): 1-9.

Khan, H.R., McDonald, G.K. and Rengel, Z. 2003. Zn fertilization improves water use efficiency, grain yield and seed Zn content in chickpea. *Plant Soil*., 249: 389-400.

Kimball, B.A., Pinter, P.J., García, R.L., LaMorte, R.L., Wall, G.W., Hunsaker, D.J., Wechsung, G., Wechsung, F. and Kartschall, T. 1995. *Productivity and water use of winter wheat under free air CO_2 enrichment. Global Change Biology*, 1: 429–442.

Kirmizi, S. and Bell, R.W. 2012. Responses of barley to hypoxia and salinity during seed germination, nutrient uptake, and early plant growth in solution culture. *J. Plant Nutr. Soil Sci.,* 175: 630–640.

Kirkby, E.A., LeBot, J., Adamowicz, S. and Römheld, V. 2009. Nitrogen in Physiology—An Agronomic Perspective and Implications for the Use of Different Nitrogen Forms. International Fertiliser Society; Cambridge, York, UK: 2009.

Kirschbaun, M.U.F. and Tompkins, D. 1990. Photosynthetic response to phosphorous nutrition in *Eucalyptus grandis* seedling. *Trop Agrie*., 64: 91-96.

Knight, H., Trewavas, A.J. and Knight, M.R. 1996.Cold calcium signaling in Arabidopsis involves two cellular pools and a change in calcium signature after acclimation . *Plant Cell*, 8: 489–503.

Kochian, L. V. 1995. Cellular mechanisms of aluminum toxicity and resistance in plants. *Annual Review of Plant Physiology and Molecular Biology,* 46: 237–260.

Kozlowski, T.T. and Pallardy, S.G. 1997. Mineral nutrition. In, "*Physiology of Woody Plants*," Ed. T.T. Kozlowski, and S.G. Pallardy, San Diego: Academic Press, pp. 210–250.

Krogmeier, M.J., McCarty, G.W., Shogren, D.R. and Bremner, J.M. 1991. Effect of nickel deficiency in soybeans on the phytotoxicity of foliar-applied urea. *Plant and Soil*. 135: 283-286.

Lambers, H., van den Boogaard, R., Veneklaas, E.J. and Villar, R.1995. Effects of global environmental change on carbon partitioning in vegetative plants of *Triticum aestivum* and closely related *Aegilops species. Global Change Biology*, 1: 397–406.

Lambers, J. and Jones, T.L. 1998. Plant physiological ecology. Spring-er-Verlag, New York.

Lee, S.H., Singh, A.P., Chung, G.C., Ahn, S.J., Noh, E.K. and Steudle, E. 2004. Exposure of roots of cucumber (*Cucumis sativus*) to low temperature severely reduces root pressure, hydraulic conductivity and active transport of nutrients. *Physiol. Plant*., 120: 413–420.

Lian, H.L., Yu, X., Ye, Q., Ding, X.S., Kitagawa, Y., Kwak, S.S., Su, W.A. and Tang Z.C. 2004. The role of aquaporin RWC3 in drought avoidance in rice. *Plant Cell Physiol.,* 45: 481-489.

Liang, Y.C., Sun, W.C., Zhu, Y.G. and Christie, P. 2007. Mechanisms of silicon-mediated alleviation of abiotic stresses in higher plants, a review. *Environ Pollut*., 147:422-428.

Lin, C.C. and Kao C.H. 2001. Abscisic acid induced changes in cell wall peroxidase activity and hydrogen peroxide level in roots of rice seedlings. *Plant Sci.,* 160: 323–329.

Lindhauer, M.G. 1985. Influence of K nutrition and drought on water relations and growth of sunflower (*Helianthus-annuus* L.) *J. Plant Nutr. Soil Sci.,* 148: 654–669.

Lindhauer, M.G. 1995. Influence of K nutrition and drought and water stressed sunflower plants differing in K nutrition. *J Plant Nutr*., 10: 1965-1973.

Liu, Z. 1996. Microelements in Soils of China. - Jiangsu Science and Technology Publishing House, Nanjing 1996. [In Chinese].

Liu, H.Y., Sun, W.N., Su, W.A., and Tang Z.C. 2006. Co-regulation of water channels and potassium channels in rice. *Physiol. Plantarum.,* 128: 58–69.

Lock, J. and Price, A.H. 1994. Evidence that disruption of cytosolic calcium is critically important in oxidative plant stress. *Proceedings of the Royal Society of Edinburgh. Section B: Biology*, 102: 261–264.

Loggini, B., Scartazza, A., Brugnoli, E. and Navari-Izzo, F 1999. Anti oxidative defense system, pigment composition, photosynthetic efficiency in two wheat cultivars subjected to drought. *Plant Physio*l., 119: 1091-1099.

Lawlor, D.W. 2002. Carbon and nitrogen assimilation in relation to yield mechanisms are the key to understanding production system s.*J Exp Bot.*, 53: 773-787.

Loren-Plucinska, G. and Ziegler, H. 1996. Changes in ATP levels in Scots pine needles during aluminium stress. *Photosynthetica*, 32: 141–144.

Lukatkin, A.S., Sharkaeva, E.Sh. and Zauralov, O.A. 1995. Lipid peroxidation in the leaves of heat-loving plants as dependent on the duration of cold stress. *Russian Journal of Plant Physiology*, 42(4):, 538–542 .

Lukatkin, A.S. and Isaikina, E.E. 1997. Calcium status and chilling injury in maize seedlings. *Russian Journal of Plant Physiology*, 44(3):, 339–342.

Lukatkin, A.S. 2003. Contribution of oxidative stress to the development of cold-induced damage to leaves of chilling-sensitive plants.3. Injury of cell membranes by chilling temperatures. *Russian Journal of Plant Physiology*, 50(2): 243-246.

Lukatkin, A.S., Brazaityte, A., Bobinas, C., Duchovskis, P. 2012. Chilling injury in chilling-sensitive plants: a review. *Þemdirbystë=Agriculture*, 99(2): 111– 124.

McKersie, B.D. and Leshem, Y.Y. 1994. Stress and Stress Coping in Cultivated Plants. KluwerAcademic Publishers; Dordrecht, The Netherlands: 1994. pp. 181–193.

Marschner, H. 1995. Mineral nutrition of higher plants. 2nd Edition, Academic Press, San Diego, pp. 889.

MacAdam, J.W., Volenec, J.J. and Nelson, C.J. 1989. Effects of nitrogen on mesophyll cell division and epidermal cell elongation in tall fescue leaf blades. *Plant Physiol.*, 89: 549-56.

Mackay, A.D. and Barber, S.A.1984. Soil temperature effects on root growth and phosphorus uptake by corn. *Soil Science Society of America Journal*, 48: 818–*823*.

Makkanen, S.K. 1995. Damage by drought stress and magnesium deficiency: experiments on young spruce clones. *Allgemeine Forst Zeitschrift,* 50(5): 263-267.

Mancuso, S. and Marras, A.M. 2006. Adaptative response of *Vitis* root to anoxia. *Plant Cell Physiol.*, 47: 401–409.

Marschner, H. and Cakmak, I. 1989. High light intensity enhances chlorosis and necrosis in leaves of zinc-, potassium- and magnesium-deficient bean *(Phaseolus vulgaris)* plants. *Journal of Plant Physiology* 134: 308-315.

Marschner H. 1995. *Mineral nutrition of higher plants,* 2nd edn. London: Academic Press.

Marschner, P. 2012. Marschner's Mineral Nutrition of Higher Plants. 3rd ed. Academic Press; London, UK: 2012. pp. 178–189.

Marschner, H., E. A. Kirkby2, E.A. and I. Cakmak, I. 1996. Effect of mineral nutritional status on shoot-root partitioning of photoassimilates and cycling of mineral nutrients. *Journal of Experimental Botany,* 47(Special Issue):,1255-1263,

Marschner, H., Kirkby, E.A. and Cakmak, I. 1996. Effect of mineral nutritional status on shoot-root partitioning of photoassimilates and cycling of mineral nutrients. *J. Exp. Bot.*, 47: 1255–1263.

Maurel, C. and Chrispeels, M.J. 2001. Aquaporins: A molecular entry into plant water relations. *Plant Physiol.*, 125: 135-138.

Mehrotra, S.C. and Gupta, P. 1990. Screening of chickpea (*Cicer arietinum*) varieties under iron deficiency stress conditions using biochemical indicators. *Indian J. Agric. Sci.* 60(8): 566-568

Mengel, K. and Kirkby, E.A. 2001. Principles of plant nutrition. 5th Edition, Kluwer Academic Publishers, Dordrecht, pp. 848.

Mian, A., Oomen, R.J., Isayenkov, S., Sentenac, H., Maathuis, F.J. and Very A.A. 2011. Over-expression of an Na^+-and K^+-permeable HKT transporter in barley improves salt tolerance. *Plant J.*, 68: 468–479.

Minorsky, P.V. 1985. A heuristic hypothesis of chilling injury in plants: a role for calcium as the primary physiological transducer of injury. *Plant, Cell and Environment*, 8: 75–94.

Moustakas, M. and Ouzounidou, G. 1994. Increases non-photochemical quenching in leaves of aluminum-stressed wheat plants is due to Al-induced elemental loss. *Plant Physiology and Biochemistry,* 32: 527–532.

Moustakas, M., Eleftheriou, E.P. and Ouzounidou, G. 1997. Short-term effects of aluminium at alkaline pH on the structure and function of the photosynthetic apparatus. *Photosynthetica,* 34: 169–177.

Muchow, R.C. 1988. Effect of nitrogen supply on the comparative productivity of maize and sorghum in semi arid tropical environment I. Leaf growth and leaf nitrogen. *Field Crops Res.*, 18: 1-16.

Miao, M., Zhang, Z., Xu, X., Wang, K., Cheng, H. and Cao, B. 2009. Different mechanisms to obtain higher fruit growth rate in two cold-tolerant cucumber (*Cucumis sativus* L.) lines under low night temperature. *Sci. Hortic.*, 119: 357–361.

Mihailovic, N., Drazic, G. and Vucinic, Z. 2008. Effects of aluminium on photosynthetic performance in Al-sensitive and Al-tolerant maize inbred lines. *Photosynthetica* 46: 476–480.

Mart′ýnez-Estevez, M., Racagni-Di Palma, G., Munoz-Sanchez, J.A., Brito-Arg aez, L., Loyola-Vargas, V.M. and Hernandez-Sotomayor, S.M.T. 2003. Aluminum differentially modifies lipid metabolism from ′ the phosphoinositide pathway in *Coffea arabica* L. cells. *Journal of Plant Physiology*, 167: 1297–1303.

Mitchell, R.A.C., Mitchell, V.J., Driscoll, S.P., Franklin, J. and Lawlor, D.W. 1993. *Effects of increased* CO_2 concentration and tempertaure on growth and yield of winter wheat at two levels of nitrogen *application. Plant, Cell and Environment*, 16: 521–*529.*

Mitchell, R.A.C., Lawlor, D.W., Mitchell, V.J., Gibbard, C.L., White, E.M. and Porter, J.R. 1995. *Effects of* elevated CO_2 concentration and tempertaure on growth and yield of winter wheat: test of ARCWHEAT1 simulation model. Plant, *Cell and Environment*, 18: 736–748.

Mittler, R. 2002. Oxidative stress, antioxidants and stress tolerance. *Trends Plant Sci.,* 7: 405–410.

Mugnai, S., Marras, A.M. and Mancuso S. 2011. Effect of hypoxic acclimation on anoxia tolerance in *Vitis* roots: Response of metabolic activity and K^+ fluxes. Plant Cell Physiol.,52:1107–1116.

Munns, R., 2008. Tester M. Mechanisms of salinity tolerance. *Annu. Rev. Plant Biol.,* 59: 651–681.

Nachegowda, V., Sulladmath, U.V., Vijyakumar, N and Subhadrabandu, S. 1992. Visual nutrient deficiency symptoms in sapota (*Manilkara achras* Forsberg). *Acta Horticulturae.* 321: 566-573.

Nautiyal, N., Chaterjee, C. and Sharma, C.P. 1999. Copper stress affects grain filling in rice. *Commun. Soil Sci. & Plant Analysis.* 30(11-12): 1625-1632.

Neumann, D. and Niede, U. 2001. Silicon and heavy metal tolerance of higher plants. *Phytochem.,* 56: 685-692.

Niedziela, C.E. Jr,, Kim, S.H., Nelson, P.V. and De Hertogh, A.A. 2008. Effects of N-P-K deficiency and temperature regime on the growth and development of *Lilium longiflorum* 'Nellie White' during bulb production under phytotron conditions. *Sci. Hortic.*, 116: 430–436.

Nybe, E.V. and Nair, P.C.S. 1986. Nutrient deficiency in black pepper (*Piper nigrum* L.). 1. Nitrogen, phosphorus and potassium. *Agri. Res. J. Kerala.* 24(2): 132-150.

Nybe, E .V. and Nair, P.C.S. 1987. Nutrient deficiency in black pepper (*Piper nigrum* L.). 2. Calcium, magnesium and sulfur. *Agri. Res. J. Kerala.* 25(1): 52-65.

Oerke, E.C. and Dehne, H.W. 2004. Safeguarding production-losses in major crops and the role of crop protection. *Crop Prot.,* 23: 275–285.

O'Sullivan, J.N., Asher, C.J., Blamey, F.P.C. and Edwards, D.G. 1993. Mineral nutrient disorder of root crops of the Pacific: Preliminary observations on sweet potato (*Ipomea batatas*). *Plant and Soil*.155/156: 263-267.

Owuoche, J.O., Briggs, K.G., Taylor, G.J and Penney, D.C. 1995. Response of eight Canadian spring wheat (*Triticum aestivum* L.) cultivars to copper: copper content in leaves and grain. *Canadian J. Plant Sci.*, 75(2): 405-411.

Palta, J.P. 2000. Stress interactions at the cellular and membrane levels. *Hort. Sci.*, 25(11): 1377.

Platten, J.D., Cotsaftis, O., Berthomieu, P., Bohnert, H., Davenport, R.J., Fairbairn, D.J., Horie,T., Leigh, R.A., Lin, H.X., Luan, S., Mâser, P., Pantoia, D., Rodriquez-Navarno, A., Schachtman, O.P., Schroeder, J., Sentenac, H., Uozumi, N., Very, A.A., Zhu, J.K., Dennis, E.S. and Tester, M. 2006. Nomenclature for HKT transporters, key determinants of plant salinity tolerance. *Trends Plant Sci.*, 11: 372–374.

Pang, J.Y., Newman, I., Mendham, N., Zhou, M. and Shabala, S. 2006. Microelectrode ion and O_2 fluxes measurements reveal differential sensitivity of barley root tissues to hypoxia. *Plant Cell Environ.*, 29: 1107–1121.

Pang, J.Y., Cuin, T., Shabala, L., Zhou, M.X., Mendham, N. and Shabala, S. 2007. Effect of secondary metabolites associated with anaerobic soil conditions on ion fluxes and electrophysiology in barley roots. *Plant Physiol.*, 145: 266–276.

Payne, W.A., Hossner, L.R., Onken, A.B. and Wedt, C.W. 1995. Nitrogen and phosphorous uptake in pearl millet and its relation to nutrient and transpiration efficiency. *Agron J* 87: 425-431.

Peng, Y.Y. and Dang, Q.L. 2003. Effects of soil temperature on biomass production and allocation in seedlings of four boreal tree species. *For. Ecol. Manage.*, 180: 1–9.

Pervez, H., Ashraf, M. Makhdum, M.I. 2004. Influence of potassium nutrition on gas exchange characteristics and water relations in cotton (*Gossypium hirsutam L.). Photosynthetica,* 42: 251–255.

Peuke, A.D., Hartung, W. and Jeschke, W.D. 1994. The uptake and flow of C, N and ions between roots and shoots in *Ricinus communis* L. II. Grown with low or high nitrate supply. *Journal of Experimental Botany,* 45: 733-740.

Peuke, A.D., Jeschke, W.D. and Hartung W. 2002. Flows of elements, ions and abscisic acid in *Ricinus communis* and site of nitrate reduction under potassium limitation. *J. Exp. Bot.*, 53: 241–250.

Pilbeam, D.J., Cakmak, I., Marschner, H. and Kikby, E.A. 1993. Effect of withdrawl of phosphorous on nitrate assimilation and PEP carboxylase activity in tomato. *Plant soil* 154: 111-117.

Pinkerton, A. and Simpson, J.R. 1986. Interactions of surface drying and subsurface nutrients affecting plant-growth on acidic soil profiles from an old pasture. *Aust J Exp Agric.*, 26: 681-689.

Price, A.H., Taylor, A., Ripley, S.J., Griffiths, A., Trewavas, A.J. and Knight, M.R..1994. Oxidative signals in tobacco increase cytosolic calcium *Plant coll.,* 6:, 1301–1310.

Pinton, R., Cakmak, I. and Marschner, H. 1994. Zinc deficiency enhanced NAD(P)H-dependent superoxide radical production in plasma membrane vesicles isolated from roots of bean plants. *Journal of Experimental Botany*, 45: 45-50.

Polle, A. 1996. Mehler reaction: friend or foe in photosynthesis? *Botanica Acta*, 109: 84-89.

Pollock, F.T. and Cairns, M.D. 1991. Nitrogen transfer and dry mat-ter production in soybean and sorghum mixed cropping sys-tems at different population densities. *Soil Sci Plant Nutri*, 36: 233-241.

Poovaiah, B.W. and Reddy, A.S.N. 2000. Calcium Messenger Systems in Plants. *CRC Crit Rev Plant Sci.*, 6: 47-102

Preeti, J., Mehrotra, S.C. and Jain, P. 1994. A critical evaluation of some nutrient ratios as indicators of iron chlorosis in plants. *New Botanist*. 21(1-4): 167-168.

Premachandra, G.S., Saneoka, H. and Ogata S. 1991. Cell membrane stability and leaf water relations as affected by potassium nutrition of water-stressed maize. *J. Exp. Bot.,* 42: 739–745.

Qu, C.X., Liu, C., Ze, Y.G., Gong, X.L., Hong, M.M., Wang, L. and Hong F.S. 2011. Inhibition of nitrogen and photosynthetic carbon assimilation of maize seedlings by exposure to a combination of salt stress and potassium-deficient stress. *Biol. Trace. Elem. Res.*, 144: 1159–1174.

Qu, C.X., Liu, C., Gong, X.L., Li, C.X., Hong, M.M., Wang, L. and Hong F.S. 2012. Impairment of maize seedling photosynthesis caused by a combination of potassium deficiency and salt stress. *Environ.* Exp. Bot., 75: 134–141.

Quartacci, M.F., Sgherri, C.L.M., Pinzino, C. and Navariizzo, F. 1994. Superoxide radical production in wheat plants differently sensitive to drought. *Proc. Royal Soc. Edinburgh*,Section B, 102: 287-290.

Rawson, H.M. 1995. Yield responses of two wheat genotypes to carbon dioxide and temperature in field studies using temperature gradient tunnels. *Australian Journal of Plant Physiology*, 22: 23–*32*.

Raun, W.R. and Johnson, G.V. 1999. Improving nitrogen use efficiency for cereal production. *Agron J.*, 9(1): 357-363.

Razmjoo, K., Imada, T., Sugiura, J. and Kanek, S. 1997. Seasonal variations in nutrient and carbohydrate levels of tall fescue cvs. in Japan. *J. Plant Nut.* 20(12): 1667-1679.

Reddy, K.B. ; Ashalatha, M. and Venkaiah, K. 1993. Differential response of groundnut genotypes to iron stress. *J. Plant Nut.* 16(3): 523-531.

Reddy, A.R., Chaitanya, K.V. and Vivekanandan, M. 2004. Drought-induced responses of photosynthesis and antioxidant metabolism in higher plants. *J. Plant Physiol.*, 161: 1189–1202.

Riesmeier, J.W., Willmitzer, L. and Frommer, W.B. 1994. Evidence for an essential role of the sucrose transporter in phloem loading and assimilate partitioning. *EMBO Journal,* 13: 1-7.

Roberts, S. and Rhee, J.K. 1990. Boron utilization by potato in nutrient cultures and in field plantings. *Commun Soil Sci. Plant Analysis*. 21(11-12): 921-932.

Romheld, V. and Kirkby, E.A. 2010. Research on potassium in agriculture: Needs and prospects. *Plant Soil.,* 335: 155–180.

Russelle, M.P. and McGraw, R.L. 1986. Nutrient stress in birdsfoot trefoil. *Can. J. Pl. Sci.* 66: 933-944.

Sabour, I., Merah, O., El Jaafari, S., Paul, R. and Monneveux, P.H. 1997. Leaf osmotic potential, relative water content and leaf excised water loss variations in oasis wheat landraces in response to water deficit. *Arch Int Physiol Biochem Biophys*., 105: 14.

Sahi, H.P.S.; Sharma, P.K. and Sachan, P.L. 1992. Check *khaira* disease of rice with zinc sulfate spray. *Indian Farming*. 41(11): 26.

Saebo, A., Haland, A., Skre, O. and Mortensen, L.M. 2001. Influence of nitrogen and winter climate stresses on *Calluna vulgaris (Hull). Anal of Botany,* 88: 823-828.

Sangakkara, U.R., Frehner, M., and Nosberger, J. 2000. Effect of sou moisture and potassium fertilizer on shoot water potential, photosynthesis and partitioning of carbon in mungbean and cowpea. *J Agron Crop Sci.*, 185: 201-207.

Sawwan, J., Shibli, R.A., Swaidat, L. and Tahat, M. 2000. Phosphorus regulates osmotic potential and growth of African violet under *in vitro*-induced water deficit. *J Plant Nutr.*,23: 759-771.

Saxena, M.C. Mehrotra, R.S. and Singh, K.B. 1990. Iron deficiency in chickpea in the Mediterranean region and its control through resistant genotypes and nutrient application. *Plant and Soil.* 123(2): 251-254.

Schinozaki, K. 2000. Molecular responses to dehydration and low temperature: *Curr Opinion in Plant Biol.*, 217-223 p.

Sen Gupta, A., Berkowitz, G.A. and Pier, P.A. 1989. Maintenance of photosynthesis at low leaf water potential in wheat. *Plant Physiol.*, 89: 1358-1365.

Shabala, S.N. and Lew, R.R. 2002. Turgor regulation in osmotically stressed Arabidopsis epidermal root cells. Direct support for the role of inorganic ion uptake as revealed by concurrent flux and cell turgor measurements. *Plant Physiol.*, 129: 290–299.

Shabala, S. and Cuin T.A. 2008. Potassium transport and plant salt tolerance. *Physiol. Plant.*, 133: 651–669.

Shabala, S. 2009. Salinity and programmed cell death: Unravelling mechanisms for ion specific signalling. *J. Exp. Bot.*, 60: 709–711.

Shabala, S. and Pottosin I.I. 2010. Potassium and potassium-permeable channels in plant salt tolerance. *Signal. Commun. Plants.*, 2010:87–110.

Shabala, S. 2011. Physiological and cellular aspects of phytotoxicity tolerance in plants: The role of membrane transporters and implications for crop breeding for waterlogging tolerance. *New Phytol.*, 190: 289–298.

Sharma, C.P., Mehrotra, S.C., Sharma, P.N. and Bisht, S.S. 1984. Water stress induced by zinc deficiency in cabbage. *Curr Sci.*, 53: 44-45.

Shaul, O. 2002. Magnesium transport and function in plants: the tip of the iceberg. *Biometals.*, 15: 309-323.

Shelp, B.J.; Shattuck, V.I. and Proctor, J.T.A. 1987. Boron nutrition and mobility and its relation to the elemental composition of greenhouse grown root crops.2.Radish. *Commun. in Soil Sci. and Plant Analysis.* 18(2): 203-219.

Shi, X.H., Zhang, C.H., Wang, H. and Zhang, F.S. 2005. Effect of Si on the distribution of Cd in rice seedlings. *Plant Soil.*, 272: 53-60.

Shin, R. and Schachtman, D.P. 2004. Hydrogen peroxide mediates plant root cell response to nutrient deprivation. *Proc. Natl. Acad. Sci. USA.*, 101: 8827–8832.

Silva, H. and Rodrigue, J. 1996. Physiological disorders of Pomaceae fruit induced by calcium deficiency. An integrated focus on the problem. *Revista Fruticola,* 17(1): 5-17.

Simon, J.E., Wilcox, G.E.., Simini, M., Elamin, O.M. and Decoteau, R.D. 1986. Identification of manganese toxicity and magnesium deficiency on melons grown in low-pH soils. *Hort Science*, 21(6): 1383-1386.

Smart, L.B., Moskal, W.A., Cameron, K.D. and Bennett, A.B. 2001. MIP genes are down-regulated under drought stress in *Nicotiana glauca. Plant Cell Physiol.*, 42: 686–693.

Smienoff, N. 1998. Plant resistance to environmental stress. *Curr. Opin Biotech.*, 9: 214-219.

Smith, G.S., Clark, C.J. and Holland, P.T. 1987. Chlorine requirement of kiwifruit (*Actinidia deliciosa). New Phytologist.* 106(1): 71-80.

Smith, F.W., Jackson, W.A. and van den Berg, P.J. 1990. Internal phosphorus flows during development of phosphorus stress in *Stylosanthes hamata. Australian Journal of Plant Physiology* 17: 451-464.

Sonneveld, C. and Voogt, W. 1991. Effects of Ca-stress on blossom-end rot and Mg-deficiency in rockwood grown tomato. *Acta Hort.* 294: 81-88.

Spehar, C.R. and Galwey, N.W. 1997. Screning soybean (*Glycine max* L. Merrill) for calcium efficiency by root growth in low Ca nutrient solution. *Euphytica.* 94(1): 113-117.

Stuiver, C.E.E., Kok, L.J., Westerman, S. and Dekok, L.J. 1997. Sulfur deficiency in *Brassica oleracea* L : Development, biochemical characterisation and sulfur/nitrogen interactions, *Russian J. Plant Physiol.* 44(4): 505-513.

Suzuki, N. and Mittler, R. 2006. Reactive oxygen species and temperature stresses: a delicate balance between signaling and destruction. *Physiologia Plantarum*, 126(1): 45–51.

Swietlik, D. 1995. Interactions between zinc deficiency and boron toxicity on growth and mineral nutrition of sour orange seedlings. *J. Plant Nut.* 18(6): 1191-1207.

Subbarao, G.V., Ito, O., Berry, W.L..and Wheeler, R.M. 2003. Sodium—A Functional Plant Nutrient, *Critical Reviews in Plant Sciences* 22(5): 391–416.

Suzuki, N. and Mittler, R. 2006. Reactive oxygen species and temperature stresses: A delicate balance between signaling and destruction. *Physiol. Plantarum.*, 126: 45–51.

Tamada, T. 1989. Nutrient deficiency of rabbiteye and highbush blue barries. *Acta Hort.* 241: 132-138.

Tanaka, Y., Sano, T., Tamaoki, M., Nakajima, N., Kondo, N. and Hasezawa, S. 2005. Ethylene inhibits abscisic acid-induced stomatal closure in *Arabidopsis. Plant Physiol.,* 138: 2337-2343.

Tanaka, Y., Sano, T., Tamaoki, M., Nakajima, N., Kondo, N. and Hasezawa, S. 2006. Cytokinin and auxin inhibit abscisic acid-induced stomatal closure by enhancing ethylene production in *Arabidopsis. J. Exp. Bot.,* 57: 2259–2266.

Teakle N.L., Bazihizina N., Shabala S., Colmer T.D., Barrett-Lennard E.G., Rodrigo-Moreno, A. and Läuchli, A.E. 2013. Differential tolerance to combined salinity and O_2 deficiency in the halophytic grasses *Puccinellia ciliata* and *Thinopyrum ponticum*: The importance of K^+ retention in roots. *Environ. Exp. Bot.,* 87: 69–78.

Teroaka, T., Kaneko, M., Mori, S. and Yoshimura, E. 2002. Aluminum rapidly inhibits cellulose synthesis in roots of barley and wheat seedlings. *Journal of Plant Physiology*, 159: 17–23.

Tomemori, H., Hamamura, K. and Tanabe, K. 2002. Interactive effects of sodium and potassium on the growth and photosynthesis of spinach and komatsuna. *Plant Prod. Sci.,* 5: 281-285.

Toth, V.R., Meszkaros, L. Veres, S. and Nagy, J. 2002. Effect of the available nitrogen on the photosynthetic activity and xanthophylls cycle pool of maize in field. *J Plant Physiol.*, 159: 627-634.

Tsonev, T., Velikova, V., Yildiz-Aktas, L., Gurel, A. and Edreva, A. 2011. Effect of water deficit and potassium fertilization on photosynthetic activity in cotton plants. *Plant Biosyst.,* 145: 841–847.

Trull, M.C., Guitinan, M.J., Lynch, J.P. and Deikman, J. 1997. The response of wild type and ABA mutant *Arabidiopsis thaliana* plants to phosphorus starvation. *Plant cell and environment.* 20(1): 85-92.

Tyerman, S.D., Niemietz, C.M. and Bramley, H.2002. Plant aquaporins: Multifunctional water and solute channels with expanding roles. *Plant Cell Environ.,* 25: 173–194.

Verhoeven, A.S., Demming-Adams, B. and Adams III, W.W. 1997. Enhanced employment of the xanthophylls cycle and thermal energy dissipation in spinach exposed to high light and N stess. *Plant Physiol.*, 113: 817-824.

Vitorello, V.A., Capaldi, F.R. and Stefanuto. V.A. 2005. Recent advances in aluminium toxicity and resistance in higher plants. *Brazilian Journal of Plant Physiology*, 17: 129–143.

Vranova, E., Inze, D. and van Breusegem, F. 2002. Signal transduction during oxidative stress. *J. Exp. Bot.,* 53: 1227–1236.

Walker, D.J., Leigh, R.A. and Miller A.J. 1996. Potassium homeostasis in vacuolate plant cells. *Proc. Natl. Acad. Sci. USA.* 93: 10510–10514.

Wang, Z.L. and Huang, B.R. 2004. Physiological recovery of kentucky bluegrass from simultaneous drought and heat stress. *Crop Sci.,* 44: 1729–1736.

Wang, X.M., Li W.Q., Li M.Y. and Welti R. 2006. Profiling lipid changes in plant response to lowtemperatures. *Physiol. Plantarum.,* 126: 90–96.

Wang, M., Zheng, Q., Shen, Q. and Guo, S. 2013. The critical role of potassium in plant stress response.*Int J Mol Sci.*, 14(4): 7370–7390.

Webster, D.E. and Ebdon, J.S. 2005. Effects of nitrogen and potassium fertilization on perennial ryegrass cold tolerance during deacclimation in late winter and early spring. *Hortscience,* 40: 842–849.

Wegner, L. 2010. Oxygen Transport in Waterlogged Plants. In, "*Waterlogging Signalling and Tolerance in Plants*", Eds. S. Mancuso, S. Shabala, Springer; Berlin/Heidelberg, Germany: pp. 3–22.

Waraich, E.A., Amad, R., Ashraf, M.Y., Saifullah, Ahmad, M. 2011. Improving agricultural water use efficiency by nutrient management. *Acta Agri Scandi - Soil & Plant Sci.*, 61(4): 291-304.

Wilkinson, S., Clephan, A.L. and Davies, W.J. 2001. Rapid low-temperature induced stomatal closure occur in cold tolerant *Commelina communis* leaves but not in cold-sensitive tobacco leaves, via a mechanism that involves apoplastic calcium but not abscisic acid. *Plant Physiol.*, 126: 1566-157.

Winter, H., Lohaus, G. and Heldt, A.W. 1992. Phloem transport of amino acids in relation to their cytosolic levels in barley leaves. *Plant Physiology,* 99:, 996-1004.

White P. and Karley A. 2010. Potassium. In, "*Cell Biology of Metals and Nutrients*", Ed. R. Hell, R.R. Mendel, Springer; Berlin/Heidelberg, Germany: 2010. pp. 199–224.

Whitehead, D.C. 1985. Chlorine deficiency in red clover grown in solution culture. *J. Plant Nutr.* 8(2): 193-198.

Woods, C.M., Reid, M.S. and Patterson, B.D. 1984. Response to chilling stress in plant cells. I. Changes in cyclosis and cytoplasmic structure. *Protoplasma*, 121(1): 8–16.

Wood, A.J., Saneoka, H., Rhodes, D., Joly, R.J. and Gildsbrough, P.B. 1996. Betaine aldehyde dehydrogenase in sorghum. Molecular cloning and expression of two related genes. *Plant Physiol.*, 110: 1301-1308.

Xiong, L.M., Schumaker, K.S. and Zhu, J.K. 2002. Cell signaling during cold, drought, and salt stress. *Plant* Cell, 14: S165–S183.

Xu, Q and Huang, B. 2006. Seasonal changes in root metabolic activity and nitrogen uptake for two cultivars of creeping bentgrass. *Hort Sci.*, 41: 822–826.

Yamamoto, Y. Kobayashi, Y. and Matsumoto, H. 2001. Lipid peroxidation is an early symptom triggered by aluminum, but not the primary cause of elongation inhibition in pea roots. *Plant Physiology*, 125: 199–208.

Yan, Q., Duan, Z., Mao, J., Li, X. and Dong, F. 2012. Effects of root-zone temperature and N, P, and K supplies on nutrient uptake of cucumber (*Cucumis sativus* L.) seedlings in hydroponics. *Soil Science and Plant Nutrition*, 58(6): 707-717.

Yu, X., Bell, P.F. and Yu, X.O. 1998. Nutrient deficiency symptoms and boron uptake mechanism of rice. *J. Plant Nut.* 21(10): 2077-2088.

Yu, Q. and Rengel, Z. 1999. Micronutrient deficiency influences plant growth and activities of super oxide dismustases in narrow-leafed lupines. *Ann. Bot.* 83(2): 175-182.

Zaiter, H.Z. and Ghalayini, A. 1994. Iron deficiency in lentils in the Mediterranean regions and its control through resistant genotypes and nutrient application. *J. Plant Nut.* 17(6): 945-952.

Zhao, Z.G., Chen, G.C. and Zhang, C.L. 2001. Interaction between reactive oxygen species and nitric oxide in drought-induced abscisic acid synthesis in root tips of wheat seedlings. *Aus. J. Pl. Physiol.*, 28: 1055-1061.

Zhou, M. 2010. Improvement of Plant Waterlogging Tolerance. In, "*Waterlogging Signalling and Tolerance in Plants*", Eds. S. Mancuso, S. Shabala., Springer; Berlin/Heidelberg, Germany: pp. 267–285.

Zhu, Z., Gerendas, J., Bendixen, R., Schinner, K., Tabrizi, H., Sattelmacher, B. and Hansen, U.P. 2000. Different tolerance to light stress in NO_3 and NH_4^+ - grown *Phaseolus vulgaris. Plant Physiol.,* 2: 558-570.

Zhu, J. K. 2001. Plant salt tolerance. *Trends in Plant Science,* 6(2): 66–71.

Zhu, J.K. 2001a. Cell signaling under salt, water and cold stresses. *Curr. Opin. Plant Biol.,* 4:401–406.

Zhu, X., Gong, H., Chen, G., Wang, S. and Zhang, C. 2005. Different solute levels in two spring wheat cultivars induced by progressive field water stress at different developmental stages. *J Arid Environ*., 62: 1-14.

Zia, M.S., Salim, M., Aslam, M., Gill, M.A. and Rahmatullah. 1994. Effect of low temperature of irrigation water on rice growth and nutrient uptake. *The Journal of Agricultural Science.*, 173(1): 22–31.

Zocchi, G. and Hanson, J.B. 1982. Calcium influx into corn roots as a results of cold shock. *Plant Physiology*, 70: 318–319.

7

Transgenic Approaches for Enhanced Drought Tolerance (EDT) in Plants

Alok Das[1], Shallu Thakur[1], P S Basu[2] and N P Singh[1]

The global climate change present new challenges that threats sustainability of crop production. Drought (water limiting) alone accounts for higher productivity losses as compared to other factors in rain fed agriculture. Hence, enhancing the ability of plants to tolerate drought situation is crucial for agricultural production. Genetic engineering for enhanced drought tolerant (EDT) plants, utilizing genes known to be involved in stress response and putative tolerance has been demonstrated. Candidate gene includes transcription factors, protective proteins, osmolyte metabolism, ROS-scavenging proteins and signaling factors. Several lines of evidence implicate that tailoring of genes has the potential to overcome a number of limitations in creating EDT transgenic plants. Novel strategies including post-translational modifications, small RNAs, epigenetic control of gene expression and hormonal networks are currently being envisaged. The development and application of plant synthetic biology tools and approaches will add new functionalities and perspectives to genetic engineering programs for enhancing drought tolerance.

In agriculture, water shortage and climate change are the major challenges worldwide. The global water shortage threatens sustainable farming as 75% of global water is consumed in agricultural activities. The Intergovernmental Panel on Climate Change (IPCC) has concluded that subtropics will dry due to increase in concentration of green house gases by the end of century that will lead to drought stress in Agriculture. Abiotic stresses like drought, temperature extremes, and salinity are the most common stresses that plant encounter which in turn affects growth and productivity. It is estimated that approximately 22% of the

[1]*Division of Plant Biotechnology, ICAR, Indian Institute of Pulses Research, Kanpur - 208024, Uttar Pradesh*

[2]*Division of Basic Science, ICAR, Indian Institute of Pulses Research Kanpur - 208024, Uttar Pradesh*

global agriculture land is saline (FAO 2004), and the areas under drought are expanding and expected to increase further. Crops are often exposed to multiple stresses, and the way in which a plant senses and responds to different stresses are overlapping. Expression profiling of barley plants under either drought or salt-stressed indicated that various genes were differentially regulated in response to different stress but they possibly induce a similar defense response.

The most important is to understand the cellular, biochemical and molecular changes occurs in cell in response to stress, is the key to progress towards breeding better crops under stress. Modern molecular techniques to enhance breeding program involve identification and use of markers. However, the introgression of genomic portions (QTLs) linked to stress tolerance often brings along undesirable agronomic characteristics due to lack of precise knowledge of the key genes underlying the QTLs. Development of genetic engineered plants by introduction/ over-expression of desired gene would be a faster way to insert beneficial genes than conventional or molecular breeding. Till now, various transgenic technologies have been used to improve stress tolerance in plants.

Stress induced gene expression can be broadly categorized into three groups: (1) genes encoding proteins with known enzymatic or structural functions, (2) proteins with as yet unknown functions, and (3) regulatory proteins. Initial attempts to develop transgenics (mainly tobacco) for abiotic stress tolerance involved "single action genes" *i.e.* genes responsible for modification of a single metabolite that would confer increased tolerance to salt or drought stress. Stress-induced proteins with known functions such as water channel proteins, key enzymes for osmolyte (proline, betaine, suger such as trehalose, and polyamines) biosynthesis, detoxification enzymes, and transport proteins were the initial targets of plant transformation. In fact, metabolic traits, especially pathways with relatively few enzymes, have been characterized genetically and appear more amenable to manipulations than structural and developmental traits. However, that approach has overlooked the fact that abiotic stress tolerance is likely to involve many genes at a time, and that single-gene tolerance is unlikely to be sustainable. Therefore, a second "wave" of transformation attempts to transform plants with the third category of stress induced genes, namely, regulatory proteins has emerged. Through these proteins many genes involved in stress response can be simultaneously regulated by a single gene encoding stress inducible transcription factor, thus offering possibility of enhancing tolerance towards multiple stresses including drought salinity and freezing. It is interesting to note that this "second wave" has also coincided with a better integration of genetic engineering and plant physiology.

Further, genetic engineering also allows control the timing, tissue-specificity, and expression level of the introduced genes for their optimal function. In recent years, the basic findings on stress promoters have led to a major shift in the paradigm for genetically engineering stress-tolerant crops. Constitutively expressed promoters are widely used in generating transgenic plants. However, these promoters are not suitable where gene expression of stress related genes need at specific time or in specific organ. The major drawback with constitutive expression of stress induced genes is possible deleterious effects on the plant. The most recent approach is to use gene cassettes driven by stress-induced promoters in generating transgenic plants. With an increasing number of stress genes becoming available and genetic transformation becoming more or less a routine procedure, characterization of stress-induced promoters (particularly those induced by anaerobic, low or high temperature and salt stresses) has taken a firm footing. Drought Gard maize contains the gene for "cold shock protein B" (*cspB*) from *Bacillus subtilis*. Cold shock proteins were discovered (and named) due to their rapid accumulation in cold shocked bacterial cells. Some CSPs such as CSPB act as RNA chaperones, which help to maintain normal physiological performance during stress events by binding and unfolding tangled RNA molecules so that they can function normally.

This chapter highlights potential candidate genes used in drought tolerance, their evaluation, and its application to other crop species and recent progress in breeding for abiotic stress tolerant crops using transgenic technology.

7.1. Single Action Gene

7.1.1. Osmoprotectants

Osmoprotectants are compounds produced in plants during osmotic stress condition. Chemically, these are small, electrically neutral molecules, plays important roles in protection and stabilization of proteins and membranes against abiotic stresses in plants without disrupting metabolism. Osmoprotectants are classified into three groups; (1) amino acids (*e.g.* proline), (2) polyol/sugars (*e.g.* mannitol, trehalose, fructans), and (3) quaternary amines (*e.g.* glycine betaine and polyamines). Osmoregulation would be the best approach for abiotic stress tolerance, especially if osmoregulatory genes could be triggered in response to drought, salinity and high temperature. Various strategies are being pursued to genetically engineering osmoprotection in plants. The first step involved in engineering of genes that encode enzymes for the synthesis of selected osmolytes. This has resulted in profusion of reports involving osmoprotectants such as glycine-betaine (Ishitani *et al.*, 1997) and proline (Delauney and Verma, 1993). Also, a number of "sugar alcohols" (mannitol, trehalose, myo-inositol and sorbitol) have been targeted for the engineering of compatible- solute over production, thereby protecting the membrane and protein complexes during stress (Gao *et al.*, 2000). Similarly, transgenics engineered for the over expression of polyamines have also been developed (Capell *et al.*, 2004). Studies conducted on transgenic petunia and pigeon pea (*Cajanus cajan*) expressing pyrroline-5-carboxylate synthetase (P5CS) accumulated more proline in transgenic plants compared to non-transgenic leading to drought and salt tolerance, respectively (Yamada *et al.*, 2005; Surekha *et al.*, 2013).

In another study, tobacco plants were transformed with *imtl* gene encoding myo-inositol-o-methyl-transferase, an enzyme that plays role in the inositol biosynthetic pathway and resulted plants were found to be more tolerant to salt and drought stress compared to wild. In another study, same gene was used to create potato transgenic plants that showed significantly enhanced drought tolerance.

The *Escherichia coli otsA* and *otsB* genes were introduced into potato and tobacco under the control of either a constitutive CaMV35S promoter or the tuber specific promoter. However, very low level of trehalose accumulation was observed in the leaves of the obtained transgenic tobacco plants, whereas transgenic potato tubers showed no trehalose accumulation. Same genes were expressed in rice under both tissue-specific and stress-inducible promoters. Compared to control plants, the transgenic rice plants showed vigorous growth under salt, drought, temperature and osmotic stress. The TPS1 gene was expressed in potato under a drought responsive StDS2 promoter. In transgenic plants, the TPS1 expression was not induced; however the very low level of constitutive expression was sufficient to alter their response to drought stress. The transgenic accumulation of trehalose seems to be species specific as variable concentrations were found with genes of different origins. Recently, a trehalose biosynthetic gene, trehalose synthase (TSase), from *Grifola frondosa* was expressed in tobacco, and the transgenic plants accumulated significantly higher trehalose content than previously reported for *E. coli* TPS, TPP, TPSP, and yeast TPS1 gene expressed in tobacco and rice plants. Transgenic tobacco plants also showed enhanced tolerance to salt and drought stress. In some other studies, expression of the trehalose biosynthetic genes conferred tolerance to abiotic stresses in transgenic Arabidopsis and rice plants.

Fructans are polymers of fructose and in many plant species serve as storage of carbohydrates. Fructans get accumulated in vacuoles of plants and known to play a role in abiotic stress tolerance. Several fructans biosynthetic genes were isolated from both bacteria and higher plants. *SacB* (fructan biosynthetic gene) encoding levan sucrase enzyme was expressed in tobacco and potato under constitutive promoter. Transgenic tobacco plants showed increased tolerance to salt, drought and chilling stress.

The results of transgenic modifications of biosynthetic and metabolic pathways in most of the above mentioned cases indicate that higher stress tolerance and the accumulation of compatible solutes may protect plants against damage by scavenging of reactive oxygen species (ROS), and by their chaperone like activities in maintaining protein structures and functions. However, pleiotropic effects (*e.g.* necrosis and growth retardation) have been observed due to disturbance in endogenous pathways of primary metabolisms. Also, there are also some reports showing a negative effect of osmotic stress on yield potential. Genetic manipulations of compatible solutes do not always lead to a significant accumulation of the compound, thereby suggesting that the function of compatible solute is not restricted to osmotic adjustment, and that osmoprotection may not always confer drought tolerance. Another recent study with chickpea has also shown that osmotic adjustment provided no beneficial effect on yield under drought stress. Besides, the results of simulation modeling also suggest that changes in a given metabolic process may end up with little benefit for actual yield under stress. For agricultural practices, over-synthesis of compatible solutes should not account for the primary metabolic costs and hence to minimize the pleiotropic effects, over production of compatible solutes should be stress inducible and/or tissue specific.

7.1.2. Detoxifying Genes

In most of the aerobic organisms, there is a need to effectively eliminate reactive oxygen species (ROS) generated as a resulted of environmental stresses. Depending on the nature of the ROS, some are highly toxic and need to be rapidly detoxified. In order to control the level of ROS and protect the cells form oxidative injury, plants have developed a complex antioxidant defense system to scavenge the ROS. These antioxidant defense include various enzymes and non enzymatic metabolites that may also play a significant role in ROS signalling in plants. A number of transgenic improvements for abiotic stress tolerance have been achieved through detoxification strategy. These include transgenic plants over expressing enzymes involved in oxidative protection, such as glutathione peroxidase, superoxide dismutase, ascorbate peroxidases and glutathione reductases. Transgenic tobacco over expressing SOD in the chloroplast, mitochondria and cytosol have been generated and these have been shown to enhance tolerance to oxidative stress induced by methyl viologen (MV) in leaf disc assays. Over-expression of chloroplast Cu/Zn SOD showed a dramatic improvement in the photosynthetic performance under chilling stress condition in transgenic tobacco and potato plants. Tobacco transgenic plants over-expressing MnSOD rendered enhanced tolerance to oxidative stress only in the presence of other antioxidant enzymes and substrate, thereby, showing that the genotype and the isozyme composition also have profound effect on the relative tolerance of the transgenic plants to abiotic stress. While transgenic alfalfa (*Medicago sativa*) plants cv. RA3 over-expressing MnSOD in chloroplasts showed lower membrane injury, transgenic plants overproducing alfalfa aldose redutase gene showed lower concentrations of reactive aldehydes and increased tolerance against oxidative agents and drought stress.

7.1.3. Late Embryogenesis Abundant (LEA) Protein

LEA proteins represent another category of high molecular weight proteins that are abundant during late embryogenesis and accumulate during seed desiccation and in response to water stress. Most LEA proteins identified till date belongs to hydrophilins, a group of proteins characterized by a high content of charged amino acid residues, as well as glycine or other small amino acids such as alanine, serine, or threonine. Moreover, most hydrophilins lack tryptophanes and cysteines. Analyses of LEA protein amino acid sequences recognize seven different groups, each one showing distinctive motifs (LEA1 – LEA7). No significant sequence similarity was detected between the seven LEA protein groups. Those belonging to group 3 are predicted to pay a role in sequestering ions that are concentrated during cellular dehydration. These proteins have 11-mer amino acid motifs with the consensus sequence TAQAAKEKAGE repeated as many as 13 times. The group 1 LEA proteins are predicted to have enhanced water-binding capacity, while the group 5 LEA proteins are thought to sequester ions during water loss. Constitutive over-expression of the HVA 1, a group 3 LEA protein from barley conferred tolerance to soil water deficit and salt stress in transgenic rice plats. Constitutive or stress induced expression of the HVA 1 gene resulted in the improvement of growth characteristics and stress tolerance in terms of cell integrity in wheat and rice under salt and water-stress conditions. Although, the reported water use efficiency (WUE) was extremely low when compared to other data reported in wheat cultigens, transgenic rice (TNG67) plants expressing a wheat LEA group 2 protein (PMA80) gene or the wheat LEA group 1 protein (PMA1959) gene resulted in increased tolerance to dehydration and salt stresses. Besides, protective chaperone like function of LEA proteins acting against cellular damage has been proposed, indicating the role of LEA proteins in anti-aggregation of enzymes under desiccation and freezing stresses. LEA proteins have been detected in different stages of plant development in response to different water deficit conditions (salinity, freezing, and drought) in common bean, soybean or *Medicago truncatula* seedlings. Their accumulation during late embryogenesis has also been reported in *M. truncatula* , *G. max* , *Arachis hypogaea* and *P. vulgaris*. In case of *M. truncatula*, microarray data showed that under water deficit conditions along with NaCl treatments (200 mM) induce the transcript accumulation groups 2, 3, 4, 6, and 7 LEA genes.*LEA1* transcripts were mostly detected during late embryogenesis in *Arabidopsis*, cotton, and wheat.

7.1.4. Transporter Genes

An important strategy for achieving greater tolerance to abiotic stress is to help plants to re-establish homeostasis under stressful environments, restoring both ionic and osmotic homeostasis. This has been and continues to be a major approach to improve salt tolerance in plants through genetic engineering, where $^+$ excretion out of the root, or their storage in the vacuole. A number of abiotic stress tolerant transgenic plants have been produced by increasing the cellular levels of proteins (such as vacuolar antiporter proteins) that control the transport functions. A vacuolar chloride channel, AtCLCd gene, which is involved in cation detoxification, and AtNHXI gene which is homologous to NhxI gene of yeast have been cloned and over-expressed in *Arabidopsis* to confer salt tolerance by compartmentalizing Na^+ ions in the vacuoles. Transgenic *Arabidopsis* and tomato plants that over express AtNHXI accumulated abundant quantities of the transporter in the tonoplast and exhibited substaintially enhanced salt tolerance. Salt Oerly Sensitive 1 (SOS 1) locus in *A. thaliana*, which is similar to plasma membrane NA^+/H^+ antiporter form bacteria and fungi, was cloned and over expressed using CaMV 35S promoter. The up-regulation of SOS 1 gene was found to be consistent with its role in Na^+ tolerance, providing a greater proton motive force that is necessary for elevated Na^+/H^+ antiporter activities.

7.1.5. Multifunctional Genes for Lipid Biosynthesis

Transgenic approaches also aim to improve photosynthesis under abiotic stress condition through changes in the lipid biochemistry of the membranes. Adaptation of living cells to chilling temperatures is a function of alteration in the membrane lipid composition by increased fatty acid unsaturation. Genetically engineered tobacco plants over-expressing chloroplast glycerol-3-phosphate acyltransferase (GPAT) gene (involved in phosphatidy1 glycerol fatty acid desaturation) form squash (*Cucubita maxima*) and *A. thaliana* showed an increase in the number of unsaturated fatty acids and a corresponding decrease in the chilling sensitivity. Besides, transgenic tobacco plants with silenced expression of chloroplast ω3-fatty acid desaturase (Fad7, which synthesizes trienoic fatty acids) were able to acclimate to high temperature as compared to the *wild type*.

7.1.6. Heat -shock Protein Genes

The heat shock response, the increased transcription of a set of genes in response to heat or other toxic agent exposure is a highly conserved biological response, occurring in all organisms. The response is mediated by heat shock transcription factor (HSF) which is present in a monomeric, non-DNA binding form in unstressed cells and is activated by stress to a trimeric form which can bind to promoters of heat shock genes. The induction of genes encoding heat shock proteins (HSPs) is one of the most prominent responses observed at the molecular level of organisms exposed to high temperature. Genetic engineering for increased thermo-tolerance by enhancing heat shock protein synthesis in plants has been achieved in a number of plant species. Positive correlations between the level of heat shock proteins and stress tolerance were also observed. Although the precise mechanism by which these heat shock proteins confer stress tolerance is not known, a recent study demonstrated that *in vivo* function of thermo protection of small heat shock proteins is achieved *via* their assembly into functional stress granules (HSG).

7.2. Regulatory Gene

Genes respond to multiple stresses like dehydration and low temperature and the transcriptional level are also induced by ABA, which protects the cell form dehydration. In order to restore the cellular function and make plants more tolerant to stress, transferring a single gene encoding a single specific stress protein may not be sufficient to reach the required tolerance levels to overcome such constraints, enhancing tolerance towards multiple stresses by a gene encoding a stress inducible transcription factor that regulates a number of other genes is a promising approach. Therefore, a second category of genes of recent preference for crop genetic engineering are those that switch on transcription factors regulating the expression of several genes related to abiotic stresses.

7.2.1. Transcription Factors

Plants have evolved several mechanisms to respond abiotic stresses by induction of stress-responsive and stress-tolerance genes. In plant genomes small portion (approximately 7%) of coding sequences are assigned to transcription factors (TFs). TFs are an attractive target category for manipulation and gene regulation, bind to promoter regulatory elements in genes which is regulated by abiotic stresses. The transcription factors activate cascades of genes that act together in enhancing tolerance towards multiple stresses. TFs may be attractive targets for genetic engineering to enhance stress tolerance in plants. Dozens of transcription factors are involved in the plant response to drought stress. Most of this fall into several large transcription factor families, such as AP2/ERF, bZIP, NAC, MYC, Cys_2His_2 zinc-finger and WRKY. Individual members of the same family, respond differently to various stresses. On the other hand, some stress responsive genes may share the same transcription factors, as indicated by the significant overlap of the gene expression profiles that are induced in response to different stresses. Two families, bZIP and MYB, are involved in ABA signaling and its gene activation. Many ABA inducible genes share the (C/T) ACGTGGC consensus, *cis*-acting ABA-responsive element (ABRE) in their promoter regions. Introduction of transcription factors in the ABA signaling pathway can also be a mechanism of genetic improvement of plant stress tolerance. Constitutive expression of ABF3 or ABF4 demonstrated enhanced drought tolerance in *Arabidopsis,* with altered expression of ABA/stress-responsive genes, *e.g.* rd29B, rab18, ABI1 and ABI2. Several stress induced *cor* genes such as rd29a, cor15A, kin 1 and cor6.6 are triggered in response to cold treatment, ABA and water deficit stress. There have been numerous efforts in enhancing tolerance towards multiple stresses such as cold, drought and salt stress in crops other than the model plants like *Arabidopsis*, tobacco and alfalfa. An increased tolerance to freezing and drought in *Arabidopsis* was achieved by over-expressing CBF4, a close CBF/DREB 1 homolog whose expression is rapidly induced during drought stress and by ABA treatment, but not by cold. Similarly, a *cis*-acting element, dehydratrion responsive element (DRE) identified in *A. thaliana*, is also involved in ABA independent gene expression under drought, low temperature and high salt stress conditions in many dehydration responsive genes like *rd29A* that are responsible for dehydration and cold induced gene expression. Several cDNA encoding the DRE binding proteins, DREB1A and DREB2A have been isolated from *A. thaliana* and shown to specifically bind and activate the transcription of genes containing DRE sequences. DREB1/CBFs are thought to function in cold-responsive gene expression, whereas DREB2s are involved in drought-responsive gene expression. The transcriptional activation of stress-induced genes has been possible in transgenic plants over-expressing one or more transcription factors that recognize regulatory elements of these genes. In *Arabidopsis*, the transcription factor, DREB1A specifically interacts with the DRE and induces expression of stress tolerance genes. DREB1A cDNA under the control of CaMV 35S promoter in transgenic plants elicits strong constitutive expression of the stress inducible genes and brings about increased tolerance to freezing, salt and drought stress. Strong tolerance to freezing stress was observed in transgenic *Arabidopsis* plants that over-express CBF/(DREB1b) cDNA under the control of the CaMV 35S promoter. Subsequently, the over expression of DREB1A has been shown to improve the drought and low temperature stress tolerance in tobacco, wheat and groundnut. The use of stress inducible rd29A promoter minimized the negative effects in growth in these crop species. However, over expression of DREB2 in transgenic plants did not improve stress tolerance, suggesting involvement of post-translational activation of DREB2 proteins. Recently, an active form of DREB2 was shown to trans-activate target stress-inducible genes and improve drought tolerance in transgenic

Arabidopsis. DREB2 protein is expressed under normal growth conditions and activated by osmotic stress through post-translational modification in the early stages of the osmotic stress response.

Relatively few bZIP regulators have been explored as potential candidates for application in the improvement of drought tolerance in crops. However, the Group A TF *Os*ABF1 from rice (*Oryza sativa*) and *Sl*AREB from tomato (*Solanum lycopersicum*) both enhanced tolerance to drought and salt stress, and *Sl*AREB was found to bind to and activate transcription from the *Arabidopsis* RD29B and the tomato leucine amino peptidase (LAP) promoters in an ABA-dependent manner. Furthermore, *Os*bZIP23 over-expression conferred ABA hypersensitivity, and increased salinity and drought tolerance of rice, and over-expression of *Os*bZIP46 increased ABA sensitivity. However, positive regulation of drought and osmotic stress tolerance by OsbZIP46 was dependent on its activation state. The soybean (*Glycine max*) AREB/ABF *Gm*bZIP1 was highly induced by ABA, drought, high salt, and low temperature, and its over-expression enhanced the response of transgenic plants to ABA and triggered stomatal closure under stress conditions. Over-expression of *Gm*bZIP1 also enhanced the drought tolerance of transgenic wheat, suggesting that *Gm*bZIP1 may be useful for engineering stress tolerance for crops in general.

Another ABA-independent, stress-responsive and senescence-activated gene expression involves ERD gene, the promoter analysis of which further identified two different novel *cis* acting elements involved with dehydration stress induction and in dark induced senescence. Similarly, transgenic plants developed by expressing a drought responsive AP2-type TF, SHN1-3 or WXP1, induced several wax related genes resulting in enhanced cuticular accumulation and increased drought tolerance. Thus, clearly the over-expression of some drought-responsive transcription factors can lead to the expression of downstream genes and the enhancement of abiotic stress tolerance in plants. The regulatory genes/factors reported so far not only play a significant role in drought and salinity stresses, but also in submergence tolerance. More recently, an ethylene-response-factor like gene *Sub1a*, one of the cluster of three genes at the Sub1 locus have been identified in rice and the over-expression of Sub1A-1 lead to submergence tolerance in plants.

Transcription factors play an important role in plant stress adaptation by regulating the downstream genes involved in stress tolerance. These stress-related genes and pathways may also include expression of genes encoding osmoprotectants. In one study, transformed rice with the OsWRKY11 gene under the control of HSP11 promoter has enhanced heat and drought tolerance. Further studies were conducted to examine the contents of sucrose, glucose, fructose and raffinose, and their relationship to desiccation tolerance in the OsWRKY11-transgenic plants. Up-regulation of the expression of two genes, one encoding raffinose synthase and the other encoding galactinol synthase induced enhanced raffinose accumulation and desiccation tolerance in transgenic plants. These studies suggest a transcriptional control of the stress-induced metabolic alterations.

7.2.2. Signal Transduction Genes

Genes involved in stress signal sensing and a cascade of stress signaling in *A. thaliana* has been of recent research interest. Components of the same signal transduction pathway may also be shared by various stress factors such as drought, salt and cold. Although there are multiple pathways of signal-transduction systems operating at the cellular level for gene regulation, ABA is known component acting in one of the signal transduction pathways, while other act independently of ABA. The early response genes have been known to encode transcription factors that activate downstream delayed response genes. Although, specific branches and components exist, the signaling pathways for salt, drought, and cold stresses all interact with ABA, and even converge at multiple steps. Abiotic stress signaling in plants involves receptor-coupled phospho-relay, phospho-ionositol-induced Ca^{2+} changes, mitogen activated protein kinase (MAPK) cascade, and transcriptional activation of stress responsive genes. A number of signaling components are associated with the plant response to high temperature, freezing, drought and anaerobic stresses.

One of the merits for the manipulation of signaling factors is that they can control a broad range of down-stream events that can result in superior tolerance for multiple aspects. Alteration of these signal transduction components is an approach to reduce the sensitivity of cells to stress conditions, or such that a low level of constitutive expression of stress genes is induced. Over-expression of functionally conserved At-DBF2 (homolog of yeast DBf2 kinase) showed striking multiple stress tolerance in *Arabidopsis* plants. Salt stress-tolerant transgenic plants were also reported by over expressing calcineurin (a Ca^{2+}/ Calmodulin dependent serine/threonine protein phosphatase), a protein known to be involved in salt-tress signal transduction in yeast. Transgenic tobacco plants produced by altering stress signaling through functional reconstitution of activated yeast calcineurin not only opened up new routes for study of stress signaling, but also for engineering transgenic crops with enhanced stress tolerance. Over-expression of an osmotic stress activated protein kinase, SRK2C resulted in a higher drought tolerance in *A. thaliana*, which coincided with the upregulation of stress responsive genes. Similarly, a truncated tobacco mitogen activated protein kinase, NPK1, activated an oxidative signal cascade resulting in cold, heat salinity and drought tolerance in transgenic plants. However, suppression of signaling factors could also effectively enhance tolerance to abiotic stress. This hypothesis was based on previous reports indicating that a and b subunits of farnesyltransferase ERA1 functions as negative regulator of ABA signalling. Conditional antisense downregulation of a or b subunits of protein faernesyltransferase resulted in enhanced drought tolerance of *Arabidopsis* and canola plants.

7.3. Choice of Promoters

An important aspect of transgenic technology is the regulated expression of transgene. Tissue specificity of transgene expression is also an important while deciding on the choice of the promoter so as to increase the level of expression of the gene. Thus, the strength of the promoter and the possibility of using stress inducible, developmental-stage or tissue-specific promoters have also proved to be critical for tailoring plant response to these stresses. Some gene products are needed in large amounts, such as LEA3, thereby necessitating the need for a very strong promoter. With other gene products, such as enzymes for polyamine biosynthesis, it may be better to use an inducible promoter of moderate strength. The promoters that have been most commonly used in the production of abiotic stress tolerant plants so far, include the CaMV 35S, ubiquitin 1 and actin promoter. These promoters being constitutive in nature, by and large express the downstream transgenes in all organs and at all the stages. However, constitutive over production of molecules, such as trehalose or polyamines cause abnormalities in plants grown under normal conditions. Also, the production of the above described molecules can be metabolically expensive. In this case, the use of a stress inducible promoter may be more desirable. In plants, various types of abiotic stresses induce a large number of well characterized and useful promoters. An ideal inducible promoter should not only be devoid of any basal level of gene expression in the absence of inducing agents, but the expression should be reversible and dose-dependent. The transcriptional regulatory region of the drought induced and cold induced genes have been analyzed to identify several *cis* acting and *trans* acting elements involved in the gene expression that is induced by abiotic stress. Most of the stress promoter contains an array of stress specific *cis*-acting elements that are recognized by the requisite transcription factors; for example, the transcriptional regulation of *hsp* genes is mediated by the core "heat shock element" (HSE) located in the promoter region of these genes, 5' of the TATA box. All the plant *hsp* genes sequenced so far have been shown to contain partly overlapping multiple HSEs proximal to TATA motif. Apart from these *hsp* promoters, rd29 are stress responsive genes, but are differentially induced under abiotic stress conditions. The rd29A promoter includes both DRE and ABRE elements, where dehydration, high salinity and low temperatures induce the gene while the rd29B promoter includes only ABREs and the induction is ABA-dependent. Over-expression of DREB1A transcription factors under the control of stress inducible promoter form rde29A showed a better phenotypic growth of the transgenic plants than the ones obtained using the constitutive CaMV 35S promoter. A stress inducible expression of *Arabidopsis* CBF1 in transgenic tomato was achieved using the ABRC1 promoter form barley HAV22. Gene expression is induced by cold and water stress, to a *cis* acting DRE element in the promoters of genes such as rd29A, rd17, cor6.6, cor15A,erd10, and kin1, thereby, initiating synthesis of gene products imparting tolerance to low temperatures and water stress in plants. The regions of respiratory alcohol dehydrogenase *adh1* gene promoter in maize and rice that are required for anaerobic induction include a string of bases called anoxia response element (ARE) with the consensus sequence of its core element as TGGTTT. Besides, other stress responsive *cis* acting promoter sequences like low temperature responsive elements (LTRD) with a consensus sequences of A/GCCGAC have been identified in genes such as Cor6.6, Cor 15 and Cor 78. These basic findings on stress promoters have led to a major shift in the paradigm for genetically engineering stress tolerant crops.

7.4. Physiological Evaluation of Stress Effect

A large number of studies have evaluated different transgenic constructs in different plant species, and to different stresses such as a drought, salinity and cold. The expression of the genes inserted as well as altered levels of metabolites have been reported in great detail. The assessment of stress tolerance of transgenic needs to be done with respect to its cross talk with other stress related genes/mechanisms and where the effects of stress need to be observed over longer periods/ conditions. This is particularly important, in order to emulate the life span of most crops under recurrent cycles of stress, rather than brief exposure to stresses, although short exposures to stress are certainly adequate if the purpose is to assess gene expression only. Two major issues that typically need to be addressed in stress response evaluation of transgenic include: (1) Mean of stress imposition, details about the stress, and growth conditions (including the intensity, timing, and quickness of imposition, *etc.*), and (2) Data on the response of tested materials to support conclusions (comparison within the same species).

7.5. Conclusions

Drought tolerance is a complex trait and the use of transgene(s) to improve the tolerance of crops to abiotic stress remains and attractive option. Targeting multiple gene regulation appears better than targeting single genes. Most importantly, how specific abiotic stress is assessed, and whether the achieved tolerance compares to existing tolerance. The cost of production of different metabolites to cope with stress and their effect on yield needs be properly evaluated. Synergistic multi-pronged approach combining the molecular, physiological and metabolic aspects of tolerance is required for bridging the knowledge gaps between short and long term effect of the genes and their products, and between the molecular or cellular expression of the genes and the whole plant phenotype under stress. A thorough understanding of the physiological processes in response to different abiotic stresses can efficiently/ successfully drive the choice of a given promoter or transcription factor to be used for transformation.

References

Abebe, T., Guenzi, A.C., Martin, B. and Cushman, L.C. 2003. Tolerance of mannitol-accumulating transgenic wheat to water stress and salinity. *Plant Physiol*, 131: 1748-1755.

Aharoni, A., Dixit, S., Jetter, R., Thoenes, E., van Arkel, G. and Pereira, A. 2004. The SHINE clade of AP2 domain transcription factors activates wax biosynthesis, alters cuticle properties, and confers drought tolerance when over expressed in *Arabidopsis. Plant Cell*, 16: 2463-2480.

Alia, H., Kondo, Y., Sakamoto, A., Nonaka, H., Hayashi, H., Pardha Saradhi, P., Chen, T.H.H. and Norio, M. 1999. Enhanced tolerance to light stress of transgenic *Arabidopsis* plants that express the *codA* gene for a bacterial choline oxidase. *Plant Mol Biol*, 40: 279-288

Alia, H., Sakamoto, A. and Murata, N. 1998. Enhancement of tolerance of *Arabidopsis* to high temperature by genetic engineering of the synthesis of glycine betaine. *Plant J*, 16:155-161.

Allen, R.D. 1995. Dissection of oxidative stress tolerance using transgenic plants. *Plant Physiol.*, 107: 1049-1054.

Anderson, S.E., Bastola, D.R. and Minocha, S.C. 1998. Metabolism of polyamines in transgenic cell of carrot expressing a mouse ornithine decaroxylase cDNA. *Plant Physiol*, 116: 299-307.

Apse, M.P., Aharon, G.S., Snedden, W.A. and Blumwald, E. 1999. Salt tolerance conferred by over expression of a vascoular Na^+/H^+ antoport in *Arabidopsis*. *Science*, 285: 1256-1258

Bajaj, S,, Tagolli, J,, Liu, L.-F., Ho., T-H. D. and Wu, R. 1999. Transgenic approaches to increase dehydration stress tolerance in plants. *Mol Breed.*, 5: 493-503.

Bartels, D. and Sunkar, R 2005. Drought and salt tolerance in plants, *Crit. Rev. Plant Sci.*, 21: 1-36.

Behnam, B., Kkkuchi, A., Celebi-Toprak, F., Yamanaka, S., Kasuga, M., Yamaguchi- Shinozaki, K. and Watanabe, K.N. 2006. The *Arabidopsis* DREB1A gene driven by the stress-inducible rd29a promoter increases salt-stress tolerance in proportion to its copy number in tetrasomic tetraploid potato (*Solanum tuberosum*). *Plant Biotech.*, 23:169-177.

Bhatnagar-Mathur, P., Devi, M.J, Serraj, R., Yamaguchi-Shinozaki, K., Vadez, V. and Sharma, K.K. 2006. Over expression of *Arabidopsis thaliana* DREB1A in transgenic peanut (*Arachis hypogaea* L.) for improving tolerance to drought stress (poster presentation) In, "*Sackler colloquia on From Functional Genomics of Model Organism to Crop Plants for Global Health*" Ed. M. Arthur, April 3-5, 2006. National Academy of Science, Wasington, DC

Bohnert, H.J., Nelson, D.F. and Jenson, R.G. 1995 Adaption to environmental stresses. *Plant Cell*, 7: 1099-1111.

Bohnert, H.J. and Shen, B. 1999. Tranformation and compatible solutes *Sci Hort*, 78: 237-260

Bordas, M., Montesinos, D., Dabauza, M., Savador, A., Riog Lam Serrano, R., Moreno, V. 1997. Transfer of the yeast salt tolerance gene HAL1 to *Cucumis melo* L. cultivars and *in vitro* evaluation of salt tolerance of salt tolerance. *Transgenic Res.*, 5: 1-10.

Bowler, C, Slooten, L., Vandenbranden, S., Rycke, R.D., Botteerman, J., Sybesma, C., van Montagu, M. and Inze, D. 1991. Manganese superoxide dismutase can reduce cellular damage mediated by oxygen radicals in transgenic plant. *EMBO J.*, 10: 1723-1732.

Bray, E.A. 1993. Molecular responses to water deficit. *Plant Physiol.*, 103: 1035-1040.

Burke, E.J., Brown, S.J. and Christidis, N. 2006. Modeling the recent evolution of global drought and projections for the twenty-first century with the Hardley centre climate model. *J Hydometeor.*,7: 113-1125.

Capell, T., Bassie, L. and Christou, P. 2004. Modulation of the polyamine biosynthetic pathway in transgenic rice confers tolerance to drought stress. *Proc Natl Acad Sci USA.*, 101: 9909-9914.

Capell, T., Escobar, C., Lui, H., Burtin, H., Lepri, O. and Christou, P. 1998. Over expression of the oat arginine decarboxylase cDNA in transgenic rice (*Oryza sativa* L.) affects normal development patterns *in vitro* and results in puterescine accumulation in transgenic plants. *Theor Appl Genet.*, 97: 246-254.

Chen, T.H. and Murata, N. 2002. Enhancement of tolerance of abiotic stress by metabolic engineering of betaines and other compatible solutes. *Curr Opin Plant Biol.*, 5: 250-257.

Cheng, W.H., Endo, A., Zhou, L., Penny, J., Chen, H.C., Arroyo Leon, P., Nambara, E., Asami, Seo, M. 2002. A unique short-chain dehydrogenase/reductase in *Arabidopsis* glucose signaling and abscisic acid biousynthesis and funtions. *Plant Cell*, 14: 2723-2743.

Chinnusamy, V., Jagendorf, A. and Zhu, J.K. 2005. Understanding and improving salt tolerance in plants. *Crop Sci.*, 45: 437-448.

Cortina, C., Francisco, A. and Culiàñez-Macia.Tomato abiotic stress enhanced tolerance by trehalose biosynthesis. *Plant Sci.*, 169: 75-82.

Cutler, S., Ghassemian, M., Bonetta, D., Cooney, S. and McCourt, P. 1996. A protein farnesyl transferase involved in abscisic acid signal transduction in *Arabidopsis*. *Science* , 273: 1239-1241

Delauney, A.J. and Verma, D.P.S. 1993. Proline biosynthesis and osmo-regulation in plants. *Plant J.*, 4: 215-223

Diamant, S., Elihu, N., Rosenthal, D., Golouninoff, P. 2001. Chemical chaperones regulate molecular chaperones *in vitro* and in cells under combined salt and heat stresses. *J Biol Chem.*, 276: 39586-39591.

Dure, L. 1993. A repeating 11-mer amino acid motif and plant desiccation. *Plant J.*, 3: 363-369

Dure, L., Crouch, M., Harada, J., Ho, T.-H.D. and Mundy, J. 1989. Common amino acid sequence domains among the LEA proteins of higher plants. *Plant Mol Biol.*, 12: 475-486.

FAO (Food, Agriculture Organization of the United Nations) 2004. FAO production yearbook FAO, Rome.

Fukai, S. and Cooper, M. 1995. Developing resistant cultivars using physio-morphological traits in rice. *Field Crop Res.*, 40: 67-86.

Galau, G.A., Bijaisoradat, N. and Huges, D.W. 1987. Accumulation kinetics of cotton late embryogenesis-abundant (LEA) mRNAs and storage protein mRNAs coordinate regulation during embryogenesis and role of abscisic acid. *Dev Biol.*, 123: 198-212.

Gao, M., Sakamoto, A., Miura, K., Murata, N., Sugiura, A. and Tao, R. 2000. Transformation of Japanese persimmon (Diospyros kaki Thunb.) with a bacterial gene for choline oxidase. *Mol Breed.*, 6: 501-510.

Garg, A.K., Kin, J.K., Owens, T.G., Ranwala, A.P., Choi. Y.C, Kochian, L.V. and Wu, R.J. 2002. Trehalose accumulation in rice plants pmfers high tolerance levels to different abiotic stresses. *Proc Natl Acad Sci USA*, 99: 15898-15903.

Gisbert, C., Rus, A.M., Bolarin, M.C., Lopez-Caronado, M., Arrillaga, I., Montesinos, C., Caro, M., Serrano, R. and Moreno, V. 2000. The yeast HAL1 gene improves salt tolerance of transgenic tomato. *Plant Physiol.*, 123: 393-402.

Goyal, K., Walton, L.J. and Tunacliffe, A. 2005. LEA Proteins prevent protein aggregation due to water stress. *Biochem J.*, 388: 151-157.

Grover, A. and Minhas, D. 2000. Towards production of abiotic stress tolerant transgenic rice plants: issues, progress and future research needs. *Pro Indian Natl Sci Acad B Rev Tracts. Biol Sci.,* 66: 13-32.

Grover, A., Kapoor, A., Satya Lakshmi, O., Agrawal, S., Sahi, C., Katiyar- Agrawal, S., Agarwal, M. and Dubey, H. 2001. Understanding molecular alphabets of the plant abiotic stress responses. *Curr Sci.*, 80: 206-216.

Grover, A., Sahi, C. Samam, N. and Grover, A. 1999.Taming abiotic stresses in plants through genetic engineering: current strategies and perspective. *Plant Sci.*, 143: 101-111.

Guiltinan, M.J., Marcotte, W.R. and Quatrano, R.S. 1990. A Plant leucine zipper protein that recognizes an abscisic acid response element. *Science*, 250: 267-271.

Haake, V., Cook, D., Riechmann, J.L., Pineda, O., Thomashow, M.F. and Zhang, J.Z. 2002. Transcription factor CBF4 is a regulator of drought adaptation in *Arabidopsis. Plant Physiol.,* 130: 639-648.

Hare, P.D., Cress, W.A. and Van Staden, J. 1998. Dissecting the roles of osmolyte accumulation during stress. *Plant Cell Environ.*, 21: 535-553.

Hayashi, H., Alia, Sakamoto, A., Nonaka, H., Chen, T.H.H. and Murata, N. 1988. Enhanced germination under high salt conditions of seeds of transgenic *Arabidopsis* with a bacterial gene (codA) for choline oxidase. *J Plant Res.*, 111: 357-362.

Hayashi, H., Mustardy, L., Deshnium, P., Ida, M. and Murata, N. 1997. Transformation of *Arabidopsis thaliana* with the codA gene for choline oxidase: accumulation of glycine betaine and enhanced tolerance to salt land cold stress. *Plant J*.12: 133-142.

Holmstorm, K.O., Manty, E., Welin, B. and Palva, E.T. 1996. Drought tolerance in tobacco. *Nature*, 379 :683-684.

Holmstrom, K.O., Somersalo, S., Mandal, A., Palva, E.T. and Welin, B. 2000. Improved tolerance to salinity and low temperature in transgenic tobacco producing glycine betaine. *J Expt Bot.*, 51: 177-185.

Hsieh, T.H., Lee, J.T., Chang, Y.Y. and Chan, M.T. 2002. Tomato plants ectopically expressing *Arabidopsis* CBF1 show enhanced resistance to water deficit stress. *Plant Physiol., 130:* 618-626 Ishitani, M., Xiong, L., Stevenson, B. and Zhu, J.K. 1997. Genetic analysis of osmotic and cold stress signal transduction in *Arabidopsis* interaction and convergence of abscisic acid dependent and abscisic acid independent pathways. *Plant Cell*, 9: 1935-1949.

Iwasaki, T., Kiyosue, T. and Yamaguchi-Shinozaki, K. 1997. The dehydration inducible rd17 (cor47) gene and its promoter region in *Arabidopsis thaliana. Plant Physiol.*, 115: 1287.

Jaglo, K.R., Kleff, K.L., Amundsen, X., Zhang, V., Haake, J.Z., Zhang, T. and Thomashow, M.F. 2001. Components of the *Arabidopsis* C-repeat/dehydration responsive element binding factor cold response pathway are conserved in *Brassica napus* and other plant species. *Plant Physiol.*, 127: 910-917.

Jagloo-Ottosen, K.R., Gilmour, S.J., Zarka, D.G., Schabenberger, O. and Thomashow, M.F. *Arabidopsis* CBF1 over expression induces cor genes and enhances freezing tolerance. *Science*, 280: 104-106.

Kagaya, Y., Hobo, T., Murata, M., Ban, A. and Hattori, T. 2002. Abscisic acid induced transcription is mediated by phosphorylation of an abscisic acid response element binding factor, TRAB1. *Plant Cell*, 14: 3177-3189.

Kasuga, M., Liu, Q., Miura, S., Yamaguchi-Shinozaki, K. and Shinozaki K 1999. Improving plant drought, salt and freezing tolerance by gene transfer of a single stress inducible transcription factor. *Nat Biotechnol*, 17: 287-291.

Kasuga, M., Miura, S., Shinozaki, K. and Yamaguchi-Shinozaki, K. 2004. A combination of the *Arabidopsis* DREB1A gene and stress inducible rd29A promoter improved drought- and low temperature stress tolerance in tobacco by gene transfer. *Plant Cell Physiol.*, 45: 346-350.

Katiyar-Agrawal, S., Agarwal, M. and Grover, A. 1999. Emerging trends in agricultural biotechnology research: use of abiotic stress induced promoter to drive expression of a stress resistance gene in the transgenic system leads to high level stress tolerance associated with minimal negative effects on growth. *Curr Sci.*, 77: 1577-1579.

Katiyar-Agrawal, S., Agarwal, M., Grover, A. 2003. Heat-Tolerance basmati rice engineered by over-expression of hsp 101. *Plant Mol Biol.*, 51: 677-686.

Kimpel, J.A. and Key, J.L. 1985. Heat shock in plants. *Trends Biochem Sci.*, 10: 353-357.

Kovtun, Y., Chiu W.L., Tena, G. and Sheen, J. 2000. Functional analysis of oxidative stress-activated mitogen-activated protein kinase cascade in plants. *Pro Natl Acad Sci USA.*, 97: 2940-2945.

Kumria , R. and Rajam, M.V. 2002. Ornithine decarbosylase transgene in tobacco affects polyamines, *in vitro*-morphogenesis and response to salt stress. *J Plant Physiol.*, 159: 983-990.

Lee, H., Xiong, L., Gong, Z., Ishitani, M., Stevenson, B. and Zhu, J.K. 2001. The *Arabidopsis* HOS1 gene negatively regulated cold signal transduction land encodes a RING finger protein that displays cold regulated nucleo-cytoplasmic partitioning. *Gene Dev.*, 15: 912-924.

Lee, J.H., van Montagu, M. and Verbruggen, N. 1999. A highly sponsored kinase is an essential component for stress tolerance in yeast and plant cells. *Proc Natl Acad Sci USA,* 96: 5873-5877.

Lee, S.S., Cho, H.S,, Yoon, G.M., Ahn, J.W., Kim, H.H. and Pai, H.S. 2003. Interaction of Nt CDPK1 calcium-dependent protein kinase with NtRpn3 regulatory subunit of the 26S proteasime in *Nicotiana tabacum. Plant J.*, 33: 825-840.

Li, H.Y., Chang, C.S., Lu, L.S., Liu, C.A., Chan, M.T. and Chang, Y.Y. 2003. Over-expression of *Arabidopsis thaliana* heat shock factor gene (AtHsfA1b) enhances chilling tolerance in transgenic tomato. *Bot Bull Acad Sin.*, 44:129-140.

Lilius, G., Holmberg, N. and Bulow, L. 1996. Enhanced NaCl stress tolerance in transgenic tobacco expressing bacterial choline dehydrogenase. *Biotechnol.*, 14: 177-180.

Lindquist, S. 1986. The heat-shock response. *Annu Rev Biochem* ,55: 1151-1191.

Liu, Q., Kasuga, M., Sakuma, Y., Abe, H., Miura, S., Yamaguchi-Shinozaki, K. and Shinozaki, K. 1998. Two transcription factors, DREBI and DREB2, with an EREBP/AP2 DNA Binding domain seprate two cellular signal transductional pathways in drought and low temperature responsive gene expression respectively, in Arabidopsis. Plant Cell. 10:1391-1406.

Malik, M.K., Solvin, J.P., Hwang, C.H. and Zimmerman. J/L/ 1999. Modified expression of carrot small heat shock protein gene, Hsp 17.7 results in increased or decreased thermo tolerance. *Plant J.*, 20: 89-99.

Maliro, M.F.A., McNeil, D., Kallmorgen, J., Pittock, C. and Redden, B. 2004. Screening chickpea (*Cicer arietinum* L) and wild relatives germplasm form diverse sources for salt tolerance. New directions for a diverse planet: Proceeding of the 4th International Crop Science Congress, Brisbane, Australia 26 Sep to 1 Oct 2004.

Mckersue, B.D., Bowley, S.R., Harjanto, E. and Leprince, O. 1996. Water deficit tolerance and field performance of transgenic alfalfa over expressing superoxide dismutase. *Plant Physiol.,* 111: 1177-1181.

McNeil, S.D., Nuccio, M.L., Rhodes, D., Shachar, Hill. Y., Hanson, A.D. 2000. Radiotracer and computer modeling evidence that posphobase methylatonis the main route of choline synthesis in tobacco. *Plant Physiol.*, 123: 371-380.

Mironshnichenko, S., Tripp, J., Nieden, U.Z., Neumann, D., Conrad, U. and Manteuffel, R. 2005. Immuno modulation of function of small heat shock proteins prevents their assembly into heat stress granules and results in cell death at sub-lethal temperatures. *Plant J.*, 41: 269-281.

Mundy, J., Yamaguchi-shinzaki, K., Chua, N.H. 1990. Nuclear proteins bind conserved elements in the abscisic acid-responsive promoter of a rice rab gene. *Proc Natl Acad Sci USA* , 87: 406-410.

Mundy, R. and Chua, N.-H. 1988. Abscisic acid and water stress induce the expression of a novel rice gene. *EMBO J.*, 7: 2279-2286.

Munns, R., Hussain, S., Rivelli, A.R., James, R.A., Condon, A.G 2002. Avenues for increasing salt tolerance of crops, and the role of physiologically based selection traits. *Plant Soil,* 247: 93-105.

Murakami, Y., Tsuyama, M., Kobayshi, Y., Kodama, H. and Iba K 2000. Trienoic fatty acids and plant tolerance of high temperature. *Science*, 287: 476-479.

Murata, N., Ishizki-Nishizawa, O., Higashi, S., Hayashi, S., Tasaka, Y. and Nishida, I. 1992. Genetically engineered alteration in the chilling sensitivity of plants. *Nature*, 356: 710-713.

Nanjo, T., Kobayashi, M., Yoshiba, Y., Kakubari, Y., Yamaguchi-Shinozaki, K. and Shinozaki, K. 1999 Antisense suppression of proline degradation improves tolerance to freezing and salinity in *Arabidopsis thaliana. FEBS Lett.*, 461: 205-210.

Nordin, K., Heino, P. and Palva, E.T. 1991. Separate signal pathways regulate the expression of a low temperature induced gene in *Arabidopsis thaliana* (L) Heynh. *Plant Mol Biol*, 115: 875-879.

Oberschall, A., Deak, M., Torok, K., Sass, L., Vass, I., Kovacs, I., Feher, A., Dudits, D.M. and Hovarth, G.V. 2000. A novel aldose/aldehyde reductase protects transgenic plants against lipid peroxidation under chemical and drought stress. *Plant J.*, 24: 437-446.

Ozuturk, Z.N., Talame, V., Dehholos, M., Michalowski, C.B., Galbraith, D.W., Gozukirmizi, N., Tuberosa, R. and Bohner, H.J. 2002. Monitoring large-scale changes in transcript abundance in drought and salt stressed barley. *Plant Mol Biol.*, 48: 551-571.

Pardo, J.M., Reddy, M.P. and Yang, S. 1998. Stress signaling though Ca^{2+}/Camodulin dependent protein phosphatase calcineurin mediates salt adaptation in plants. *Proc Natl Acad Sci USA,* 95: 9681-9683.

Passioura, J. 1977. Physiology of grain yield in wheat growing on stored water. *Aust J Plant Physiol* .3: 559-565.

Passioura, J. 2007. The drought environment: physical, biological and agricultural perspectives. *J Exp Bot.*, 58: 113-117.

Pei, Z.M., Ghassemian, M., Kwak, C.M., McCourt, P. and Schroeder, J.I. 1998. Role of farnesyl transferase in ABA regulation of gurard cell anion channels and plant water loss.. *Science*, 282: 287-290.

Pellegrineschi, A., Reynolds, M., Pacheco, M., Brito, R.M., Almeraya, R., Yamaguchi-Shinozaki, K. and Hoisington, D. 2004. Stress-induced expression in what of the *Arabiidopsi thaliana* DREB1A gene delays water stress symtoms under green house conditions. *Genome,* 47: 493-500.

Perl, A., Perl-Treves, R., Galili, S., Aviv, D., Shalgi, E., Malkin, S. and Galun, E. 1993. Enhanced oxidative stress defense om transgenic potato expressing tomato Cu, Zn superoxide *Theor Appl Genet.*, 85: 568-576.

Pilon-Smiths, E.A.H., Ebskamp, M.J.M., Jeuken, M.J.W., van der meer, IM., Visser, R.G.F., Weisbeek, P.J. and Smeekens, J.C.M. 1996. Microbial fructan production in transgenic potato plants and tubers. *Ind Crop Prod* ,5: 35-46.

Pilon-Smiths, E.A.H., Terry, N., Sears, T., Kim, H., Zayed, A., Hwang, S., Van Dun, K., Voogd E., Verwoerd, T.C., Krutwagen, R.W.H.H. and Giddinjn, J.M. 1998. Trehalose-producing transgenic tobacco plants show improved growth performance under drought stress. *J Plant Physiol.,* 152: 525-532.

Pilon-Smiths, E.A.H., Ebskamp, M.J.M., Jeuken, M.J.W., Paul, M.J., Jeuken, J.W., Weisbeek Smeekens, S.C.M. 1995. Improved performance of transgenic fructan –accumulating tobacco under drought stress. *Plant Physiol.*, 107: 125-130.

Pilon-Smiths, E.A.H., Terry, N., Sears, T. and van Dun, K. 1999. Enhanced drought resistance in fructan producing sugar beet. *Plant Physiol Biochem.* 37: 313-317.

Quimlo, C.A., Torrizo, L.B., Setter, T.L., Ellis, M., Grover, A., Abrigo, E.M., Oliva,, N.P., Ella, E.S., Carpena, A.L., Ito, O., Peacock, W.J., Dennis, E. and Datta, S.K. 2000. Enhancement of submergence tolerance in transgenic rice plants over producinjg pyruvate decarboxylase. *J Plant Physiol.*, 156: 516-521.

Quintero, F.J., Blatt, M.R. and Pardo, J.M. 2000. Functional conservation between yeast and plant 3 endosomal Na^+/H^+ antiporters. *FEBS Lett.*, 471: 224-228.

Ritchie, G.A. 1982. Carbohydrate reserves and root growth potential in Douglas-fir seedling before and after cold storage. *Can J Res.*, 12: 905-912.

Rohila, J.S., Jain, R.K. and Wu, R. 2002. Genetic improvement of Basmati rice for salt and drought tolerance by regulated expression of a barley Hva1 cDNA. *Plant Sci.*, 163: 525-532.

Romeo, C., Belles, J.M., Vaya, J.L., Serrano, R. and Culianez-Macia, F.A. 1997. Expression of the yeast trehalsoe-6-phosphate synthase gene in transgenic tobacco plants: pleiotropic phenotypes include drought tolerance. *Planta*, 201: 293-297.

Roxas, V.P., Smith, R.K Jr., Allen, E.R. and Allen, R.D. 1997. Over expression of glutathione S-transferase/glutathione peroxidase enhances the growth of transgenic tobacco seedlings during stress. *Nat Biotechnol.*, 15: 988-991.

Roy, M. and Wu, R. 2001. Arginine decarboxylase transgene expression and analysis of environmental stress tolerance in transgenic rice plant. *Science,* 160: 869-875.

Roy, M., Wu, R. 2002. Over expression of S-adenosylmethionine decarboxylase gene in rice increases polyamine level and enhances sodium chloride stress tolerance. *Plant Sci.,* 163: 987-992.

Rubio, M.C., Gonzalez, E.M., Minchin, F.R., Webb, K.J., Arrese-Igor, C., Ramos, J. and Becana, M. 2002. Effects of water stress on antioxidant enzymes of leaves and nodules of transgenic alfalfa over expressing superoxide dismutase. *Physiol Plant*, 115: 531-540.

Sakamoto, A., Alia, Murata, N. 1998. Metabolic engineering of rice leading to biosynthesis of glycine betaine and tolerance to salt and cold. *Plant Mol Biol.*, 38: 1011-1019.

Sakamoto, A., Valverde, R., Alia, Chen, T.H. and Murata, N. 2000. Transformation of *Arabidopsis* with the codA gene for choline oxidase enhances freezing tolerance of plants. *Plant J* .,22: 449-453.

Sakuma, Y., Maruyama, K., Oskabe, Y., Qin, F., Seki, M., Shinozaki, K. and Yamaguchi-shinozaki, K. 2006. Functional analysis of an *Arabidopsis* transcription factor, DREB2a, involved in drought-responsive gene expression. *Plant Cell,* 158: 1292-1309.

Schobert, B. Tschesche, H. 1978. Unusual solution properties of proline and its interaction with proteins. *Biochem Biophys Acta.*, 541: 270-277.

Seki, M., Narusaka, M., Abe, H., Kasuga, M., Yamaguchi-Shinozaki, K., Carninci, P., Hayashizaki,

Y. and Shinozaki, K. 2001. Monitoring the expression pattern of 1300 *Arabidopsis* genes under drought and cold stresses by using a full-length cDNA Microarray. *Plant Cell*, 13: 61-72.

SenGupta, A., Heinen, J.L., Holady, A.S., Burke, J.J. and Allen, R.D. 1993. Increased resistance to oxidative stress in transgenic plants that over express chloroplastic Cu/Zn superoxide dismutase. *Proc Nat Acad Sci USA*, 90: 1629-1633.

Serraj, R. and Sinclair, T.R. 2002. Osmolyte accumulation: can it really help increase crop under drought condition? *Plant Cell Environ*.,25: 333-341.

Shen, B., Jensen, R.G and Bohnert, H.J. 1997. Increased resistance to oxidative stress in transgenic plants by targeting mannitol biosynthesis to chloroplast. *Plant Physiol.,* 113: 1177-1183

Shi, H. Ishitani, M., Kim, C. and Zhu, J.K. 2000. The *Arabidopsis thaliana* salt tolerance gene SOS1 encodes a putative Na+/H+ antiporter. *Proc Nat Acad Sci USA*, 97: 6896-6901.

Shinozaki, K. and Yamaguchi-Shinozaki, K. 1999. Molecular responses to drought stress. In, "*Molecular response to cold, drought, heat and salt stress in higher plants*", Eds. K. Shinozaki, Yamaguchi-Shinozaki, R.G Lands Co. Austin, pp 11-28

Shinwari, Z.K. 1999. Function and regulation of genes that are induced by dehydration stress. *Biosci Agric*, 5: 39-47.

Shinwari, Z.K., Nakashima, K., Miura, S., Kasuga, M., Seki, M., Yamaguchi-Shinozaki, K., and Shinozaki, K. 1998. An *Arabidopsis* gene family encoding DRE/CRT binding proteins involved in low-temperature-reponsive gene expression. *Biochem Biophys Res Commun*.50: 161-170.

Shou, H., Bordallo, P. and Wang, K. 2004. Expression of the *Nicotiana* protein kinase (NPK1) enhanced drought tolerance in transgenic maize. *J Exp Bot.,* 55: 1013-1019.

Simpson, S.D., Nakashima, K., Narusaka, Y., Seki, M., Shinozaki, K. and Yamaguchi Shinozaki, K. 2003. Two different novel *cis* acting elements of erd1, a clp A homologous *Arabidopsis* gene function in induction by dehydration stress and dark-induced senescence. *Plant J.,* 33: 329-341.

Sinclair, B.J., Jaco Klok, C. and Chown, S.L. 2004. Metabolism of the sun-Antarctic caterpillar P*ringleophaga marioni* during cooling. freezing and thawing. *J Exp Biol*., 207: 1287-1294.

Sivamani, E., Bahieldin, A., Wraith, J.M., Al-Niemi, T., Dyer, W.E., Ho, T.H.D., Qu, R. 2000. Improved biomass productivity and water use efficiency under water deficit conditions in transgenic wheat constitutively expressing the barley HVA 1 gene. *Plant Sci*., 155:1-9.

Skriver, K. and Mundy, J. 1990. Gene expression in response to abscisic acid and osmotic stress. *Plant Cell*, 5: 503-512.

Slooten, L., Capiau, K., Van Camp, W., Montagu, M.V., Sybesma, C. and Inze, D. 1995. Factors affecting the enhancement of oxidative stress tolerance in transgenic tobacco over expressing manganese superoxide dismutase in the chloroplasts. *Plant Physiol.,* 107: 737-775.

Sun, W., Bernard, C., van de Cotte, B., Montagu, M.V., Verbruggen, N. 2001. At-HSP17.6A encoding a small heat shock protein in *Arabidopsis*, can enhance osmotolerance upon over expression. *Plant J*., 27: 407-415.

Two transcription factors, DREB1 and DREB2, with an EREBP/AP2 DNA binding domain separate two cellular signal transductional pathways in drought and low temperature responsive gene expression respectively, in *Arabidopsis. Plant Cell,* 10: 1391-1406.

Tarczynski, M.C., Jensen, R.G and Bohnert, H.J. 1993. Stress protection of transgenic tobacco by production of osmolyte mannitol. *Science*, 259: 508-510.

Thomashow, M.F. 1998. Role of cold responsive genes in plant freezing tolerance. *Plant Physiol.,* 118: 1-7.

Turner, N.C., Shahal, A., Berger, J.D., Chaturvedi, S.K., French, R.J., Ludwig, C., Mannur, D.M., Singh. S.J. and Yadav, H.S. 2007. Osmotic adjustment in chickpea (*Cicer arietinum* L.) results in no yield benefit under terminal drought. *J Exp Bot* .,58: 187-194.

Umezawa, T., Fujita, M., Fujita, Y., Yamaguchi-Shinozaki, K. and Shinozaki, K. 2006. Engineeringdrought tolerance in plants: discovering and tailoring genes to unlock the future. *Curr Opin Biotechnol.*, 17: 113-122.

Umezawa, T., Yoshida, R., Maruyama, K., Ymaguchi-Shinozaki, K. and Shinozaki, K. 2004. SRK2C, a SNF1- related protein kinase 2, improves drought tolerance by controlling stress responsive gene expression in *Arabidopsis thaliana*. *Pro Natl Acad Sci USA.*, 101: 17306-17311.

Vadez, V., Krishnamurthy, L., Gaur, P.M., Upadhyaya, H.D., Hiosington, D.A., Varshney, R.K., Turner, N.C. and Siddique, K.H.M. 2006. Tapping the large genetic variability for salinity tolerance in chickpea. *Proc Aust Soc Agron Meet.*, 10-14 Sept 2006.

Vadez, V., Krishnamurthy, L., Serraj, R., Gaur, P.M., Upadhyaya, H.D., Hoisington, D.A., Varshney, R.K., Turner, N.C. and Siddique, K.H.M. 2007. Large variation in salinity tolerance in chickpea is explained by difference in sensitivity at the reproductive stage. *Field Crop Res.*, 100(1-3): 123-129.

Van Camp, W., Capiau, M., van Montagu, M., Inze, D. and Slooten L 1996. Enhancement of oxidative stress tolerance in transgenic tobacco plants overproducing Fe superoxide dismutase in chloroplasts. *Plant Physiol.,* 112: 1703-1714.

Vierling, E. 1991. The roles of heat shock proteins in plants. *Annu Rev Plant Physiol Plant Mol Biol.*, 42: 579-620.

Vincour, B. and Altman, A. 2005. Recent advances in engineering plant tolerance to abiotic stress: achievements and limitations, *Curr Opin Biotechnol.*, 16: 123-132.

Voetberg, G.S. and Sharp, R.E. 1991. Growth of the maize primary root tip at low water potentials. III. Role of increased proline deposition in osmotic adjustment. *Plant Physiol.,* 96: 1125-1130.

Vranova, E., Inze, D. and van Breusegem, F. 2002. Signal transduction during oxidation stress. *J Exp Bot.*, 53: 1227-1236.

Waie, B. and Rajam, M.V. 2003. Effect of increased polymine biosynthesis on stress responses in trans-genic tobacco by introduction of human S-adenosyl methionine gene. *Plant Sci.*, 164: 727-734.

Wang, Y., Ying, J., Kuzma, M., Chalifoux, M., Sample, A., McArthur, C., Uchacz, T., Sarvas, C., Wan, J., Dennis, D.T., McCourt, P. and Huang, Y. 2005. Molecular tailoring of farnesylation for plant drought tolerance and yield protection. *Plant J.*, 43: 413-424.

Water, E.R., Lee, G.J. and Vierling, E. 1996. Evolution, structure and function of the small heat shock proteins in plants. *J Exp Bot.*, 47: 325-338.

Winicov, I. and Bastola, D.R. 1999. Transgenic over expression of the transcription factor Alfin 1 enhances expression of the endogenous MsPRP2 gene in alfalfa and improves salinity tolerance of the plant. *Plant Physiol.*, 120: 473-480.

Winicov, I. and Bastola, D.R. 1997. Salt tolerance in crop plants: new approaches through tissue culture and gene regulation. *Acta Physiol Plant.*, 19: 435-449.

Xiong, L., Ishitani, M. and Zhu, J.K. 1999. Interaction of osmotic stress, temperature and abscisic acid in the regulation of gene expression in *Arabidopsis*. *Plant Physiol.*119: 205-211.

Xiong, L. and Zhu, J.K. 2001. Plant abiotic stress signal transduction: molecular and genetic perspectives. *Physiol Plant*, 112: 205-211.

Xu, D., Duan, X., Wang, B., Hong, B., Ho, T.-H.D. and Wu, R. 1996. Expression of a late embryogenesis abundant protein gene, HVA 1, from barley confers tolerance to rice. *Nature*, 422: 705-708.

Yamada, M., Morishita, H., Urano, K., Shinozaki, N., Yamaguchi-Shinozaki, K., Shinozaki, K., Yoshiba, Y. 2005. Effects of free proline accumulation in petunias under drought stress. *J Exp Bot.*, 56: 1975-1981.

Yamaguchi-Shinozaki, K. and Shinozaki, K. 1993. Characterization of the expression of a desiccation responsive rd29 gene of *Arabidopsis thaliana* and analysis of its promoter in transgenic plants. *Mol Gen Genet* 236: 331-340.

Yamaguchi-Shinozaki, K., Urao, T., Iwasaki, T., Kiyosue, T. and Shinozaki, K. 1994. Function and regulation of genes that are induced by dehydration stress in *Arabidopsis thaliana*. *JIRCAS J.,* 1: 69-79.

Yang, G., Rhodes, D. and Joly, R.J. 1996. Effects of high temperature on membrane stability and chlorophyll fluorescence in glycine betaine deficient and glycine betaine containing maize lines. *Aust J Plant Physiol.*, 23: 437-443.

Zhang, H.-X. and Blemwald, E. 2001. Transgenic salt tolerant tomato plants accumulate salt in foliage but not in fruit. *Nat Biotechnol.*,19: 765-768.

Zhang, J.Y., Broeckling, C.D., Blancaflor, E.B., Sledge, M.K., Sumner, L.W. and Wang, Z.Y.2005. Over expression of WXO1, a putative *Medicago truncatula* AP2 domain-containing transcription factor gene, increases cuticular wax accumulation and enhances drought tolerance in transgenic alfalfa (*Medicago sativa*). *Plant J.*, 42: 689-707.

Zhang, J.Z., Creelman, R.A. and Zhu, J.K. 2004. From laboratory to field. Using information from *Arabidopsis* to engineer salt, cold, and drought tolerance in crops. *Plant Physiol.*, 135: 615-621.

Zhao, H.W., Chen, Y.J., Hu, Y.L., Gao, Y. and Lin, Z.P. 2000. Construction of a trehalose-6-phosphate synthase gene driven by drought responsive promoter and expression of drought resistance in transgenic tobacco. *Acta Bot Sinica.*, 42: 616-619.

Zhu, B.M., Su, J., Chang, M., Verma, D.P.S., Fan Y.L. and Wu, R. 1998. Over expression of delta 1-pyrroline-5-carboylate synthase gene and analysis of tolerance to water and salt stress in transgenic rice. *Plant Sci.*, 199: 41-48.

Zhu, J.K. 2002. Salt and drought stress signal transduction in plants. *Annu Rev Plant Physiol Plant Mol Biol.*, 53: 247-273.

Zhu, L., Tang, G.S., Hazen, S.P., Kim, H.S., War, R.W. 1999. RFLP based genetic diversity and its development in Shaanxi wheat lines. *Aca Bot Boreali Occident Sin*, 19: 13.

8

Abiotic Stresses and Respiration in Plants

A. Bhattacharya

Abiotic stress affects plant physiology at whole plant as well as at cellular levels through osmotic and ionic adjustments that result in reduced biomass production. The adverse effect of abiotic stress appears on whole plant level at almost all growth stages including germination, seedling, vegetative and maturity stages. Despite causing osmotic and ionic stress, salinity causes ionic imbalances that may impair the selectivity of root membranes and induce potassium deficiency.

The current state of the art on the regulation of photosynthesis and respiration in response to water stress has been summarized. It was suggested that in all but very severe stress situations, diffusional limitations to CO_2 *(stomatal plus mesophyll) account for most of the observed decreases in photosynthesis during water stress development. At severe stress, metabolic impairment of photosynthesis also occurs in some cases, probably due to secondary oxidative stress. The response of respiration to severe water stress consists in increased electron partitioning to alternative oxidase and decreased cytochrome pathway, with little effect on total respiration rates.*

Respiration is the set of metabolic reactions and processes that take place in the cells of organisms to convert biochemical energy from nutrients into adenosine triphosphate (ATP), and then release waste products. The reactions involved in respiration are catabolic, which break large molecules into smaller ones, releasing energy in the process, as weak so-called "high-energy" bonds are replaced by stronger bonds in the products. Respiration is one of the key ways a cell gains useful energy to fuel cellular activity. Cellular respiration is considered an exothermic redox reaction which releases heat. The overall reaction occurs in a series of biochemical steps, most of which are redox reactions themselves. Although technically, cellular respiration is a combustion reaction, it clearly does not resemble one when it occurs in a living cell due to slow release of energy from the series of reactions. The non-photorespiratory carbon release in leaves is called mitochondrial respiration—a central metabolic process that produces energy (ATP, NADPH) and carbon skeletons for cellular maintenance and growth. It also contributes to significant carbon losses—especially under stress conditions—altering the net carbon gain (van Oijen *et al.*, 2010).

Respiration involves the participation of different processes responsible for the oxidation of glucose molecules for energy and C structures, either in the presence (aerobic) (Millar *et al.*, 2011; van Dongen *et al.*, 2011) or absence (anaerobic) of oxygen (Gupta *et al.*, 2009). In the latter case, the most affected organ is the root, inducing partial oxidation strategies of substrates in order to continue to generate energy without oxygen (O_2). These strategies are called fermentation, which differentiate themselves by their end products: ethanol, lactic acid and alanine (Sousa and Sodek, 2002).

In the presence of O_2, substrates are completely oxidized to CO_2 and H_2O (van Dongen *et al.,* 2011). This is done through three metabolic processes: glycolysis, the TCA cycle and the oxidative phosphorylation (Fernie *et al.,* 2004). To these is added a fourth process; transport of the products of respiration. This corresponds to the movement of substrates and cofactors to facilitate the release of products throughout the cell (Millar *et al.*, 2011). The operation of these processes is the most efficient way to obtain energy from complete oxidation of hydrocarbon substrates, both in plants and animals (Plaxton, 1996).

It has been reported that there are differences in respiration according to species and plant tissue (Millar *et al.*, 2011); for example, spinach leaf respiration is preferably performed at night, because during the day it is inactivated at low solar radiation intensities (10-50/μ mol m^{-2} s^{-1}) (Atkin *et al.*, 1998), presumably cause excess ATP from photosynthesis in chloroplasts (Atkin *et al.*, 2004). However, for many years the effects of radiation on leaf respiration were studied without considering the effect of temperature. Thus, a study in eucalyptus plants showed that leaf respiration is highly dependent on both radiation and temperature, showing a high degree of inhibition of respiration at high temperatures and high radiation levels, which reduces the CO_2 ratio (provided by photosynthesis) that is respired (Atkin *et al.,* 2000).

Likewise, root respiration can be altered by a variety of factors such as temperature (Rachmilevitch *et al.*, 2006), salinity (Bernstein *et al.*, 2013), heavy metals (Moyen and Roblin, 2013), drought (Jimenez *et al.*, 2013), waterlogging and flooding (Liao and Lin, 2001); however, the availability of O_2 is what most affects root respiration (Gupta *et al.*, 2009). This factor is key in respiration metabolism, because oxygen is the final electron acceptor in oxidative phosphorylation (Moller, 2001). Most of the energy (ATP) produced by root respiration is used for processes such as growth (Thongo M'Bou *et al.*, 2010), nitrate reduction, symbiotic N fixation (in legumes), the absorption of nitrate and other ions absorption by the roots (Poorter *et al.*, 1991), protein turnover (Scheurwater *et al.*, 2000), maintenance of the ion gradient and membrane potential (Bouma and De Visser, 1993) and waste mechanisms and production of heat through alternative pathways (Cannell and Thornley, 2000).

8.1. Glycolysis in Plants

Glycolysis is an anaerobic pathway responsible for oxidizing sucrose (glucose in animals) to generate ATP, a reductant (NADH) and pyruvate (Millar *et al.*, 2011; van Dongen *et al.*, 2011). The universality of glycolysis is associated with its importance in adaptations to different environmental stressors, such as nutritional stress, temperature, drought, and anoxia, among others (Plaxton, 1996). In general, the transformation of glucose to pyruvate is performed through a series of reactions catalyzed by numerous enzymes (Figure 1) (Plaxton, 1996; van Dongen *et al.*, 2011), which not only act as catalysts and energy metabolism regulators (Camacho-Pereira *et al.*, 2009), but also as signal transducers in response to changes in the environment. For instance, it has been observed that the activity level of hexokinase would correspond to a key component in the sugar signal detection. .

Nutrients that are commonly used by animal and plant cells in respiration include sugar, amino acids and fatty acids,, and the most common oxidizing agent (electron acceptor) is molecular oxygen (O_2). The chemical energy stored in ATP (its third phosphate group is weakly bonded to the rest of the molecule and is cheaply broken allowing stronger bonds to form, thereby transferring energy for use by the cell) can then be used to drive processes requiring energy, including biosynthesis, locomotion or transportation of molecules across cell membranes.

Aerobic respiration requires oxygen (O_2) in order to generate ATP. Although carbohydrates, fats and proteins are consumed as reactants, it is the preferred method of pyruvate breakdown in glycolysis and requires that pyruvate enter the mitochondria in order to be fully oxidized by the Krebs cycle. The products of this process are carbon dioxide and water, but the energy transferred is used to break strong bonds in ADP as the third phosphate group is added to form ATP (adenosine triphosphate), by substrate-level phosphorylation NADH and $FADH_2$.

Simplified reaction: $C_6H_{12}O_6$ (s) + 6 O_2 (g) → 6 CO_2 (g) + 6 H_2O (l) + heat

$$\Delta G = -2880 \text{ kJ per mol of } C_6H_{12}O_6$$

The negative ΔG indicates that the reaction can occur spontaneously.

Respiration is a major sink for carbohydrates. Carbon lost through respiration can account for up to 50% of the daily carbon gain by photosynthesis (Morgan and Austin, 1983; Poorter *et al.*, 1995). Respiration generally increases with temperatures with a short-term Q10 of about 2 (Lambers, 1985). Shortage of assimilates due to high respiratory losses has long been proposed to be a primary factor responsible for root growth inhibition and dysfunction (Youngner and Nudge, 1968). Heat-tolerant roots may be able to control their respiration rate or use more efficient respiratory pathways. Increasing temperature by 4°C (from 16.8°C to 20.8°C) caused a 30% increase in root respiration rates in three species - two perennial grasses, *Dactylis glomerata*, *Poa annua*, and *Bellis perennis*. However, *B. perennis*, that was better acclimated to high temperature conditions, maintained a slower respiration rate than the other two species (Gunn and Farrar, 1999). The critical question remains how plants adapted to high soil temperatures control root respiration rates to survive high temperatures for prolonged periods of time.

The rate of respiration depends on three major energy requiring processes: maintenance of biomass, growth, and ion uptake transport and assimilation (Poorter *et al.*, 1991). High respiratory carbon consumption has been found to be mainly due to high specific respiratory maintenance costs (Scheurwater *et al.*, 1998). Maintenance respiration increases with increasing temperatures (McCree and Amthor, 1982; Kase and Catsky, 1984), and its response is similar to the increased response of total respiration. The specific respiratory costs for growth, however, are usually thought to be relatively unaffected by temperature (Johnson and Thornley, 1985); *i.e.* as long as the substrate and product compositions do not change in response to temperature. In addition, even in a comparison of 24 species with a 3.5-fold variation in relative growth rate (RGR), growth respiration changed <20% (Poorter *et al.*, 1991). While specific respiratory costs for maintenance and growth have frequently been determined (Amthor, 1984; Lambers, 1985), the specific respiratory costs for ion uptake have been little investigated (Scheurwater *et al.*, 1998). Scheurwater *et al.* (1998) calculated costs of 6 mmol O^2 g^{-1} DW for ion uptake by assuming costs for growth respiration similar to values found by Poorter *et al.* (1991). Their calculations were made for plants for which NO^{-3} was the sole nitrogen source; therefore, the major costs of ion uptake were for NO^{-3} uptake and assimilation. They concluded that the major cause for the relatively slow rates of root respiration in fast-growing grasses was lower specific respiratory costs for ion uptake.

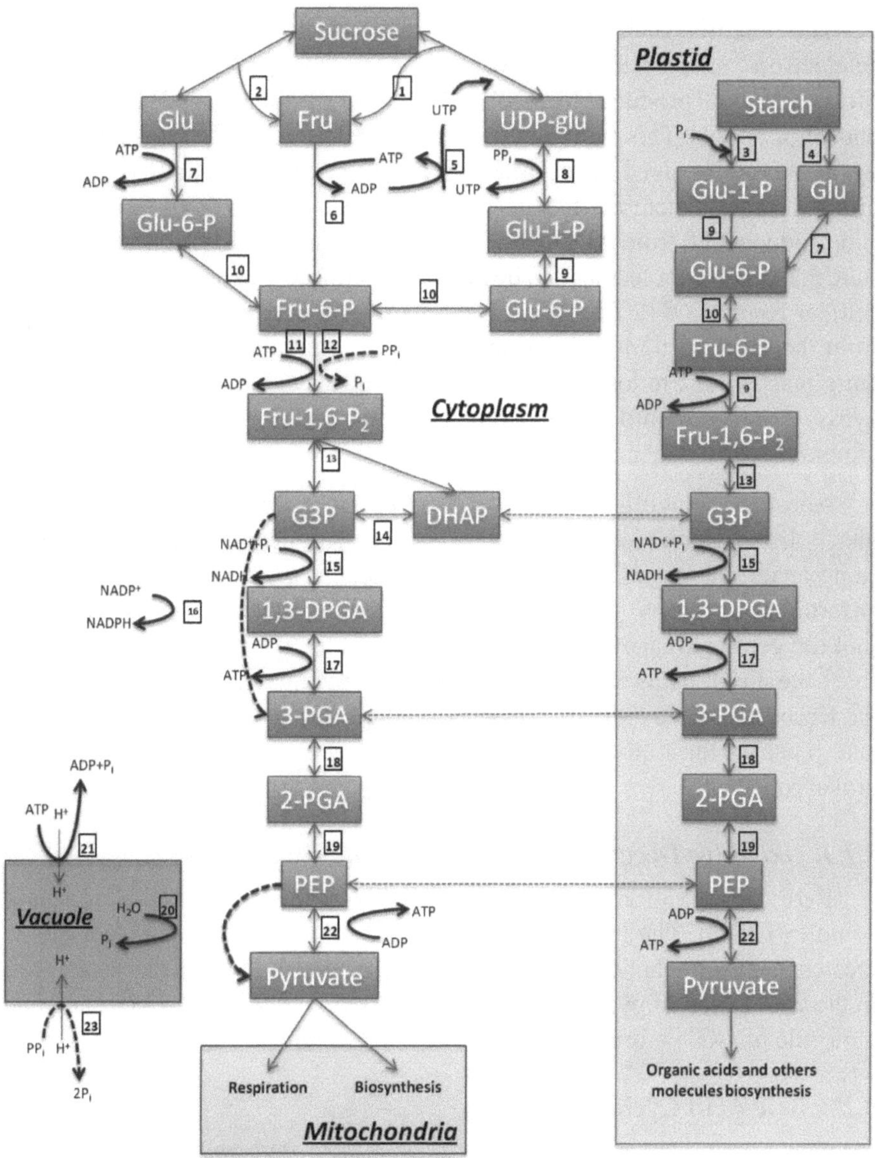

Source: After Plaxton (1996) and Plaxton and Podesta (2006)

Fig. 1: Schematic representation of the glycolysis pathway. Continuous line represents glycolytic flux and broken lines represent alternatives. Enzymes involved in each reaction are as follows: α and β-amylase, 5) NDP kinase, 6) fructokinase, 7) hexokinase, 8) UDP-glucose pyrophosphorylase, 9) phosphoglucomutase, 10) phosphoglucose isomerase, 11) ATP-PFK, 12) PP_i-PFK, 13) aldolase, 14) triose phosphate, 15) and 16) glyceraldehyde-3-phosphate phosphorylated and non-phosphorylated, respectively, 17) phosphoglycerate kinase, 18) phosphoglyceromutase, 19) enolase, 20) PEP phosphatase, 21) H^+-ATPasa, 22) piruvate kinase, and 23) H^+-PP_iasa.

The potential of NADH and $FADH_2$ is converted to more ATP through an electron transport chain with oxygen as the "terminal electron acceptor". Most of the ATP produced by aerobic cellular respiration is made by oxidative phosphorylation. This works by the energy released in the consumption of pyruvate being used to create a chemiosmotic potential by pumping protons across a membrane. This potential is then used to drive ATP synthase and produce ATP from ADP and a phosphate group. Biology textbooks often state that 38 ATP molecules can be made per oxidised glucose molecule during cellular respiration (2 from glycolysis, 2 from the Krebs cycle, and about 34 from the electron transport system). However, this maximum yield is never quite reached due to losses (leaky membranes) as well as the cost of moving pyruvate and ADP into the mitochondrial matrix, and current estimates range around 29 to 30 ATP per glucose.

Aerobic metabolism is up to 15 times more efficient than anaerobic metabolism (which yields 2 molecules ATP per 1 molecule glucose). However some anaerobic organisms, such as methanogens are able to continue with anaerobic respiration, yielding more ATP by using other inorganic molecules (not oxygen) as final electron acceptors in the electron transport chain. They share the initial pathway of glycolysis but aerobic metabolism continues with the Krebs cycle and oxidative phosphorylation. The post-glycolytic reactions take place in the mitochondria in eukaryotic cells, and in the cytoplasm in prokaryotic cells.

8.1.1. Oxidative Decarboxylation of Pyruvate

Pyruvate is oxidized to acetyl-CoA and CO_2 by the pyruvate dehydrogenase complex (PDC). The PDC contains multiple copies of three enzymes and is located in the mitochondria of eukaryotic cells and in the cytosol of prokaryotes. In the conversion of pyruvate to acetyl-CoA, one molecule of NADH and one molecule of CO_2 is formed.

8.2. Citric Acid Cycle

This is also called the *Krebs cycle* or the *tricarboxylic acid cycle*. When oxygen is present, acetyle-CoA is produced from the pyruvate molecules created from glycolysis. When oxygen is present, the mitochondria will undergo aerobic respiration which leads to the Krebs cycle. However, if oxygen is not present, fermentation of the pyruvate molecule will occur. In the presence of oxygen, when acetyle-CoA is produced, the molecule then enters the citric acid cycle (Krebs cycle) inside the mitochondrial matrix, and is oxidized to CO_2 while at the same time reducing NAD to NADH. NADH can be used by the rlectron transport chain to create furtherr ATPas part of oxidative phosphorylation. To fully oxidize the equivalent of one glucose molecule, two acetyl-CoA must be metabolized by the Krebs cycle. Two waste products, H_2O and CO_2, are created during this cycle.

The citric acid cycle is an 8-step process involving different enzymes and co-enzymes. During the cycle, acetyl-CoA (2 carbons) + oxaloacetate (4 carbons) yields citrate (6 carbons), which is rearranged to a more reactive form called isocitrate (6 carbons). Isocitrate is modified to become α-ketoglutarate (5 carbons), succinyl-CoA, succinate, fumarate, malate, and, finally, oxaloacetate. The net gain of high-energy compounds from one cycle is 3 NADH, 1 $FADH_2$, and 1 GTP; the GTP may subsequently be used to produce ATP. Thus, the total yield from 1 glucose molecule (2 pyruvate molecules) is 6 NADH, 2 $FADH_2$, and 2 ATP.

8.3. Oxidative Phosphorylation

In eukaryotes, oxidative phosphorylation occurs in the mitochondrial cristae. It comprises the electron transport chain that establishes a proton gradient (chemiosmotic potential) across the boundary of inner membrane by oxidizing the NADH produced from the Krebs cycle. ATP is synthesized by the ATP synthase enzyme when the chemiosmotic gradient is used to drive the phosphorylation of ADP. The electrons are finally transferred to exogenous oxygen and, with the addition of two protons, water is formed.

8.3.1. Efficiency of ATP Production

The table below describes the reactions involved when one glucose molecule is fully oxidized into carbon dioxide. It is assumed that all the reduced coenzymes are oxidized by the electron transport chain and used for oxidative phosphorylation.

Although there is a theoretical yield of 38 ATP molecules per glucose during cellular respiration, such conditions are generally not realized due to losses such as the cost of moving pyruvate (from glycolysis), phosphate, and ADP (substrates for ATP synthesis) into the mitochondria. All are actively transported using carriers that utilize the stored energy in the proton electromagnetic gradient.

Photorespiration (also known as the oxidative photosynthetic carbon cycle or C_2 photosynthesis) is a process in plant metabolism which attempts to ameliorate the consequences of a wasteful oxygenation reaction by the enzyme RuBisCO. The desired reaction is the addition of carbon dioxide to RuBP (carboxylation), a key step in the Calvin-Benson cycle, however approximately 25% of reactions by RuBisCO instead add oxygen to RuBP (oxygenation), producing a product that cannot be used within the Calvin-Benson cycle. This

Step	Co-enzyme yield	ATP yield	Source of ATP
Glycolysis preparatory phase		(-) 2	Phosphorylation of glucose and fructose-6-phosphate used two ATP from the cytoplasm
Glycolysis pay off phase		4	Substrate level phosphorylation
	2 NADH	3 or 5	Oxidative phosphorylation. Each NADH ptoduces net 1.5 ATP (instead of usual 2.5) due to NADH transport over the mitochondrial membrane
Oxidative decarboxylation of pyruvate	2 NADH	5	Oxidative phosphorylation
Krebs cycle		2	Substrate-level-phosphorylation
	6 NADH	15	Oxidative phosphorylation
	2 $FADH_2$	3	Oxidative phosphorylation
Total yield	30 or 32 ATP		From the complete oxidation of one glucose molecule to carbon di-oxide and oxidation of all reduced co-enzymes.

process reduces efficiency of photosynthesis, potentially reducing photosynthetic output by 25% in C_3 plants. Photorespiration involves a complex network of enzyme reactions that exchange metabolites between chloroplasts, leaf peroxisomes and mitochondria.

The oxygenation reaction of RuBisCO is a wasteful process because 3-phosphoglycerate is created at a reduced rate and higher metabolic cost compared with RuBP carboxylase activity. While photorespiratory carbon cycling results in the formation of G3P eventually, there is still a net loss of carbon (around 25% of carbon fixed by photosynthesis is re-released as CO_2) and nitrogen, as ammonia. The ammonia must be detoxified at a substantial cost to the cell. Photorespiration also incurs a direct cost of 2ATP and one NAD(P)H.

8.4. Photorespiration

While it is common to refer to the entire process as photorespiration, technically the term refers only to the metabolic network which acts to rescue the products of the oxygenation reaction (phosphoglycolate). Addition of molecular oxygen to ribulose-1,5-bisphosphate produces 3-phosphoglycerate (PGA) and 2-phosphoglycolate (2PG, or PG). PGA is the normal product of carboxylation, and productively enters the Calvin cycle. Phosphoglycolate, however, inhibits certain enzymes involved in photosynthetic carbon fixation (hence is often said to be an inhibitor of photosynthesis). It is also relatively difficult to recycle: in higher plants it is salvaged by a series of reactions in the peroxisome, mitochondria, and again in the peroxisome where it is converted into glycerate. Glycerate reenters the chloroplast and by the same transporter that exports glycolate. A cost of 1 ATP is associated with conversion to 3-phosphoglycerate (PGA) (Phosphorylation), within the chloroplast, which is then free to re-enter the Calvin cycle.

There are several costs associated with this metabolic pathway; one being the production of hydrogen peroxide in the peroxisome (associated with the conversion of glycolate to glyoxylate). Hydrogen peroxide is a dangerously strong oxidant which must be immediately broken down into water and oxygen by the enzyme catalase. The conversion of 2x 2 carbon glycine to 1 C3 serine in the mitochondria by the enzyme glycine-decarboxylase is a key step, which releases CO_2, NH_3, and reduces NAD to NADH. Thus, 1 CO_2 molecule is produced for every 3 molecules of O_2 (two deriving from the activity of RuBisCO, the third from peroxisomal oxidations). The assimilation of NH_3 occurs via the GS-GOGAT cycle, at a cost of one ATP and one NADPH.

Cyanobacteria have three possible pathways through which they can metabolise 2 phosphoglycolate. They are unable to grow if all three pathways are knocked out, despite having a carbon concentrating mechanism that should dramatically reduce the rate of photorespiration.

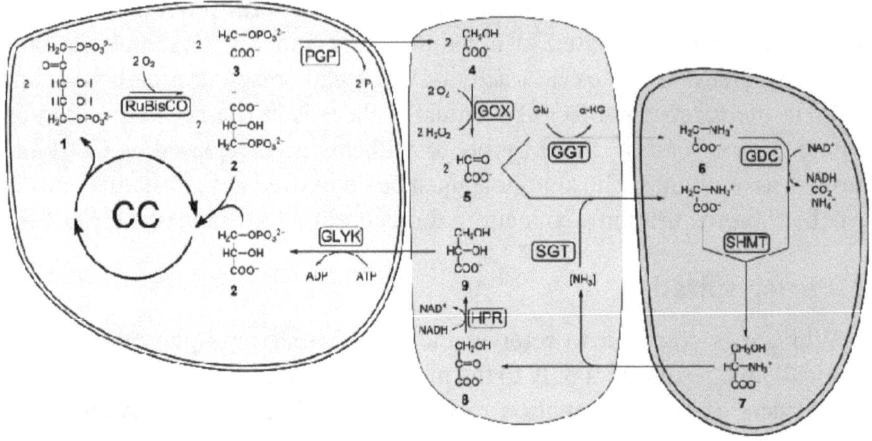

8.5. Effect of Abiotic Stresses

Abiotic stresses such as drought, salinity, flood, and cold vastly effect plant growth and metabolism that ultimately disturbs plant life (Bray *et al.*, 2000; Ahmad and Prasad, 2012a, b). This has a negative impact on global crop production since majority of world's arable lands are exposed to these abiotic stress conditions (Rockström and Falkenmark, 2000). Up to 50–70% decline in major crop productivities have been attributed to abiotic stresses on several occasions (Mittler, 2006). For their survival under these stress conditions, plants respond by modifying several aspects in their metabolic cascade (Dos Reis *et al.*, 2012). These response mechanisms help plants to survive during the stress period as well as to recover following cessation of the stress.

When the ATP and NADPH production by photochemical processes exceed
$_2$ fixation, plants can use several processes to dissipate energy and avoid or minimised photoinhibition. These processes include alternative electron sinks dependent of O_2 such as the oxygenase reaction catalised by ribulose-1,5-bisphosphate carboxylase/ oxigenase which initiates photorespiration (Foyer and Noctor, 2000). The light-dependent O_2 uptake by photorespiration not only use ATP and reducing power from photosynthetic electron transport system but also cause a loss of the CO_2 fixed by Calvin cycle. Even in plants under no photoinhibitory conditions, photorespiration occur due to the capacity of Rubisco to catalise the carboxylation and oxygenation of ribulose-1,5-bisphosphate, depending on the CO_2/ O_2 ratio. At 25°C, photorespiration increases the cost of carbon fixation to 4.75 ATP and 3.5

NADPH per carbon fixed under atmospheric CO_2 and O_2 concentrations, which compares to three ATP and two NADPH per carbon fixed under no photorespiration conditions, *e.g.* only 2% O_2 instead of atmospheric 21% O_2 (Keys, 1986). In plants submitted to drought, a reduction of photosynthesis and photorespiration is observed as a result of the lower CO_2 and O_2 availability in the chloroplast. However, in this situation, the photorespiratory pathway is less decreased than photosynthesis (Lawlor, 1976, 1981). In fact, despite the much higher affinity of Rubisco for CO_2 than O_2, the CO_2 concentration is almost at the sub saturating level in C3 plants. Thus any decrease in stomatal conductance or in the gases solubility limits the carboxylase activity while the oxigenase activity is unaffected or less affected (Hall and Keys, 1983). In C4 plants, the higher CO_2 concentration at the Rubisco level allows a lower decrease in the photosynthesis / photorespiration ratio under water deficit (Carmo-Silva *et al.*, 2008) than the one observed in C3 plants, despite the C4 pathway having *per se* specific energy costs. The less efficient light use for CO_2 fixation caused by photorespiration lowers the quantum yields of photosynthesis in C3 plants under drought (Nunes *et al.*, 2009) or high temperature but this was not observed in C4 plants (Ehleringer *et al.*, 1977). Since photorespiration is the major cause of a lower bioenergetic balance in photosynthetic tissues of C3 plants, increasing plant growth by overcoming the limitation of photosynthesis imposed by Rubisco is still an important target of research and plant improvement (Bogorad, 2003; Zhu *et al.*, 2004; Long *et al.*, 2006; Parry *et al.*, 2007).

In C3 and C4 plants under stress, the photosynthetic rate decreases with the leaf relative water content and water potential (Kramer and Boyer, 1995; Cornic and Massacci, 1996; Carmo-Silva *et al.*, 2007; Nunes *et al.*, 2008). This decrease is frequently correlated to the impairment of photochemical processes in C3 plants (Flexas and Medrano, 2002; Parry *et al.*, 2002) including inhibition of ATP synthesis (Tezara *et al.*, 1999; Lawlor, 2002). It is still unclear if photosynthesis is primarily limited by water deficit through the restriction of CO_2 supply to metabolism (stomatal limitation) (Chaves, 1991) or by the impairment of other processes which decrease the potential rate of photosynthesis (non-stomatal limitation) (Duque *et al.*, 2013). Photosynthesis light curves allow the determination of the relative contribution of respiration, photosynthesis and photorespiration to the light energy dissipation (Bjorkman, 1981). More often, under stress, no change or a decrease in respiration is observed in leaves but the variations were always minor comparing to photosynthesis, despite the interdependence of the two processes through photorespiration (Dutilleul *et al.*, 2003). However, at the whole-plant level, the contribution of respiration to the plant bioenergetics status is relevant because respiration can account for a release of 30-70% of the carbon fixed daily in well-watered plants, whereas in drought plants the proportion of carbon lost increases, mainly due to the decrease observed on photosynthesis (Flexas *et al.*, 2005; Duque *et al.*, 2013).

Abiotic stress responses in plants occur at various organ levels among which the root specific processes are of particular importance. Under normal growth condition, root absorbs water and nutrients from the soil and supplies them throughout the plant body, thereby playing pivotal roles in maintaining cellular homeostasis. However, this balanced system is altered during the stress period when roots are forced to adopt several structural and functional modifications. Examples of these modifications include molecular, cellular, and phenotypic changes such as alteration of metabolism and membrane characteristics, hardening of cell wall and reduction of root length (Gowda *et al.*, 2011; Atkinson and Urwin, 2012). These changes are often caused by single or combined effects of several abiotic stress responsive pathways that can be best explored at the global level using high-throughput approaches such as proteomics (Petricka *et al.*, 2012).

Distinct light regimes, both in terms of quantity and composition, influence almost all physiological processes like photosynthesis and respiration rates, affecting also variables like plant height, fresh and dry mass and water content of the plant (Pystina and Danilov, 2001). Water content, on the other hand, shifts both the stem length and leaf area of the plant, in a way to adapt the plant to the amount or quality of light intercepted (Aspiazú *et al.*, 2008). During mixed cropping, inter-specific and intra-specific plant competition affects the amount and the quality of the final product, as well as its efficiency in utilization of environmental resources (VanderZee and Kennedy, 1983; Melo *et al.*, 2006). This is noted when assessing physiological characteristics associated to photosynthesis, such as concentration of internal and external gases (Kirschbaum and Pearcy, 1988), light composition and intensity (Merotto Jr. *et al.*, 2009) and mass accumulation by plants under different conditions. Although gas exchange capability by stomata is considered a main limitation for photosynthetic CO_2 assimilation (Hutmacher and Krieg, 1983), it is unlikely that gas exchange will limit the photosynthesis rate when interacting with other factors. However, photosynthetic rate is directly related to the photosynthetically active radiation composition of light, to water availability and gas exchange (Naves-Barbiero *et al.*, 2000). Plants have specific needs for light, predominantly in bands of red and blue (Messinger *et al.*, 2006). When plants do not receive these wave lengths in a satisfactory manner, they need to adapt themselves in order to survive (Attridge, 1990). When under competition for light, the red and far-red ratio affected by shading is also important (Merotto Jr. *et al.*, 2009) and influences photosynthetic efficiency (Da Matta *et al.,* 2001).

Salinity, cold and drought stress are all osmotic stresses, they cause a primary loss of cell water, and, therefore, a decrease of cell osmotic potential (Duque *et al.,* 2013). However, the elicitor of cell water loss differs between stresses: (a) salinity stress decreases cell water content due to the decrease of external water potential, caused by the increased ion concentration (mainly Na^+ and Cl^-), turning more difficult water uptake by roots and water translocation to metabolically active cells; (b) cold stress decreases cell water content due to the so-called physiological drought, *i.e.,* the inability to transport the water available at the soil to the living cells, mainly the ones of the leaf mesophyll; (c) the decrease of the cell water content under drought stress is due to water shortage in soil or/and in the atmosphere. Anyway, dehydration triggers the biosynthesis of the phytohormone abscisic acid (ABA) and it has been known for a long time that a significant set of genes, induced by drought, salt, and cold stresses, are also activated by ABA (Boudsocq and Laurière, 2005).

As a consequence of water loss and decreased cell volume, cell sap solute concentrations increase and thereby cell osmotic potential decreases. As cell turgor also decreases, an early effect common to these stresses is a sharp decrease in leaf expansion rate and overall plant growth rate. Furthermore, an additional active decrease of the cell sap osmotic potential is observed, as an attempt to keep cell hydration. In fact, at the metabolic level, a common feature to these three stresses is the osmotic adjustment by synthesis of low-molecular weight osmolytes [carbohydrates (da Silva *et al.*, 2004), betain (Wyn Jones and Storey, 1981) and proline (Aspinall and Paleg, 1981) that can counteract cellular dehydration and turgor loss (Beck *et al.*, 2007). On the other hand, differences between these stresses do also exist. While drought stress is mainly osmotic, ion toxicity, namely Na^+, is a distinctive feature of salinity stress. Cold stress, behinds physiological drought, has an impact on the rate of most biochemical reactions, including photosynthetic carbon metabolism reactions, as enzyme activities are extremely temperature-dependent. Also water stress and salinity stress decrease photosynthesis, which create conditions to increased photoinhibition, particularly under high irradiances.

The effect of common salt (NaCl) on ion contents, Krebs cycle intermediates and its regulatory enzymes was investigated in growing mungbean (*Vigna radiata* L. Wilczek, B 105) seedlings. Concentration of sodium and chloride were tested on ion contents, Kerbs cycle intermediates and its regulatory enzymes were studied in growing mungbean (*Vigna radiata* L. Wilczek) seedlings and it was reported that sodium and chloride ion contents increased in both root and shoot whereas potassium ion content decreased in shoot of test seedlings with increasing concentrations of NaCl. Organic acids like pyruvate and citrate levels increased whereas malate level decreased under stress in both roots and shoots.

Salt stress also variedly affected the activities of different enzymes of respiratory chain. The activity of pyruvate dehydrogenase (E.C. 1.2.4.1) decreased in 50 mM NaCl but increased in 100 mM and 150 mM concentrations, in both root and shoot samples. Succinate dehydrogenase (E.C. 1.3.5.1) activity was reduced in root whereas stimulated in shoot under increasing concentrations of salt. The activity of isocitrate dehydrogenase (E.C. 1.1.1.41) and malate dehydrogenase (E.C. 1.1.1.37) decreased in both root and shoot samples under salt stress. On the contrary, pretreatment of mungbean seeds with sublethal dose of NaCl was able to overcome the adverse effects of stress imposed by NaCl to variable extents with significant alterations of all the tested parameters, resulting in better growth and efficient respiration in mungbean seedlings. Thus, plants can acclimate to lethal level of salinity by pretreatment of seeds with sublethal level of NaCl, which serves to improve their health and production under saline condition, but the sublethal concentration of NaCl should be carefully chosen (Saha *et al.*, 2012). Mitova and Iquamberdiev (2000) studied the activity of NADP-dependent isocitrate dehydrogenase, malate dehydrogenase, succinate dehydrogenase, catalase, and peroxidase as well as the rate of $^{14}CO_2$ release after introduction of labeled substrates for glycolysis and citrate acid cycle within 24 h after salt stress (1% NaCl) in 10-14 days old germinants of wheat (*Triticum aestivum* L.) and maize (*Zea mays* L.) as well as thallus of small duckweed (*Wolffia arrhiza* (L.) Hork ex Wimmer). It was reported that oscillations in the enzymes activity with 4-6 h period have been revealed under stress conditions. Activity of glycolysis decreased in wheat and maize and increased in duckweed under the influence of stress stimulus. Six hours after NaCl action decarboxylation of exogenous citrate and succinate was enhanced in all three plants while the rate of exogenous malate decarboxylation was decreased. It was concluded that adaptation of higher plants to salinization is accompanied by rearrangements in oxidative metabolism reflected by oscillations in activity of the enzymes involved in oxidative metabolism.

8.6. Effect of Drought on Respiration

Warmer and drier conditions are expected globally under current climate change scenarios and particularly in the Mediterranean region (Somot *et al.*, 2008; Friend, 2010; IPCC, 013). Seasonal reoccurring drought is the main natural environmental factor in the Mediterranean region limiting plant growth and yield (Spechet *et al.*, 1969; Di Castri, 1973). Several studies have reported drought-induced impacts in the Mediterranean region (Peñuelas *et al.*, 2001; Martinez-Vilata and PiEol, 2002; Raftovannis *et al.*, 2008; Allen *et al.*, 2010; Camicer *et al.*, 2011; Matusick *et al.*, 2013), as well as shifts in vegetation composition (Jump and Peñuelas, 2013; Anderegg *et al.*, 2013). Seasonal summer drought limits plant growth and productivity most strongly through reductions in the plant carbon budget, which depends on the balance between photosynthesis and respiration (Flexas *et al.*, 2006). The Mediterranean region is characterized by a high variability in temperature and precipitation regimes, especially in mountainous areas (Sperlich *et al.*, 2015). Climate extremes combined with high interannual variability complicate the scaling of carbon dynamics from one year to another (Reynolds *et al.*, 1996; Moraleset *et al.*, 2005; Gullias *et al.*, 2009).

Although the drought responses of Mediterranean vegetation have been investigated extensively, most studies concern *photosynthetic* responses (Flexas *et al.*, 2014), whereas *respiratory* responses in leaves have largely been neglected (Ninemets, 2014). Also, it is not clear how seasonality and other abiotic stressors affect the balance of night respiration (R_n) and day respiration (R_d) in the leaves. This is partly owing to measurement difficulties; R_n can easily be measured by darkening the leaf, but R_d is harder to obtain and is traditionally estimated from carbon-response curves with the Laisk method, from light-response curves with the Kok method, or with an amended version of the Kok method with chlorophyll fluorescence developed by Yin *et al.* (2009) (reviewed by Yin *et al.*, 2011). Measurement constraints and lacking research priorities can account for the dearth of data on respiratory responses to abiotic stress, particularly drought (Atkin and Macherel, 2009; Heskel *et al.*, 2014). Wright *et al.* (2006) provided evidence that irradiance, temperature, and precipitation affect respiration in a wide range of woody species around the world; Mediterranean species, however, were not covered. Catoni *et al.* (2013) provided evidence that temperature, and monthly rainfall to a lesser extent, could explain the seasonal variation of R_d in several Mediterranean maquis species. Seasonal acclimation of respiration is believed to be more important in sclerophyllic perennial leaves (Galmìs *et al.*, 2007; Zaragoza-Castells *et al.*, 2007, 2008) than in plants with short-lived leaves (Atkin and Macherel, 2009). A better characterization of the respiratory responses to drought relative to carbon gain is vital for elucidating the overall effects on carbon exchange dynamics in water-limited environments. Rainfall-exclusion experiments in natural ecosystems are laborious and expensive but are highly valuable to simulate more realistically long-term drought. Some studies have addressed the photosynthetic limitations under long-term drought in natural ecosystems comprising stomatal, mesophyll, and biochemical components (Limousin *et al.*, 2010; Martin-StPaul *et al.*, 2012). The effects of long-term experimental drought on photosynthesis in parallel with night and day respiration has not been investigated so far on mature species in natural ecosystems.

Water scarcity is considered as the main environmental factor limiting plant growth and yield worldwide (Chaves *et al.*, 2003). Drought has alone adverse effects on the plant growth, affecting mainly leaf and root growth, stomatal conductance, photosynthetic rate and biomass production (Blum, 1998). As drought stress mainly affects the plant carbon balance, in particular, photosynthesis and respiration, adjustments at the leaf level are of primary importance, while long-term adjustments at the whole plant level may then follow (Chaves *et al.*, 2003; Flexas *et al.*, 2006). Changes in plant growth elicited by low water availability have also been related to modulation of plant cell carbon metabolism, which are dependent on the balance between photosynthesis and respiration (Martin *et al.*, 2009). Although photosynthesis may decrease up to 100% becomes completely impaired under severe drought, respiration rate may either increase (Bartoli *et al.*, 2005) or decrease (Martin *et al.*, 2009). Respiration is an essential metabolic process that generate ATP but several other metabolites that are used in many synthetic processes essential for growth and maintenance of the cell homeostasis, including under stress conditions (MacCabe *et al.*, 2000; Bartoli *et al.*, 2000).

Drought tolerance is a cost-intensive phenomenon, as a considerable quantity of energy is spent to cope with it. The fraction of carbohydrate that is lost through respiration determines the overall metabolic efficiency of the plant (Davidson *et al.*, 2000). The root is a major consumer of carbon fixed in photosynthesis and uses it for growth and maintenance, as well as dry matter production (Lambers *et al.*, 1996). Plant growth and developmental processes as well as environmental conditions affect the size of this fraction (*i.e.* utilized in respiration). However, the rate of photosynthesis often limits plant growth when soil water availability is reduced (Huang and Fu, 2000). A negative carbon balance can occur as a result of diminished photosynthetic capacity during drought, unless simultaneous and proportionate reductions in growth and carbon consumption take place. In wheat, depending on the growth stage, cultivar and nutritional status, more than 50% of the daily accumulated photosynthates were transported to the root, and around 60% of this fraction was respired (Lambers *et al.*, 1996). Drought sensitive spring wheat (cv. Longchun, 8139–2) used a relatively greater amount of glucose to absorb water, especially in severe drought stress (Liu *et al.*, 2004). Severe drought reduced the shoot and root biomass, photosynthesis and root respiration rate. Limited root respiration and root biomass under severe soil drying can improve growth and physiological activity of drought-tolerant wheat, which is advantageous over a drought-sensitive cultivar in arid regions (Liu and Li, 2005). There are two mitochondrial electron transport pathways from ubiquinone to oxygen in plants. The alternative pathway branches from the cytochrome pathway and donates electrons to oxygen directly by alternative oxidase (Moore and Siedow, 1991). The alternative pathway is not coupled with adenosine tri phosphate synthesis, but can be induced in response to stress or inhibition of the main electron transfer pathway (Wagner and Moore, 1997). When plants are exposed to drought stress, they produce reactive oxygen species, which damage membrane components (Blokhina *et al.*, 2003). In this regard, alternative oxidase activity could be useful in maintaining normal levels of metabolites and reduce reactive oxygen species production during stress. Oxygen uptake by sugar beet was characterized by a high rate, distinct cytochrome oxidase dependent terminal oxidation and up to 80% inhibition of respiration in the presence of 0.5 mM potassium cyanide. At an early drought stage (10 days), a decrease in the activity of the cytochrome-mediated oxidation pathway was largely counterbalanced by the activation of mitochondrial alternative oxidase, whereas long-term dehydration of plants was accompanied by activation of additional oxidative systems insensitive to both potassium cyanide and salicylhydroxamate (Shugaeva *et al.*, 2007). In summary, water deficit in the rhizosphere leads to an increased rate of root respiration, leading to an imbalance in the utilization of carbon resources, reduced production of adenosine triphosphate and enhanced generation of reactive oxygen species.

A special feature of plant cell respiration is the presence of an alternative pathway which drains electrons from the ubiquinone pool without involvement of the cytochrome oxidase (Brownleader *et al.*, 1997). The mitochondrial alternative oxidase (AOX) apparently reduces molecular oxygen to water in a single four-electron transfer step (Day *et al.*, 1991; Moore and Siedow, 1991). This alternative pathway is non-phosphorylating, resistant to cyanide and antimicine and inhibited by salicilhydroxamic acid (SHAM) and n-propyl galate (Siedow and Grivin, 1980). It has been the focus on many studies in plant respiratory metabolism under several environmental stresses, mainly because an increased AOX capacity might contribute to controlling the formation of reactive oxygen species (ROS) (Wanger and Krab, 1995; Popov *et al.*, 1997; Maxwl *et al.*, 1999; Umbach *et al.*, 2005).

Water stress invariably decreases the photosynthetic rate and the intensity of this effect influences the capacity of different species to cope with the drought, which also depends on the duration of stress and plant genetic background (Chaves, 1991; Chaves *et al.*, 2002). Generally, when respiration rate decreases upon drought conditions the photosynthesis and growth requirements are further affected. Nevertheless, this behavior seems to be somewhat species dependent, and respiratory rates can also increase, particularly under severe drought (Gashshaie *et al.*, 2001).

Suppression of photosynthesis during drought stress due to the closure of stomata and the contribution of leaf-internal limitations to CO_2 diffusion, particularly mesophyll conductance (g_m), has been determined in numerous studies and plant species (Flexas *et al.*, 2004, 2008; Ninemets *et al.*, 2005; Waren and Adams, 2006). Although plant species respond differently to varying drought stress intensities, photosynthetic limitation is firstly and predominantly driven by stomata, in particular, by a decline in stomatal conductance. Further, when stomatal conductance drops below a certain threshold (<50 mmol H_2O m^{-2} s^{-1}) limitations of non-stomatal processes become more important, in particular, decreased g_m and impaired photo-biochemistry (Flexas and Medrano, 2002; Chevas *et al.*, 2003; Flexas *et al.*, 2004). Adjustment of leaf diffusion components for CO_2 is one way for plants to cope with situations of limited water supply and concurrently to improve their water use efficiency. As well as leaf internal adjustments of diffusion components during drought stress, increased thermal dissipation of excess energy and other photo-protective processes (*e.g.* an enhanced xanthophyll cycle) may contribute to improved stress tolerance and adaptation (Havaux and Niyogi, 1999; Mittler, 2002).

In parallel to these changes inside the chloroplast, respiratory pathways in mitochondria might also be altered, because of their interaction with the photosynthetic pathway. The respiratory chain is thought to dissipate excess reductants originated from chloroplasts (Raghvendra and Padmasree, 2003). Moreover, the non-phosphorylating pathways, which involve the cyanide-resistant alternative oxidase (AOX) and the type II NAD(P)H dehydrogenases, are considered to be efficient dissipation systems for these reductants, because electron flow through these pathways is not limited by adenylate control (Noctor *et al.*, 2007). Thus, the non-phosphorylating pathways may function as a mechanism of plant photo-protection, while the components of this mechanism have not been characterized in detail. Indeed, several studies have highlighted that different mutants with some impaired mitochondrial function also have a lower photosynthetic capacity (Juszczuk *et al.*, 2007; Nunes-Nesi *et al.*, 2007; Giraud *et al.*, 2008), and this includes the CMSII mutant of *Nicotiana sylvestris* (Sabar *et al.,* 2000; Priault *et al.,* 2206a, b).

Several studies on the effect of severe drought stress on respiratory pathways have revealed contrasting results, as respiration remained unaltered in soybean (Ribas-carbo *et al.*, 2005), increased in wheat (Bartoli *et al.*, 2005), and decreased in bean and pepper (Gonzalez-Meler *et al.*, 1997). However, changes in the *in vivo* activities of the cytochrome oxidase (COX) and AOX pathways, measured with the oxygen isotope fractionation technique that has been demonstrated to be the most reliable technique for the studies of electron partitioning between the two main respiratory pathways (Ribas-Carbo *et al.*, 1995; Dey *et al.*, 1996), have been reported by Ribas-Carbo and colleagues (Flexas *et al.*, 2005; Ribas-carbo *et al.*, 2005). In their study on soybean (Ribas-carbo *et al.*, 2005), a decrease in COX activity was detected in leaves during severe drought stress, while AOX activity increased. Whether, and to what extent, plant species-specific factors and/or experimental conditions affect *in vivo* respiratory pathways under drought stress awaits further studies.

To examine the effect of stress-induced changes on respiratory pathways, in particular with relation to photosynthesis, plants with modified AOX expression have been intensively studied (Noctor *et al.*, 2007). Among these, the *Nicotiana sylvestris* cytoplasmic male-sterile CMSII mutant, which lacks a functional mitochondrial complex I (Gutierres *et al.*, 1997) and possesses high amounts of AOX transcript and protein (Sabar *et al.*, 2000), has received increased attention. The observed differences in activities of the photosynthetic and respiratory pathways as compared with wild-type plants have been proposed to result from alterations in the cellular redox balance (Dutilleul *et al.*, 2003b; Priault *et al.*, 2006a; Vidal *et al.*, 2007) and/or limitations of CO_2 diffusion inside the leaf (Priault *et al.*, 2006b). Moreover, loss of complex I function might be compensated by enhanced non-phosphorylating NAD(P)H dehydrogenases, resulting in the maintenance of cell redox balance and cross-talk between mitochondria and chloroplasts (Dutilleul *et al.*, 2003a, b; Vidal *et al.*, 2007).

Thus, the tobacco CMSII mutant offers great possibilities to study the interaction between respiratory and photosynthetic pathways, in particular, under contrasting environmental conditions (*e.g.* high light, temperature, and drought stress). With regard to the increasing importance of understanding plant responses to drought stress, particularly within the context of climate changes, more studies are needed to examine the plant carbon balance, especially the relationship between the carboxylation and oxygenation processes. Due to the scarcity of *in vivo* studies on photosynthesis and respiration under drought stress (Gonzalez-Meler *et al.*, 1997; Bartoli *et al.*, 2005; Ribas-Carbo *et al.*, 2005; Giraud *et al.*, 2008), more research is necessary to understand the interrelation between chloroplasts and mitochondria.

8.7. Effect of Waterlogging on Respiration

Heavy or continuous rainfall in areas with poorly drained soil causes flood, one of the most severe environmental stresses affecting plants, in particular at their early growth stages. Soybean, wheat, barley, and maize are categorized as flood-sensitive whereas rice is an examplc of flood-tolerant species (Komatsu *et al.*, 2012). The hypoxic environment formed due to submerged state during flooding stress, affects aerobic respiration (Bailey-Serres and Voesenek, 2008), which leads to a boosted production of ATP and regeneration of NAD^+ through anaerobic respiration (Gibbs and Greenway, 2003). Under this oxygen deprived condition, protein synthesis is hampered as well, since translation is a tremendously energy-intensive process (Nanjo *et al.*, 2010). In order to cope with this stress, plants need to adopt several changes in their gene expression profiles as well as at cellular protein levels.

Waterlogging is a severe problem, which affects crop growth and yield in those areas where the concentration of rain is low. The main cause of damage under water logging is oxygen deficiency, so the plants show wilting even when enclosed by excess of water, which affect nutrient and water uptake (Sairam *et al.*, 2008). Water logging causes a condition of hypoxia (low oxygen concentrations) in soils, because of the low solubility of oxygen in water. Water logging can also cause the accumulation of ethylene and products of root and bacterial anaerobic metabolism (Barrett-Lennard, 2003). Plants tolerant to water logging stress exhibit certain adaptation, such as, formation of aerenchyma and adventitious roots. Furthermore, due to interaction of plant hormones, auxin and ethylene the formation of adventitious roots take place (Ashraf, 2012). Rice is one of the important flooding tolerant crops. The rice has ability to germinate in the complete absence of oxygen and grow in standing water. Rice and some associated weed species, such as, barnyard grass (*Echinochloa* spp.), will germinate quickly under anoxia once the seeds are imbibed (Drawford, 2003). But in the case of drought some species deep root systems to tap deep water tables while some has large surface root systems to quickly absorb rainfall. Some plants avoid drought by quickly regrowing new leaves when environmental condition are favorable and by improving dropping their leaves (Knox, 2005). Some species are very sensitive to water logging at the germination stage. A decrease in germination ability is due to oxygen deficiency in water logging soil. Respiration, electron transport and ATP formation are inhibited during germination when oxygen is short (Hsu *et al.*, 2000). Plants under drought and water logging stresses exhibit growth reduction, low SLA (specific leaf area), photosynthesis declination, stomatal closure, decrease in respiration and biomass production and protein degradation (Juan *et al.*, 2009). Plants could get only two ATP per glucose molecule, in fermentation, whereas, 36 ATP the molecules are produced per glucose molecule in aerobic respiration. Flood-tolerant plants are able to retain their energy status by using fermentation. In waterlogged plants, initial decline in cytosolic pH attributed to the production of lactic acid during fermentation. This initial decrease in pH helps the plant to change from lactate to ethanol fermentation by establishment of alcohol dehydrogenase and inhibition of lactate dehydrogenase grow.

Hypoxia and Anoxia: Low oxygen concentration in soil (hypoxia) or complete absence of oxygen (anoxia) affect nutrient uptake, synthesis and translocation of growth regulators, respiration and carbohydrate partitioning, as well as photosynthesis decreasing the yield of crops grown in soil with insufficient drainage (Ferreira *et al.*, 2008).

According to Jackson and Colmer (2005), diffusion of oxygen through water is 104-fold slower than in air. Excess water also leads to other changes in the soil that influence plants, due to the threat of oxygen deficiency. Oxygen deficiency in roots causes a change from aerobic to anaerobic respiration, which is much less efficient in consumption of organic compounds as energy sources and effect in accumulation of toxic products *e.g.* ethanol (Halcomb, 2003). The accumulation of starch in leaves of flooded plants is a reflection of reduced phloem transport. Furthermore, flooding reduces phloem transport and causes reduction of carbohydrates in roots. The deleterious effects linked with hypoxia and anoxia, which are responsible for the slowed growth and reduced yield of many agriculturally important crops, include drop in cytoplasmic pH and decrease in cellular energy charge and the accumulation of toxic metabolites and reactive oxygen species (ROS) (Liao and Lin, 2000).

Excess water causes a sharp decrease in soil redox potential, resulting in very significant changes to the soil elemental profile. As soon as the free oxygen is depleted, nitrate is used by soil microorganisms as an alternative electron acceptor in respiration. Manganese oxides are the next electron acceptors, followed by iron and sulphate. This results in a dramatic build up in the amount of soluble iron (Fe^{2+}) and Mn^{2+} in the soil solution, often to above toxic levels (Marschner, 1991). It should be also kept in mind that the solubility of potentially toxic metal cations is strongly dependent on soil pH and the toxicity of some of these elements may increase even further in some soil types (such as Mn^{2+} in acid soils). It should be noted in this context that waterlogging usually results in a drift towards neutrality, thus causing changes in nutrient solubility. As waterlogging stress progresses, a further decrease in redox potential results in the reduction of sulphate to hydrogen sulphide. Finally, methane formation, from the reduction of carbon dioxide and certain organic acids, is initiated at redox potentials of 200 mV (at pH7). Impeded gas exchange during soil waterlogging also leads to high partial pressure of CO_2 (pCO_2) in the root zone, with some serious consequences for root growth and metabolism. High root zone (pCO_2 will lead to high concentrations of HCO_3) in the cytoplasm, with implications for metabolic regulatory systems. Rice and other wetland plants can tolerate high pCO_2 but non wetland species suffer rapid damage. Also, in waterlogged soils, the main form of plant-available nitrogen (N) is NH_4^+, and plant adaptations to NH_4^+ Vs NO_3^- nutrition may be important under prolonged waterlogging. Some wetland species are particularly efficient at absorbing NO_3^-, possibly to scavenge NO_3^- formed in the rhizosphere by nitrifiers depending on O_2 released from roots. Numerous volatile (short-chain) fatty acids and phenolics also accumulate in soils during prolonged waterlogging (Patel *et al.*, 2014).

Soil flooding usually influences plant growth in a negative way as nutritional point of view. Waterlogging reduces the endogenous levels of nutrient in different *et al.*, 2011). Oxygen deficiency in the root zone causes a marked decline in the selectivity of K^+/Na^+ uptake and impedes the transport of K^+ to the shoots. The early senescence of leaves and the retarded growth of shoots in flooded plants are caused by the inhibition of nitrogen uptake and the consequent redistribution of nitrogen within the shoot. Moreover, the hampered efficiency of photosystem (PS) II is attributed to the deficiencies of nitrogen, phosphorus, potassium, magnesium, and calcium. It is evident from the literature that adverse effects of waterlogging are not due to the toxic levels of Na and Fe but reduced concentrations of nitrogen, phosphorus, potassium, calcium and magnesium are the major contributors (Smethurst *et al.*, 2005).

8.7.1. Plant Metabolism and Energetic Alteration Under Waterlogging

Aerobic respiration by roots and microorganisms reduces the oxygen concentration in the rhizosphere leading to hypoxia/anoxia. Therefore, aerobic respiration gets hampered, oxidative phosphorylation is negligible and plants shift their metabolism to anaerobic mode. One of the most important consequences of energy limitation under anoxia is altered redox state of the cell. Due to unavailability of oxygen, the intermediate electron carriers in electron transport chain become reduced, affecting redox active metabolic reactions. Therefore, cells require maintaining their redox characteristics, *i.e.* NADH/ NAD^+ ratio unaffected under waterlogging (Chirkova *et al.*, 1992). Saturated electron transport components, the highly reduced intracellular environment and low energy supply are the factors favorable for reactive oxygen species (ROS) generation. The consequences of ROS formation depend on the intensity of the stress as well as on the physicochemical conditions in the cell (*i.e.* antioxidant status, redox state and pH. Generation of ROS under stress condition may cause lipid peroxidation, enzyme inactivation, and oxidative damage to DNA (Shewfelt and Purvis, 1995). Elevated cytoplasmic Ca^{2+} concentration, cytoplasmic acidification and change in cell membrane permeability are the most common features observed under waterlogged conditions. Species tolerant to soil waterlogging are generally considered those able to maintain their energy level via fermentation. In addition to keep on appropriate energy level maintenance of cytosolic pH is critical. When hypoxia or anoxia occurs the pH of the cytoplasm shows an early decrease that is attributed to an initial production of lactic acid by fermentation. According to the "Davis- Roberts pH–state theory" the decline in pH permits the switch from lactate to ethanol fermentation by inhibition of lactated dehydrogenase (LDH) and activation of ADH (Chang *et al.*, 1983). Because acidosis increases cell necrosis, the switch taking place maintains pH at approximately 6.8 thus allowing cell survival.

Reduction of root respiration is one of the earliest responses of plants under anoxia, regardless of whether the plants are flooding-tolerant or intolerant (Carpenter and Mitchell, 1980; Hsu *et al.*, 1999; Kuo and Chen, 1980; Lambers, 1980; Lin and Lin, 1992; Liao and Lin, 1995; Su and Lin, 1996; McNamara and Mitchel, 1989). The respiratory capacity of bitter melon fell to 28% that of the non-flooded bitter melon level after 5 days of flooding (Liao and Lin, 1995). This phenomenon may be valuable for survival and may help flooded roots scavenge any oxygen available around the roots. A high oxygen consumption rate in root tips is associated with respiration, which is required for related metabolic activities, such as generation of ATP. Under flooded conditions, plant roots are in a state of hypoxia, their metabolic activity is inhibited and ATP production decreased (Sagilo *et al.*, 1980). The decreased ATP production restricts the supply of energy for root growth, thus reducing vegetative growth.

It was found that respiration of leaves increased significantly during flooding (Liao and Lin, 1994). Flooding has been reported to cause stomatal closure directly, without affecting the photosynthetic capacity, in *V. ashei* (Davies and Flore, 1986a, b), thus decreasing *Ci*. It has been suggested that the stomatal aperture regulates the decline of CER. However, in flooded bitter melon seedlings, *Ci* was observed to increase (Liao and Lin, 1994). Other plants, in which *Ci* was found to remain unchanged with flooding, are *C. illinoensis* (Smith and Ager, 1988), *C. sinensis* grafted onto *C. jamthiri* and *C. aurantium* rootstocks (Vu and Yelenosky, 1991) and bitter melon grafted onto luffa rootstocks (Liao and Lin, 1996). These observations suggest that stomatal aperture is not the only limiting factor for CER but is partly responsible for the decrease in the photosynthetic capacity of mesophyll tissue.

The activation level of Rubisco in flooded bitter melon increased above the control value after 1 day of flooding and subsequently declined to a lower level (Liao and Lin, 1994). Changes in the level of activation of Rubisco reflect the level of carbamylation of Rubisco (Miziorko and Lorimer, 1983), which in turn is regulated by *Ci* and light intensity (Sage *et al.*, 1990). In bitter melon seedlings, Rubisco activity was found to average 59% of the control value on the 7th day after flooding (Liao and Lin, 1994). In as much as Rubisco catalyzes the initial reaction during the assimilation of atmospheric CO_2 (Andrews and Lorimer, 1987), the activation level of Rubisco should be positively correlated with the rate of photosynthesis. The leaf soluble protein of flooded bitter melon was found to decline slowly and to reach an average of 75% of the control level on the 7th day of flooding, but the Rubisco content was reduced significantly.

Based on the positive correlation between Rubisco activity and Rubisco content ($R^2 = 0.89$) as well as on the decreased activation level of Rubisco, it can be inferred that a reduced quantity of Rubisco protein as well as reduced activity of existing enzymes may cause Rubisco activity to decline during flooding. Furthermore, it was reported that phloem transport of photosynthate was blocked and that the demand for sucrose loading was lowered. This may lead to an accumulation of starch in the chloroplasts (Liao and Lin, 1994; Wample and Davis, 1983). It can, thus, be suggested that feedback inhibition of starch accumulation may result in a reduction of CER in flooded plants. The many physiological responses observed indicate that both stomatal and metabolic factors are responsible for the reduction of CER during flooding stress.

8.8. Effect of Low Oxygen on Respiration

A major factor that affects plant respiration is O_2 depletion, in the rhizosphere or directly in tissues (Toro and Pinto, 2015). The O_2 availability in tissues and cells depends on the plant's age and especially on the O_2 supply from the environment, tissue O_2 having a great influence on the central pathways of carbohydrate synthesis (Millar *et al.,* 1998). In plants, the major external factors affecting O_2 availability are flooding and waterlogging of soil (Liao and Lin, 2001). Under these conditions one can distinguish three broad categories of oxygen status induced by water: normoxia, hypoxia, and anoxia. Excess water in soil decreases the O_2 diffusion rate, because the diffusion of O_2 is 10^4 times slower in water than in air (Liao and Lin, 2001; Jackson and Colmer, 2005). The reason for this is the high level of interaction between O_2 and H_2O through hydrogen bonds (Mommer *et al.*, 2004). In flooded soils not only is the diffusion of O_2 reduced, but also the diffusion of several other gases such as CO_2 and ethylene (Wegner, 2010). It is important to consider that low O_2 availability in cells can even occur in some situations under normoxia, due mainly to a high resistance to O_2 diffusion between different plant tissues (Gibbs and Greenway, 2003) such as roots and stems (van Dongen *et al.*, 2011).

Sometimes, to reduce the anaerobic state in shoots, roots regulate frequency of absorption (Zabalza *et al.*, 2009) and oxygen consumption (Mancuso and Marras, 2006). Morphological adaptations also play a key supporting role in promoting the O_2 flux from the roots through the rest of the plant, *e.g.* aerenchyma formation due to induction by ethylene (Greenway and Gibbs, 2003). One of the main effects of anaerobiosis on metabolism is a decreased adenylate energy charge (AEC), which also reduces the ATP:ADP ratio (Drew *et al.*, 2000).

Under hypoxic conditions the TCA cycle pathway gives more flexibility to the overall metabolism. For instance, in *Lotus japonicus,* it was shown that alanine aminotransferase (AlaAT) generates a link between glycolysis and the TCA cycle through the conversion of 2-oxoglutarate to succinate. This generates NADH that is used in the transformation of OAA into malate; together with succinate CoA ligase, both contribute to the generation of ATP under conditions of oxygen deficit (Geigenberger, 2003; Bailey-Serres and Voesenek, 2008).

Moreover, glycolysis is affected when O_2 concentration falls below 1.5-2.0 mg O_2 L^{-1} in the bulk solution (hypoxia). This decreases glycolysis and increases the activities of enzymes involved in fermentation (Li *et al.*, 2010). In maize subjected to hypoxia, pyruvate decarboxylase activity increases 5-to-9-fold compared to that under normal conditions (Gupta *et al.*, 2009). It is known that there are two major control points for electron transfer under oxygen deficit, corresponding to alternative oxidase and cytochrome c oxidase. Some studies suggest that the response to oxygen deficiency in roots is driven by the terminal oxidation of respiration (Kennedy *et al.*, 1992). Under some conditions, when the availability of O_2 in roots is decreased, adaptation to stress may occur through two strategies; the first is a decrease in ATP consumption which leads to a metabolic crisis at the cellular level (Igamberdiev *et al.*, 2010), and the second is an increase in the glycolytic flux (Mancuso and Marras, 2006). The latter is called the "Pasteur effect" (Gibbs *et al.*, 2000), consisting of a progressive acceleration in carbohydrate metabolism that allows the plants to maintain their energy level, especially during the early phase of acclimation to oxygen deficiency (Greenway and Gibbs, 2003; Camacho-Pereira *et al.*, 2009; Wegner, 2010).

Although many have emphasized the importance of the "Pasteur effect" in acclimatization processes to compensate for the energy inefficiency caused by oxidative phosphorylation inhibition (Gibbs and Greenway, 2003; Huang *et al.*, 2008), this can only generate ATP at about 37.5% of the rate produced in tissues under optimal oxygen conditions (Geigenberger, 2003).

The O_2 is a molecule that participates as a final electron acceptor in the transport chain complexes (Gibbs and Greenway, 2003); therefore, when oxygen availability decreases dramatically in the rhizosphere (anoxia), oxidative phosphorylation is inactivated (Moller, 2001) and terminal oxidase (COX) is inhibited (Liao and Lin, 2001; Greenway and Gibbs, 2003; Mancuso and Marras, 2006; Mailloux *et al.*, 2007). Under these conditions, plants have developed alternative pathways to aerobic respiration in order to maintain glycolysis through the regeneration of NAD+ (Gupta *et al.*, 2009) and thus obtain the energy needed to maintain key processes of metabolism such as membrane integrity and thus ion selectivity (Tadege *et al.*, 1999), protein synthesis and turnover and regulation of cytosolic pH, among others (Blokhina *et al.*, 2003). These alternative pathways are known as fermentations.

Three fermentative pathways are known (Greenway and Gibbs, 2003) that are active under conditions of oxygen deficiency alcoholic fermentation, lactic fermentation (Sousa and Sodek, 2002) and the alanine pathway; the last one is specific to plants and its final product is alanine from glutamate and the pyruvate *et al.*, 1999). There are also other final products of fermentation such as succinate, malate, and γ-aminobutynic acid (Dennis *et al.*, 2000). From the energy point of view, fermentative pathways can produce between 5% and 10% of the ATP per mole of glucose oxidized in aerobic respiration (Sousa and Sodek, 2002). In the roots of grapevines, for example, under normal conditions about 22.5 nmol ATP g^{-1} fresh weight are obtained, but in anoxia only 2 nmol ATP g^{-1} fresh weight are produced (Geigenberger, 2003).

The activation of a fermentation pathway is usually initiated when an anaerobic event occurs (Tadege *et al.*, 1999; Bailey-Serres and Voesenek, 2008); later, the pyruvate is reduced to lactic acid by lactate dehydrogenase in the cell's cytoplasm (Tadege *et al.*, 1999). Lactate dehydrogenase acts at a pH near 7.4 (Kato-Noguchi and Morokuma, 2007) and is responsible for maintaining the redox balance without the loss of carbon associated with alcoholic fermentation. However, the accumulation of this enzyme leads to a decrease in cytosolic pH to levels between 6.4 and 6.8 (Tadege *et al.*, 1999). Acidification of the cytoplasm inactivates lactate dehydrogenase and activates pyruvate decarboxylase (Roberts *et al.*, 1984), which decarboxylates pyruvate and forms acetaldehyde, which in turn is finally reduced to ethanol by alcohol dehydrogenase (Felle, 2005). In *Vitis riparia* Michx. and *V. rupestri* Scheele plants subjected to anoxia during 24 h, both species showed high levels of ethanol as the principal component of fermentation, yielding 225 and 175 μmol g^{-1} fresh weight (*V. riparia* and *V. rupestri,* respectively) at 20 h of treatment. Under normal conditions the level was 40 μmol g^{-1} fresh weight (Tadege *et al.*, 1999; Kato-Noguchi and Morokuma, 2007). The importance of alcoholic fermentation in tolerance to oxygen deficiency has also been shown in rice, where the response of different cultivars to 48 h of anoxia has been demonstrated. This study showed that at the end of the treatment lactate dehydrogenase activity was insignificant.

Evaluation in *Arabidopsis* mutants has indicated that overexpression of pyruvate decarboxylase and alcohol dehydrogenase promotes stress tolerance during oxygen deficiency, because the carbon flux is controlled through alcoholic fermentation and through the lactate and alanine pathways (Kato-Noguchi and Morokuma, 2007). Nevertheless, additional information about the special features associated with fermentative pathways, and specifically alcoholic fermentation, where the absence of oxygen would not be responsible for inducing enzymes such as pyruvate decarboxylase and alcohol dehydrogenase, is highly desirable (Ismond *et al.*, 2003; Kursteiner *et al.*, 2003).

8.9. Effect of Higher Day and Night Temperature on Respiration

Photosynthesis and respiration of plants and microbes increase with temperature, especially in temperate latitudes. As respiration increases more with increased temperature than does photosynthesis, global warming is likely to increase the flux of carbon dioxide to the atmosphere which would constitute a positive feedback to global warming. In tomato (*Lycopersicon esculentum*) (Behboudian and Lai, 1994) and cotton (*Gossypium hirsutum*) (Thomas *et al.,* 1993), elevated CO_2 increased dark respiration possibly because of increased carbohydrate accumulation in tissues. The latter has been shown to increase alternative pathway respiration as well (Amthor, 1991).

Apparent dark respiration may decline under elevated CO_2 if there is dark CO_2 fixation or if elevated CO_2 directly inhibits or inactivates respiratory enzymes as may occur through increased formation or carbamate (Wullschleger *et al.,* 1994). Few studies have successfully partitioned the effects of elevated CO_2 on growth and maintenance respiration. Both components appear to decline, probably because of decrease in leaf protein levels which results in reduced construction and maintenance costs (Wullschleger *et al.,* 1994). Elevated CO_2 reduced maintenance respiration of *Medicago sativa* and *Dactylis glomerata* at lower temperatures (15 to 20°C), whereas elevated CO_2 reduced growth respiration of *M. sativa* at 20 to 30°C and *D. glomerata* at 15 to 25°C (Ziska and Bunce, 1993).

It has been projected that the global temperature is likely to increase by 1.4 to 5.8°C because of projected increase in the concentrations of all greenhouse gases by the end of the 21st century (IPCC, 2001; Section 9.3.3, p. 555). Night time temperatures are projected to increase more than daytime temperatures (Alward *et al*., 1999). A rise in night time temperature by 1°C can reduce rice grain yield by 10% (Peng *et al*., 2004). Thus, identifying and developing management practices, such as the application of agrochemicals to prevent or negate the negative effects of high night time temperatures, can be beneficial for worldwide rice production and food stability (Mohammad and Tarpley, 2009). A projected increase in plant respiration in response to climate warming is of serious concern, as respiratory processes could consume a larger portion of total photosynthates (Paembonan *et al*., 1992).

Respiration is typically partitioned into the functional components of construction (growth) and of maintenance and ion uptake to facilitate our understanding of the impact of the environment on respiratory processes (Lambers, 1985; Amthor, 1986). Maintenance respiration is mainly associated with turnover of proteins and lipids and maintenance of ion concentration gradients across membranes (Penning de Vries, 1975) and is the most responsive to environmental changes (Ryan, 1991). High night time temperatures (HNT) are considered to be disadvantageous because they can stimulate respiration, *et al.*, 2002). Increased respiration rates as a result of high temperatures can lead to production of reactive oxygen species (ROS) in many plants (McDonald and Vanlerberghe, 2005). Under normal physiological conditions, the toxic effects of ROS are minimized by enzymic and non- enzymic antioxidants; however, under stress conditions oxidant levels can overwhelm the antioxidant levels, leading to cell damage (Kreiner *et al.*, 2002). Increased cell damage as a result of ROS can decrease membrane thermal stability (MTS), thereby disrupting water, ion, and organic solute movement across plant membranes, thus affecting carbon production, consumption, transport, and accumulation (Christiansen, 1978). A common method of evaluating damage to membranes is by examining MTS, which measures electrolyte leakage from tissues, such as leaves, subjected to stresses such as drought (Blum and Ebercon, 1981), heat (Sullivan, 1972), and freezing (Dexter, 1956). Membrane thermal stability was positively associated with yield performance in wheat (*Triticum esculentum* L.) under heat-stressed conditions (Reynolds *et al.*, 1994). The antioxidant concentration of a plant is closely associated with its stress tolerance in some circumstances (Smirnoff, 1995). The severity of ROS-induced damage depends on the antioxidant status of the plant. Plants pretreated with α-tocopherol (vitamin E), glycine betaine (GB), or salicylic acid (SA), the potential preventive exogenous effectors (chemicals) used in this study, showed induced thermal tolerance and protection against oxidative damage (Diaz-Zorita *et al.*, 2001; Larkindale and Knight, 2002). The α-tocopherol is one of the most effective single-oxygen quenchers and is a strong antioxidant, whereas GB enhances tolerance to high temperatures by protecting certain enzymes (*e.g.*, RuBiSCo and citrate synthase) against heat-induced inactivation (Caldas *et al.*, 1999; Makela *et al.*, 2000). Salicylic acid plays an essential role in preventing oxidative damage in plants by detoxifying superoxide radicals (Bowler *et al.*, 1992) and stabilizing trimers of heat shock transcription factors (Larkindale and Knight, 2002). Salicylic acid is also involved in calcium signaling (Kawano *et al.*, 1998) and induces thermal tolerance (Larkindale and Knight, 2002). Despite the importance of antioxidants in stress tolerance, little is known about the response of rice thermal tolerance to these preventative exogenous effectors (vitamin E, GB, and SA).

High night temperature decreases crop production by decreasing photosynthetic function, sugar and starch content (Loka and Oosterhuis, 2010; Turnbull *et al.*, 2002), increasing respiration rate (Mohammed and Tarpley, 2009b), suppressing floral bud development (Ahmed and Hall, 1993), causing male sterility and low pollen viability and hastening crop maturity (Mohammed and Tarpley, 2009a). Under normal physiological conditions, the toxic effects of reactive oxygen species (ROS) are minimized by enzymatic and non-enzymatic antioxidants. Under stress conditions, oxidant levels can overwhelm the antioxidant levels leading to cell damage. The increased production of ROS [oxide radical (O_2^-), H_2O_2, and the hydroxyl radical ($^{\cdot}OH$)], or the plant's decreased ability to neutralize ROS, as a result of heat stress negatively affects many physiological processes in plants, thus decreasing yield Razak *et al.*, 2011. Increase in respiration from climate warming is of serious concern as respiratory processes can consume a larger portion of total photosynthates (Paembonan *et al.*, 1992). On average, the carbon lost from respiratory metabolism within an individual plant ranges between 30 and 70% of the carbon gained through photosynthesis (Peterson and Zelitch, 1982). Respiration is typically partitioned into the functional components of construction (growth), maintenance and ion uptake to facilitate our understanding of the impact of the environment on respiratory processes (Amthor, 1986; Farrar, 1985; Lambers, 1985). Maintenance respiration is mainly associated with turnover of proteins and lipids and maintenance of ion concentrations across membranes (Penning de Vries, 1975). It is the most responsive to environmental changes among the functional components of respiration (Ryan, 1991). At high temperatures, the cost of maintenance increases to support protein turnover and to maintain active ion fluxes across the membranes (Penning de Vries, 1975), thereby increasing maintenance respiration. An increase in night temperature from 27°C to 32°C increased respiration rates by 40% in rice leaves. Earlier it was shown that an increase in maintenance respiration with warmer nights (Frantz *et al.,* 2004; Loka and Oosterhuis, 2010, Mohammed and Tarpley, 2009b) using rice, cotton, lettuce (*Lactuca sativa* L.), tomato and soybean indicated. Hence,, hugh night temperature can stimulate respiration rates, thereby negatively affecting the yield (Zheng *et al.*, 2002). The other consequence of increased respiration is increased production of ROS.

The production of ROS is an unavoidable consequence of aerobic respiration, with the majority of the ROS produced in photosynthetic tissue (mostly leaves) in the dark by mitochondrial electron transport chain activity (McDonald and Vanlerberghe, 2005). The production of ROS by mitochondria has been shown to increase in many plants as a result of biotic and abiotic stresses (McDonald and Vanlerberghe, 2005). An increase in night temperature increases maintenance respiration and thus increases production of ROS. Physiological injury due to abiotic stress has been associated with increases in oxidative damage to the membrane in plant species (Larkindale and Knight, 2002). Plants increase maintenance respiration to support repair mechanisms of the membranes due to oxidative damage (Amthor and McCree, 1990). Thus, an increase in respiration occurs with an increase in temperature (Huang *et al.*, 1998) to support repair mechanisms of the membranes due to oxidative damage (Amthor and McCree, 1990). In addition, ROS interferes with photosynthesis and respiration by disrupting water, ion, and organic solute movement across plant membranes by affecting membrane stability (Christiansen, 1978).

Destabilized membranes are leaky membranes, thus interpreting the amount of electrolytic leakage from the membrane can be used as an indicator of cell membrane stability (Sullivan and Ross, 1979). Many studies have examined stability loss of the membranes subjected to environmental stresses by measuring electrolytic leakage from the membranes (Ibrahim and Quick, 2001; Ismail and Hall, 1999). An increase in night temperature from 27^{0}C to 32^{0}C increased injury to the membrane by 60% in rice leaves. Similar results of increased injury to the membrane as a result of heat stress were seen in many crop species (Ibrahim and Quick, 2001; Ismail and Hall, 1999; Mohammed and Tarpley, 2009b). Mohammed and Tarpley (2009b) and Reynolds *et al.* (1994) positively associated membrane stability with yield performance under heat-stressed conditions in rice and wheat.

8.10. Respiration Under Low Temperature Stress

Photosynthetic process is a thermosensitive function in higher plants (Havaux, 1993) and green algae (Gaevskell *et al.*, 1986). The primary effect of chilling and none freezing temperatures (*e.g.* 0-13°C), on photosynthesis are apparent within some important processes. Some of these effects were observed as inactivation of Calvin-cycle enzymes through oxidation in the stroma (Wise, 1995), delay in circadian rhythm of sucrose phosphate synthase activity in the cytosol (Jones *et al.*, 1998) and oxidative damage to PSI and PSII reaction centers (Ortiz Lopez *et al.*, 1990; Kudoh and Sonoike, 2002).

Upon ideal conditions, high irradiance (full sunlight) may be required to cause an accelerated damage to photosynthesis, but under environmental conditions, a stress, which reduces CO_2-fixing capacity (*e.g.* chilling), will result in a hypersensitivity to light (Kyle and Ohad, 1986). Under normal turnover conditions, oxidative damage to D1 (reaction center protein of PSII) comes along with the repair process. When the rate of repair can keep pace with the rate of damage, no inhibition of electron flow is observed. Under such conditions there is a maximum potential for photosynthesis (light-saturated rates). Chilling can also decrease the D1 repair rates. Any decrease in CO_2-fixing or in the repair rate of D1 protein can shift the maximum light intensities needed for photosynthesis saturation to a much lower intensities (Kyle and Ohad, 1986).

Respiration rates of plant cells are affected by temperature (Rasion, 1980) and there is a linear increase in O_2 uptake with temperature (Buetow, 1962). In addition to short-term responses, species grown at low temperatures often show higher rates of respiration than species grown at higher temperatures when measured at the same temperature (Collier and Cummins, 1990). This stimulation of respiration was reported as adaptive behavior of plants grown at cold conditions compared with related species from warmer conditions (McNulty and Cummins, 1987). The increased rate of respiration at low temperatures in mitochondria involves two energy dissipating systems, Alternative Oxidase (AOX) pathway (Purvis and Shewfelt, 1993) and Plant Uncoupling Mitochondrial Protein (PUMP) (Vercesi, 2001). This stimulation of respiration causes a decrease in the yield of ATP synthase which is linked to an increase in heat dissipation (Calegario *et al.*, 2003). These systems are also involved in removal of Reactive Oxygen Species (ROS), because they do not generate a proton electrochemical gradient (Möller, 2001).

At chilling temperatures, plants show many changes including in the process of respiration (Lukatkin *et al.*, 2012). There is evidence of decline in temperature, occurring as a result of destruction of the mitochondria structure, the general lowering of kinetic energy, and the inhibition of some enzymes (Prasad *et al.*, 1994; Lawrence and Holaday, 2000; Munro *et al.*, 2004). Others have observed that an increase in respiratory activity during chilling and prolonged elevation of the respiration rate after cold exposure may indicate irreversible metabolic dysfunction and accumulation of incompletely oxidized intermediates (Steward *et al.*, 1990; Yadegari *et al.*, 2008). The mechanism of stimulation is unknown, but it is possible to assume that it was the result of uncoupling of oxidative phosphorylation (Wang, 1982). It is also possible that the increased respiration reflects a reaction to the transfer of plants from chilling temperatures to the higher temperatures (Zauralov and Lukatkin, 1997). As a result of decreased respiration and increased consumption of energy-rich phosphates at chilling temperatures is a reduction of ATP levels (Takeda *et al.*, 1995; Lawrence and Holaday, 2000). Cold-tolerant crop species have greater temperature homeostasis of leaf respiration than cold-sensitive species (Yamori *et al.*, 2009). Chilling reduces the cytochrome path of the electron transport in seedlings (Prasad *et al.*, 1994; Reyes and Jennings, 1997) and enhances alternative respiratory pathways (Ordentlich *et al.*, 1991; Purvis and Shewfeld, 1993; Gonzalez-Meier *et al.*, 1999; Ribas-Carbo *et al.*, 2000). Perhaps these alternative pathways play an important role in plant adaptation to chilling (Steward *et al.*, 1990). They are triggered at the chilling period and increase with decreasing temperature (Ordentlich *et al.*, 1991). These alternative pathways induced by chilling caused a decrease in superoxide generated in mitochondria (Purvis and Shewfelt, 1993; Hu *et al.*, 2008).

Exposure of coffee plants to non-freezing temperature has a depressing effect on growth, photosynthetic performance and yield. Besides that, low temperatures can damage root system, with a consequent leaf death. Plants have mechanisms of cold acclimation that promote quantitative and qualitative modifications of membrane lipids, increased activity of antioxidant enzymes and increased ability to dissipate excess energy. In fact, the existence of different acclimation abilities within the *Coffea* genus seems to be related to the tolerance of the photosynthetic apparatus in face of oxidative stress conditions (Ramalho *et al.*, 2003). The presence of these mechanisms is essential for the implementation and maintenance of plant species in a region, influencing plant performance and agricultural practices, as well as the economical viability of the crop. Thus, this knowledge is important for the selection and management of the genus *Coffea* (Praxedes *et al.*, 2006).

8.11. Effect of Salt Stress on Respiration

Salt stress is one of the major abiotic stress factors that affect almost every aspect of physiology and biochemistry of a plant, resulting in a reduction in its yield (Foolad, 2004). Thus it is a serious threat to agricultural productivity especially in arid and semi-arid regions (Parvaiz *et al.*, 2008). Soil contaminated with salts (ECe > 4 dS m-1 or 40 mM NaCl or osmotic potential < 0.117 MPa) are with defined as saline land, which directly affects plant growth and development in vegetative stage prior to reproductive stage, especially crop species (Sairam and Tyagi, 2004; Chinnusamy *et al.*, 2005; Ashraf *et al.*, 2008; Ashraf, 2009). Salinity in soil and water is irrevocably associated with irrigated agriculture throughout the world and salt management becomes an integral part of the production system and cause slow growth and injure leaf cells (Hasegawa *et al.*, 2000). The salinity-induced crop yield reduction takes place due to a number of physiological and biochemical dysfunctions and have been listed in a number of comprehensive reviews on salinity effects and tolerance in plants (Ashraf *et al.*, 2008; Munns and Tester, 2008; Jamil *et al.*, 2011). The salinity effect on the water stress of the plant, its gaseous exchanges and its metabolism has been analyzed over short periods. Some results mentioned that salinity decrease of sugars store and as a result it can distort the respiration metabolism of embryo growth (Kazerouni *et al.*, 2005). Reduced availability of water, increased respiration rate, altered mineral distribution, membrane instability, failure in the maintenance of turgour pressure are some of the events that prevails during this stress episode. Plants try to withstand these stresses either by tolerating it or by adopting a dormant stage (Cuartero *et al.*, 2006).

Salt stress affects many physiological aspects of plant growth. Shoot growth was reduced by salinity due to inhibitory effect of salt on cell division and enlargement in growing point (Mccue, and Hanson, 1990). Toxic effect of salts may change enzymatic activity and hormonal balance of plants (Moud and Maghsoudi, 2008). Net photosynthesis is decreased due to the reduction in photosynthesis and increase in respiration per unit leaf area. Studies have reported that increasing salt concentration in growing medium of pea plants increased root and stem respiration (Nieman, 1962; Livne and Levin, 1966). Soaking seeds before germination in ascorbic acid and thiamin solutions have been suggested as a remedy to increase seedling growth due to partly to reduction in dark respiration under saline condition (Hamada, 1998).

Under high salt concentration stress, net carbon assimilation rate is not affected, although stomatal conductance and, consequently, transpiration, were lower at higher salinity levels. The lower biomass of salt-stressed plants was thus not associated with lower net carbon assimilation (Houle *et al.*, 2001). This suggests that respiration, particularly root and maintenance respiration were significantly higher for the salt-stressed than for the control plants, resulting in lower C-accumulation (Wang *et al.*, 1977; Epron *et al.*, 1999). Water-use efficiency is reported to be higher for plants exposed to higher salinities, suggesting an adjustment within the plant, such as stomatal control of water losses, to conserve water (Ayala and O'Leary, 1995). Environmentally induced changes in specific leaf area (leaf surface/leaf mass) may also lead to changes in water use efficiency (Stanhill, 1986); however, the higher water use efficiency of salt-stressed plants was not associated with a lower specific leaf area. Plant growth was reduced, but net carbon assimilation rate was not affected by high salinity levels. Increased root respiration and respiratory costs associated with salt tolerance might have contributed to lower carbon accumulation at higher salinity levels (Houle *et al.*, 2001). High concentrations of salinity have often been reported to increase in respiration. This increase in respiration is greater in salt sensitive than salt tolerant species (Sabbagh *et al.*, 2014).

High salt levels do not only lead to damaging effects on plants but also increase the pH level of soil. Most crop plants do not grow well under high pH levels. Salt stress also leads to deterioration of soil structure and hinders desirable air-water balance essential for biological processes occurring at plant roots. As a result of all the detrimental effects of salinisation, crop yields are decreasing, while arable land is being lost irreversibly (Supper, 2003). Salt stress causes various effects on plant physiology such as increased respiration rate, ion toxicity, changes in plant growth, mineral distribution, and membrane instability resulting from calcium displacement by sodium (Marschner, 1986), membrane permeability (Gupta *et al.*, 2002), and decreased photosynthetic rate (Hasegawa *et al.*, 2000; Munns, 2002; Ashraf and Shahbaz, 2003; Kao *et al.*, 2003; Sayed, 2003). Salt stress affects plant physiology at whole plant as well as cellular levels through osmotic and ionic stress (Hasegawa *et al.,* 2000; Muranaka *et al.,* 2002a; Ranjbarfordoei *et al.,* 2002; Murphy and Durako, 2003). Despite causing osmotic and ionic stress, salinity causes ionic imbalances that may impair the selectivity of root membranes and induce potassium deficiency (Gadallah, 2000).

The accumulation of high amounts of toxic salts in the leaf apoplasm leads to dehydration and turgor loss, and eventually death of leaf cells and tissues (Marschner, 1995). As a result of these changes, the activities of various enzymes and plant metabolism are affected (Lacerda *et al.,* 2003). At high rates of transpiration, the xylem of all species contains much lower chloride and sodium concentrations than those in the external saline medium. Salt stress enhances the accumulation of sodium chloride in chloroplasts of higher plants, affects growth rate, and is often associated with decrease in photosynthetic electron transport activities (Kirst, 1989). In higher plants, salt stress inhibits photosystem PS-II activity (Kao *et al*., 2003; Parida *et al*., 2003), although some studies showed contrary results (Brugnoli and Björkman, 1992; Morales *et al*., 1992). The reduction of plant growth and dry-matter accumulation under saline conditions has been reported in several important grain legumes (Tejera *et al*., 2006).

8.11.1. Reactive Oxygen Species

Exposure of plants to salt stress can up-regulate the production of reactive $_2O_2$ (hydrogen peroxide), O_2^- (superoxide), $1O_2$ (singlet oxygen) and .OH (hydroxyl radical). Excess ROS causes phytotoxic reactions such as lipid peroxidation, protein degradation and DNA mutation (McCord, 2000, Wang *et al*., 2003; Vinocur and Altman, 2005; Pitzschke *et al*., 2006). In plant cells, ROS, mainly H_2O_2, superoxide anion (O_2^-), and hydroxyl radical (OH) are generated in the cytosol, chloroplasts, mitochondria, and the apoplastic space (Bowler and Fluhr, 2000; Mittler, 2002), while ROS have the potential to cause oxidative damage to cells during environmental stress. Recent studies have shown that ROS play a key role in plants as signal transduction molecules involved in mediating responses to pathogen infection, environmental stresses, programmed cell death and developmental stimuli (Mittler *et al.,* 2004; Torres and Dangl, 2005).

Membrane injury induced by salt stress is related to an enhanced production of highly toxic ROS (Shalata *et al*., 2001). A rise in ROS production may result from stomatal closure, causing a decrease in CO_2 concentration inside the chloroplasts. This in turn causes a decrease in $NADP^+$ concentration with the concomitant generation of ROS (Foyer and Noctor, 2003). The increased concentration of ROS damages the D1 protein of PS II leading to photoinhibition.

Stress enhanced photorespiration and NADPH activity also contributes to increase in H_2O_2 accumulation, which may inactivate enzymes by oxidizing their thiol groups. This toxicity of H_2O_2 is not due to its reactivity alone, but requires the presence of a metal reductant to form the highly reactive hydroxyl radical (OH), which has the ability to react with all biological molecules (Halliwell and Gutteridge, 1989). Salinity-associated reductions in elongation in the expansion zone of maize leaves are associated with reduced ROS levels and could be alleviated by the addition of ROS (Rodrýguez *et al.*, 2004).

8.11.2. Physiological and Biochemical Processes

Soil salinity affects various physiological and biochemical processes which result in reduced biomass production. This adverse effect of salt stress appears on whole plant level at almost all growth stages including germination, seedling, vegetative and reproductive stages. However, tolerance to salt stress at different plant developmental stages varies from species to species. For example, it has been observed that the degree of salt tolerance at different developmental growth stages varies in rice (Akbar and Yabuno, 1977), barley (Norlyn, 1980) and wheat (Ashraf and Khanum, 1997). In contrast, salt tolerance in some other crops such as *Medicago sativa*, *Trifolium alexandranium* and *Trifolium pratense* examined at the seedling stage was also confirmed at the later growth stages (Ashraf *et al.*, 1986). Similarly, while working with safflower, Ashraf and Fatima (1995) also found that salt tolerance does not vary at different plant growth stages in these plants.

There are reports for variation in salt tolerance in a number of crop species depends on the extent of Na^+ exclusion at root level or ability to compartmentalize salts in the vacuole (Munns, 2002, 2005; Ashraf, 2004). For example, Wyn Jones *et al.* (1984) found that the higher salt tolerance of *Agropyron junceum* to that of *Agropyron intermedium* was related to its efficient exclusion of both Na^+ and Cl^-. In another study, Carden *et al.* (2003) found that the salt tolerant variety maintained a 10-fold lower cytosolic Na^+ in the root cortical cells than the more sensitive variety. It is well established that high accumulation of Na^+ in shoots inhibits enzyme activity, and other metabolic processes such as protein synthesis and photosynthesis (Ashraf, 2004; Munns, 2005) thereby reducing leaf growth or causing leaf death. Thus, in most plant species, particularly glycophytes, Na^+ exclusion from the shoot and retention in the root is a general trend and hence an important component of salt tolerance (Ashraf, 2004). However, Mansour *et al.* (2005) found that salt induced increase in Na^+ accumulation compared with a decrease in K^+ and Ca^{2+} was higher in salt tolerant maize cultivar Giza 2 compared with that in salt sensitive Trihybrid 321. A number of studies have shown that photosynthetic capacity of different species is reduced due to salinity (Ashraf, 2004; Dubey, 2005). It is evident that higher photosynthetic capacity causes increased plant growth under normal or stress conditions as has earlier been observed in a number of plant species *e.g.*, in cotton (Pettigrew and Meredith, 1994), *Zea mays* (Crosbie and Pearce, 1982), *Brassica* spp. (Nazir *et al.*, 2001) and wheat (Raza *et al.*, 2007). Furthermore, salt-induced reduction in photosynthesis could be due to stomatal and non-

$^{+}$ and Cl^{-} in

the leaves also reduces photosynthetic capacity and Na^{+} content in the leaves of rice (Yeo, 1998), and wheat (James *et al.*, 2002), while high Cl^{-} contents in the citrus (Walker *et al.*, 1981) and in the chloroplast of *Phaseolous vulgaris* Seemann and Critchley, 1985) were found to be detrimental to photosynthesis. In view of all these reports, it can be concluded that growth inhibition may occur due to both osmotic and toxic effects. However, osmotically induced reduction in growth occurs at early growth stages under salt stress. Furthermore, photosynthesis is also one of the main contributing factors in salt-induced reduction in plant growth and yield. Tolerance of photosynthetic system to salinity depends on how effectively plant excludes or compartmentalizes the toxic ions. However, the extent of the adverse effects of salt stress on photosynthesizing tissue or on growth varies with the type of species, level of stress and duration of stress.

There has been some controversy regarding the main physiological targets responsible for photosynthetic impairment under drought and/or salinity (Flexas and Medrano, 2002, Lawlor and Cornic, 2002). However, there is now substantial consensus that reduced CO_2 diffusion from the atmosphere to the site of carboxylation is the main cause for decreased photosynthesis under most water stress conditions (Chaves and Oliveira, 2004, Flexas *et al.*, 2004a) and were due to at least two components that are regulated almost simultaneously: stomatal closure and reduced mesophyll conductance (Flexas *et al.*, 2006). Although the former has been known for a long time to be one of the first responses of plants to soil water shortage, the latter has only recently been recognized as an equally important cause for reduced CO_2 diffusion, under drought (Flexas *et al.* 2002, Warren *et al.*, 2004) and salinity (Centritto *et al.*, 2003).

Although restricted CO_2 diffusion across leaves is likely to be the most usual cause for decreased photosynthesis rates under water and salinity stress, metabolic impairment may also occur, particularly under severe stress. Generally, the respiration rate decreases during water and salinity stress, due to reduced photosynthate assimilation and growth needs. However, this behaviour may be somewhat species dependent, and respiration rate can also increase, particularly under severe water stress (Flexas *et al.,* 2005, Ghashghaie *et al.,* 2001). Nevertheless, the total respiration rates are usually kept within a narrower range than those of photosynthesis during water stress development, resulting in a progressively increased respiration to photosynthesis ratio, *i.e.* a decreased carbon balance (Flexas *et al.*, 2005). Plants possess two respiratory pathways: the energy-conserving, cyanide-sensitive, cytochrome pathway and the energy-wasteful, cyanide-resistant, alternative pathway (Lambers *et al.*, 2005). In soybean it has been demonstrated that severe water stress induces a sharp decrease in the cytochrome respiration rate concomitant with a similar increase in the alternative respiration rate, so that total respiration rate remains quite constant (Ribas-Carbo *et al.*, 2005). Moreover, these changes are not accompanied by increases in alternative oxidase protein content, indicating that these changes may be regarded as a biochemical regulation. The most surprising aspect is that the observed changes in mitochondrial electron partitioning during water stress only occur when stomatal conductance drops below 0.1 mol H_2O m^{-2} s^{-1}, thus in coincidence with the observed threshold for photosynthesis metabolic impairment.

8.12. Effect of High CO_2 Concentration on Respiration

With increase in global atmospheric CO_2 concentrations of 43% from the μ mol mol^{-1} in 1750 to the present level of 400 μ mol mol^{-1} (an annual increase of 1.35%), the global CO_2 concentration has increased by about 1.55 ppm CO_2 per year over the past 55 years (Xu *et al.*, 2015). It continues to be elevated at an unprecedented pace of $<$–1.0 μ mol mol^{-1} per year, as a result of the further increase in the cumulative emissions of CO_2 to the atmosphere during the 21st century (400 μ mol mol^{-1} in 2011 vs.936 μ mol mol^{-1} in 2100; IPCC, 2013; NASA, 2014). Meanwhile, the global mean surface temperature is expected to increase by 2.6–4.8°C by the end of the 21st century (2081–2100), relative to the 1986–2005 level (IPCC, 2013). The climate changes, such as elevated CO_2, rising temperature, and altered precipitation, have resulted in drastic impacts on the natural ecosystems, such as in vegetation function, sustainable food production, and crop yields (Lobell *et al.*, 2011; Peñuelas *et al.*, 2013; Ruiz-Vera *et al.*, 2013; Xu *et al.*, 2013a, 2014; Lavania *et al.*, 2015), leading to more profound impacts when the climate changes are combined with other environmental constraints, such as air pollution, nutrition limitation, and their interactions (Gillespie *et al.*, 2012; Peñuelas *et al.*, 2012; Xu *et al.*, 2013a, b; Wang *et al.*, 2015). Photosynthesis and respiration are two fundamental physiological processes of plants, because the former involves initial carbon fixation, light energy transfer, and oxygen release, and the latter works on carbon efflux, energy production, and the relevant substrate metabolisms, such as those providing the carbon skeleton. They play a critical role in balancing the carbon budget and maintaining the carbon sink in terrestrial ecosystems, as well as in the response and feed back to climate change (Prentice *et al.*, 2001; Sage, 2004; Long *et al.*, 2006; Atkin *et al.*, 2010; Atkin, 2015).

Rising CO_2 has affected almost all crucial biological processes, including photosynthesis, respiration, and antioxidant systems, as well as other key secondary metabolisms in plants (Long *et al.*, 2004; Matros *et al.*, 2006; Peñuelas *et al.*, 2013; Singh and Agrawal, 2015). All other effects of elevated CO_2 on individual plants and ecosystems may be partly derived from these fundamental biological responses (Long *et al.*, 2004; Ainsworth and Rogers, 2007; Peñuelas *et al.*, 2013; Zinta *et al.*, 2014). Many reports have reviewed the biological responses to CO_2 enrichment, and their interactions with environmental change, including photosynthesis and stomatal behavior (Long *et al.*, 2004; Ainsworth and Long, 2005; Ainsworth and Rogers, 2007).

In a changing climate, rising atmospheric CO_2 and increasing temperature may be considered as the two main factors affecting plant respiration (Frantz *et al.*, 2004; Gonzalez-Meler *et al.*, 2004; He *et al.*, 2005). However, the response of plant respiration to elevated CO_2 concentration is subject to dispute (Liu *et al.*, 2011). Many investigator believe that plant respiration is reduced at high CO_2 concentration (Amthor, 1995; Reuveni and Bugbee, 1997; Wang *et al.*, 2002). However, Gonzalez-Meler *et al.* (2004) reviewed studies of plant respiration on elevated atmospheric CO_2 concentration and concluded that specific respiration rates are generally not reduced when plants are grown at elevated CO_2 concentration. Bunce (2005) also testified that elevated CO_2 concentration had no effect on night-time respiration through an experiment of soybeans grown in field plots at the current ambient and elevated CO_2 concentration in open top chambers. In contracts, studies on the effects of temperature on plant respiration are more conclusive. Bunce (2004) showed that low temperature reduced rates of respiration, increased temperature increased rates of respiration. Bunce (2007) concluded that short-term increase in temperature early in the dark period led to exponential increases in rates of respiration. For rapidly growing plants, Frantz *et al.* (2004) found that the whole-plant respiration was slightly sensitive to temperature and that sensitivity does not change among species tested, even after 20 days of treatment. This can be explained by the classical respiration theory, that is, whole plant respiration R (g CH_2O m^{-2} d^{-1}) can be separated into growth (R_g) and maintenance respiration (R_m) (McCree, 1974). Growth respiration is proportional to the total assimilation (P) and maintenance respiration is proportional to the dry mass (W). This gives:

$$R = R_g + R_m = kP + cW$$

Where k is the coefficient for growth respiration and c is the coefficient for maintenance respiration. The growth respiration is the cost of converting the immediate products of photosynthesis into plant materials. The rate of the metabolic process of conversion will obey the rule of temperature dependence.

The maintenance respiration refers to the CO_2 that results from protein breakdown, plus the CO_2 produced in the respiratory process that provide energy for maintenance processes including cellular structure and gradients of ions and metabolites and also the processes of physiological adaptation that maintain cells and active units in changing environments (Penning de Vries, 1975). The coefficient c varied with many biotic and environmental factors including temperature, nitrogen status, water status (Amthor, 1985), salt stress (Moud and Maghsoudi, 2008) and different plant organs (Ploschuk and Hall, 1997).

References

Abd Elgawad, H., and Asard, H. 2013. Elevated CO_2 attenuates oxidative stress caused by drought and elevated temperature in four C3 plant species. *Biotechnologia* 94: 156–202.

Ahmad, P. and Prasad, M.N..V. 2012a. Abiotic stress responses in plants: Metabolism, productivity and sustainability. New York, NY: Springer Science Business Media.

Ahmad, P. and Prasad, M.N.V. 2012b. Environmental adaptations and stress tolerance in plants in the era of climate change. New York, NY: Springer Science Business Media, LLC;

Akbar, M. and Yabuno, T. 1977. Breeding saline-resistant varieties of rice. IV. Inheritance of delayed type panicle sterility induced by salinity. *Jpn. J. Breed.*, 27: 237-240.

Allen, C.D., Macalady, AK., Chenchouni, H., Bachelet, D., McDowel, l N., Vennetier, M., Kitzberger, T., Rigling, A., Breshears, D.D. and Hogg, E.H. (Ted) 2010. A global overview of drought and heat-induced tree mortality reveals emerging climate change risks for forests. *Forest Ecology and Management*, 259: 660–684.

Alward, R.D., Detling, J.K. and Milchunas, D.G. 1999. Grassland vegetation changes and nocturnal global warming. *Science*, 283: 229–231.

Amthor, J.S. 1984. The role of maintenance respiration in plant growth. *Plant, Cell and Environment,* 7: 561–569.

Amthor, J.S. 1986. Evolution and applicability of a whole plant respiration model. *J. Theor. Biol.* 122: 473–490.

Amthor, J.S., 1989. Respiration and Crop Productivity. Springer-Verlag, New York.

Amthor, J.S. and McCree, K.J. 1990. Carbon balance of stress plants: A conceptual model for integrating research results, In, "*Stress Response in Plants: Adaptation and Acclimation Mechanism*", Eds. R.G. Alscher and J.R. Cumming, Alan R. Liss, New York, USA. pp. 1-15.

Amthor. J.S. 1991. Respiration in a future higher CO_2 world. *Plant Cell Environ.* 14: 13-20.

Amthor, J.S., 1995. Terrestrial higher-plant response to increasing atmospheric $[CO_2]$ in relation to the global carbon cycle. *Global Change Bio.*, 1: 243-274.

Anderegg, W.R.L. Kane, J.M. and Anderegg, L.D.L. 2013. Consequences of widespread tree mortality triggered by drought and temperature stress. *Nature Climate Change*, 3: 30–36.

Andrews, T.J. and Lorimer, G.H. 1987. Rubisco: structure, mechanism and prospects for improvement. In, "*The Biochemistry of Plants*", Eds. M.D. Hatch, N.K. Boardman, Academic press, New York, USA, pp. 131-218.

Ashraf, M., McNeilly, T. and Bradshaw, A.D. 1986. Response and ion uptake of selected salt tolerant and unselected lines of three legume species. *New Phytol.*, 104: 463-472.

Ashraf, M. 1994. Breeding for salinity tolerance in plants. *Crit. Rev. Plant Sci.*, 13: 17-42.

Ashraf, M. and Fatima, H. 1995. Responses of some salt tolerant and salt sensitive lines of safflower (*Carthamus tinctorius* L.). *Acta Physiol. Plant*, 17: 61-71.

Ashraf, M. and Khanum, A. 1997. Relationship between ion accumulation and growth in two spring wheat lines differing in salt tolerance at different growth stages. *J. Agron. Crop Sci.*, 178: 39-51.

Ashraf, M. 2004. Some important physiological selection criteria for salt tolerance in plants. *Flora*, 199: 361-376.

Ashraf, M.A., Ahmad, M.S.A., Ashraf, M., Al-Qurainy, F. and Asheaf, M.Y. 2011. Allevation of waterlogging stress in upland cotton (*Gossipium hirsutam* L.) by exogenous application of potassium in soil and as foliar spray. *Crop Pasture Sci*, 62(1): 25-38.

Aspinall, D. and Paleg, L.G 1981. Proline accumulation: Physiological aspects. In, "*The Physiology and Biochemistry of Drought Resistance in Plants*", Eds. L.G Paleg and D. Aspinall, Sidney: Academic Press, pp. 205-241.

Aspiazú, I., Sediyama, T., Ribeiro Jr., J.I., Silva, A.A., Concenço, G, Galon, L.V., Ferreira, E.A., Silva, A.F., Borges, E.T. and Araujo, W.F. 2010. Eficiência fotosintética y de uso del agua por malezas. *Planta Daninha*, 28(1): 87-92.

Atkin, O.K., Evans, J.R. and Siebke, K. 1998. Relationship between the inhibition of leaf respiration by light and enhancement of leaf dark respiration following light treatment. *Australian Journal of Plant Physiology*, 25: 437-443.

Atkin, O.K., Evans, J.R., Ball, M.C., Lambers, H. and Pons, T.L. 2000. Leaf respiration of snow gum in the light and dark. Interactions between temperature and irradiance. *Plant Physiology*, 122: 915-924.

Atkin, O., Millar, A. Gardestrom, P. and Day, D. 2004. Photosynthesis, carbohydrate metabolism and respiration in leaves of higher plants. In, "*Photosynthesis*", Eds. R. Leegood, T. Sharkey, and S. Caemmerer, Springer, Dordrecht, The Netherlands, pp. 153-175. .

Atkin, O.K. and Macherel, D. 2009. The crucial role of plant mitochondria in orchestrating drought tolerance. *Annals of Botany*, 103: 581–597.

Atkin, O., Millar, H.A. and Turnbull, M.H. 2010. Plant respiration in a changing world. *New Phytol.* 187: 268–272.

Atkin, O. 2015. New Phytologist and the 'fate' of carbon in terrestrial ecosystems. *New Phytol.* 205: 1–3.

Atkinson, N.J. and Urwin, P.E. 2012. The interaction of plant biotic and abiotic stresses: from genes to the field. *J. Exp. Bot.*, 63: 3523–3544.

Attridge, T.H. 1990. The natural environment. In, "*Light and Plant Responses*", Ed. T.H. Attridge, Edward Arnold, London, England, pp. 1-5.

Ayala, F. and O'Leary, J.W. 1995. Growth and physiology of *Salicornia bigelovii* Torr. at suboptimal salinity. *International Journal of Plant Sciences,* 156: 197–205.

Behboudian, M.H. and Lai, R. 1994. Carbon dioxide in 'Virosa' tomato plants; responses to enrichment duration and to temperature. *Hort. Sci.* 29: 1456-1459.

Ashraf, M., Athar, H.R., Harris, P.J.C. and Kwon, T.R. 2008. Some prospective strategies for improving crop salt tolerance. *Adv. Agron.*, 97: 45-110.

Ashraf, M. 2009. Biotechnological approach of improving plant salt tolerance using antioxidants as markers. *Biotech. Adv.*, 27: 84-93.

Ashraf, M.A. 2012. Waterlogging stress in plants. *African Journal of Agricultural Research,* 7(13): 1976-1981.

Bailey-Serres, J., Voesenek, L.A.C.J. 2008. Flooding stress: acclimations and genetic diversity *Annu. Rev. Plant Biol.*, 59: 313–339.

Barrett-Lennard, E.G., 2003. The interaction between waterlogging and salinity in higher plants causes, consequences and implications. *Plant and Soil,* 253: 35-54.

Bartoli, C.G., Pastori, G.M. and Foyer, C.H. 2000. Ascorbate biosynthesis in mitochondria in linked to the electron transport chain between complexes III and IV. *Plant Physiol.*, 123(1): 335-344.

Bartoli, C.G., Gomez, F., Gergoff, G., Gulaìt, J.J. and Puntarulo, S. 2005. Up-regulation of the mitochondrial alternative oxidase pathway enhances photosynthetic electron transport under drought conditions. *Journal of Experimental Biology*, 56: 1269-1276.

Beck, E.H., Fettig, S., Knake, C., Hartig, K. and Bhattarai, T. 2007. Specific and unspecific responses of plants to cold and drought stress. *Journal Biosciences*, 32(3): 501–510.

Bernstein, N., Eshel, A. and Beeckman, T. 2013. Effects of salinity on root growth. CRC Press, Boca Raton, Florida, USA.

Björkman, O. 1981. Response to different quantum flux densities. In, "*Encyclopedia of Plant Physiology, New Series*", Eds. O.L. Lange, P.S. Nobel, C.B. Osmond and H., Zeigler, Vol 12A. Berlin: Springer; pp. 57-107.

Blokhina, O., Virolainen, E. and Fagerstedt, K.V. 2003. Antioxidants, oxidative damage and oxygen deprivation stress: a review. *Ann. Bot.*, 91: 179–194.

Blum, A., and Ebercon. A. 1981. Cell membrane stability as a measure of drought and heat tolerance in wheat. *Crop Sci.*, 21: 43–47.

Blum, A. 1998. Improving wheat grain filling under stress by stem reserve mobilization. *Euphytica*, 100: 77-83.

Bouma, T.J., and De Visser, R. 1993. Energy requirements for maintenance of ion concentrations in roots. *Physiologia Plantarum*, 89: 133-142.

Bowler, C., Montagu, M.V. and Inze, D. 1992. Superoxide dismutase and stress tolerance. *Annu. Rev. Plant Physiol. Mol. Biol.*, 43: 83–116.

Bowler, C. and Fluhr, R. 2000. The role of calcium and activated oxygen as signals for controlling cross-tolerance. *Trends Plant Sci.*, 5: 241-246.

Boudsocg, M. and Lauriere, C. 2005. Osmotic signaling in plants: Multiple pathways mediated by emerging kinase families. *Plant Physiol.*, 138(3): 1185-1194.

Bogorad, L. 2003. Photosynthesis research: advances through molecular biology-the beginnings, 1975-1980s and on. *Photosynthesis Research,* 76(1-3): 13-33.

Bray, E.A., Bailey-Serres J. and Weretilnyk, E. 2000. Responses to abiotic stresses, In, "*Biochemistry and Molecular Biology of Plants*", Eds W. Gruissem, B. Buchnnan, and R. Jones, M.D. Rockville, American Society of Plant Physiologists, pp. 1158–1249

Brownleader, M.D., Harbone, J.B. and Dey, P.M. 1997. Carbohydrate metabolism: Primary metabolism of mitochondria. In, "*Plant Biochemistry*", Ed. P.M. Dey and J.B. Harbone, Academic Press, California, USA, pp. 111-140.

Brugnoli, E. and Bjorkman, O. 1992. Growth of cotton under continuous salinity stress: influence on allocation pattern, stomatal and non stomatal components and dissipation of excess light energy. *Planta*, 187: 335-347.

Bunce, J.A., 2004. A Comparison of the effects of carbon dioxide concentration and temperature on respiration, translocation and nitrate reduction in darkened soybean leaves. *Ann. Bot.*, 93: 665-669.

Bunce, J.A., 2005. Response of respiration of soybean leaves grown at ambient and elevated carbon dioxide concentrations to day-to-day variation in light and temperature under field conditions.*Ann. Bot.*, 95: 1059-1066.

Bunce, J.A., 2007. Direct and acclimatory responses of dark respiration and translocation to temperature. *Ann. Bot.*, 100: 67-73.

Buetow,D.E., 1962. Differential effects of temperature on the growth of *Euglena gracilis*. *Experimental Cell Research*, 27: 137-142.

Caldas, T., Demont-Caulet, N., Ghazi, A. and Richarme, G. 1999. Thermo protection by glycine betaine and choline. *Microbiol.*, 145: 2543–2548.

Calegario, F.F., Coso, R.G., Fagian, M.M., Almeida, V., Jardim, W.F., Jezek, P., Arruda, P. and Vercesi, A.E. 2003. Stimulation of potato tuber respiration by cold stress is associated with an increased capacity of both Plant Uncoupling Mitochondrial Protein (PUMP) and alternative oxidase. *Journal of Bioenergetics Biomembranes,* 35: 211-220.

Camacho-Pereira, J., Meyer, L.E., Machado,, L.B., Oliveira, M.F. and Galina, A. 2009. Reactive oxygen species production by potato tuber mitochondria is modulated by mitochondrially bound hexokinase activity. *Plant Physiology*, 149: 1099-1110.

Cannell, M.G.R., and Thornley, J.H.M. 2000. Modelling the components of plant respiration: Some guiding principles. *Annals of Botany*, 85: 45-54.

Carnicer, J., Coll, M., Ninyerola, M., Pons, X., Sánchez, G. and Peñuelas, J. 2011.Widespread crown condition decline, food web disruption, and amplified tree mortality with increased climate change-type drought. *Proceedings of the National Academy of Sciences, USA*, 108: 1474–1478.

Carden, D.E., Walker, D.J., Flowers, T.J. and Miller, A.J. 2003. Single-cell measurements of the contributions of cytosolic Na^+ and K^+ to salt tolerance. *Plant Physiol.*, 131: 676-683.

Carmo-Silva, A,.E., Soares, A.S., Marques, Da Silva J., Bernardes, Da Silva A., Keys, A.J. and Arrabacca, M.C. 2007. Photosynthetic responses of three C4 grasses of different metabolic subtypes to water deficit. *Functional Plant Biology*, 34(1): 1-10.

Carmo-Silva, A.E., Power, S.J., Keys, A.J., Arrabaca, M.C. and Parry, M.A.J. 2008. Photorespiration in C4 grasses remains slow under drought conditions. *Plant Cell and Environment,* 31(7): 925-940.

Catoni, R., Varone, L. and Gratani, L. 2013. Variations in leaf respiration across different seasons for Mediterranean evergreen species. *Photosynthetica*, 51: 295–304.

$_2$] to estimate diffusional and non-diffusional limitations of photosynthetic capacity of salt-stressed olive saplings. *Plant Cell Environ*, 26: 585–594.

Chang, L.A., Hammett, L.K. and Pharr, D.M. 1983. Carbon dioxide effect on ethanol production, pyruvate decarboxylase and alcohol dehydrogenase activities in aerobic sweet potato roots. *Plant Physiol*,71(1): 59-62.

Chaves, M.M. 1991. Effects of water deficite in carbon assimilation. *Journal of Experimental Botany*, 42: 1-16.

Chaves, M.M., Pereira, J.S., Rodrigues, M. Ricardo, C.P.P., Osonio, M.L., Carvalho, I., Faria, T. and Pinheiron, C. 2002. How plants cope with water stress in he field: Photosynthesis and growth. *Annals of Botany*, 89: 907-916.

Chaves, M.M., Maroco, J.P. and Pereira, J.S. 2003. Understanding plant response to drought: from genes to the whole plant. *Functional Plant Biology*, 30: 239-264.

Chaves, M.M. and Oliveira, M.M. 2004 Mechanisms underlying plant resilience to water deficits: prospects for watersaving agriculture. *J Exp Bot*, 55: 2365–2384.

Chinnusamy, V., Jagendorf, A. and Zhu, J.K. 2005. Understanding and improving salt tolerance in plants. *Crop Sci.*, 45: 437-448.

Chirkova, T.V., Zhukova, T.M. and Bugrova, M.P. 1992. Redox reaction of plant cells in response to short-term anaerobiosis. *Vestnik SPBGU*, 3: 82-86.

Christiansen, M.N. 1978. The physiology of plant tolerance to temperature extremes. In, "*Crop tolerance to suboptimal land conditions*," Ed. G.A. Jung, ASA Madison, WI, pp. 173–191.

Collier, D.E. and Cummins, W.R.1990. The effect of low growth and measurement temperature on the respiratory properties of five temperate species. *Annals of Botany*, 65: 533-538.

Comic, C. and Massacci, A. 1996. Leaf photosynthesis under drought stress. In, "*Photosynthesis and Environment*", Ed. N.R. Baker, Dordrecht: Kluwer Academic Publs.

Crawford, R.M.M. 2003. Seasonal differences in plant responses to flooding and anoxia. *Can. J. Bot.*, 81: 1224-1246.

Crosbie, T.M. and Pearce, R.B. 1982. Effects of recurrent phenotypic selection for high and low photosynthesis on agronomic traits in two maize populations. *Crop Sci.*, 22: 809-813.

Cuartero, J,, Bolarin, M.C., Asins, M.J. and Moreno, V. 2006. Increasing salt tolerance in tomato. *J. Exp. Bot.*, 57: 1045-1058.

Da Matta, F.M., Loos, R.A., Rodrigues, R. and Barros, R. 2001. Actual and potential photosynthetic rates of tropical crop species. *Revista Brasileira de Fisiologia Vegetal*, 13(1): 24-32.

Davidson, E.A., Verchot, L.V., Cattanio, J.H., Ackerman, I.L. and Carvalho, H.M. 2000. Effects of soil water content on soil respiration in forests and cattle pastures of eastern Amazonian, *Biogeochemistry*, 48: 53–69.

Davies, F.S. and Flore, J.A. 1986a. Short-term flooding effects on gas exchange and quantum yield of rubbiteye blueberry (*Vaccinium ashe* Reade). *Plant Physiol.*, 81(1): 289-292.

Davies, F.S. and Flore, J.A. 1986b. Gas exchange and flooding stress of rubbiteye blueberry. *J. Am. Soc. Hort. Sci.*, 111: 565-571.

Da Silva, J. M. and Arrabaça, M.C. 2004. Contributions of soluble carbohydrates to the osmotic adjustment in the C4 grass *Setaria sphacelata*: a comparison between rapidly and slowly imposed water stress. *Journal of Plant Physiology*, 161(5): 551-555.

Day, D.A., Dry, I.B., Soole, K.L.., Wiskich, J.T. and Moore, A.L. 1991. Regulation of alternative pathway activity in plant mitochondria. *Plant Physiology*, 95: 948-953.

Day, D.A., Krab, K., Lambers, H., Moore, A.L., Siedow, J.N., Wagner, A.M., Wiskich, J.T. 1996. The cyanide-resistant oxidase: to inhibit or not to inhibit, that is the question. *Plant Physiology*, 110: 1-2.

Dennis, E.S., Dolferus, R., Ellis, M., Rahman, M., Wu, Y., Hoeren, F.U., Grover, A., Jsmond, K.P., Good, A.G and Peacock, W.J. 2000. Molecular strategies for improving waterlogging tolerance in plants. *Journal of Experimental Botany*, 51(342): 89-97.

Dexter, S.T. 1956. Evaluation of crop plants for winter hardiness. Adv. Agron. 8:203–209.

Diaz-Zorita, M., Fernandez-Canigia, M.V. and Grosso, G.A. 2001. Application of foliar fertilizers containing glycinebetaine improved wheat yields. *J. Agron. Crop Sci.*, 186: 209–215.

Di Castri, F. 1973. Climatographical comparison between Chile and the western coast of North America. In, "*Mediterranean-type ecosystems: origin and structure*", Eds. F. Di Castri and H.A. Mooney, Springer, Berlin, pp. 21–36.

Dos Reis, S.P., Lima, A.M. and De Souza, C.R.B. 2012. Recent molecular advances on downstream plant responses to abiotic stress. *Int. J. Mol. Sci.*, 13: 8628–8647.

Drew, M.C. 1997. Oxygen deficiency and root metabolism: Injury and acclimation under hypoxia and anoxia. *Annual Review of Plant Physiology and Plant Molecular Biology*, 48: 223-250.

Drew, M.C., He, C.J. and Morgan, P.W. 2000. Programmed cell death and aerenchyma formation in roots. *Trends in Plant Science*, 5: 123-127.

Dubey, R.S. 2005. Photosynthesis in plants under stressful conditions. In, "*Handbook of photosynthesis*", 2nd edition, Ed. M. Pessarakli,. CRC Press. Florida, pp. 479-497.

Duque, A.S., de Almeida, A.M., da Silva, A.B., da Silva, J.M., Farinha, A.P., Santos, D., Fevereiro, P. and de Sousa Araújo, S. 2013. Abiotic stress responses in plants: unraveling the complexity of genes and networks to survive. In, "*Abiotic Stress—Plant Responses and Applications in Agriculture*," Eds. K. Vahdati and C. Leslie, InTech: Rijeka, Croatia, pp. 3–23.

Dutilleul, C., Driscoll, S., Cornic, G., de Paepe, R., Foyer, C.H. and Noctor, G. 2003a. Functional mitochondrial complex I is required by tobacco leaves for optimal photosynthetic performance in photorespiratory conditions and during transients. *Plant Physiology*, 131: 264-275.

Dutilleul, C., Garmier, M., Noctor, G., Mathieu, C., Chetrit, P., Foyer, C.H. and de Paepe, R. 2003b. Leaf mitochondria modulate whole cell redox homeostasis, set antioxidant capacity, and determine stress resistance through altered signaling and diurnal regulation. *The Plant Cell*, 15: 1212-1226.

Epron, D., Toussaint, M.-L. and Badot, P.-M. 1999. Effects of sodium chloride salinity on root growth and respiration in oak seedlings. *Annals of Forest Science,* 56: 41–47.

Ehleringer, J.R. and Björkman, O. 1977. Quantum yield for CO_2 uptake in C3 plants. *Plant Physiology*, 59(1): 86-90.

Farooq, M., Wahid, A., Kobayashi, N., Fujita, D. and Basra, S.M.A. 2009. Plant drought stress: effects, mechanisms and management. *Agron. Sustain. Dev.*, 29: 185–212.

Farrar, J.F. 1985. The Respiratory Source of CO_2. *Plant Cell and Environment*, 8: pp. 427-438.

Felle, H.H. 2005. pH regulation in anoxic plants. *Annals of Botany*, 96: 519-532.

Fernie, A.R., Carrari, F. and Sweetlove, L.J. 2004. Respiratory metabolism: Glycolysis, the TCA cycle and mitochondrial electron transport. *Current Opinion in Plant Biology*, 7: 254-261.

Ferreira, J.L., Coelho, C.H.M., Magalhães, P.C., Sant'ana, G.C. and Borém, A. 2008. Evaluation of mineral content in maize under flooding. *Crop Breed. and Applied Biotech.*, 8: 134-140.

Flexas, J. and Medrano, H. 2002. Drought-inhibition of photosynthesis in C_3 plants: stomatal and non-stomatal limitations revisited. *Annals of Botany*, 89: 183-189.

Flexas, J., Bota, J., Escalona, J.M., Sampol, B. and Medrano, H. 2002. Effects of drought on photosyn-thesis in grapevines under field conditions: an evaluation of stomatal and mesophyll limitations. *Funct Plant Biol*, 29: 461–471.

Flexas, J. and Medrano, H. 2002. Drought inhibition of photosynthesis in C3 plants: Stomatal and non stomatal limitations revisited. *Annals of Botany,* 89(2): 183-189.

Flexas, J., Bota, J., Loreto, F., Cornic, G. and Sharkey, T.D. 2004. Diffusive and metabolic limitations to photosynthesis under drought and salinity in C3 plants. *Plant Biol*, 6: 269–279.

Flexas, J., Galmes, J., Ribas-Carbo, M. and Medrano, H. 2005. The effects of drought in plant respiration. In, "*Advances in Photosynthesis and Respiration 18. Plant Respiration: from Cell to Ecosystem*", Eds. H. Lambers and M. Ribas-Carbo, Kluwer Academic Publishers, Dordrecht, pp 85–94.

Flexas, J., Bota, J., Galmés, J., Medrano, H. and Ribas-carbo, M. 2006. Keeping a positive carbon balance under adverse conditions: responses of photosynthesis and respiration to water stress. *Physiologia Plantarum*, 127: 343–352.

Flexas, J., Ribas-Carbo, M., Diaz-Espejo, A., Galmes, J., Medrano, H. 2008. Mesophyll conductance to CO_2: current knowledge and future prospects. *Plant, Cell and Environment*, 31: 602-621.

Flexas, J., Bota, J., Loreto, F., Cornic, G. and Sharkey, T.D. 2004. Diffusive and metabolic limitations to photosynthesis under drought and salinity in C_3 plants. *Plant Biol.*, 6: 269-279.

Flexas, J., Diaz-Espejo, A., Gago, J., Gallé, A., Galmés, J., Gulías, J. and Medrano, H. 2014. Photosynthetic limitations in Mediterranean plants: a review. *Environmental and Experimental Botany,* 103: 12–23.

Foolad, M.R. 2004. Recent advances in genetics of salt tolerance in tomato. *Plant Cell Tiss. Org.*, 76: 101-119.

Foyer, C.H. and Noctor, G. 2000. Oxygen processing in photosynthesis: Regulation and signaling. *New Phytologist*, 146(3): 359-388.

Foyer, C.H. and Noctor, G. 2003. Redox sensing and signaling associated with reactive oxygen in chloroplasts, peroxisomes and mitochondria. *Physiol. Plant*, 119: 355-364.

Frantz, J.M., Cometti, N.N. and Bugbee, B. 2004. Night temperature has a minimal effect on respiration and growth in rapidly growing plants. *Ann. Bot.*, 94: 155-166.

Friend, A.D. 2010. Terrestrial plant production and climate change. *Journal of Experimenta Botany,* 61: 1293–1309.

Gaevskell, N.A., Soroklna, G.A. Gold, V.M., Ladygln, V.G. and Gekhman, A.V. 1986. Use of *Chlamydomonas reinhardtii* mutants to study heat induced changes in chlorophyll fluorescence. *Fiziol. Restenii,* 32: 674-680.

Galmés, J., Ribas-Carbó, M., Medrano, H. and Flexas, J. 2007. Response of leaf respiration to water stress in Mediterranean species with different growth forms. *Journal of Arid Environments,* 68: 206–222.

Ghashghaie, J., Duranceau, M., Badeck, P.W., Comic, G., Adeline, M.T. and Deleens, E. 2001. $\Delta^{13}C$ of CO_2 respired in the dark in relation to $\Delta^{13}C$ of C of leaf metabolites comparison between *Nicotiana sylvestris* and *Helianthus annuus* under drought. *Plant Cell and Environment,* 28: 834-849.

Gibbs, J., Morrell, S., Valdez, A. Setter, T.L. and Greenway, H. 2000. Regulation of alcoholic fermentation in coleoptiles of two rice cultivars differing in tolerance to anoxia. *Journal of Experimental Botany,* 51: 785-796.

Gibbs, J. and Greenway, H. 2003. Mechanisms of anoxia tolerance in plants. I. Growth, survival and anaerobic catabolism. *Funct. Plant Biol.,* 30: 1–47.

Geigenberger, P. 2003. Response of plant metabolism to too little oxygen. *Current Opinion in Plant Biology,* 6: 247-256.

Gillespie, K.M., Xu, F., Richter, K.T., McGrath, J.M., Markelz, R.C., Ort, D.R., Andrew, D.B., Leakey and Euzabeth, and Ainsworth, A. 2012. Greater antioxidant and respiratory metabolism in field-grown soybean exposed to elevated O_3 under both ambient and elevated CO_2. *Plant Cell Environ.* 35: 169–184.

Giraud, E., Ho, L.H. Clifton, R., Carroll, A., Estavillo, G., Tan, Y.F., Howell, K.A., Ivanova, A., Pogson, B.J., Miller, A.H. and Whelan, J. 2008. The absence of ALTERNATIVE OXIDASE1a in Arabidopsis results in acute sensitivity to combined light and drought stress. *Plant Physiology,* 147: 595-610.

Gonzalez-Meler, M.A., Matamala, R. and Penuelas, J. 1997. Effects of prolonged drought stress and nitrogen deficiency on the respiratory O_2 uptake of bean and pepper leaves. *Photosynthetica,* 34: 505-512.

Gonzalez-Meler, M.A., Ribas-Carbo, M., Giles, L. and Siedow J.N. 1999. The effect of growth and measurement temperature on the activity of the alternative respiratory pathway. *Plant Physiology,* 120(3): 765–772.

Gonzalez-Meler, M.A., Taneva, L. and Trueman, R.J. 2004. Plant respiration and elevated atmospheric CO_2 concentration: cellular responses and global significance. *Ann. Bot.,* 94: 647-656.

Gowda, V.R.P., Henry, A., Yamauchi, A., Shashidhar, H.E. and Serraj, R. 2011. Root biology and genetic improvement for drought avoidance in rice. *Field Crops Res.,* 122: 1–13.

Greenway, H. and Gibbs. L. 2003. Review: Mechanisms of anoxia tolerance in plants. II. Energy requirements for maintenance and energy distribution to essential processes. *Functional Plant Biology,* 30: 999-1036.

Gunn, S. and Farrar, J.F. 1999. Effects of a 4°C increase in temperature on partitioning of leaf area and dry mass, root respiration and carbohydrates. *Functional Ecology,* 13: 12–20.

Gulías, J., Cifre, J., Jonasson, S., Medrano, H. and Flexas, J. 2009. Seasonal and inter-annual variations of gas exchange in thirteen woody species along a climatic gradient in the Mediterranean island of Mallorca. *Flora—Morphology, Distribution, Functional Ecology of Plants,* 204: 169–181.

Gupta, N.K., Meena, S.K., Gupta, S. and Khandelwal, S.K. 2002. Gas exchange, membrane permeability, and ion uptake in two species of Indian jujube differing in salt tolerance. *Photosynthetica,* 40: 535-539.

Gupta, K.J., Zabalza, A. and van Dongen, J.T. 2009. Regulation of respiration when the oxygen availability changes. *Physiologia Plantarum*, 137: 383-391.

Gutierres, S., Sabar, M., Lelandais, C., Chetrit, P., Diolez, P., Degand, H., Boutry, M., Vedal, F., de Kouchkovsky, Y. and de Paepe, R. 1997. Lack of mitochondrial and nuclear-encoded subunits of complex I and alteration of the respiratory chain in*Nicotiana sylvestris* mitochondrial deletion mutants. *Proceedings of the National Academy of Sciences, USA* 1997;94:3436-3441.

Halcomb, M., 2003. Too much Water Kills More Plants than too little Water. The University of Tennessee Extension.

Hall, N.O. and Keys, A.J. 1983. Temperature dependence of the enzymatic carboxylation and oxygenation of RuBP in relation to effects of temperature on photosynthesis. *Plant Physiology*, 72(4): 945-948.

Halliwell, B. and Gutteridge, J.M.C. 1989. Free Radicals in Biology and Medicine, Clarendon Press. Hasegawa, P., Bressan, R.A., Zhu, J.K. and Bohnert, H.J. 2000. Plant cellular and molecular responses to high salinity. *Annu Rev Plant Physiol Plant Mol Biol*, 51: 463–499.

Heskel, M.A., Bitterman, D., Atkin, O.K., Turnbull, M.H. and Griffin, K.L. 2014. Seasonality of foliar respiration in two dominant plant species from the Arctic tundra: response to long-term warming and short-term temperature variability. *Functional Plant Biology*, 41: 287–300.

Havaux,M., 1993. Characterization of thermal damage to the photosynthetic electron transport system in potato leaves. *Plant Science,* 94: 19-33.

Havaux, M. and Niyogi, K.K. 1999. The violaxanthin cycle protects plants from photooxidative damage by more than one mechanism. *Proceedings of the National Academy of Sciences, USA*, 96: 8762-8767.

He, J., Wolfe-Bellin, K.S. and Bazzazy, F.A. 2005. Leaf-level physiology, biomass and reproduction of *Phytolacca americana* under conditions of elevated CO_2 and altered temperature regimes. *Int. J.Plant Sci.*, 166(4): 615-622.

Houle, G., Morel, L., Reynolds, C.E. and Siégel, J. 2001. The effect of salinity on different developmental stages of an endemic annual plant, *Aster laurentianus* (Asteraceae). *Am. J. Bot.* 88(1): 62-67.

Hsu, Y.M., Tesng, M.J. and Lin, C.H. 1999. The fluctuation of carbohydrate and nitrogen compounds in flooded wax-apple trees. *Bot. Bull. Acad. Sin.,* 40: 193-198.

Hsu, F.S., Lin, J.B. and Chang, S.R. 2000. Effects of water logging on seed germination, electric conductivity of seed leakage and development of hypocotyls and radical in sudangrass. *Bot. Bull. Acad. Sin.*, 41: 267-273.

Hu, W.H., Song, X.S., Shi, K., Xia, X.J., Zhou, Y.H. and Yu J.Q. 2008. Changes in electron transport superoxide dismutase and ascorbate peroxidase isoenzymes in chloroplasts and mitochondria of cucumber leaves as influenced by chilling. *Photosynthetica*, 46(4): 581–588.

Huang, B., Liu, X. and Fry J.D. 1998. Shoot physiological responses of two bentgrass cultivars to high temperature and poor soil aeration. *Crop Science*, 38: 1219-1224. .

Huang, B.R. and Fu, J. 2000. Photosynthesis, respiration, and carbon allocation of two cool-season perennial grasses in response to surface soil drying, *Plant Soil*, 227: 17–26.

Huang, S., Colmer, T.D. and Millar, A.H. 2008. Does anoxia tolerance involve altering the energy currency towards PPi? *Trends in Plant Science*, 13: 221-227.

Hutmacher, R.B. and Krieg, D.R. 1983. Photosynthetic rate control in cotton. *Plant Physiology*,73(3): 658-661.

Ibrahim, A.M.H. and Quick, J.S. 2001. Heritability of heat tolerance in winter and spring wheat. *Crop Science*, 41: 1401-1405.

Igamberdiev, A.U., Bykova, N.V., Shah, J.K. and Hill, R.D. 2010. Anoxic nitric oxide cycling in plants: participating reactions and possible mechanisms. *Physiologia Plantarum*, 138: 393-404.

IPCC. 2001. Climate change 2001. In, "*Scientific Basis*", Eds. J.T. Houghton, Y. Ding, D.J. Griggs, M. Noguer, P.J. Van der Linden, X. Da, K. Maskell, and C.A. Johnson, Cambridge University Press, New York, pp. 555.

IPCC. 2013. Summary for Policy makers. In, "*Climate Change 2013: The Physical Science Basis. Contribution of Working Group to the Fift hAssessment Report of the Intergovernmental Panel on Climate Change*", Eds. T.F. Stocker, D. Qin, G.K. Plattner, M. Tignor, S.K. Allen, J. Boschung, *et al.,* Cambridge University Press, New York, NY.

Ismail, A.M. and Hall, A.E. 1999. Reproductive-stage heat tolerance, leaf membrane thermostability and plant morphology in cowpea. *Crop Science*, 39: 1762-1768.

Ismond, K.P., Dolferus, R., de Pauw, M., Dennis, E.S. and Good, A.G. 2003. Enhanced low oxygen survival in Arabidopsis through increased metabolic flux in the fermentative pathway. *Plant Physiology*, 132: 1292-1302.

Jackson, M.B. and Colmer, T.D. 2005. Response and adaptation by plants to flooding stress. *Annals of Bot.*, 96: 501-505.

James, R.A., Rivelli, A.R., Munns, R. and Caemmerer, S.V. 2002. Factors affecting CO_2 assimilation, leaf injury and growth in salt-stressed durum wheat. *Func. Plant Biol.*, 29: 1393-1403.

Jamil, A., Riaz, S., Ashraf, M. and Foolad, M.R. 2011. Gene expression profiling of plants under salt stress. *Crit Rev Plant Sci*, 30: 435-458.

Jimenez, S., Dridi, J., Gutierrez, D. Moret, D. Irigoyen, J.J., Moreno, M.A. and Gogorcena, Y. 2013. Physiological, biochemical and molecular responses in four Prunus rootstocks submitted to drought stress. *Tree Physiology*, 33: 1061-1075.

Johnson, I.R. and Thornley, J.H.M. 1985. Temperature dependence of plant and crop processes. *Annals of Botany*, 55: 1–24.

Jones, T.L., Tucker, D.E. and Ort, D.R. 1998. Chilling delays circadian pattern of sucrose phosphate synthase and nitrate reductase activity in tomato. *Plant Physiology*, 118: 149-158.

Juan, A., González, M. Gallardo, A. Hilal, M. Rosa and Prado, F.E. 2009. Physiological responses of quinoa (*Chenopodium quinoa* Willd.) to drought and water logging stress: dry matter partitioning. *Botanical Studies*, 50: 35-42.

Juszczuk, I.M., Flexas, J., Szal, B., Dabrowska, Z., Ribas-Carbo, M. and Rychter, A.M. 2007. Effect of mitochondrial genome rearrangement on respiratory activity, photosynthesis, photorespiration and energy status of MSC16 cucumber (*Cucumis sativa*) mutant. *Physiologia Plantarum*, 131: 527-541.

Kao, W.Y., Tsai. T.T. and Shih, C.N. 2003. Photosynthetic gas exchange and chlorophyll a fluorescence of three wild soybean species in response to NaCl treatments. *Photosynthetica*, 41: 415-419.

Kato-Noguchi, H., and Morokuma, M. 2007. Ethanolic fermentation and anoxia tolerance in four rice cultivars. *Journal of Plant Physiology*, 164: 168-173.

Kase, M. and Catsky, J. 1984. Maintenance and growth componenets of dark respiration rate in leaves of C3 and C4 plants as affected by leaf temperature. *Biologia Plantarum*, 26: 461–470.

Kawano, T., Sahashi, N., Takahashi, K., Uozumi, N. and Muto, S. 1998. Salicylic acid induces extracellular superoxide generation followed by an increase in cytosolic calcium ion in tobacco suspension culture: The earliest events in salicylic acid signal transduction. *Plant Cell Physiol.*, 39: 721–730.

Kazerouni, E., Akramian, M., Tokassi, S. and Eghbali, S. 2005. Physiological effects of salt and drought stresses on germination and seedling growth of lentil (*Lens culinaris*) and mungbean (*Vigna radiate* L.). *Horticultural. J.*, 445-448.

Kennedy, R.A., Rumpho, M.E. and Fox, T.C. 1992. Anaerobic metabolism in plants. *Plant Physiology*, 100: 1-6.

Keys, A.J. 1886. Rubisco in photorespiration. *Philosophical Transction of the Royal Society of London, Series B-Biological Sciences,* 313(1162): 325-336.

Kirst, G.O. 1989. Salinity tolerance of eukaryotic marine algae. *Annu. Rev. Plant Physiol. Plant Mol. Biol.*, 40: 21-53.

Knox, G. 2005. Drought-Tolerant Plants for North and Central Florida. University of Florida Cooperative Extension Service.

Komatsu, S., Hiraga, S. and Yanagawa, Y. 2012. Proteomics techniques for the development of flood tolerant crops. *J. Proteome Res.*, 11: 68–78.

Kramer, P.J. andBoyer, J.S. 1995. Water relation of plants and soils. Academic Press, London.

Kreiner, M., Harvey, L.M. and McNeil, B. 2002. Oxidative stress response of a recombinant *Aspergillus niger* to exogenous menadione and H_2O_2 addition. Enzyme Microb. *Technol.* 30: 346–353.

Kirschbaum, M.U.F. and Pearcy, R.W. 1988. Gas exchange analysis of the relative importance of stomatal and biochemical factors in photosynthetic induction in *Alocasia macrorrhiza. Plant Physiology*, 86(3): 782-785.

Kudoh, H. and Sonoike, K. 2002. Irreversible damage to photosystem I by chilling in the light: cause of the degradation of chlorophyll after returning to normal growth temperature. *Planta,* 215: 541-548.

Kuo, C.C. and Chen, B.W. 1980. Physiological responses of tomato cultivars to flooding. *J. Am.Soc. Hort. Sci.*, 105: 751-755.

Kursteiner, O., Dupuis, I. and Kuhlemeier, C.K. 2003. The pyruvate decarboxylase1 gene of *Arabidopsis* is required during anoxia but not other environmental stresses. *Plant Physiology*, 132: 968-978.

Kyle, D.J. and Ohad, I. 1986. The mechanism if photoinhibition in higher plants and green algae. In, "*Encyclopedia of plant physiology*," Eds., L.A. Staehelin, and C.J. Arntzen, Vol 19, Springer-Verlag, New Series, Berlin, pp: 468-475.

Lambers, H. 1976. Respiration and NADH oxidation of the roots of flood intolerant *Senecia* species as affected by anaerobiosis. *Physiol. Plant.*, 37(2): 117-122.

Lambers, H. 1985. Respiration in tack plants and tissues: Its regulation and dependence on environmental factors, metabolism and invaded organism. In, "*Higher plant cell respiration. Encyclopedia of plant physiology. New Ser",* Eds. R. Douce and D.A. Day, Vol. 18. Springer, Berlin.

Lambers, H., Atkin, O.K. and Scheureater, I. 1996. Respiratory patterns in roots in relation to their function. In, "*Plant Roots, The Hidden Half*", Ed. Y. Waisel, Marcel Dekker, New York.

Lambers, H., Robinson, S.A. and Ribas-Carbo, M. 2005. Regulation of respiration in vivo. The effects of drought in plant respiration. In, "*Advances in Photosynthesis and Respiration 18. Plant Respiration: from Cell to Ecosystem*", Eds. H. Lambers and M. Ribas-Carbo, Kluwer Academic Publishers, Dordrecht, pp 1-15.

Larkindale, J., and Knight, M.R. 2002. Protection against heat stress-induced oxidative damage in *Arabidopsis* involves calcium, abscisic acid, ethylene, and salicylic acid. *Plant Physiol.,* 128: 682–695.

Lavania, D., Dhingra, A., Siddiqui, M.H., Al-Whaibi, M.H., and Grover, A. 2015. Current status of the production of high temperature tolerant transgenic crops for cultivation in warmer climates. *Plant Physiol. Biochem.* 86: 100–108.

Lawlor, D.W. 1976. Water stress induced changes in photosynthesis, photorespiration, respiration and CO compensation of wheat. *Photosynthetica*, 10(3): 378-387.

Lawlor, D.W. and Pearlman, J.G. 1981. Compartmental modeling of photorespiration and carbon metabolism of water stressed leaves. *Plant Cell and Environment,* 4!1): 37-52.

Lawlor, D.W. 2002. Limitation to photosynthesis in water-stressed leaves: stomata vs metabolism and the role of ATP. *Annals of Botany,* 89(7): 871-885.

Lawlor, D.W. and Cornic, G. 2002 Photosynthetic carbon assimilation and associated metabolism in relation to water deficits in higher plants. *Plant Cell Environ,* 25: 275–294

Lawrence, C. and Holaday, A.S. 2000. Effects of mild night chilling on respiration of expanding cotton leaves. *Plant Science*, 157(2): 233–244.

Li, C., Bai, T., Ma, F. and Han, M. 2010. Hypoxia tolerance and adaptation of anaerobic respiration to hypoxia stress in two *Malus* species. *Scientia Horticulturae* ,124: 274-279.

Liao, C.T. and Lin, C.H. 1994. Effects of flooding stress on photosynthetic activities of *Momordica charantia. Plant Physiol. Biochem.*, 32: 1-5.

Liao, C.T. and Lin, C.H. 1995. Effects of flooding stress on morphology and anaerobic metabolism of *Momordia charantia. Environ. Exp. Bot.*, 35(1): 105-113.

Liao, C.T. and Lin, C.H. 1996. Photosynthetic responses of grafted bitter melon seedlings to flooding stress. *Environ. Expt. Bot.*, 36(2): 167-172.

Liao, C.T. and Lin, C.H. 2000. Physiological adaptation of crop plants to flooding stress. *Proc. Natl. Sci. Counc. ROC(B)*, 25(3): 148-157.

Liao, C.T., and C.H. Lin. 2001. Physiological adaptation of crop plants to flooding stress. *Proceedings of the National Science Council, Republic of China. Part B, Life Sciences* 25: 148-157.

Limousin, J.-M., Misson, L., Lavoir, A.-V., Martin, N.K. and Rambal, S. 2010. Do photosynthetic limitations of evergreen *Quercus ilex* leaves change with long-term increased drought severity? *Plant, Cell and Environment*, 33: 863–875.

Lin, C.H. and Lin, C.H. 1992. Physiological adaptation of wax apple to waterlogging. *Plant Cell Environ.*, 15(3): 321-328.

Liu, H.P., Dong, B.H., Zhang, Y.Y., Liu, Z.P. and Liu, Y.L. 2004. Relationship between osmotic stress and the levels of free, soluble conjugated and insoluble-conjugated polyamines in leaves of wheat seedlings. *Plant Sci.*, 166: 1261–1267.

Liu, H.S. and Li, F.M. 2005. Root respiration, photosynthesis and grain yield of two spring wheat in response to soil drying. *Plant Growth Regul.*, 46: 233–240.

Liu, D.L., Yang, Y., Mo, J. and Scott, B.J. 2011. Simulation of maintenance respiration in wheat (*Triticum aestivum* L.). *World Journal of Agricultural Sciences*, 7(6): 777-784.

Livne, A. and Levin, N. 1966. Energy metabolism of salt-affected pea seedlings. *Israel J. Botany*, pp. 15.

Lobell, D.B., Schlenker, W., and Costa-Roberts, J. 2011. Climate trends and global crop production since 1980. *Science,* 333: 616–620.

Loka, D.A. and Oosterhuis, D.M., 2010. Effect of high night temperatures on cotton respiration. ATP levels, and carbohydrate content. *Environmental and Experimental Botany*, 68: 258-263.

Long, S.P., Zhu, X.-G., Naidu, S., and Ort, D.R. 2006. Can improvement in photosynthesis increase crop yields? *Plant Cell Environ.*, 29: 315–330.

Lukatkin, A.S., Brazaityte, A., Bobinas, C. and Duchovskis, P. 2012. Chilling injury in chilling-sensitive plants: a review. *Þemdirbystë-Agriculture*, 99(2): 111 124.

Mailloux, R.J., Beriault, R., Lemire, J., Singh, R., Chenier, D.R., Hamel, R.D. and Appanna, V.D. 2007. The tricarboxylic acid cycle, an ancient metabolic network with a novel twist. *PLoS ONE,* 2(8): :e690.

MacCabe, T.C., Daley, D. and Whelan, J. 2000. Regulatory, developmental and tissue aspects of mitochondrial biogenesis in plant cell. *Plant Biology,* 2: 121-135.

McNamara, S.T. and Mitchel, C.A. 1989. Differential flood stress resistance of two tomato genotypes. *J. Am. Soc. Hort. Sci.*, 114(6): 976-980.

Makela, P., Karkkainen, J. and Somersalo, S. 2000. Effect of glycine betaine on chloroplast ultra structure, chlorophyll and protein content, and RUBPCO activities in tomato grown under drought or salinity. *Biol. Plant.* 3: 471–475.

Mancuso, S. and Marras, A.M. 2006. Adaptive response of *Vitis* root to anoxia. *Plant and Cell Physiology*, 47: 401-409.

Mansour, M.M.F., Salama, K.H.A., Ali, F.Z.M. and Abou Hadid, A.F. 2005. Cell and plant responses to NaCl in *Zea mays* cultivars differing in salt tolerance. *Gen. Appl. Plant Physiol.*, 31: 29-41.

Marschner, H. 1986. Mineral Nutrition in Higher Plants. Acad. Press, London. pp. 477-542.

Marschner, H. 1991. Mechanism of adaptation of plants in acid soil. In, "*Plant soil interaction at low pH*", Eds. R.J. Wright, V.C. Baliger and R.P. Moorman, *Proceedings of the Second International Symposium on Plant-Soil Interaction at Low pH, 1990, Beckley, West Virginia, USA,* Kluwer Academic Publisher, pp. 663-702.

Martin, S.A., Santos, M.P., Pecanha, A.L., Pommer, C., Campostrini, E., Viana, A.P., Facanha, A.R. and Bressan-Smith, R. 2009. Photosynthesis and cell respiration modulated by water deficit in grape vine (*Vitis vinifera* L.) cv. Cabernet Sauvignon. *Braz. J. Plant Physiol.*, 21(2): 95-102.

Martin-StPaul, N.K., Limousin, J.-M., Rodríguez-Calcerrada, J., Ruffault, J., Rambal, S., Matthew, L.G. and Misson, L. 2012. Photosynthetic sensitivity to drought varies among populations of *Quercus ilex* along a rainfall gradient. *Functional Ecology*, 39: 25–37.

Martínez-Vilalta, J. and Piñol, J. 2002. Drought-induced mortality and hydraulic architecture in pine populations of the NE Iberian Peninsula. *Forest Ecology and Management*, 161: 247–256.

Matusick, G., Ruthrof, K.X., Brouwers, N.C., Dell, B. and Hardy, G.S.J. 2013. Sudden forest canopy collapse corresponding with extreme drought and heat in a Mediterranean-type eucalypt forest in southwestern Australia. *European Journal of Forest Research*, 132: 497–510.

Maxwell, D.P., Wang, Y. and Mclntosh, L. 1997. The alternative oxidase lowers mitochondrial reactive oxygen production in plant cell. *PNAS*, 96: 8271-8276.

Mccue, K.F. and Hanson, A.D. 1990. Drought and salt tolerance: towards understanding and application. *Trends Biotechnol*, 8: 358-362.

McCord, J.M. 2000. The evolution of free radicals and oxidative stress. *Am. J. Med.*, 108: 652-659.

McCree, K.J., 1974. Equations for dark respiration of white clover and grain sorghum, as functions of dry weight, photosynthetic rate and temperature. *Crop Sci.*, 14: 509-514.

McCree, K.J. and Amthor, M.E. 1982. Effects of diurnal variation in temperature on the carbon balances of white clover plants. *Crop Science*, 22: 822–827.

McDonald, A.E., and Vanlerberghe, G.C. 2005. Alternative oxidase and plastoquinol terminal oxidase in marine prokaryotes of the *Sargasso Sea. Gene*, 349: 15–24.

McNulty, A.K. and Cummins, W.R. 1987. The relationship between respiration and temperature in leaves of the arctic plant *Saxifraga cernua. Plant Cell and Environment*, 10: 319-325.

Melo, P.T.B.S., Schuch, L.O.B., Assis, F. and Concenço, G. 2006. Comportamento de populações de arroz irrigado em função das proporções de plantas originadas de sementes de alta e baixa qualidade fisiológica. *Revista Brasileira de Sementes*, 12(1): 37-43.

Merotto Jr., A., Fischer, A.J. and Vidal, R.A. 2009. Perspectives for using light quality knowledge as an advanced ecophysiological weed management tool. *Planta Daninha*, 27(2): 407-419.

Messinger, S.M., Buckley, T.N. and Mott, K.A. 2006. Evidence for involvement of photosynthetic processes in the stomatal response to CO_2. *Plant Physiology*, 140(2): 771-778.

Millar, A.H., Atkin, O.K., Ian Menz, Henry, R.B. Farquhar, G. and Day, D.A. 1998. Analysis of respiratory chain regulation in roots of soybean seedlings. *Plant Physiology*, 117: 1083-1093.

Millar, A.H., Whelan, J., Soole, K.L. and Day, D.A. 2011. Organization and regulation of mitochondrial respiration in plants. *Annual Review of Plant Biology*, 62: 79-104.

Mitova, V.O. and Iquamberdiev, A.U. 2000. Effect of salt stress on respiration in higher plants. *Izy Akad Nauk Ser Biol.*, 3: 322-328.

Mittler, R. 2002. Oxidative stress, antioxidants and stress tolerance. *Trends in Plant Science, 7:* 405-410.

Mittler, R., Vanderauwera, S., Gollery, M. and Breusegem, F.V. 2004. Reactive oxygen gene network of plants. *Trends Plant Sci.*, 9: 490-498.

Mittler, R. 2006. Abiotic stress, the field environment and stress combination. *Trends Plant Sci.*, 11: 15–19.

Miziorko, H.M. and Lorimer, G.H. 1983. Rubulose-1-5-bisphosphate carboxylase-oxygenase. *Ann. Rev. Biochem.,* 52: 507-535.

Mohammed, A.R. and Tarpley, L. 2009a. High nighttime temperatures affect rice productivity through altered pollen germination and spikelet fertility. *Agricultural and Forest Meteorology*, 149: 999-1008.

Mohammed, A.R. and Tarpley, L. 2009b. Impact of high nighttime temperature on respiration, membrane stability, antioxidant capacity, and yield of rice plants. *Crop Science*, 49: 313-322.

Mommer, L.., Pedersen, O. and Visser, E.J.W.. 2004. Acclimation of a terrestrial plant to submergence facilitates gas exchange under water. *Plant, Cell and Environment* 27: 1281-1287.

Möller, B.M., 2001. Plant mitochondria and oxidative stress: Electron transport, NADPH turnover and metabolism of reactive oxygen species. *Annual Review of Plant Physiology and Plant Molecular Biology,* 52: 561-591.

Moller, I.M. 2001. Plant mitochondria and oxidative stress: Electron transport, NADPH turnover, and metabolism of reactive oxygen species. *Annual Review of Plant Physiology and Plant Molecular Biology,* 52: 561-591.

Moore, A.L. and Siedow, J.N. 1991. The regulation and nature of cyanide resistant alternative oxidase in mitochondria. *Biochemistry and Biophysical Acta*, 1058: 121-140.

Morales, F., Abadía, A., Gómez-Aparisi, J. and Abadía, J. 1992. Effect of combined NaCl and $_2$ salinity on photosynthetic parameters of barley grown in nutrient solution. *Physiol. Plant*, 86: 419-426.

Morales, P., Sykes, M.T., Prentice, I.C., Smith, P., Smith, B., Bugmann, H., Zierl, B., Friedlingstein, P., Viovy, N., Sabaté, S., Sánchez, A., Pla, E., Gracia, C.A., Sitch, S., Arneth, A. and Ogee, J. 2005. Comparing and evaluating process-based ecosystem model predictions of carbon and water fluxes in major European forest biomes. *Global Change Biology*, 11: 2211–2233.

Morgan, C.L. and Austin RB. 1983. Respiratory loss of recently assimilated carbon in wheat. *Annals of Botany*, 51:, 85–95.

Moud, A.M. and Maghsoudi, K. 2008. Salt stress effects on respiration and growth of germinated seeds of different wheat (*Triticum aestivum* L.) cultivars. *World Journal of Agricultural Sciences*, 4(3): 351-358,

Moyen, C., and Roblin, G. 2013. Occurrence of interactions between individual Sr^{2+}- and Ca^{2+}-effects on maize root and shoot growth and Sr^{2+}, Ca^{2+} and Mg^{2+} contents, and membrane potential: Consequences on predicting Sr^{2+}-impact. *Journal of Hazardous Materials*, 260: 770-779.

Munro, K.D., Hodges, D.M., DeLong, J.M., Forney, C.F. and Kristie, D.N. 2004. Low temperature effects on ubiquinone content, respiration rates and lipid peroxidation levels of etiolated seedlings of two differentially chilling-sensitive species. *Physiologia Plantarum,* 121(3): 488–497.

Munns, R. and Termaat, A. 1986. Whole-plant responses to salinity. *Aust. J. Plant Physiol.*, 13: 143-160.

Munns, R. 2002. Comparative physiology of salt and water stress. *Plant Cell Environ.*, 25: 239-250.

Munns, R. 2005. Genes and salt tolerance: bringing them together. *New Phytol.*, 167: 645-663.

Nanjo, Y., Skultety, L., Ashraf, Y. and Komatsu, S. 2010. Comparative proteomic analysis of early-stage soybean seedlings responses to flooding by using gel and gel-free techniques. *J. Proteome Res.*, 9: 3989–4002.

NASA. 2014. Global Climate Change: Vital Signs of the Planet. Available at:http:// climate.nasa.gov/ 400ppmquotes[accessedDecember4,2014.

Naves-Barbiero, C.C., Franco, A.C., Bucci, S.J. and Goldstein, G. 2000. Fluxo de seiva e condutância estomática de duas espécies lenhosas sempre-verdes no campo sujo e cerradão. *Revista Brasileira de Fisiologia Vegetal,* 12(1): 119-134.

Nazir, N., Ashraf, M. and Rasul, E. 2001. Genomic relationships in oilseed *Brassica* with respect to salt tolerance-photosynthetic capacity and ion relations. *Pak. J. Bot.*, 33: 483-501.

Nieman, R.H., 1962. Some effects of sodium chloride on growth, photosynthesis and respiration of twelve crop plants. *Bot.. Gaz,* ,123: 279-85.

Niinemets, U., Cescatti, A., Rodeghiero, M. and Tosens, T. 2005. Leaf internal diffusion conductance limits photosynthesis more strongly in older leaves of Mediterranean evergreen broad-leaved species. *Plant, Cell and Environment*, 28: 1552-1566.

Niinemets, U. 2014. Improving modeling of the 'dark part' of canopy carbon gain. *Tree Physiology*, 34: 557–563.

Noctor, G., de Paepe, R. and Foyer, C.H. 2007. Mitochondrial redox biology and homeostasis in plants. *Trends in Plant Science*, 12: 125-134.

Norlyn, J.D. 1980. Breeding salt tolerant crop plants. In, "*Genetic engineering of osmoregulation: Impact on plant productivity for pood, chemicals and energy*", Eds. D.W. Rains, R.C. Valentine, and A. Hollaender, Plenum Press, New York. pp. 293-309.

Nunes, C., Araújo, S.S., Silva, M.M., Fevereiro, M.P.S. and Silva, A.B. 2008. Physiological responses of the legume model *Medicago truncatula* cv. Jemalong to water deficit. *Environmental and Experimental Botany,* 63(1-3): 289-296.

Nunes, C., Araújo, S.S., Silva, M.M., Fevereiro, M.P.S. and Silva, A.B. 2009. Photosynthesis light curves: A method for screening water deficit rResistance in the model legume *Medicago truncatula. Annals of Applied Biology,* 155(3): 321-332.

Nunes-Nesi, A., Carrari, F., Gibon, Y., Sulpice, R., Lytovchenko, A., Fisahn, J., Graham, J., Ratcliffe, R.G., Sweetlove, L.J. and Fernie, A.R. 2007. Deficiency of mitochondrial fumarase activity in tomato plants impairs photosynthesis via an effect on stomatal function. *The Plant Journal*, 50: 1093-1106.

Ordentlich, A., Linzer, R.A. and Raskin, I. 1991. Alternative respiration and heat evolution in Plants. *Plant Physiology*, 97(4): 1545–1550.

Ortiz Lopez, A., Nie, G.Y., Ort, D.R. and Baker, N.R. 1990. The involvement of the photo-inhibition of photosystem II and impaired membrane energization in the reduced quantum yield of carbon assimilation in chilled maize. *Planta,* 181: 78-84.

Paembonan, S.A., Hagihara, A. and Hozumi, K. 1992. Long-term respiration in relation to growth and maintenance processes of the aboveground parts of a hinoki forest tress. *Tree Physiol.*, 10: 101–110.

Parida, A.K., Das, A.B. and Mittra, B. 2003. Effects of NaCl stress on the structure, pigment complex composition, and photosynthetic activity of mangrove *Bruguiera parviflora* chloroplasts. *Photosynthetica*, 41: 191-200.

Parry, M.A.J., Andralojc, P.J., Shahnaz, K., Lee P.J. and Keys, A. J. 2002. Rubisco activity: effect of drought stress. *Annals of Botany,* 89(7): 833-839.

Parry, M.A.J., Madgwick, P.J., Carvalho, J.F.C. and Andralojc, P.J. 2007. Prospects for increasing photosynthesis by overcoming the limitations of Rubisco. *Journal of Agricultural Science,* 145(1): 31-43.

Parvaiz, A. and Satyawati, S. 2008. Salt stress and phyto-biochemical responses of plants-a review. *Plant Soil Environ.*, 54: 89–99.

Patel, P.K., Singh, A.K., Tripathi, N., Yadav, D. and Hemanteranjan, A. 2014. Flooding abiotic constraint limiting vegetable productivity. *Adv. Plant Agric. Res.* 1(3): 00016.

Peng, S., Huang, J., Sheehy, J.E., Laza, R.C., Visperas, R.M., Zhong, X., Centeno, G.S., Khush, G.S. and Cassman, K.G. 2004. Rice yields decline with higher night temperature from global warming. *Proc. Natl. Acad. Sci. USA* , 101: 9971–9975.

Penning de Vries, F.W.T., 1975. The cost of maintenance processes in plant cells. *Ann. Bot.*, 39: 77-92.Peñuelas, J., Lloret, F. and Montoya, R 2001. Severe drought effects on Mediterranean woody flora in Spain. *Forest Science*, 47: 214–218.

Peñuelas, J., Sardans, J., Rivas-Ubach, A. and Janssess, I. 2012. The human- induced imbalances between C, N and P in Earth's life system. *Global Change Biol.* 18: 3–6.

Peñuelas,J., Sardans, J., Estiarte,M., Ogaya, R., Carnicer, J., Coll, M., Barbeta, A., Ubach, A.R., Llusià, J., Garbulsky, M., Filella, I. and Jump, A.S. 2013. Evidence of current impact of climate change on life: a walk from genes to the biosphere. *Global Change Biol.* 19: 2303–2338.

$_2$ assimilation and dry weight accumulation in field-grown tobacco. *Plant Physiology*, 70: 677-685.

Petricka, J.J., Schauer, M.A., Megraw, M., Breakfield, N.W., Thompson, J.W., Georgiev, S., Soderblom, E.J., Ohler, U., Moseley, M.A., Grossniklaus, U. and Benfey, P.N. 2012. The protein expression landscape of the *Arabidopsis* root. *Proc. Natl. Acad. Sci.*, U.S.A. 109: 6811–6818.

Pettigrew, W.T. and Meredith, W.R. 1994. Leaf gas exchange parameters vary among cotton genotypes. *Crop Sci.*, 34: 700-705.

Pitzschke, A., Forzani, C. and Hirt, H. 2006. Reactive oxygen species signaling in plants. *Antioxidants and Redox Signaling*, 8: 1757-1764.

Plaxton, W.C. 1996. The organization and regulation of plant glycolysis. *Annual Review of Plant Physiology and Plant Molecular Biology*, 47: 185-214.

Ploschuk, E.L. and Hall, A.J. 1997. Maintenance respiration coefficient for sunflower grains in less than that for entire capitulam. *Field Crops Res.*, 49: 147-157.

Poorter, H., Van Der Werf, A., Atkin, O.K. and Lambers, H. 1991. Respiratory energy requirements of roots vary with the potential growth rate of plant species. *Physiologia Plantarum*, 83: 469–475.

Poorter, H., Vandevijver, C., Boot, R.G.A. and Lambers, H. 1995. Growth and carbon economy of a fast-growing and a slow-growing grass species as dependent on nitrate supply. *Plant and Soil*, 171: 217–227.

Popov, V.M., Simonia, R.A., Skulachev, V.P. and Starkov, A,.A. 1997. Inhibition of alternative oxidase stimulates H_2O_2 production in plant mitochondria. *FEBS Letters*, 416: 87-90.

Prasad, T.K., Anderson, M.D., Martin, B.A. and Stewart, C.R. 1994. Evidence for chilling-induced oxidative stress in maize seedlings and a regulatore role for hydrogen peroxide. *Plant Cell*, 6: 65–74.

Praxedes, S.C., DaMatta, F.M., Loureiro, M.E., Ferrão, M.A.G. and Cordeiro, A.T. 2006. Effects of long-term soil drought on photosynthesis and carbohydrate metabolism in mature robusta coffee (*Coffea canephora* Pierre var. *kouillou*) leaves. *Environmental and Experimental Botany*, 56: 263-273.

Priault, P., Fresneau, C., Noctor, G., de Paepe, R., Cornic, G. and Streb, P. 2006a. The mitochondrial CMSII mutation of *Nicotiana sylvestris* impairs adjustment of photosynthetic carbon assimilation to higher growth irradiance. *Journal of Experimental Botany,* 57: 2075-2085.

Priault, P., Tcherkez, G, Cornic, G., de Paepe, R., Naik, R., Ghashghaie, J. and Streb, P. 2006b. The lack of mitochondrial complex I in a CMSII mutant of *Nicotiana sylvestris* increases photorespiration through an increased internal resistance to CO_2 diffusion. *Journal of Experimental Botany*, 57: 3195-3207.

Prentice, I.C., Farquhar, G.D., Fasham, M., Goulden, M.L., Heimann, M., Jaramillo, V.J., Kheshgi, H.S., Quere, C.Le., Scholes, R.J. and Walace, D.W.R. 2001. The carbon cycle and atmospheric carbon dioxide. In, "*Climate Change 2001: The Scientific Basis. Contribution of Working Group to the Third Assessment Report of the Intergovernmental Panel on Climate Change", Eds.* J. Houghton, Y. Ding, and D. Griggs, Cambridge University Press, New York, NY, pp. 185–237.

Purvis, A.C. and Shewfelt, R.L. 1993. Does the alternative pathway ameliorate chilling injury in sensitive plant tissues? *Physiology of Plant*, 88: 712-718.

Pystina, N.V. and Danilov, R.A. 2001. Influence of light regimes on respiration, activity of alternative respiratory pathway and carbohydrates content in mature leaves of *Ajuga reptans* L. *Revista Brasileira de Fisiologia Vegetal*, 13(3):, 285-292,

Rachmilevitch, S., Lambers, H. and Huang, B. 2006. Root respiratory characteristics associated with plant adaptation to high soil temperature for geothermal and turf-type *Agrostis* species. *Journal of Experimental Botany*, 57: 623-631.

Raghavendra, A.S. and Padmasree, K. 2003. Beneficial interactions of mitochondrial metabolism with photosynthetic carbon assimilation. *Trends in Plant Science*, 8: 546-553.

Raftoyannis, Y., Spanos, I. and Radoglou, K 2008. The decline of Greek fir (*Abies cephalonica* Loudon): relationships with root condition. *Plant Biosystems*, 142: 386–390.

Ramalho, J.C., Quartin, V.L., Leitão, E., Campos, P.S., Carelli, M.L.C., Fahland, J.I. and Nunes, M.A. 2003. Cold acclimation ability and photosynthesis among species of the tropical *Coffea genus*. *Plant Biology*, 5: .631-641.

Rasion, J.K., 1980. Effect of low temperature on respiration. In, "*The Biochemistry of plants: Metabolism and Respiration*," Ed. D.D. Davies, Academic Press, New York, 2: 613-626.

Raza, S.H., Athar, H.R., Ashraf, M., Hameed, A. 2007. Glycine betaine induced modulation of antioxidant enzymes activities and ion accumulation in two wheat cultivars differing in salt tolerance. *Environ. Exp. Bot.,* 60: 368-378.

Razack, A., Mohammed and Tarpley, L. 2011. Effects of high night temperature on crop physiology and productivity: Plant growth regulators provide a management option. In, "*Global Warming Impacts - Case Studies on the Economy, Human Health, and on Urban and Natural Environments*", Ed. S. Casalegno, Intech Europe, pp. 153-172.

Reuveni, J. and Bugbee, B. 1997. Very high CO_2 reduces photosynthesis, dark respiration and yield in wheat. *Ann. Bot.*, 80: 539-546.

Reyes, L. and Jennings, P.H. 1997. Effects of chilling on respiration and induction of cyanide-resistant respiration in seedling roots of cucumber. *Journal of the American Society for Horticultural Science,* 122(2): 190–194.

Reynolds, M.P.; Balota, M.; Delgado, M.I.B.; Amani, I. and Fisher, R.A. 1994. Physiological and morphological traits associated with spring wheat yield under hot irrigated conditions. *Australian Journal of Plant Physiology*, 21: 717–730.

Reynolds, J.F. Kemp, P.R., Acock, B., Chen, J. and Moorhead, D.L. 1996. Progress, limitations $_2$ on plants and ecosystems. In, "*Carbon dioxide and terrestrial ecosystems*", Eds. G.W. Koch and H.A. Mooney, San Diego: Academic Press, 347–380.

Ribas-Carbo, M., Berry, J.A., Yakir, D., Giles, L., Robinson, S.A., Lennon, A.M. and Siedow, J.N. 1995. Electron partitioning between the cytochrome and alternative pathways in plant-itochondria. *Plant Physiology*, 109: 829-837.

Ribas-Carbo, M., Aroca, R., Gonzalez-Meler, M.A., Irigoyen, J.J. and Sanchezdiaz, M. 2000. The electron partitioning between the cytochrome and alternative respiratory pathways during chilling recovery in 2 cultivars of maize differing in chilling sensitivity. *Plant Physiology,* 122(1): 199–204.

Ribas-Carbo, M., Taylor, N.L., Giles, L., Busquets, S., Finnegan, P.M., Day, D.A., Lambers, H., Medrano, H., Berry, J.A. and Flexas, J. 2005. Effects of water stress on respiration in soybean leaves. *Plant Physiology*, 139: 466-473.

Roberts, J.K., Callis, J., Jardetzky, O., Walbot, V. and Freeling, M. 1984. Cytoplasmic acidosis as a determinant of flooding intolerance in plants. *Proceedings of the National Academy of Sciences of United States of America,* 81: 6029-6033.

Rockström, J. and Falkenmark, M. 2000. Semiarid crop production from a hydrological perspective: gap between potential and actual yields. *CRC*, 19: 319–346.

Rodriguez, A.A., Llicia, R.A., Cordoba, R., Ortega, L. and Taleisnik, E. 2004. Decreased reactive oxygen species concentration in the elongation zone contributes to the reduction in maize leaf growth under salinity. *J. Exp. Bot.*, 55: 1383-1390.

Ruiz-Vera, U.M., Siebers, M., Gray, S.B., Drag, D.W., Rosenthal, D.M., Kimball, B. A., Ort, D.R. and Bemacchi, C.J. 2013. Global warming can negate the expected CO_2 stimulation in photosynthesis and productivity for soybean grown in the Midwestern United States. *Plant Physiol.,* 162: 410–423.

Ryan, M.G. 1991. Eff ects of climate change on plant respiration. *Ecol. Appl.*, 1: 157–167.

Sabar, M., de Paepe, R. and de Kouchkovsky, Y. 2000. Complex I impairment, respiratory compensation and photosynthetic decrease in nuclear and mitochondrial male sterile mutants of *Nicotiana sylvestris. Plant Physiology*, 124: 1239-1249.

Sabbagh. E., Lakzayi, M., Keshtehgar, A. and Rigi, K. 2014. The effect of salt stress on respiration, PSII function, chlorophyll, carbohydrate and nitrogen content in crop plants. *Intl J Farm and Alli Sci.*, 3(9): 988-993.

Sage, R.F. 2004.The evolution of C4 photosynthesis. *New Phytol.* 161: 341–370.

Sage, R.F., Sharkey, T.D. and Seemann, J.R. 1990. Regulation of ribulose-1,5-bisphosphate carboxylase activity in responses light intensity and CO_2 in the C3 annuals *Chenopodium album* L. and *Phaseolus vulgaris. Plant Physiol.*, 94(1): 1735-1742.

Saglio, P.H., Raymond, P. and Pradet, A. 1980. Metabolic activity and energy charge of excised maize root tips under anoxia. *Plant Physiol.*, 66(6): 1053-1057.

Saha, P., Kunda, P. and Biswas, A.K. 2012. Influence of sodium chloride on the regulation of Krebs cycle intermediates and enzymes of respiratory chain in mungbean (*Vigna radiata* L. Wilczek) seedlings. *Plant Physiol. Biochem.*, 60: 214-222.

Sairam, R.K. and Tyagi, A. 2004. Physiology and molecular biology of salinity stress tolerance in plants. *Curr. Sci.*, 86: 407-421.

Sairam, R.K., Kumutha, D., Ezhilmathi, K., Deshmukh, P.S. and Srivastava, G.C. 2008. Physiology and biochemistry of waterlogging tolerance in plants. *Biologia Plantarum*, 52(3): 401-412.

Scheurwater, I., Cornelissen, C., Dictus, F., Welschen, R. and Lambers, H. 1998. Why do fast- and slow-growing grass species differ so little in their rate of root respiration, considering the large differences in rate of growth and ion uptake? *Plant, Cell and Environment*, 21: 995–1005.

Scheurwater, I., Dunnebacke, M., Eising, R. and Lambers, H. 2000. Respiratory costs and rate of protein turnover in the roots of a fast-growing (*Dactylis glomerata* L.) and a slow-growing (*Festuca ovina* L.) grass species. *Journal of Experimental Botany*, 51: 1089-1097.

Seemann, J.R. and Critchley, C. 1985. Effects of salt stress on the growth, ion content, stomatal behaviour and photosynthetic capacity of a salt sensitive species. *Phaseolus vulagris* L. *Plant Physiol.*, 82: 555-560.

Shalata, A., Mittova, V., Volokita, M., Guy. M. and Taj, M. 2001. Response of cultivated tomato and its wild salt-tolerant relative *Lycopersicon pennellii* to salt-dependent oxidative stress: the root antioxidative system. *Physiol. Plant*, 112: 487-494.

Shewfelt, R.L. and Purvis, A.C. 1995. Lipid peroxidation in plant tissue disorder. *Hort. Sci.*, 30(2): 213- 218.

Shugaeva, N., Vyskrebentseva, E., Orekhova, S. and Shugaev, A. 2007. Effect of water deficit on respiration of conducting bundles in leaf petioles of sugar beet. *Russ. J. Plant Physiol.*, 54: 329–335.

Siedow, J.N. and Grivin, M.E. 1980. Alternative respiratory pathway. Its role in seed respiration and its inhibition by propyl galalate. *Plant Physiology*, 65: 669-674.

Smethurst, C.F., Garnet, G and Shabala, S. 2005. Nutrition and chlorophyll fluorescence response of Lucerne (*Medicago sativa* L.) to waterlogging subsequent recovery. *Plant Soil*, 270(1-2): 31-45.

Smirnoff, N. 1995. Antioxidant systems and plant response to the environment. In, "*Environment and plant metabolism: Flexibility and acclimation*", Ed. N. Smirnoff, Bios Scientific, Oxford, UK, pp. 217–243.

Smith, M.W. and Ager, P.L. 1988. Effect of flooding on leaf gas exchange of seedling pecan trees. *Hort. Sci.*, 23(2): 370-372.

Somot, S., Sevault, F., Déqué, M. and Crépon M. 2008. 21st Century climate change scenario for the Mediterranean using a coupled atmosphere–ocean regional climate model. Global *and Planetary Change*, 63: 112–126.

Sousa, C.A.F. de, and Sodek, L. 2002. The metabolic response of plants to oxygen deficiency. *Brazilian Journal of Plant Physiology*, 14: 83-94.

Specht, R.L. 1969. A comparison of the sclerophyllous vegetation characteristics of Mediterranean type climates in France, California, and southern Australia. I. Structure, morphology, and succession. *Australian Journal of Botany*, 17: 277–292.

Sperlich, D., Barbeta, A., Ogava, R., Sabate, S. and Peñuelas, J. 2015. Balance between carbon gain and loss under long-term drought: impacts on foliar respiration and photosynthesis in *Quercus ilex* L. *J. Exp. Bot.* 66(21): 6863-68750.

Stanhill, G. 1986. Water use efficiency. *Advances in Agronomy,* 39: 53–85.

Steward, Ñ.R., Martin, B.A., Reding, L. and Cerwick, S. 1990. Respiration and alternative oxidase in corn seedling tissues during germination at different temperatures. *Plant Physiology*, 92(3): 755–760.

Su, P.H. and Lin. C.H. 1996. Metabolic response of luffa roots to long term flooding. *J. Plant Physiol.*, 148(6): 735-740.

Sullivan, C.Y. 1972. Mechanisms of heat and drought resistance in grain sorghum and methods of measurement. In, "*Sorghum in the seventies*", Eds. N.G.P. Rao and L.R. House, Oxford and IPH, New Delhi, India. Pp. 247-264.

Sullivan, C.Y. and Ross, W.M. 1979. Selecting for drought and heat resistance in grain sorghum. In, "*Stress Physiology in Crop Plants*", Eds. H. Mussell and R.C. Staples, John Wiley and Sons, New York, USA, pp. 263-281,

Supper, S. 2003. Verstecktes Wasser. Sustainable Austria, Nr. 25 December.

Tadege, M., Dupuis, I.I. and Kuhlemeier, C. 1999. Ethanolic fermentation: New functions for an old pathway. *Trends in Plant Science*, 4: 320-325.

Takeda, Y., Ogawa, Ò., Nakamura, Y., Kasamo, K., Sakata, M. and Ohta, E. 1995. ^{31}P-NMR study of the physiological conditions in intact root cells of mung bean seedlings under low temperature stress. *Plant and Cell Physiology*, 36(5): 865–871.

Tejera, N., Ortega, E., Rodes, R. and Lluch, C. 2006. Nitrogen compounds in the apoplastic sap of sugarcane stem: Some implications in the association with endophytes. *J. Plant Physiol.*, 163: 80-85.

Tezara, W., Mitchell, V.J., Driscoll, S.D. and Lawlor, D.W. 1999. Water stress inhibits plant photosyn-thesis by decreasing coupling factor and ATP. *Nature,* 401: 914-917.

Thomas, R.B., Reid, C.D., Ybema, R. and Strain, B.R. 1993. Growth and maintenance components of leaf respiration of cotton grown in elevated carbon dioxide partial pressure. *Plant Cell Environ.,* 16: 533-546.

Thongo M'Bou, A., Saint-Andre, L., de Grandcourt, A. Nouvellon, Y., Jourdan, C., Mialoundama, F. and Epron, D. 2010. Growth and maintenance respiration of roots of clonal *Eucalyptus* cuttings: scaling to stand-level. *Plant and Soil*, 332(1): 41-53.

Toro, G. and Pinto, M. 2015. Plant respiration under low oxygen. *Chilean Journal of Agricultural Research*, 75(Suppl. 1): 57-70.

Torres, M.A. and Dangl, J.L. 2005. Functions of the respiratory burst oxidase in biotic interactions, abiotic stress and development. *Curr. Opin. Plant Biol.*, 8: 397-403.

Turnbull, M.H.; Murthy, R. and Griffin, K.L. 2002. The relative impacts of daytime and night-time warming on photosynthetic capacity in *Populus detoides*. *Plant Cell and Environment*, 25: 1729-1737.

Umbach, A.L., Fiorani, F. and Siedow, J.N. 2005. Characterization of transformed *Arabidopsis* with altered alternative oxidase levels and analysis of effects on reactive oxygen species in tissues. *Plant Physiology*, 139: 1806-1820.

van Dongen, J.T., Gupta, K.J., Rami'rez-Aguilar, S.J., Araujo, W.L., Nunes-Nesi, A. and Fernie, A.R. 2011. Regulation of respiration in plants: A role for alternative metabolic pathways. *Journal of Plant Physiology*, 168: 1434-1443.

van Oijen, M., Schapendonk, A. and Höglind, M. 2010. On the relative magnitudes of photosynthesis, respiration, growth and carbon storage in vegetation. *Annals of Botany,* 105: 793–797.

Vercesi, A.E., 2001. The discovery of an uncoupling mitochondrial protein in plants. *Bioscience Reports*, 21: 195-200.

Vidal, G., de Paepe, R. and Ribas-Carbo, M. 2007. Leaf age-related changes in respiratory pathways are dependent on complex I activity in *Nicotiana sylvestris*. *Physiologia Plantarum,* 129: 152-162.

Vinocur, B. and Altman, A. 2005. Recent advances in engineering plant tolerance to aboitic stress: achievements and limitations. *Curr. Opin. Biotechnol*, 16: 123-132.

Vanderzee, D. and Kennedy, R.A. 1983. Development of photosynthetic activity following anaerobic germination in rice-mimic grass (*Echinochloa crus-galli* var. *oryzicola*). *Plant Physiology*, 73(2): 332-339,

Vu, C.V. and Yelenosky, G. 1991. Photosynthetic responses of citrus trees to soil flooding. *Physiol. Plant.* 81(1): 7-14.

Walker, R.R., Törökfalvy, E., Scott, N.S.S. and Kriedemann, P.E. 1981. An analysis of photosynthetic response to salt treatment in *Vitis vinifera*. *Aust. J. Plant Physiol.*, 8: 359-374.

Wample, R.L. and Davis, R.W. 1983. Effect of slooding on starch accumulation in chloroplasts of sunflower (*Helianthus annus* L.). *Plant Physiol.*, 73(1): 195-198.

Wang, C.Y. 1982. Physiological and biochemical responses of plants to chilling stress. *Hort. Science.* 17(2): 173–186.

Wang, L.-W., Showalter, A.M. and Ungar, I.A. 1997. Effect of salinity on growth, ion content, and cell wall chemistry in *Atriplex prostrata* (Chenopodiaceae). *American Journal of Botany,* 84: 1247–1255.

Wang, K.Y., Zha, T. and Kellomaki, S. 2002. Measuring and simulating crown respiration of Scots Pine with increased temperature and carbon dioxide enrichment. *Ann. Bot.*, 90: 325-335.

Wang, W.X., Vinocur, B. and Altaman, A. 2003. Plant responses to drought, salinity and extreme temperatures: towards genetic engineering for stress tolerance. *Planta*, 218: 1-14.

Wang, X., Taub, D.R., and Jablonski, L.M. 2015. Reproductive allocation in plants as affected by elevated carbon dioxide and other environmental changes: a synthesis using meta-analysis and graphical vector analysis. *Oecologia* 177: 1075–1087.

Wanger,A.M. and Karb, K. 1995. The alternative respiratory pathway in plant? Role and regulation. *Physiologia Plantarum*, 95: 318-325.

Wagner, A.B. and Moore, A.L. 1997. Structure and function of the plant alternative oxidase: its putative role in the oxygen defense mechanism. *Bioscience Rep.*, 17: 319–333.

Warren, C.R., Livingston, N.J. and Turpin, D.H. 2004. Water stress decreases the transfer conductance of Douglas-fir (*Pseudotsuga menziensii*) seedlings. *Tree Physiol*, 24: 971–979.

Warren, C.R. and Adams, M.A. 2006. Internal conductance does not scale with photosynthetic capacity: mplications for carbon isotope discrimination and the economics of water and nitrogen use in photosynthesis. *Plant, Cell and Environment*, 29: 192-201.

Wegner, L.H. 2010. Oxygen transport in waterlogged plants, waterlogging signalling and tolerance in plants. In, "*Waterlogging signalling and tolerance in plants*", Eds. S. Mancuso, and S. Shabala, Springer-Verlag, Berlin, Germany, pp. 3-22.

Wise, R.R., 1995. Chilling-enhanced photooxidation: the production, action and study of reactive oxygen species produced during chilling in the light. *Photosynthesis Research*, 45: 79-97.

Wright, I.J., Reich, P.B., Atkin, O.K., Lusk, C.H., Tjoelker, M.G. and Westoby, M. 2006. Irradiance, temperature and rainfall influence leaf dark respiration in woody plants: evidence from comparisons across 20 sites. *New Phytologist*, 169: 309–319.

Wullschleger, S.D., Ziska, L.H. and Bunce, J.A. 1994. Respiratory responses of higher plants to atmospheric CO_2 enrichment. *Physiol. Plant* 90: 221-229.

Wyn Jones, R.G. and Storey, R. 1981. Betaines. In, "*The Physiology and Biochemistry of Drought Resistance in Plants*", Eds. L.G. Paleg and D. Aspinall, Sidney: Academic Press, pp. 171–204.

Wyn Jones, R.G., Gorham, J. and McDonnell, E. 1984. Organic and inorganic solute contents as selection criteria for salt tolerance in the Triticeae. In, "*Salinity Tolerance in Plants: Strategies for Crop Improvement*", Eds. R. Staples and G.H. Toennissen, Wiley and Sons, New York, pp. 189-203.

Xu, Z.Z., Shimizu, H., Yagasaki, Y., Ito, S., Zheng, Y.R., and Zhou, G.S. 2013a. Interactive effects of elevated CO_2, drought, and warming on plants. *J. Plant Growth Regul.,* 32: 692–707.

Xu, Z.Z., Yu, Z.W., and Zhao, J.Y. 2013b.Theory and application for the promotion of wheat production in china: past, present and future. *J. Sci. Food Agrc.,* 93: 2339–2350.

Xu, Z., Shimizu, H., Ito, S., Yagasaki, Y., Zou, C., Zhou, G., Zheng, Y. 2014. Effects of elevated CO_2, warming and precipitation change on plant growth, photosynthesis and peroxidation in dominant species from North China grassland. *Planta,* 239: 421–435.

Xu, Z., Jiang, Y. and Zhou, G. 2015. Response and adaptation of photosynthesis, respiration, and antioxidant systems to elevated CO_2 with environmental stress in plants, *Frontiers of Plant* Science, 6: 701.

Yadegari, L.Z., Heidari, R. and Carapetian, J. 2008. Chilling pretreatment causes some changes in respiration, membrane permeability and some other factors in soybean seedlings. *Research Journal of Biological Sciences*, 3(9): 1054–1059.

Yamori, W., Noguchi, Ê., Hikosaka, K. and Terashima, I. 2009. Cold-tolerant crop species have greater temperature homeostasis of leaf respiration and photosynthesis than cold-sensitive species. *Plant and Cell Physiology*, 50(2): 203–215.

Yeo, A. 1998. Molecular biology of salt tolerance in the context of whole-plant physiology. *J. Exp. Bot.*, 49: 915-929.

Yin, X., Struik, P.C., Romero, P., Harbinson, J., Evers, J.B., Van Der Putten, P.E.L. and Vos, J. 2009. Using combined measurements of gas exchange and chlorophyll fluorescence to estimate parameters of a biochemical C photosynthesis model: a critical appraisal and a new integrated approach applied to leaves in a wheat (*Triticum aestivum*) canopy. *Plant, Cell and Environment,* 32: 448–64.

Yin, X., Sun, Z., Struik, P.C. and Gu, J. 2011. Evaluating a new method to estimate the rate of leaf respiration in the light by analysis of combined gas exchange and chlorophyll fluorescence measurements. *Journal of Experimental Botany*, 62: 3489–99.

Youngner, V.B. and Nudge, F.G. 1968. Growth and carbohydrate storage of three *Poa pratensis* L. strains as influenced by temperatures. *Crop Science*, 8: 455–456.

Zabalza, A., van Dongen, J.T., Froehlich, A., Oliver, S.N., Faix, B., Gupta, K.J., Schmälzlin, E., Igal, M., Orcarray, L., Royuela, M. and Geigenberger, P. 2009. Regulation of respiration and fermentation to control the plant internal oxygen concentration. *Plant Physiology*, 149: 1087-1098.

Zaragoza-Castells, J., Sánchez-Gómez, D., Valladares, F., Hurry, V. and Atkin, O.K. 2007. Does growth irradiance affect temperature dependence and thermal acclimation of leaf respiration? Insights from a Mediterranean tree with long-lived leaves. *Plant, Cell and Environment,* 30: 820–33.

Zaragoza-Castells, J., Sánchez-Gómez, D., Hartley, I.P., Matesanz, S., Valladares, F., Lloyd, J. and Atkin, O.K. 2008. Climate-dependent variations in leaf respiration in a dry-land, low productivity Mediterranean forest: the importance of acclimation in both high-light and shaded habitats. *Functional Ecology*, 22: 172–184.

Zauralov, O.A. and Lukatkin, A.S. 1997. After effect of low temperatures on the respiration of heat-loving plants. *Russian Journal of Plant Physiology*, 44(5): 640–644.

$_2$ as

modified by growth temperature. *Physiol. Plant.* 87: 459-466.

Zheng, S.H., Nakamoto, H., Yoshikawa, K., Furuya, T. and Fukuyama, M. 2002. Influence of high night temperature on flowering and pod setting in soybean. *Plant Prod. Sci.*, 5: 215–218.

Zhu, X.-G., Portis, A.R.Jr. and Long, S.P. 2004. Would transformation of C3 crop plants with foreign Rubisco increase productivity? A computational analysis extrapolating from kinetic properties to canopy photosynthesis. *Plant Cell and Environment,* 27(2): 155-165.

Zeitfracht Medien GmbH
Ferdinand-Jühlke-Straße 7
99095 Erfurt, Deutschland
produktsicherheit@kolibri360.de